Life Histories of Cascadia Butterflies

David G. James and David Nunnallee

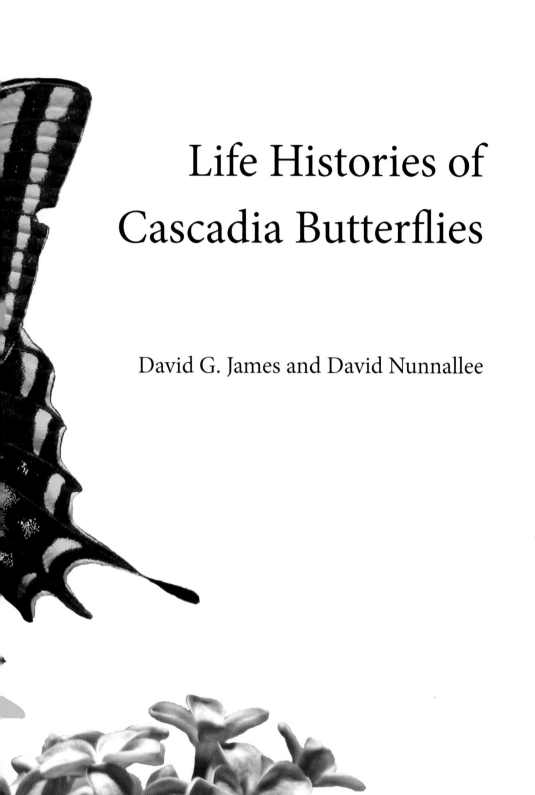

Life Histories of
Cascadia Butterflies

David G. James and David Nunnallee

Oregon State University Press
Corvallis

Library of Congress Cataloging-in-Publication Data

James, David G.
  Life histories of Cascadia butterflies / David G. James, David Nunnallee ; foreword by Robert Michael Pyle.
      p. cm.
  Includes bibliographical references and index.
  ISBN 978-0-87071-626-3 (alk. paper) -- ISBN 978-0-87071-648-5 (ebook)
  1. Butterflies--Washington (State) 2. Butterflies--Northwest, Pacific. 3. Butterflies--Washington (State)--Pictorial works. 4. Butterflies--Northwest, Pacific--Pictorial works. I. Nunnallee, D. II. Pyle, Robert Michael. III. Title.
  QL551.W2J36 2011
  595.78'909795--dc23
                          2011020404

Oregon State University Press
121 The Valley Library
Corvallis, OR 97331-4501
541-737-3166 • fax 541-737-3170
http://oregonstate.edu/dept/press

Dedication

To my father, Alan James, and mother, Doreen James, for supporting and encouraging my interest in butterflies as an 8-year-old, which led to a lifelong passion for insects and their biology. Also to my entomologist wife, Tanya, and our gorgeous daughters, Jasmine Vanessa and Rhiannon Vanessa, for making this endeavor a truly enjoyable and family affair!

—David G. James

To my wife, Jo, for her unwavering support, companionship, and assistance, both in the field and at home.

—Dave Nunnallee

Table of Contents

Foreword

Robert Michael Pyle
(author of *The Butterflies of Cascadia*, founder of The Xerces Society)

When I set out to encounter as many species of U.S. butterflies as I could in the calendar year 2008 for a book called *Mariposa Road*, I made the decision right up front to count caterpillars—and eggs, and chrysalides—as well as adults, as long as their specific identity was unequivocal. Some saw this as a liberty: "I thought you were counting *butterflies*?" they would say. And when the annual Fourth of July Butterfly Counts were launched by the Xerces Society in 1975, they too counted all life stages, though later the rules were changed so that the counts (now run by the North American Butterfly Association) tally only adults. To me, this seems not only silly, but also downright nonintuitive: after all, the flying butterfly is but one-quarter of the overall animal! And the other three-quarters are just as interesting, arresting, and, in their own way, beautiful, as the insects' imagos themselves.

In forty years of writing and teaching about butterflies, three questions have been asked of me more often than all the others combined: What is the difference between a butterfly and a moth? How can I tell their caterpillars apart? and How long do butterflies live? Well, there is no rule for distinguishing moth larvae from those of butterflies, unlike the adults with their differing antennae. But by learning to recognize different *groups* of caterpillars, you will usually be able to tell if a particular "worm" (as most people still call them) is a candidate for a butterfly. And as for telling those apart, this book is just the ticket.

When people wonder how long butterflies live, they are usually thinking of the adult stage only. Very often they are amazed to hear that the free-flying butterfly—for most species but not all—occupies far and away the briefest portion of a given butterfly's life on earth. Depending on which stage overwinters, the bulk of the lifespan may be spent as an egg, as a larva, or as a pupa; or, for the anglewings, tortoiseshells, and the migratory generation of the monarch, it is the adult form in which the organism spends most of its days. But most species by far live for months in their early metamorphic stages, and only for days as a flying adult. This being the case, the necessity for a volume such as this one becomes apparent. Except that there is, really, no other book like this.

Some years ago, early in the life of the Washington Butterfly Association, I became aware that Dave Nunnallee was laboriously and successfully rearing many Washington butterflies and photographing them beautifully in each stage of their development. Soon thereafter, I learned that David James shared this enthusiasm and was doing the same on the eastern side of the Cascades. It was obvious that they needed to be in touch. If they were productive on their own, their synergy (which some of us call Dave Squared) can be compared only to wildfire. A few years later, the result of their collaboration is what you are now cradling in your hands and thrilling to with your eyes. Thanks to OSU Press, these authors' vision of depicting the butterflies of Cascadia in the fullness of their amazing metamorphoses has come true.

This ambitious undertaking honors not only the old-time, painstaking kind of natural history that distinguished an earlier era, but also the fine publishing tradition that grew out of it. In many ways, this book is a throwback to a much-lamented era of closely observed and reported animal studies—but with full benefit of modern tools, techniques, and notions, so even better. While the authors' clear, engaging text is worthy of that earlier era's best examples, their stunning images rival, even surpass, the most beautiful and accurate portraits from nature study's Golden Age.

I can't say enough about the distinction of *Life Histories*. In the first place, it represents an enormous advance in our knowledge. When I wrote *Watching Washington Butterflies* (1974), many of our species were next to unknown in their early stages; and by the time *Butterflies of Cascadia* was published (2000), things weren't that much better. In the mere decade since, our entire Washington fauna (save one), including most of those found in the Pacific Northwest, has been reared by these two dynamos, as well as studied and exquisitely photographed in every stage. In the second place, this book is the apex of life history treatments to date. In the whole world, no other comparable region enjoys a work of this scale, ambit, and acuity for its butterfly fauna. Only Great Britain, Japan, and one or two other hyper-studied Old World countries come close; and none of them to my knowledge match the completeness, depth of detail, precision, and presentation of this book.

In recent years, some excellent larval guides have appeared—Wagner for eastern butterflies and moths; Allen, Brock, and Glassberg for selected American butterflies; Thomas Allen again for West Virginia (see bibliography)—but nothing that covers its butterfly young as completely as this volume. The lot of the naturalist in search of an identity (and understanding) of immatures has greatly improved around the country, but now, especially so in Cascadia. The publication of James and Nunnallee, or "the Daves" as we know them, is a matter for unreserved celebration, not only for lepidopterists and nature lovers of all stripes, but for anyone who cares about our butterflies' lives, futures, conservation management, and the plants with which they have co-evolved.

So go ye forth now out-of-doors, this book in hand, on the dash, or in the day pack, and enjoy the rich array of our Cascadian butterflies even more—and not just the flying ones as usual. Examine the minute, reticulated beadwork on a hairstreak egg; contemplate the sublimely cryptic coloration of a pine white larva or the pastel caterpillar-camo of a buckwheat blue; ponder the means of making a Painted Lady's chrysalis as brilliantly metallic as any suit of armor. In short, let this wonderful book help you to discover butterflies all over again—as their whole selves!

Introduction

This book is the fulfillment of the authors' long-time dream to document all the life stages of all the butterflies of a significant geographic region, that of Cascadia. For David James the dream of comprehensively documenting butterfly life histories extends back to the 1960s in England, when he first began rearing butterflies—at the age of 8!

We define Cascadia for the purposes of this book as Washington State and the adjacent areas of northern Oregon, southern British Columbia, and the Idaho panhandle. Our coverage is virtually complete for Washington and nearly complete for these adjacent areas.

Area of coverage.

This work is the result of approximately twenty years of combined effort by the two authors to produce the information and photographs for this book. Initially both authors began this project independently, working separately and unbeknownst to each other for several years. The authors met early in 2005 and, recognizing that they had very similar goals, pooled their resources and efforts and have been working together ever since. David James is an associate professor in the Department of Entomology at Washington State University, Prosser, located in the Yakima Valley of eastern Washington. David Nunnallee is an engineer by occupation and a naturalist by avocation who resides in western Washington State near Seattle.

Life Histories of Cascadia Butterflies is a book about the immature stages—the eggs, larvae, and pupae—of the butterflies of the Pacific Northwest. The vast majority of the photographs and information were obtained by diligent and, usually, multiple rearings of individual species. This book is unique in the Americas in describing and illustrating with high-quality photographs all the immature stages of virtually all the butterfly species within a substantial geographic area. While there are other American publications that deal with butterfly larvae, there is no other publication in the New World, or to our knowledge in the Western world, with such a complete scope. Other publications either have complete life histories for a few butterfly species, or cover only the final instar (stage) of caterpillars for a wider selection of species; however, because many butterfly larvae change radically during their development, an illustration of only the final stage is often inadequate for identification or comparative studies.

Several converging factors underscore the need for a book of this kind. First, there has been a great surge of public interest in butterflies over the past decade or so. As members of the birding community have expanded their interests in recent years to butterflies, native plants, and dragonflies, they continue to search for interesting new pursuits, not to replace their natural history interests but to expand them. The study of butterfly larvae is a "natural" for this audience. Now that many people are becoming comfortable with identification of adult butterflies, they are ready to move on to immature stages. They want to be able to identify caterpillars they find, to locate larvae to rear with their children, or simply to expand their knowledge of the natural world. In addition, growing numbers of gardeners and horticulturists want not only to attract butterflies but also to provide a complete life history habitat for them and to be able to identify the larvae they find in their gardens.

Youngsters are our future. Rhiannon James enjoys her "butterfly nose."

There is also a great deal of interest in the immature stages of butterflies within the scientific community as a taxonomic tool for clarifying species relationships, for comparison of geographically separated populations, and as indicators of habitat degradation and climate change. Larvae are currently recognized, perhaps even more now than in earlier decades, for their value in differentiating butterfly species. In this book we hope that the considerable new information we present will help solve longtime problems in butterfly taxonomy and species identification. Professional and amateur entomologists are increasingly interested in the biology and ecology of butterflies because they provide excellent subjects for laboratory and field research.

Butterflies are dependent on their environment, most particularly on plants, and as human encroachment compromises their environments, many butterfly species are struggling to survive. As a result, in Cascadia several butterfly species have been listed as endangered within the past few years, and still others have been placed on candidate species lists. Such listings typically require recovery plans that in turn may include captive rearing programs. State agencies, zoos, universities, and conservation organizations are currently cooperating to rear some of the listed species for reintroduction to the wild. We cannot protect what we do not understand. We hope this book will increase our understanding of butterfly life histories and that this will lead to more effective preservation programs.

People have been fascinated with adult butterflies for centuries; however, the adults of many butterfly species live for only a week or so, with 98 percent of their life history hidden or unknown in immature stages, so a full understanding of the factors controlling populations must necessarily include the study of immature stages. Resources for understanding life histories are scattered and fragmentary. We hope that our comprehensive approach to showing all the life stages in our complete butterfly fauna will provide a reference point for others, to assist in regional and perhaps even international comparisons with similar faunas elsewhere.

We present a considerable body of new biological and ecological information on butterfly life histories, much of it gleaned from our multiple rearings of individual species. Data on egg laying, rates of development, instar number, feeding behavior, shelters, host-plant acceptance, defense strategies, natural enemies, and pupation are presented for the first time for many species. These data are combined with previously known but difficult to locate information, to provide a succinct overview of the biology and ecology of the immature stages of each species. For most species we highlight aspects of life history that need further research and include pertinent references to previously published research.

Prior to the species accounts we present overviews of fundamental themes relevant to immature stages of butterflies, like defense, natural enemies, dormancy, and habitats. We also include chapters on rearing, photography, and field techniques.

We hope that scientists and butterfly enthusiasts alike will be stimulated to conduct additional research and investigation of our fascinating butterfly fauna, or at least to record and share their observations, not only to satisfy individual curiosity and learn more about our natural history but also to help fill the many large gaps in our knowledge.

The Life History of a Butterfly

In this, the twenty-first century, the life history of a butterfly features in the curricula of most preschool and nearly all elementary school students. Today's children often know more about the fantastic metamorphosis through which butterflies pass than their parents do. Today, many 5-year-olds experience firsthand in the classroom the thrill and excitement of rearing their own "pet" caterpillar, thanks to companies providing butterfly livestock. Rearing companies in North America supply and ship eggs and caterpillars of Painted Lady butterflies along with artificial diet, so that nearly every schoolchild has the chance to see for him or herself the miracle that is the basis of this book. Observing butterfly development from egg to caterpillar to chrysalis is a tremendous life experience as well as learning experience for children, one they can carry with them the rest of their lives. It is hoped that it will also stimulate some of them to learn more about the butterflies in their area and to look at other life cycles.

Life Histories of Cascadia Butterflies documents individual life histories, each one unique in major or minor details of biology and ecology; however, all these life histories have the same underlying physiology of metamorphosis. All butterflies court, mate, and lay eggs that hatch into caterpillars (larvae). Larvae feed and grow, molting a number of times before becoming chrysalids (pupae) within which the adult butterflies develop and from which they ultimately emerge.

Freshly eclosed butterflies are in mint condition and, depending on temperature and sunshine, may take their first flights within an hour or so. All newly eclosed butterflies expel a red or orange fluid known as meconium, an accumulated waste product from the pupal stage. Nutrients are carried over from the pupal stage, and most newly eclosed butterflies do not need to seek nectar for at least a few hours and maybe a day or so, depending on activity. In most species, males develop faster and eclose before females (a phenomenon known as protandry), sometimes by up to a week or ten days. Consequently, when females eclose, males usually find them immediately and mating quickly follows. This means that

Euphyes vestris (Dun Skipper) **mature eggs with larval heads visible inside.**

for the majority of species, any female caught has usually been mated and will lay fertile eggs; there are exceptions, however, particularly during early spring, when cooler conditions may limit butterfly activity and females may remain unmated for a longer period. In some species, females eclose with mature eggs in their ovaries, but in others a period of feeding and maturation is needed before eggs are laid (e.g., most nymphalids). Butterflies that enter diapause as adults (e.g., Mourning Cloak, anglewings) may not develop their ovaries for 2–10 months, depending on the species.

Euphilotes sp. ovipositing on *Eriogonum heracleoides*.

Butterflies lay or oviposit eggs on healthy host plants chosen by the female. In some cases, eggs are laid on nonplant substrates (ground, rocks, etc.) near host plants or on senesced or dead hosts (e.g., fritillaries). In these cases, the eggs don't usually hatch until host plants reappear. Females can be extremely "choosy" when selecting oviposition sites and use visual and chemical cues to make their "decisions." Visual and/or long-distance olfactory stimuli guide females to the correct host plants, which are then chemically tested by drumming forelegs on leaves, stems, or flowers. Females are rarely satisfied with the first plant part they test and visit a number before bending their abdomen and laying an egg. Eggs are laid singly or in masses on upper or lower surfaces of leaves, on buds or flowers, or encircling twigs.

Butterfly eggs are all small but vary from the miniscule (0.4 × 0.2 mm, Pygmy Blue) to the robust (2.0 × 1.0 mm, Two-tailed Swallowtail). Species that lay eggs in batches may glue up to 300 together, covering an area of 3.0 × 1.0 cm. Small and cryptically colored, butterfly eggs are rarely found by casual observers. They develop rapidly, usually hatching within 2–10 days according to temperature. The neonate larva emerges from the egg by cutting an exit hole in the shell with its mandibles. In some species a circular "lid" is opened from the top of the egg, in others a hole is made in the wall. Neonates of some species consume the entire eggshell on the way out while others leave the empty eggshell with a telltale exit hole.

The first-instar larva immediately sets about feeding and protecting itself. It may move to a more protected location on the plant, cover itself with some strands of silk, or in the case of gregarious larvae join with its siblings in creating an extensive silk-web nest. First instars usually feed rapidly, with the larva often doubling in size from 1–2 mm at hatch to 2–4 mm within a few days. Once the larval "skin" (integument) tightens and appears stretched, the larva is nearing the first molt or ecdysis. Premolt larvae are also characterized by a swelling at the head end caused by development of the larger inelastic head capsule of the next instar. Premolt larvae spin a small pad of silk on which they attach their claspers and remain motionless for 12–48 hours. Larvae awaiting ecdysis are vulnerable to predation; thus they often select protected or hidden sites. Molting begins at the head end, with the integument splitting and moving backward along the body as the larva moves forward slightly. In most species the larva consumes the old integument except for the sclerotized head capsule. Molting takes little more than a few minutes to complete, and the new instar soon resumes feeding, seemingly making up for lost time. Newly molted larvae often show temporary coloration

The 2mm armored egg of *Adelpha californica* (California Sister) dwarfs the finely textured 0.4mm egg of *Brephidium exilis* (Western Pygmy Blue).

that disappears within 2–12 hours. Larvae may molt 3–5 times, developing through 4, 5, or 6 instars, depending on species and sometimes generation. In species that may have larval development interrupted multiple times by diapause, 7–9 instars may occur; however, most species have 5 instars with an approximate doubling of length in each successive instar. In this book we present new information for a number of species on the number of instars completed, but this is an area needing more research.

Butterfly larvae come in all shapes, colors, and sizes, as amply illustrated in our accounts; however, they all have the same purpose of consuming plant material to enable development and maturation while avoiding being eaten themselves. Larval development can be rapid in some species, taking as little as 10 days from oviposition to pupation (Western White), while larvae of other species may take 3–6 months (e.g., Mormon Metalmark). Developmental duration is strongly influenced by the timing of optimal periods for adult survival and reproduction.

Pyrgus centaureae (Grizzled Skipper) egg hatching. Larva has chewed hole in eggshell and is emerging.

Final instars consume 60–80% of the total plant mass eaten by larvae during development; consequently, final instars show the greatest increases in length and girth. The largest butterfly caterpillar in Cascadia is the final instar of the Two-tailed Swallowtail, which may reach a length of 60 mm. Contrast this with the final instar of Cascadia's smallest species, the Pygmy Blue, which reaches only 11 mm at maturity.

When nearing maturity, full-fed larvae often change color slightly or dramatically, and some enter a wandering phase. The "wanderers" seek sites away from the host plant for pupation. Species that pupate on the host plant (e.g., many lycaenids) do not wander. Examples of wanderers are common among the brushfoots and swallowtails, which may travel many meters from the host plant before choosing a site for pupation.

Pupation describes the change from active eating machine (larva) to the immobile, nonfeeding preparatory stage (pupa) that will ultimately yield the adult butterfly. Pupae are formed in one of four basic modes: loose on the ground, within a "cocoon" or shelter, hanging by the terminal end (cremaster) attached to a silk pad, or attached upright by cremaster with a supporting silk girdle. Loose pupae are common in moths but rare in butterflies, confined to a few satyrs (e.g., *Oeneis* spp.). Skippers commonly form pupae within tied-leaf or grass shelters, while some fritillary and satyr species construct less robust leaf shelters. Hanging pupae are characteristic of brushfoots and satyrs, while girdled pupae are found in swallowtails, whites, sulphurs, and lycaenids. When a pupation site is selected, the prepupal larva shrinks a little and waits motionless for the final molt to occur. Larvae that form hanging pupae adopt a characteristic J shape. After 12–48 hours, the larval skin splits behind the head, revealing not another caterpillar integument, but a fleshy, soft integument, usually green, yellow,

Atalopedes campestris (Sachem) L4 just prior to molt. Note expanded thorax; head is already withdrawn from old head capsule.

Argynnis egleis (Great Basin Fritillary) hanging in a J prior to pupation.

or orange. With much wriggling, the larval "skin" moves down the body, revealing increasingly more of the soft new pupa. Once the shed skin reaches the terminal segment, the pupal cremaster probes and seeks the silk pad spun earlier by the prepupal larva. With hanging pupae this is a critical phase; if the cremaster fails to make contact with the silk pad to which it attaches with tiny hooks, the soft pupa will fall, likely to its death. After attachment, more wriggling usually results in the shed skin dropping away from the pupa, and eventually the pupa stops moving. During the next 24 hours, the pupa hardens and assumes the coloration that allows it to blend in with the environment. In a number of species, final coloration of the pupa is dependent on the immediate environment. For example, within a single species (e.g., swallowtails, whites), pupae formed on plants tend to be greener than those formed on sticks and boulders, which are usually gray-brown.

Forms and coloration of pupae are as diverse as larvae, but unlike larvae, pupae never draw attention to themselves. A wildly colored caterpillar can thrash around and be very aggressive to potential predators, but a strikingly colored pupa would simply attract attention and end up as a snack. Even Monarch pupae that contain toxins do not advertise this fact and are colored to blend in. "Blending in" is the pupal theme, with virtually all pupae demonstrating some form of camouflage and crypsis. Consequently, pupae are usually hard to find; however, if a pupa is found and disturbed, it may respond by wriggling, presumably in an attempt to startle and dissuade consumption. The pupa of the Great Spangled Fritillary is a great "wriggler," which may dissuade small rodents on the forest floor from snacking on it.

Celestrina echo echo (Echo Blue) molting from L3 to L4.

Papilio multicaudata (Two-tailed Swallowtail) pupation series: Mature larva shedding skin to reveal pupa.

The pupa hosts the greatest magic trick in the natural world: construction of a butterfly from caterpillar soup. Some recognizable parts of a butterfly are present within a caterpillar (e.g., reproductive organs), but the genesis of most parts and assimilation take place within the pupal shell. Not only is this metamorphosis wondrous, it can also be rapid, with the transformation sometimes taking as little as 5–7 days to complete (e.g., Milbert's Tortoiseshell). In contrast, the pupae of some species oversummer and overwinter in diapause, sometimes for 2–3 years (e.g., some whites and swallowtails). A few days before a butterfly ecloses, the pupa darkens, betraying the advanced stage of development. During the final day, wing patterns and color show through the wing cases, and all parts of the body become apparent through a shell that becomes increasingly transparent. This is known as the pharate pupa stage, with eclosion just hours away.

Adult eclosion from the pupa is synchronized in many species to occur early in the morning, often soon after dawn, presumably to enable the best opportunities for successful eclosion, post-eclosion "drying" of wings, and inaugural flight. Eclosion begins with the butterfly pushing with its feet against the shell covering the legs, antennae, and proboscis. Once the legs are free, they grab hold of the shell, pulling out the rest of the body (substantially aided by gravity in hanging pupa species), until the entire butterfly is fully extricated. Hanging and girdled pupae species usually hang from the pupal shell or a nearby support, whereas butterflies eclosing from pupae on or near the ground wander for a short while to find an appropriate support site. Once a site has been chosen, the butterfly begins pumping hemolymph through the wing veins and sets about zipping together the two parts of the coiled proboscis to form a tube for sipping nectar. This vulnerable period passes quickly, with most species "pumping" wings to full size within 5–15 minutes; however, the wings remain limp and flaccid for another hour or so (depending on temperature), and the butterfly is unable to fly during this time. If a butterfly ecloses around 6 AM on a warm summer's morning, it will be capable of flight by breakfast time, ready to begin the cycle all over again.

Polygonia satyrus (Satyr Comma) pharate pupa, adult wings visible inside.

The life cycle of a butterfly may occupy as few as 28 days in summer or a whole year (or more), but the basics remain the same. The details differ in extraordinary ways, and this book offers an introduction to the lesser known parts of a butterfly's life. We have uncovered many details and secrets of the immature lives of Cascadia butterflies, but many more remain to be discovered.

Life Strategies

Despite their fragile appearance and sometimes ephemeral existence, butterflies are tough, seasoned survivors. Every delicate butterfly flitting over a flowery meadow is the survivor of millions of generations of resilient ancestors that endured, outsmarted, or avoided drought, disease, predators, parasitoids, competitors, storms, volcanoes, and even ice ages. Survival in the butterfly world requires adaptation, flexibility, and strategy, and Cascadia butterflies have become experts in all these endeavors.

Aglais milberti (Milbert's Tortoiseshell) eggs on Stinging Nettle.

Host Plants: Butterflies are inextricably linked to plants, not only as nectar sources for adults but also for larval food. Most adult butterflies will use whatever nectar source is available, provided the timing is right and they are physically able to access the nectar (proboscis length vs. flower depth); however, larvae of every butterfly species have adapted to certain host plants, some depending entirely on a single kind of plant for survival. Many other butterflies have adapted to use several closely related plant species, and a few are able to utilize a very wide range of plants. A butterfly species that overwinters as an egg needs fresh succulent food as soon as it hatches, in most cases early in the spring, so only plants that grow early in the spring are candidates as hosts. Butterflies that overwinter as pupae will eclose to adults in spring, immediately needing nectar but not requiring a larval host until sometime later, when the plant selection may be different from that available to species overwintering as eggs. Life cycle timing, or phenology, is an important determinant in host-plant selection.

Why have some butterfly species adapted by depending on only one or a few plants while others use many? There is no doubt that the ability of plant generalists, such as the Painted Lady, Gray Hairstreak, and Variegated Fritillary, to use many kinds of host plants has rendered them highly successful, allowing them to colonize vast areas, worldwide in the case of the Painted Lady. Specialization on only one or a few plants can also be a highly successful strategy if those plants are widespread and common (e.g., the case of the Cabbage White). California Tortoiseshell larvae feed almost exclusively on *Ceanothus* spp., but they are highly successful as the plants are widespread and common

Vanessa cardui (Painted Lady) larva on Canada Thistle surrounded by sharp spines.

over vast areas of the mountainous West. The Mountain Parnassian uses only stonecrops, greater fritillaries only violets, and Milbert's Tortoiseshells only Stinging Nettle, but these plants are widespread and the butterflies are very successful.

Niche exploitation occurs throughout the natural world, and Lepidoptera are no exception in a natural process of testing the boundaries and seeking a new advantage against competitors; however, it can also be a risky strategy, with some butterfly species becoming restricted to a single plant species that may be very limited in distribution and/or abundance. The Johnson's Hairstreak specializes on a species of *Arceuthobium* (dwarf mistletoe) that grows on mature hemlock trees, but with the human deforestation of old growth forests, this habitat has largely disappeared, and the Johnson's Hairstreak is now at risk. In Washington State the Golden Hairstreak feeds only on chinquapin, a broadleaf tree that barely reaches into this area, so this population is tentative and at the mercy of commercial logging operations, although it is secure farther south in Oregon.

Euphilotes on *Eriogonum thymoides*, L2, displaying its long extendable lycaenid neck useful for hollowing out host plants through a small entry hole.

Some butterflies, such as *Euphydryas editha* (Edith's Checkerspot), have adapted to using two different host plants, one for young larvae in the summer (*Castilleja*), and a different host for the spring larvae (*Collinsia* or *Plantago*) after passing the winter half-grown. Such alternate hosts can be very different, although they are usually from the same or a closely related plant family. Chemical cues undoubtedly play a strong role in selection of alternate hosts; some unrelated or introduced plants may contain chemicals similar to the preferred host, thus inducing larvae to experiment.

The part of a plant preferred by larvae can be very important, and each species of butterfly tends to specialize on only one or a few parts. Some species feed only on leaf buds, others on young leaves, others on mature leaves. Other species require flower buds, flowers, seeds, or even stems. The different parts of a plant develop at different times, so this level of specialization restricts the time period when the larvae can grow. Buds, flowers, and seeds are more nutritious (generally containing 2–10 times more nitrogen) than leaves or stems, promoting faster growth but restricting larvae to a shorter growth period (leaves usually persist longer). Use of different plant parts allows multiple species to use the same host plant without direct competition, as does the staggered timing of the different parts, another niche-exploitation strategy. Food sources such as grasses and evergreen needles are low in nutrition but hugely abundant over vast areas, so larvae exploiting these resources grow slowly but have little competition.

Voltinism and Phenology: The number of generations or broods developed over the course of one year is referred to as voltinism. The benefit of having multiple broods per year is that a species can expand its population with each brood, ultimately improving the chances of survival and success; however, late broods are often at risk of failure due to drought or early winter, so producing multiple broods may be a gamble. Voltinism is very much related to climate, and climate is related to latitude and/or elevation. A single species will often have more annual broods in its southerly range than in its northern (or higher elevation) range. Bivoltine (or "double-brooded") butterflies have two broods per year, and biennial species have one brood over two years. In Cascadia, most butterflies are univoltine, with one brood per year; however 39 of about 150 Cascadia species may be bivoltine at least on occasion, and 5 are biennial, most of these flying only in even-numbered years. Voltinism is largely a response to the growing season of host plants. When the growing season is short, larvae are unable to develop in a single season and must continue to grow the following year, overwintering as immature stages during two winters. Biennialism is common in northern Canada; all of Cascadia's biennial butterflies are northern species, most of which marginally range into our area. Those that do range farther south (*Oeneis* spp. in California) typically become univoltine. Longer growing seasons

Hesperia juba (Juba Skipper) L5 in a typical grass skipper nest.

in southerly areas produce good-quality host plants persisting well into late summer, providing an opportunity for more generations. Some butterfly species with flexible voltinism opportunistically develop a second brood when conditions are optimal, or "hedge their bets" with broods that are part univoltine and part bivoltine. This occurs in the Purplish Copper, with mid- to late-summer eggs either entering winter dormancy or hatching to produce a late summer–autumn adult generation.

Phenology refers to the timing of biological events in relation to climate and season. The phenology of butterfly life cycles in Cascadia is strongly influenced by temperature and photoperiod, with growth and development restricted to approximately nine months of the year in most species. Butterfly life cycles are adaptations to resource availability (e.g., host plant, warm temperatures). Consequently, different climatic regions and vegetation types within Cascadia produce different phenologies. For example, most plants in the arid shrub-steppe ecosystem of eastern Washington and Oregon characteristically have a brief but intense period of vegetative growth, flowering, and seed set during March–June. The phenologies of a number of butterfly species are driven by this ephemeral period of optimal environmental conditions. Development of these spring species from adults (produced by overwintering pupae) to next generation overwintering pupae typically takes as little as 4–6 weeks. Some spring species become dormant during summer in eastern Cascadia, when temperatures are excessively hot and host plants are dead or inedible, but produce a second generation of adults when temperatures ameliorate in late summer-autumn and host plants put out new growth.

Nectar and host plant–rich shrub-steppe habitat in Benton Co., Wa. during April, home to swallowtails and whites.

An example of this occurs with the Juba Skipper and some other skippers whose larvae feed and develop rapidly in mid- to late spring when host grasses are green. Once final instars are fully fed, they enter dormancy in silk-tied nests, not moving or feeding for 4–6 weeks during the heat of summer. During late August the larvae pupate, and adults eclose and lay eggs in September when grasses are showing at least some new growth. These skippers therefore avoid an unfavorable time of year by pausing development (James, 2009b).

The phenology of alpine species is strongly controlled by the timing of snowmelt; the annual appearance of adults may vary by 2–4 weeks from year to year. Most alpine butterflies overwinter as eggs or very young larvae and require an extended developmental period after the snow melts and host plants appear. Even after the snow melts, conditions are generally cool to cold and development is slow; however, most species adopt behavioral and physiological strategies to maximize body temperatures, thereby hastening development. Most larvae of alpine species are dark colored and rest either openly in sunshine or in sun-warmed refugia (under rocks, debris, etc.).

Ecosystems characterized annually by a long period of favorable climate and continued host plant availability host butterflies with flexible phenologies. Generally, species in these ecosystems are multivoltine, with generations overlapping as the season progresses, and flight periods of individual species are long. Cascadia is not well endowed with these ecosystems more characteristic of tropical and subtropical regions; however, southern coastal regions of Cascadia provide a reasonably extended favorable season for butterflies. Some irrigated urban and rural environments in eastern Cascadia provide extended seasons for some species; similarly, warm riverine environments are also favorable for prolonged periods, enabling lengthy seasons of activity for some butterfly species.

Vanessa virginiensis (American Lady) eggs, 0.7 mm, laid in the fuzzy pubescence of *Anaphalis margaritacea* (Pearly Everlasting) leaves.

The frequently overcast and cool climate of coastal Washington and British Columbia has a major impact on butterfly phenology. It reduces the hours of sunlight and holds down temperatures, restricting adult activity and development of immature stages. The importance of direct sunlight in elevating body temperatures, hastening development, and reducing exposure of caterpillars to natural enemies is considerable.

Other Strategies: Opportunism is common in the life histories of immature stages of butterflies. Some butterflies, particularly those that have evolved in semiarid regions, have the ability to opportunistically "hold over" for multiple years, typically in diapausing overwintering stages, when environmental conditions are adverse. This often happens during periods of extended drought. For example, a proportion of the pupae of some marble, white, and swallowtail species occupying shrub-steppe habitats may remain dormant for 2–5 years before eclosion. Rainfall in the shrub-steppe spring is notoriously variable, meaning that larval hosts (mustards, parsleys, etc.) do not always persist long enough for larval development to be completed. Spreading eclosion of a single brood over a number of seasons spreads the risk of a population encountering a bad spring, minimizing the possibility of local drought-mediated extinction. Research on this strategy is needed as it may be far more widespread among butterfly families and species than realized. For example, several Arrowhead Blues remained in their pupae for two years in one of our rearings.

Larvae must molt periodically, shedding their skin to make room for growth; each stage between molts is called an instar. The number of instars is generally relatively fixed for each species, with usually 4 instars in the Lycaenidae (coppers, blues, and hairstreaks),

6 in the greater fritillaries, and 5 in other families; however, larvae may have fewer or develop through more instars under some circumstances. Larvae under stress from dwindling food supply may pupate 1 instar earlier than normal. These smaller pupae produce smaller adults. Seasons with suboptimal food supplies are sometimes followed by populations with greater numbers of undersized adults. In our rearings we found that larvae of some species developing in mid- to late summer appeared to respond to declining daylengths by pupating sooner. For example, larvae of the California Sister developing in September–October, and the Western Tiger and Two-tailed swallowtails developing in August, pupated at the end of the fourth rather than the fifth instar. This perhaps is a strategy to ensure that development is completed before the onset of cooler conditions and deciduous host leaf fall. Unusual environmental conditions, such as manipulated daylengths or stressed host plants, can result in atypical extra dormancies, sometimes accompanied by extra or fewer instars (e.g, *Euphydryas* checkerspots). Extra instars may be more common than reported in some groups, particularly among checkerspots and crescents. We have occasionally observed larval molts after overwintering with no apparent change in larval size and very little change in head capsule size; such molts may go unnoticed and unreported. In one instance we observed 5 instars in a reared cohort of *Phyciodes pallida* that developed to pupae the first season, but another cohort reared from the same batch of eggs overwintered partly grown and developed through 6 instars before eclosion the following spring.

Lotus crassifolius (Big Deervetch), one of several plants hosting a suite of butterfly species; in this case at least 3 skippers, 2 hairstreaks, and a blue.

The rate of larval development varies with host-plant quality and availability as well as with temperature. It is not surprising that warm conditions and optimal food produce fast growth while a poor food supply and cool weather slow growth considerably. It is also not surprising that foods with poor nutritive value (e.g., grasses) produce slow growth. Some species, however, moderate their growth rate even in the presence of prime food and warm temperatures. Larvae of the Mormon Metalmark, which has the latest adult flight period of all Cascadian butterflies, hatch and feed in early spring. They then become sluggish but do not appear to enter aestivation, nibbling casually, frequently retreating under cover, and growing very slowly through spring and early summer, pupating and eclosing in August. Larvae of the Woodland Skipper also become active early, typically in May, but adults fly late in the season.

Pupation strategies are fixed and predictable within each species. Some pupae hang from a single attachment, others are supported upright with silk ties, and others are formed horizontally under cover. The common theme among nearly all pupae is concealment; most pupae are very well camouflaged. The different methods of attachment each have advantages and disadvantages, usually well suited to the different life histories. Butterflies that hang suspended from a single silk attachment include all the brushfoots (Nymphalidae). None of these species overwinter as pupae, and most eclose to adults within a couple of weeks of pupation, so a more secure attachment is generally unnecessary. Pupae of greater fritillaries are very poorly attached and often fall to the ground; however, pupation occurs very close to the ground under a leaf "tent," so damage is unlikely. Species that pupate upright with a cremaster silk attachment at the base plus a horizontal girdle thread around the middle include the swallowtails and pierids (whites and sulphurs). These species overwinter as pupae; thus the extra girdle support is important for stability during the long diapause, and the diagonal upright posture resembles a thorn or stick for camouflage. Most of our other species (lycaenids and skippers) pupate horizontally, many lightly anchored with a silk girdle and most under some kind of cover (rocks, leaves, sticks, nests, or in holes). The pupae of the Silver-spotted Skipper and some duskywings use an elaborate series of silk strands to suspend themselves inside a leaf shelter in such a manner that they do not touch any solid surface, protecting themselves from mold and perhaps even from small predators.

Celastrina echo (Echo Blue) L4, 14 mm, feeding on nitrogen-rich flowers of *Cornus sericea* (Red Osier Dogwood).

While this is a book about the immature stages of butterflies, one adult life strategy must be mentioned briefly, since it has a profound impact on the occurrence of immature stages. Migratory species extend their range each season into Cascadia, taking advantage of favorable spring and autumn conditions; however, they are unable to withstand winter in the Pacific Northwest and either die out or migrate back toward the south in the fall. Examples of migratory species include the Orange Sulphur, Checkered White, California Tortoiseshell, Painted Lady, West Coast Lady, and Monarch. Other species are elevational migrants, entering the mountains in late spring to spend the summer feeding where the growing season is long, then returning to the lowlands either to lay their eggs (Coronis Fritillary) or to overwinter as adults (e.g., Mourning Cloak, Milbert's Tortoiseshell). Some multiple-brooded species (Purplish Copper, Anise Swallowtail, Milbert's Tortoiseshell) move to higher elevations with each successive generation to follow the food supply, while other species remain at lower elevations, choosing instead to seek out other host plants as the season progresses (Becker's White, Orange Sulphur, Melissa Blue, Common Sootywing).

In summary, butterflies and their immature stages are enormously flexible and use a stunning variety of strategies and adaptations to enhance their survival and prosperity. We have touched on a few of these strategies, but much remains to be learned.

Overwintering, Oversummering, and Diapause

Butterflies and their immature stages, like all insects, are cold blooded and depend on environmental conditions to achieve body temperatures necessary for activity, development, and reproduction. The optimal range of body temperatures for most butterflies and their immature stages lies between 15 and 30 °C. Arctic-alpine species have lower optimal temperatures, and tropical species thrive under higher temperatures. Tropical and subtropical species depend more on ambient air temperatures, whereas butterflies living closer to the poles are generally more dependent on direct sunlight to raise body temperatures to operating levels.

When the climate is characterized by temperatures remaining below 5–10 °C with limited, low-angle sunshine, most butterflies and their immature stages are unable to develop or be active. Host-plant and nectar resources are also usually absent or minimal under these conditions, adding to the inhospitable quality of the environment. Winter in Cascadia is cold. In the mountains and lowlands, away from the coastal fringe, temperatures during December–February may remain below 0 °C for days or weeks at a time and dip as low as -35 °C. Clearly, for butterflies, eggs, larvae, or pupae to survive such an extreme environment for 2–4 months or longer, a major modification to normal insect physiology is required.

Some butterfly species, faced with an impending inhospitable winter environment, "decide" to leave Cascadia and fly many miles south to find a more forgiving climate. Painted Lady, California Tortoiseshell, and Monarch are familiar examples of migrants that escape the "frozen north" and spend winters in California or Arizona, their descendants returning north the following season. These species are robust, long lived, and powerful fliers, capable of epic journeys; however, most of Cascadia's butterflies must endure winter in the frozen north.

Chlosyne hoffmanni (Hoffmann's Checkerspot) nests on Cascade Asters, in preparation for overwintering.

Butterfly hibernation is no less impressive than hibernation of mammals like bears or groundhogs. By changing their physiology, butterflies, or their eggs, larvae, or pupae, are able to survive the winter for months in a state of suspended animation. Hibernating adult butterflies or immature stages are dormant and in a physiological state called diapause, characterized by a lowered metabolic rate and radical biochemical changes. Diapause is different from simple dormancy or inactivity as occurs in butterflies and their immature stages during cool periods in spring and autumn or overnight. It is a rigidly controlled physiological mechanism that is genetically fixed or induced by environmental cues.

Some univoltine species are genetically programmed to enter hibernal diapause in the same stage annually. These are usually species occupying short-lived optimal environments (e.g., Arctic-alpine areas) with insufficient time to produce more than one generation. For example, the diapausing, overwintering eggs of butterflies like parnassians and some lycaenids cannot be "persuaded" to hatch before winter, regardless of temperatures and daylengths to which they are exposed. These eggs are genetically programmed to remain in diapause until after exposure to cold temperatures for a certain period of time (2–8 months depending on species). Once the necessary period of chilling has elapsed (a process known as diapause development), a physiological "switch" is thrown, typically releasing certain hormones, and the egg gains competency to develop if exposed to warm temperatures and increasing daylengths. The same refractory mechanism also occurs in diapausing larvae, pupae, and adults. At some point during overwintering (usually around January), larvae gain competency to begin feeding and developing, pupae are able to develop, and adults become active when warm conditions occur.

Some species with only one annual generation in most years do not have a genetically fixed hibernal diapause. Diapause in these species, as well as species with more than two annual generations, is cued environmentally, usually by daylength, temperature, host-plant quality, or a combination of these factors; therefore, if environmental conditions are judged suitable, normally single-generation species may opportunistically "sneak in" an extra generation. Daylength, or the rate of declining daylength, is the most reliable indicator of seasonal change and is consequently the cue most often utilized by butterfly species to determine whether there is enough time for another generation. If above-average spring temperatures enable a butterfly species to complete development faster than normal, still-increasing daylengths can signal the opportunity to proceed with a second generation.

Erynnis propertius (Propertius Duskywing) L5 in its brown diapause colors.

The Purplish Copper is a good example of a species that may be single or double brooded depending on environmental cues. Females that eclose in June produce direct developing eggs and another generation of adults in August. August-eclosing females, exposed to declining daylengths, produce dispausing and nondiapausing eggs, with the percentage of diapausers greater in females eclosing in late August. By producing both types of eggs, the Purplish Copper female "hedges her bets" by placing her eggs in two baskets. If the season finishes early and larvae are unable to complete development, the diapausing eggs will ensure continuation of the lineage the following spring.

Many Cascadia species are multivoltine, with the number of generations varying annually according to seasonal conditions, or differing between lowland and highland populations. Generally warm temperatures during spring hasten development, allowing multivoltine species to produce further generations before daylengths decline. The daylength or rate of decline and/or temperature dictating diapause differs according to species; however, there are very few butterfly species for which the environmental cues causing diapause have been fully determined.

Diapausing eggs and pupae overwinter in locations chosen by females and larvae, respectively. Eggs programmed to diapause are usually laid on host plants that have senesced or on twigs, branches, and rocks. Overwintering pupae are formed in similar places, on, close to, or under the ground surface, or on branches. In all cases overwintering eggs and pupae are cryptic to minimize the chances of hungry birds or mice finding them. When larvae enter diapause in late summer or autumn, they seek

Cupido amyntula (Western Tailed Blue) larva barely visible in American Vetch pea. Larvae sometimes pupate and overwinter in such pods.

out or create refugia where they will spend the winter. Common refugia include curled leaves, seed pods/shells, crevices under rocks, and soil. In captivity, first-instar fritillary larvae commonly choose empty violet seedpods in which to overwinter. Some checkerspot and crescent larvae overwinter in groups in silken nests on or beneath host plants.

While some adult butterflies migrate south for the winter, others enter adult diapause, shutting down metabolism in the same way as diapausing eggs, larvae, and pupae. Adult diapause also incorporates reproductive

Nest of *Euphydryas colon* (Snowberry Checkerspot) L2 prior to overwintering.

diapause, with undeveloped reproductive organs in both sexes. Adult diapause is generally induced in mid- to late summer when climatic conditions are still favorable but declining daylengths signal that insufficient time is available for another generation. These butterflies (e.g., anglewings, Mourning Cloak) feed avidly on late-summer and autumn flowers as well as sap flows, rotting fruit, etc., to build up fat reserves to sustain survival during the 3–6 months of overwintering. Diapausing adults overwinter in protected sites like hollow trees, outbuildings, caves, and crannies under rocks but may venture out briefly during sunny days in winter in milder coastal areas. These species are reproductively dormant for 9–10 months, first in summer diapause (aestivation), then in winter diapause (hibernation).

Surviving the winter is the greatest challenge for most of Cascadia's butterflies, but in lowland areas east of the Cascades, surviving summer can also be a challenge. Temperatures in the arid basin areas frequently exceed 38–40 °C during July–August and may remain above 35 °C for weeks at a time. Consequently, plant life during summer in these areas is sparse and frequently dormant, with little green matter present. Ecologically, this climate and landscape is as adverse to butterfly activity and development as a winter environment. Unsurprisingly, diapause is used by many species in inland Cascadia to survive this short but highly unfavorable period. Some new examples of apparent summer diapause or aestivation in Cascadia's butterflies have been discovered during our research for this book, but this is another aspect of life history that needs further research.

Post-eclosion aestivation appears to occur in adults of some greater fritillary species, particularly those occurring at low to mid-elevations. Coronis Fritillary adults eclose in low–mid-elevation shrub-steppe areas on the western edge of the Columbia basin and eastern foothills of the Cascades in late May–early June, with mating occurring soon after; however, females do not develop their ovaries, instead beginning a well-defined upslope movement into the Cascades. Migrating females may often be seen flying strongly westward near Rimrock Lake in Yakima County, Washington in mid-June. Males do not appear to be part of this migration initially but join in later. The females oversummer on the highest ridges of the Cascades (~7,000–8,000 ft), feeding on abundant alpine flowers like phlox and asters. High-elevation

Argynnis zerene (Zerene Fritillary) L1 diapausing in an empty seed case of Viola (violet).

summer populations of the Coronis Fritillary are dominated by females, but a few males reach the oversummering grounds as well. By mid-August, the population begins moving downslope, and gravid females reappear in the shrub-steppe, ovipositing where Sagebrush

Violets grow in the spring. Great Spangled, Callippe, and possibly Zerene fritillaries appear to use the same oversummering strategy, first noted by Jonathan Pelham (pers. comm.). Fritillaries largely confined to high-elevation habitats, for example, Mormon Fritillaries, appear not to undergo adult diapause. The Common and Great Basin wood nymphs have a similar adult reproductive diapause and aestivation strategy, but instead of migrating to higher elevations, females spend 2–4 weeks after eclosion resting in shady areas before becoming flight-active and ovipositing.

Why do these species need to delay oviposition? Fritillaries and wood nymphs overwinter as diapausing unfed first-instar larvae. The extreme heat and low humidity of high summer would make survival of these tiny larvae difficult, and aestivation appears to serve as a strategy to improve survival of early-instar larvae by ensuring a less extreme environment. These butterflies are among the few species that diapause in two stages within a single generation.

Aestivation can also occur in immature stages. For example, some *Hesperia* spp. (e.g., Juba Skipper) have larvae that aestivate in the final instar, delaying pupation and adult eclosion. Larvae of the Juba Skipper develop rapidly on grass during spring, reaching final instar in early summer. The final instar builds a stronger grass nest and remains concealed within it, going without feeding for 3–6 weeks. This

Massive late-summer nests of *Euphydryas colon* (Snowberry Checkerspot) are easily spotted on penstemons.

delays adult eclosion until late August when daylengths are declining, temperatures are moderating, and grasses in basin areas of eastern Cascadia begin showing some new growth. Eggs and early-instar larvae produced by autumn Juba Skipper females encounter a less extreme environment than they would in July and August; thus larval aestivation serves to synchronize vulnerable life stages with a more optimal environment. Larvae of some other species (e.g., Mormon Metalmark, Woodland Skipper) develop extremely slowly during spring and summer and may also aestivate.

Pupae of a number of papilionid, pierid, and lycaenid species spend up to 10 months in dormancy, oversummering, then overwintering. These species fly, mate, lay eggs, and develop to pupae during 2–3 months in spring. Typically, these species occur in shrub-steppe, arid environments that provide optimal climate and host plants for only a short time each year. The remainder of the year is too hot or too cold to support these species and their necessary plant and nectar resources. Unusual adverse conditions during spring, for example, drought or prolonged cold or wet conditions, can cause serious problems for the species that depend on favorable conditions during this short time frame. Most of these spring-active species have adapted to this uncertainty by having some percentage of the population destined to remain in pupal diapause for more than one year. Pupae of Anise Swallowtails and Sara Orangetips, for example, may remain dormant for 2–3 years or more. This "bet-hedging" strategy ensures continuance of local populations should catastrophic conditions occur one year.

Using diapause as a strategy to overcome more than one period of adversity has also evolved in larvae of some species, particularly checkerspots, whites and sulphurs,

Orange-colored dormant *Hesperia colorado* (Western Branded Skipper) L1, overwintering in grass seed head.

lesser fritillaries, and satyrs. Instead of prolonging diapause, these species have the ability to reenter diapause multiple times if environmental stimuli signal the not-too-distant onset of unfavorable conditions. Postdiapause larvae of *Euphydryas* checkerspots recommence feeding in late winter–early spring on fresh host-plant growth. If host plants are moisture stressed and/or relative humidities are low, indicating the plants may senesce quickly, larvae stop feeding and reenter diapause. Larvae are capable of exiting and reentering diapause multiple times, and individuals may live for 2–3 years with only short periods of development annually. Clearly, this conservative strategy ensures persistence of populations during multiyear droughts, common in many checkerspot localities in eastern basin areas of Cascadia. Spring development of high-elevation species is dependent on timely snowmelt. In late springs, postdiapause larvae of species like the Arctic Fritillary and Pelidne Sulphur may not be able to feed soon enough to enable completion of development for the normal midsummer flight period. It is important for postdiapause larvae of these species to determine whether enough time is available to complete development, and they appear to do this by measuring and responding to daylength. If daylengths are long (more than 12 hours?) once postwintering feeding commences, Arctic Fritillary and Pelidne Sulphur larvae develop through 1 instar, then reenter diapause for a second winter. This strategy may be more common than currently realized and certainly deserves more research, particularly on the precise nature of environmental cues involved for

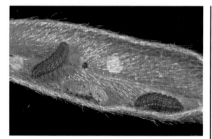

Plebejus icarioides (Boisduval's Blue) L2 stop feeding, turn reddish brown, and are dormant June–March.

different species. Some high-elevation or northerly latitude species always need two years to complete development and always overwinter twice. These species routinely diapause in two stages, usually early-instar then late-instar larvae (e.g., Great Arctic).

Butterflies, like other organisms, have evolved to survive climate-related hardships using an amazing variety of physiological and behavioral strategies. Cascadia butterflies and their immature stages must leave the Pacific Northwest in autumn or have an overwintering strategy that enables good survival under prolonged and harsh winter conditions. Many of the same species also need to deal with limited resources and excessive heat in summer. The result is a butterfly fauna that is extraordinarily tough, resilient, and well adapted to all that nature in the high latitudes can throw at it.

Natural Enemies

The immature stages of butterflies are prey to an enormous variety of natural enemies. The umbrella term "natural enemies" includes vertebrate and invertebrate predators, parasitoids, and diseases (pathogens). Some natural enemies are "generalists," feeding

Frankliniella occidentalis (Western Flower Thrips) attacking an egg of *Euchloe ausonides* (Large Marble).

on a wide range of food that may include immature stages of butterflies. Some are "specialists," specializing on a certain stage or perhaps a certain species or group of species. Eggs, larvae, and pupae of all butterflies are vulnerable to attack by natural enemies. The appearance, lifestyle, and behavior of all butterfly species are shaped to a large extent by innate strategies to avoid being eaten. Crypsis, distastefulness, gregariousness, web and shelter building, host-plant use, and adult wing patterns have all evolved largely as defenses against natural enemies. From hundreds of eggs laid by each female butterfly, only a handful will reach adulthood, largely as the result of actions of natural enemies.

Natural enemies can be a significant obstacle to butterfly rearing. Eggs, larvae, and pupae obtained from the wild may be parasitized and produce parasitic wasps or flies instead of butterflies. Some species are particularly prone to disease: in Cascadia, wild-caught eggs and larvae of the Compton Tortoiseshell (*Nymphalis l-album*) are often host to a virus that kills larvae before pupation. Host plants obtained from the wild may also harbor natural enemies, particularly predatory insects like Minute Pirate Bugs. While most of these predators can be shaken or washed off, eggs sometimes remain that hatch later, producing tiny but voracious predators.

Predators: Predators seek, catch, and consume their prey. Eggs, larvae, and pupae fall prey to many types of predators, from vertebrates like birds, mice, frogs, lizards, and snakes to a whole array of invertebrates, from predatory mites to praying mantids. The impact of predators on populations of specific butterflies is largely unknown, but the relationships are no doubt complex and variable. Very few studies have been conducted on predation of immature stages of butterflies. An exception is the agriculturally important Cabbage White (*Pieris rapae*), which is targeted by predatory and parasitic insects in early instars and by birds in later instars and as pupae.

Birds chase and consume adult butterflies, but perhaps not very commonly because of the energy involved and often limited success. Most successful bird attacks on adults probably occur early or late in the day, when conditions are cool and butterflies are lethargic or resting. Some birds specialize in searching for eggs and larvae on bushes and trees and likely make a significant impact on populations of some species. Mice, voles, and shrews search

Nymphalis californica (California Tortoiseshell) eggs sucked dry by a predatory bug.

relentlessly for food on and close to the ground, and lepidopteran larvae are a major component of their diet. Greater Fritillary larvae living on woodland floors in spring probably feature in the diets of small woodland rodents. Warren (1987) reported a population reduction of 31–40% of the European checkerspot species, *Melitaea athalia*, due to predation by "small mammals" on pupae. Other vertebrates like rats, opossums, marmots, squirrels, and others are also likely to feed on caterpillars when they find them. Even post-hibernation bears are known to search for caterpillars.

The invertebrate predators of butterfly larvae are many and likely have a much greater impact on populations overall than vertebrates do. Mites are microscopic (0.1–1.0

mm) relatives of spiders and many are carnivorous, feeding on other mites, insect eggs, and even small caterpillars. Erythraeid mites contributed to up to 80% of mortality of eggs and larvae of the checkerspot *Euphydryas gillettii* (Williams et al., 1984). Stigmaeid mites killed many early-instar *Erebia epipsodea* when we reared them on potted grass. The stigmaeids live in and on soil where the larvae rest by day. Phytoseiid mites also kill early instars of a number of species; these mites live on plants and in our studies were inadvertently introduced into larval cultures when field-collected plants were used.

Phymatidae, predatory Ambush Bug nymph, a common predator of immature Lepidoptera.

Thrips are common tiny slender insects that are mostly plant feeders; however, some are predatory and others are omnivores. They are not usually considered important predators of butterfly immature stages, but we have observed feeding by the Western Flower Thrips (*Frankliniella occidentalis*) on eggs of the Large Marble (*Euchloe ausonides*) (see photo p. 29). Thrips are also likely to prey on early-instar larvae; because of their abundance in virtually all habitats, they may cause significant mortality to populations of some butterflies.

Spiders take a great toll on butterfly larvae and are likely to be significant regulators of a number of species. A number of studies in agricultural systems have shown spiders to play a major role in reducing caterpillar populations of pestiferous moths. Hunting spiders in the families Clubionidae, Thomisidae, and Salticidae are the most important caterpillar predators. Some forage over plants looking for prey while others sit and wait for prey to come by. The population density of spiders is usually considerable in most natural habitats; thus spider-caterpillar encounters must be frequent. A single spider in a rearing cage can substantially reduce caterpillar numbers. Crab spiders, well camouflaged against flowers, often catch adult butterflies seeking nectar.

Predatory beetles include the familiar "ladybug" or ladybird beetles, which, contrary to popular notion, feed on many insects besides aphids, including butterfly eggs and caterpillars. Armed with chewing mouthparts, larval and adult lady beetles will readily consume single eggs or egg masses of butterflies, particularly when aphids are scarce. The Multicolored Asian Lady Beetle (*Harmonia axyridis*) is a particularly voracious feeder and has been shown to have a significant impact on eggs and early instars of the Monarch (*Danaus plexippus*). This introduced lady beetle is now common in Cascadia, occurring in a wide range of habitats, and likely has a major impact on egg and larval populations of many butterfly species. Ground-dwelling black-brown beetles like carabids (Carabidae) are common in most habitats and likely consume many caterpillars resting at the bases of host plants. Some species climb bushes and trees searching for caterpillars. Other beetles that feed on butterfly immature stages include tiger beetles (Cicindelidae), soldier beetles (Cantharidae), and rove beetles (Staphylinidae). Carabid and staphylinid beetles caused 2–6% mortality of pupae of the European checkerspot species *Melitaea athalia* over three years (Warren, 1987).

Orius tristicolor **(Minute Pirate Bug), a common predator of immature Lepidoptera.**

The true bugs (Hemiptera) contain at least 21 families that include predatory species, many of which target immature stages of butterflies. True bugs have tubular, piercing mouthparts used to impale prey and suck out the body contents. They develop through 3–4 nymphal stages before becoming adults, and all stages are predatory. Predatory bugs are one of the most important but likely underrated groups of natural enemies affecting butterfly populations. Most species are generalist feeders preying on a wide variety of insects, and some species are capable of developing large populations in a short time in a limited area. Some predatory bugs (Minute Pirate Bugs, damsel bugs) insert their eggs into plant tissue so they are unseen. Nymphs of these bugs often appear in caterpillar cultures when plants harboring bug eggs are used as host plants. Early-stage nymphs are very small but can kill many larvae before being discovered. In Cascadia, the most important predatory bug families that feed on eggs and caterpillars are big-eyed bugs (Lygaeidae), Minute Pirate Bugs (Anthocoridae), damsel bugs (Nabidae), stink bugs (Pentatomidae), ambush bugs (Phymatidae), mirids (Miridae), and assassin bugs (Reduviidae) (see photos). Adult ambush bugs specialize in feeding on flower-visiting insects, including adult butterflies, but nymphs forage mostly on immature insects like caterpillars. All predatory bugs are active foragers on low plants, bushes, and trees, and the winged adults disperse widely seeking new food sources. They are helped in this by plants that respond to butterfly oviposition and caterpillar feeding by sending out volatile chemical "distress signals" that predatory bugs (and other natural enemies) use as locator beacons to find the food source. Thus plants recruit natural enemies as "bodyguards" to protect them against herbivory by caterpillars. Entire egg masses laid by butterflies like the California Tortoiseshell may be sucked dry by a single predatory bug (see photo).

Hesperia juba (Juba Skipper) L1 being stalked by *Orius tristicolor* (Minute Pirate Bug) nymph.

In our rearing of butterflies in eastern Washington, the Minute Pirate Bug (*Orius tristicolor*) was the most numerous predator of eggs and larvae. Eggs, nymphs, and adults were often inadvertently introduced into cultures on host plants, and they killed many eggs and larvae, particularly among hesperiid, pierid, and lycaenid species.

Green lacewings (Neuroptera) are common predators in many habitats, preying on a wide variety of small insects, including early-instar butterfly larvae. Lacewing nymphs resemble tiny alligators with long, curved, tubular mandibles that puncture and suck the fluids out of their prey. In one of our rearings, a single lacewing nymph wiped out an entire cohort of second-instar Arctic Skipper (*Carterocephalus palaemon)* larvae. Some species are also predatory as adults, others feed on pollen or nectar.

Predatory wasps and flies remove significant numbers of caterpillars from host plants and take them back to their nests for immediate consumption or as food for developing larvae. Sand wasps, digger wasps, mud daubers (Sphecidae), hornets, paper wasps, and yellowjackets (Vespidae) are the major types of wasps preying on caterpillars. Mature larvae are often favored, providing greater food resources for the wasp colony. Some species specialize on caterpillars, sometimes specific genera or species.

Although some species of ants have symbiotic relationships with the larvae of some lycaenid butterflies, many more are predators of eggs and larvae. Local colonies of caterpillars may sometimes be extirpated by ants. On one occasion we observed *Tapinoma* sp. ants driving a tachinid fly parasitoid away from a Silvery Blue (*Glaucopsyche lygdamus*) larva on which it was attempting to lay eggs.

Mantids (Mantodea), commonly called praying mantids, are among the best known insect predators. They frequently occur in gardens, and an introduced species is now found in many natural habitats in Cascadia. They are long lived and consume many small insects, including caterpillars, from spring to autumn. They often wait near flowers, attacking bee, fly, and butterfly visitors. Mantids are often found on late summer–blooming Gray Rabbitbrush, the only late-season nectar source in the shrub-steppe and heavily visited by many insects, including butterflies. Another generalist predator, snakeflies (Raphidioptera), related to lacewings, are common in May on blooming Bitterbrush (*Purshia tridentata*) when the larvae of Behr's and California hairstreaks are present. They are also common on *Ceanothus* while the larvae of several butterfly species are present and likely utilize them as prey.

Parasitoids: Parasitoids are generally small to minute wasps (Hymenoptera) or medium-sized flies (Diptera) that develop within a host, ultimately killing it. Eggs, caterpillars, and pupae are hosts to a great number of fly and wasp parasitoids, and for most butterfly species, parasitoids are as important if not more important than predators in regulating population densities. Wild-collected eggs, larvae, and pupae often fail to hatch, develop, or eclose, instead producing single or multiple parasitoids. Many parasitoids are quite specific and will parasitize

Maggot of an unidentified wasp or fly parasitoid emerging from a *Euphilotes columbiae* (Columbia Blue) L3, killing it.

only one or a few closely related host species. The adult female parasitoid lays her egg in or on a single host stage (egg, caterpillar, pupa, adult). Some parasitoids attack and emerge from the same host life stage. Others may parasitize one host stage and emerge from another. For example, a parasitoid may lay an egg in a host egg, with the young parasitoid developing within the developing caterpillar, ultimately emerging from the host pupa, killing the host at that point. Confusingly, parasitoids may in turn also be parasitized by so-called hyperparasitoids, and these are sometimes reared out from caterpillars; however, the end result for the caterpillar is the same regardless of the type of parasitoid: death.

Cocoon of a wasp parasitoid after emerging from a *Papilio rutulus* (Western Tiger Swallowtail) L2.

The most important parasitoids of butterflies belong to the wasp families Chalcididae, Braconidae, Encyrtidae, Ichneumonidae, Pteromalidae, and Trichogrammatidae. Parasitic flies are generally in the family Tachinidae and resemble house flies. Wasps usually inject their eggs into the host, whereas tachinid flies lay them either on the host (the emerging maggot boring in) or on the host plant of the host, to be ingested during feeding. Some tachinid eggs laid on plants hatch into mobile, host-seeking maggots. The parasitoid larva consumes nonessential parts of a caterpillar's body first (e.g., fatty tissue, hemolymph), avoiding killing it until the parasitoid is ready to pupate. The parasitized caterpillar may appear quite normal until just prior to death, when colors may change, feeding ceases, and it becomes quiescent. In caterpillars of the hawkmoth *Manduca sexta*, developing parasitoid larvae interfere with the caterpillar's endocrine system, extending larval duration, thus extending the period of utilization of the host's tissues (Beckage and Riddiford, 1978). Parasitized butterfly larvae often live for an extended period before pupating, suggesting a similar situation occurs. Tachinid larvae usually complete development

Glaucopsyche lygdamus (Silvery Blue) larvae on lupine, tended by *Tapinoma sessile* ants. A parasitic tachinid fly attempts to lay eggs on the larva but is driven away by the ants. Photo by Idie Ulsh.

after the caterpillar pupates, boring out of the pupa and pupating on the ground. Trichogrammatid wasps are tiny and usually egg parasitoids, with as many as 10–20 wasps developing in and emerging from a single butterfly egg. A high percentage of eggs of the Golden Hairstreak (*Habrodais grunus*) are parasitized by tiny wasps (likely Trichogrammatidae) in southern Washington.

Parasitoids in the braconid genus *Cotesia* are commonly found parasitizing whites, swallowtails, and checkerspots. *Cotesia glomerata* is an important biological control agent of the Cabbage White (*Pieris rapae*) but may also parasitize other pierids. Caterpillars of *Euphydryas* checkerspots may be attacked by 1–3 parasitoid species, resulting in up to 67% of larvae being parasitized (Stamp, 1984). In Cascadia, swallowtail larvae appear susceptible to parasitism by braconids and ichneumonids. Most butterfly pupae are less susceptible to parasitism, perhaps because they are harder to find and have a tougher exoskeleton.

Characteristic "hanging" pose of *Nymphalis l-album* (Compton Tortoiseshell) L4 killed by a nuclear polyhedrosis virus.

Pathogens: Pathogens are infectious microorganisms that injure or kill their hosts. Bacteria, fungi, nematodes, protozoans, and viruses are the most common groups of pathogens. These diseases can have a rapid and major impact on butterfly populations, primarily affecting immature stages. Community feeding and moist conditions increase the chances that a pathogen will have a significant effect on butterfly population dynamics. Pathogens are usually a minor influence on survival of immature stages in arid environments but are probably an important factor in the reduced diversity and abundance of butterfly species in coastal areas of Cascadia.

Granulosis and nuclear polyhedrosis viruses are the most important and noticeable pathogens affecting caterpillars. Caterpillars dying from a polyhedrosis virus become limp as internal organs liquefy into an odorous soup. They also adopt characteristic hanging poses, attached by their prolegs or claspers and dripping fluid from the head end (see photo). Larvae of the Compton Tortoiseshell (*Nymphalis l-album*) are particularly prone to a polyhedrosis virus that may be responsible for the observed substantial fluctuations in abundance of this species. Rearing evidence suggests that this virus may be transovarially transmitted, infected females passing the disease to eggs and subsequent larvae. The liquefied body contents of polyhedrosis virus–affected larvae contain millions of virus particles embedded in protein crystals that are highly resistant to heat, cold, drought, etc. Caterpillars ingesting these crystals soon show symptoms of being virus affected. Once virus has occurred in caterpillar cultures it is very difficult to eliminate unless rearing equipment is disposed of or sterilized.

Bacterial diseases and fungi are probably common causes of mortality in populations of most species but rarely result in widespread outbreaks as can occur with viruses. Sublethal bacterial and fungal pathogens may have impacts on fecundity and longevity of adults. One bacterium, *Bacillus thuringiensis* (BT), has been exploited as a pest control agent. This pathogen is available commercially and used (often on a wide scale) to control pest caterpillars like larvae of the Gypsy Moth. BT is effective against all lepidopterous larvae; thus broadscale application can have substantial impacts on populations of nontarget butterflies and moths. Naturally occurring pathogens are the least understood natural enemies of butterflies and need more research.

The role and impact of natural enemies in regulating butterfly populations is an extremely fertile, productive, and underexploited field for research. We know little about the identity and impact of parasitoids, predators, and pathogens on the ecology of individual species, and there is much to be learned.

Defense

From every 100 eggs laid by a female butterfly, only a very small number, perhaps 1–5, will survive to become an adult. Biotic (natural enemies) and abiotic (mostly climatic) factors combine to ruthlessly decimate populations of eggs, larvae, and pupae. To maintain a stable population, each mated pair of butterflies needs to produce only two progeny, even though the female may lay 50–2,000 eggs in her lifetime. However, most butterfly populations are not stable, going through cyclical increases and decreases, often related to mortality pressures on immature stages.

Every species participates in an ever-evolving "arms race" that pits defense against attack. When a caterpillar evolves a new defense strategy swinging the balance toward better population survival, natural enemies in turn invariably develop improved strategies to overcome the new defenses. Multiple defense strategies are employed by virtually all species, with individual strategies often changing in importance during development. A remarkable range and diversity of defense strategies have developed in butterfly eggs, larvae, and pupae, and many more likely remain to be discovered with careful observation and study.

Concealment: *Erynnis propertius* (Propertius Duskywing) L1 in oak leaf shelters.

Concealment and Evasion: Hiding from your enemy might be the oldest and one of the most effective defense strategies used by

Camouflage: *Adelpha californica* (California Sister) L1 on a constructed "frass pier."

potential prey. This strategy is usually the first defense of butterfly immature stages. Practitioners of camouflage, subterfuge, and simple evasion can be very successful. Being active by night is an excellent way of avoiding diurnal enemies like birds and larger predatory insects. There is a risk of being found by nocturnal predators, but these are generally less common than the diurnal types in most habitats. Caterpillars of many Cascadia species are inactive by day and therefore rarely seen. Most fritillary, skipper, and satyr larvae rest by day on or in soil or at the bases of their host plants. Larvae of these groups are among the hardest to locate in the wild, and, presumably, larger diurnal predators have the same difficulty. When darkness falls, skipper and satyr larvae climb up host grasses, avoiding ground-dwelling predators like mice. Fritillary larvae likely stay closer to the ground but have a chemical weapon we discuss later.

Another concealment strategy is to hide within a refuge. In some species with relatively small larvae (e.g., blues, hairstreaks), the host plant becomes a refuge when the larva burrows into it, disappearing from the outside world, at least temporarily. Other larvae (some skippers) use silk to tie leaves together to create an individual "nest." Smaller larvae may cut a flap from a single leaf, bend it over, and tie it down, creating a bivouac.

Hiding from your enemy without actually going anywhere is the illusion known as camouflage or crypsis. Blending in with your background is probably the most common defense employed by butterfly immature stages and is exhibited by all major families and groups. Coloration of the immature stages of the majority of butterfly species is shaped by the need to be inconspicuous and not attract attention; thus, eggs, larvae, and pupae are often colored yellow-green to match their plant backgrounds. In many blues and hairstreaks, inter- and intrapopulation larval coloration is extremely variable, often based on which host plant is being used and, in some species, on which part of the host is consumed. For example, larvae of some blues and hairstreaks

Camouflage: *Euphilotes columbiae* (Columbia Blue) L4 on pink Tall Buckwheat flowers.

may feed on flowers or leaves, with the leaf-feeding larvae becoming green and flower-feeding larvae marked with red, orange, or yellow to match the flowers. Some plants have red markings on their stems (e.g., Russian Thistle, Sheep Sorrel), and the green larvae of the Western Pygmy Blue and Edith's Copper that feed on these plants also have an identical red middorsal stripe. Similarly, some clover cultivars have a red stripe on stems identical to the red stripe on the abdomen of the green pupa of some legume-feeding sulphurs (see pupa image in the *Colias philodice* account). The larvae of many grass-feeding satyrs and skippers are green with paler stripes that blend in perfectly with their grassy environment. The Pine White caterpillar living on conifer needles is similarly colored and also becomes "invisible" in its habitat. Larvae of the Western Pine

Elfin also have green and white stripes. Larvae of other whites and sulfurs feeding on broader-leaved mustards and legumes are also green but without striping, and a perfect match for their hosts. In some species living in more complex habitats, bolder markings, often white or yellow, break up the background color and disguise the caterpillar outline. Some caterpillars mimic common objects in their habitat. The best examples of this are the early instars of many swallowtail species that are black or dark brown with a white "saddle," resembling a bird dropping. They can therefore rest on upper surfaces of green leaves with impunity and not be recognized by birds as caterpillars. Swallowtail larvae grow much larger than most passerine bird droppings, and from mid-development become either leaf green or banded and dotted in black, yellow, and white to blend in with host plants. They also develop some threatening and chemical tactics.

Some larvae adopt resting postures that break up the "normal" caterpillar outline. For example, the green larvae of the California Sister rest in a Loch Ness Monster–type posture, making them disappear among the oak foliage in which they live (see *Adelpha californica* images). This posture may also be threateningly serpentlike to some predators. Pupae, because of their immobility, almost universally use crypsis for defense. Pupae formed on green hosts are invariably some shade of green or yellow, whereas pupae on inert surfaces will generally be colored gray-brown-black to match. Pupal shapes vary tremendously, again largely to break up outlines and make them inconspicuous.

Another tactic to avoid being discovered by the enemy is to reduce your odor. Most, if not all invertebrate natural enemies of caterpillars find their prey by detecting odors. A likely significant source of odor from caterpillars is frass, and some natural enemies hone in on frass odors. Disposing of your frass a distance from where you live should help put enemies off the scent. Some skipper and pierid larvae do exactly this by flinging feces for distances of up to 40 body lengths away (Weiss, 2003). Larvae of some swallowtails have been observed picking up frass with their jaws and throwing it to the ground.

Camouflage: *Callophrys gryneus* (Cedar Hairstreak) L4 on Western Red Cedar needles.

However, many larvae make no attempt to dispose of frass, and some substantially contaminate their nests with it, for example, most web-building species like crescents, checkerspots, and some brushfoots, notably including the Painted Lady. If frass attracts natural enemies to these species, they must depend on other defenses. However, it is possible that frass odor of these species is somehow neutralized or disguised and does not provide prey-location cues for predators and parasitoids. The early-instar larvae of the California Sister and to some extent the Lorquin Admiral and Viceroy actually store frass. For example, the first and second larval stages (L1, L2) of California Sister construct a "basket" from bits of chewed leaf and silk in which they carefully deposit frass pellets (James, 2009a). An Admiral species in the United Kingdom has gone one step further; it is able to throw frass pellets into the basket with great accuracy. Frass collection by admirals and sisters is associated with another strategy designed to "hide" L1 and L2 from visually searching predators. After hatching, California Sister larvae feed on a leaf tip, carefully eating away the leaf on either side of the midrib on which they rest. The dull-colored larvae blend in with their midrib substrate and are "hidden" to most predators. The apparent benefit of hiding on a "pier" or spar is substantial enough to encourage larvae to extend it. This they do by utilizing frass pellets they silk together to make a pier extension that usually doubles its length. It is possible that piers also make it difficult for ants and other predators to locate larvae.

If all the concealment and camouflage tactics fail and you are discovered, then evasion is an option. Many larvae drop from their host plants at the slightest disturbance, making it harder for the predator to find them. Getting back to the plant again may be a problem, but many larvae get around this by falling from the host suspended on a silk thread, winching themselves back up again when the threat passes. The majority of caterpillars, when removed from their host plants, respond by curling into a ball, which probably serves to protect more vulnerable parts (e.g., ventral surfaces, head) from attack.

Threats: *Adelpha californica* (California Sister) L4 "baring its fangs" in a threat display.

Threats and Scare Tactics: Another defense strategy when concealment fails is to threaten or scare your attacker. Caterpillars have developed a number of threatening tactics, but they probably only work well against more timid attackers like naïve birds. Some larvae display sudden movements when threatened, like head-jerking and thrashing the anterior part of the body from side to side. This tactic is probably most effective when a large group of larvae do it in unison, as mid-instar California Tortoiseshell caterpillars do when threatened. California Sister larvae "thrash" and also "bare their fangs" by displaying and moving their mandibles as if to bite. Larvae of the Silver-spotted Skipper and some other skippers also display open mandibles. The larvae of greater fritillaries and Compton Tortoiseshells twitch their bodies when threatened. Pupae of fritillaries and other species may also twitch when disturbed. Mature larvae of tiger swallowtails have eyespots on the thorax that are enlarged when the larva is threatened, giving the appearance perhaps of a small snake. The head capsules of some larvae and the pupal heads of many skippers have markings or modifications that give the appearance of a vertebrate face, which presumably gives pause to any potential attacker. The Silver-spotted Skipper larva has two bold orange "eye" spots, and the Propertius Duskywing larva has a slightly horned appearance. The head of the California Sister larva has two black-tipped projections apparently mimicking eyes. Larvae of some species (Monarch, admirals) have elongated horns at both ends of the body or just anteriorly that are waved around in a threatening manner when a threat approaches.

Spines: *Nymphalis antiopa* (Mourning Cloak) L4 with protective spines.

Getting Physical: Some larvae appear to be physically unpalatable, at least to vertebrate predators. The spiny armature of brushfoot larvae turns soft palatable caterpillars into prickly, tongue-stabbing specialties with which only a few predators will tangle. Spines in combination with other tactics like chemical secretions, thrashing, mandible baring, curling, and dropping are probably fairly effective deterrents for would-be predators. They also make it less easy for parasitic wasps to gain access for inserting their eggs into larvae.

Confusing the Enemy: Trying to confuse the enemy is a trick more often employed by adult butterflies, although some larvae also use it. The most famous example of enemy confusion by adults involves the use of eyespots and filaments on the hindwings of some blues and hairstreaks. The eyes and filaments (which are usually kept in motion when the butterfly is feeding or momentarily resting) resemble the head of a creature with antennae, which birds in particular are happy to attack. Such attacks result in minor wing damage, and the butterfly flies away to live another day. Pretending that the less-important rear end is actually the very important head end is a predator confusion strategy also employed by some caterpillars. Striking examples of "false heads" occur in caterpillars of some moth families (Wagner, 2005). Examples also occur in some subtropical swallowtails, but in Cascadia one of the best practitioners is the California Tortoiseshell. The two posterior abdominal segments of larvae are dark black, increasing in sclerotization and resemblance to a larval head during development. The sclerotized posterior segments most closely resemble a head in the fourth larval stage (L4). Early (1–3) instars of *N. californica* feed and rest communally, and the appearance of twice the number of "heads" in a community may reduce the risk that real heads are attacked by a predator; however, an approaching predator is usually met with a communal "head-jerking" reaction. This may be a second tier of defense with "confusion" the first.

A head at either end? A *Nymphalis californica* (California Tortoiseshell) L4 confuses the enemy.

Safety in Numbers: Aggregation or gregariousness is a behavioral strategy that reduces the odds of any single individual being attacked. Arguably the most vulnerable stages, early-instar larvae, are frequent users of communal defense, gradually shifting to other defense strategies as they mature. Examples abound in brushfoots like the Mourning Cloak, California Tortoiseshell, Milbert's Tortoiseshell, checkerspots, and crescents. Communal larvae may also build silken webs, supports, and platforms to help keep the larval community together and enhance defense.

Chemical Defense: Using toxic chemicals sequestered from host plants or producing new noxious compounds from benign chemicals, larvae of many species across many families engage in chemical defense. This is a potentially productive research area with new chemical defense strategies likely awaiting discovery. The most celebrated example of butterfly defense using plant poisons is the Monarch, whose larvae sequester cardenolides or cardiac glycosides from milkweed host plants. These plant poisons

Argynnis coronis (Coronis Fritillary) larva displaying ventral gland, which contains defensive chemicals.

make Monarch larvae, pupae, and adults unpalatable to vertebrate predators. The striking aposematic yellow, black, and white banding of Monarch larvae is recognized by educated birds as indicating distastefulness, and protection may also be conferred on a number of other similarly marked larvae (e.g., Spring White, Anise, and Old World swallowtails). Whether this is a case of mimicry to gain protection or simply convergent evolution is unknown. Other Cascadia species using distastefulness as the basis for defense include the Buckeye and checkerspots. The aposematically colored larvae of these species feed on host plants containing iridoid glycosides. *Euphydryas* checkerspot larvae are brightly colored with bright white, red, or orange spots, making them highly conspicuous to visually searching predators. Birds and some invertebrate predators tend to avoid attacking these larvae. Checkerspot larvae sequester iridoid glycosides and related iridoids produced by host plants and use them for defense. Unpalatability of checkerspot larvae for birds has been shown in cage studies, but there is much variation in unpalatability according to host plant species used and the checkerspot species. Other larvae with markings and coloration that may indicate host-derived chemical defense include those of the American Lady, Anise Swallowtail, Becker's White, and Mourning Cloak; however, studies demonstrating unpalatability of these species are lacking.

Argynnis zerene (Zerene Fritillary) larva showing ventral gland containing defensive chemicals.

Another form of chemical defense is shown by first instars of pierids (whites and sulphurs) that carry oily droplets on the tips of dorsal setae. The droplets of Cabbage White L1 consist primarily of a series of labile, unsaturated lipids, the mayolenes, which are derived from 11-hydroxylinolenic acid; in laboratory bioassays the secretion was shown to deter ants. Predation by ants in the field on Cabbage White L1 with droplets is significantly lower than that on larvae that have had their droplets removed. Interestingly, while the droplets appear to deter predators like ants, they also act as a cue for parasitic wasps searching for larvae to parasitize. The chemistry of setal droplets in other pierid species has not been researched and would make a fascinating study. In Cascadia, larvae of all crucifer-feeding whites and legume/blueberry-feeding sulphurs have setal droplets. Among the pierids only the conifer-feeding Pine White lacks the droplets. In some pierids the droplets persist to L4 or L5, but in others they are lost after L1 or L2. Setal droplets are also present in L1 of some species in other families, including lesser and greater fritillaries and some skippers. These are presumed to have a defensive function, but the chemistry remains to be elucidated.

Neophasia menapia (Pine White) L3 showing ventral gland with defensive chemicals.

The larvae of parnassians and swallowtails possess unique chemical defense in the form of an eversible forked, fleshy gland (osmeterium), colored yellow, orange, or red, that shoots out from a slit behind the head of the larva when threatened. Most developed in the swallowtails, the gland glistens with an odoriferous secretion that alters in composition (and aroma) during larval development. The strong disagreeable odor of final instars of most swallowtails is pungent and sometimes

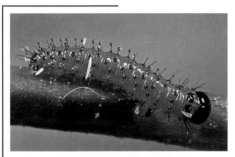

Pontia occidentalis (Western White) L1 showing droplets on setae, likely for chemical defense.

rancid, depending on the species. The odor produced by earlier instars is usually citrus-like and more pleasant. Osmeterium secretions have been demonstrated in a number of studies to repel predators like birds from swallowtail larvae, increasing their chances of survival. The chemistry of osmeterium secretions from most Cascadia swallowtails is unknown, but studies have been conducted on *Papilio machaon* and a tiger swallowtail species from the eastern United States, *Papilio glaucus*. During L1–L4, larvae of these species produce more than 50 terpene compounds in their osmeterium; however, in L5 the chemistry changes abruptly with the disappearance of terpenes and the copious production of isobutyric acid and 2-methylbutyric acid. Swallowtail larvae may alter the chemistry of osmeterium secretions during development to target different predators. The osmeterium of parnassians is yellow and relatively short, secreting a colorless liquid with a slightly offensive odor. Our species have not been examined, but the secretions of related species comprise methyl esters (as in *Papilio* L5) or monoterpene hydrocarbons.

An organ apparently homologous to the swallowtail osmeterium is found in larvae of a number of species from at least three families (Nymphalidae, Hesperiidae, Pieridae). This organ is found ventrally on the anterior margin of the first segment, essentially on the opposite side of the body from where the osmeterium is located in swallowtail larvae. Like the osmeterium it consists of a single- or bilobed eversible fleshy gland that discharges an odoriferous secretion when the larva is disturbed. Although the existence of this gland in moth caterpillars has been known for more than a century, its occurrence in butterfly larvae is less well known, and virtually nothing is known about the chemistry and function of secretions. Muyshondt and Muyshondt (1976) proposed that the gland be called "adenosma" (from the Greek *adenos*, gland, and *osme*, smell). Scott (1986a) briefly mentioned the occurrence of adenosma in some nymphalids, pierids, and hesperiids. James (2008a) described and illustrated the gland from a number of fritillary species. The existence of adenosma is noted in species accounts in this book, although we did not examine every species for this gland. Initial studies on the chemistry of adenosma secretions in fritillaries indicate that they contain saturated hydrocarbons and acetate esters. Squalene, a triterpene hydrocarbon found in the secretions, is a known ant repellent, and the acetate esters are known as "dispersal pheromones" in some moths. Thus, the adenosma secretions from fritillary larvae may serve to repel ground-dwelling invertebrate predators like ants and centipedes and may also alert nearby larvae of threats.

Bodyguards: Many larvae of blues, coppers, and hairstreaks have developed the unique defense strategy of recruiting bodyguards and depending on them to repel threats. The

A *Tapinoma* ant imbibes "honeydew" from nectary gland of *Glaucopsyche lygdamus* (Silvery Blue) L4 in exchange for protection from predators and parasitoids.

bodyguards, ants, are a significant natural enemy of most non-lycaenid butterfly larvae; however, lycaenid larvae have evolved to produce something that many ant species love: honeydew. Honeydew-collecting ants have been harnessed by a number of different insects, including aphids and scale insects, to provide defense from other insects. Most lycaenid larvae have a functional honeydew gland that ants tend. Protection is derived from the physical presence and activity of ants swarming over the larvae and substrate, preventing parasitoids and predators from attacking. If a predator is persistent, it will eventually be overpowered by the ants and discarded. Ant-lycaenid associations extend to facilitating pupation of larvae in underground ant nests in some European species, but this has not been reported in any North American species to date. Honeydew-collecting ants are less abundant in Cascadia than in warmer areas like California, and dependence by Cascadian lycaenids on ant bodyguards is less than in California; nevertheless, ant-tending of larvae is readily observed in some species, including the Silvery and Arrowhead blues, Ruddy Copper, and Gray Hairstreak. Seeking ants on lupines in May–June is a good way to find larvae of various blues without needing to search every plant. Many lycaenid-ant associations are poorly known, providing a fertile research area for naturalists and scientists.

Multifaceted Defense: We have summarized the primary defense strategies and tactics employed by the immature stages of butterflies to help reach adulthood. Other less-obvious strategies and tactics undoubtedly remain to be discovered; however, it is clear that every species employs a suite of defense tactics that usually alter with stage and development. For example, the immature stages of the California Sister use crypsis, aggression, armor, coprophagy, intimidation, and unpalatability as defense tactics during development. It is likely that the most successful species use multiple defenses and are quick to respond to changing threats. The "arms race" between butterflies and their natural enemies is dynamic, with the balance of power shifting constantly. This particular "war" is fascinating to watch.

Variation

People do not all look alike, nor do all dogs nor all butterflies, even though they may be classified as belonging to a single species. Variation has puzzled and challenged lepidopterists since the beginnings of the science. We humans prefer to classify organisms into neat, predictable boxes we can recognize and understand; however, butterflies care nothing for our predictable boxes, our names, or our classification schemes; rather, they just live, survive, reproduce, and die in the best way they can with little regard for how we may perceive them.

Variation within butterfly populations takes many forms, manifested not only in the adult insects but in the immature stages as well. In some cases striking variation can occur within the offspring of a single brood from the same parents; this is well illustrated by *Argynnis mormonia* (Mormon Fritillary), in which siblings may have reflective metallic silvered spots on their ventral wing surfaces, or they may have dull yellow nonreflective spots, imparting a strikingly different appearance. Many butterflies are multivoltine (multiple brooded), and in some species the successive broods look quite different. The spring brood of *Pieris marginalis* (Margined White) has dark olive green scales along vein lines whereas the summer brood is immaculate, entirely lacking the green lines. Temperature can affect adult butterfly appearance, with those developing in cold conditions often having darker coloration (for heat absorption) than individuals of the same species in warmer areas.

Food shortages or adverse environmental conditions during larval development can result in undersized larvae and pupae, producing small adults in the following generation. Adult flight periods can vary in response to adaptation to different host plants. Adults of *Chlosyne acastus* (Sagebrush Checkerspot) in populations where larvae feed on the early-growing *Erigeron linearis* (Linear-leaf Daisy) fly earlier than adults of populations that feed on *Chrysothamnus viscidiflorus* (Green Rabbitbrush), which grows later. Flight periods are also staggered with elevation in response to growth of the host plants.

Colors and patterns of butterfly larvae can be dramatically affected by the host plants they consume. Larvae that feed on colored flowers often concentrate their bright colors, resulting in bright stripes, bands, or spots not seen in larvae of the same species feeding on green leaves. This is especially noteworthy in the blues and hairstreaks, dramatically illustrated by *Callophrys sheridanii* (Sheridan's Hairstreak). In this species, early-season larvae feeding on *Eriogonum compositum* (Northern Buckwheat) leaves are plain green with pale yellow or white markings, whereas late-season larvae feeding on *Eriogonum elatum* (Tall Buckwheat) flowers develop bright pink or red patterned markings, imparting a dramatically different appearance. *Strymon melinus* (Gray Hairstreak) and *Glaucopsyche lygdamus* (Silvery Blue) are both renowned for the extreme variability of the color of their larvae; both species use the flowers of a wide variety of host plants, and the varied flower colors result in highly variable larval colors. In contrast, grass skippers and satyrs are two unrelated groups that utilize only grasses and sedges, virtually all of which are green; as a result, there is little variation in the colors of their larvae from one population to the next. In general, one can expect the greatest larval variability in those species that utilize a wide variety of host-plant flowers, and relatively less variability in species that feed only on leaves.

In some species the larval colors vary greatly between populations for no apparent or obvious reason. This is the kind of variation we most often encountered in our multiple rearings of the same species. In some populations of *Euphydryas anicia* (Anicia Checkerspot), the larvae are mostly black, whereas in other populations they are mostly white. There is typically relatively little variation within a given population but sometimes pronounced differences between separated populations.

Structural variations also occur in immature stages. In the eggs of nearly all butterfly species, the number of ribs and the coarseness of surface texture can be rather variable. In some species the size of spines and horns in larvae can vary fairly dramatically. In *Nymphalis antiopa* (Mourning Cloak) larvae, large dark spines are characteristic; however, the length of the spines can vary substantially in different populations at the same growth stage.

One group that stands out as unique in its level of confusion and debate among lepidopterists is the lycaenid genus *Euphilotes*. Most experts have concluded that *Euphilotes* blues are separable to species based almost entirely on the larval hosts used, each species developing on only one or in some cases two species of *Eriogonum* (buckwheat). Others have found that this separation is reinforced by phenotypical studies, by comparing collections of adults from different host plants, but most agree that adult *Euphilotes* cannot always be identified to species by observing a single adult, as they all pretty much look alike and there is considerable variation within each species. This is of course frustrating to the casual butterfly enthusiast who wants simply to be able to identify butterflies seen or photographed. In this book we present accounts for five *Euphilotes* species, four of which are described species and one a speculative possible species (*Euphilotes* on *Eriogonum sphaerocephalum*) based on its unique host plant. More research is clearly needed to sort out this difficult and variable group, and we hope the information we present will be of assistance to those intrepid souls who endeavor to do so.

Geographic separation is one of the great driving forces of evolution. Populations that are physically separated will invariably develop in different directions, over time eventually becoming so distinct that they will no longer interact or interbreed even when placed in contact with each other. At this stage some lepidopterists like to place these populations into different boxes and give them different species names, but in some areas like Cascadia, where glaciers have come and gone, separating then joining populations of butterflies over the ages, we have numerous instances of butterfly populations in some stage of separation or rejoining. Some are inexorably continuing to separate into different species, others have rejoined after long separation and are reestablishing single identities, and many others are in a jumbled and confused state, separating at one extreme of their range and joining at the other extreme. All this causes great confusion and heated disagreements among lepidopterists.

A classic example of such confusion occurs in *Callophrys gryneus* (Cedar Hairstreak), a taxon historically split into numerous species, then rejoined, and endlessly debated. Within Cascadia, *C. gryneus* is found in two separate populations, the cedar-feeding populations of forested areas and the juniper-feeding population of the Columbia Basin. These populations are nowhere in contact in our immediate area, and the adults (and arguably the larvae) are separable on sight. The natural reaction is to declare these populations different species; however, when the greater range is studied, particularly in southern Oregon and California, the differences between the two populations blur and intergrade (Warren, 2005). In this book we present separate accounts for the two populations, the cedar-feeding *Callophrys gryneus plicataria* and the juniper-feeding *C. g.* nr *chalcosiva*, to provide additional information for future debates on this subject.

In the species accounts that follow, we have given double-account coverage to several other subspecies-level taxa, typically those in which there has been some level of disagreement or debate or simply a lack of available information. These include *Satyrium sylvinus sylvinus* and *S. s. nootka*, *Celastrina echo echo* and *C. e. nigrescens*, *Argynnis hesperis dodgei* and *A. h. brico*, and *Argynnis mormonia erinna* and *A. m. washingtonia*. *Callophrys perplexa* was a candidate for double coverage (ssp. *perplexa* and *oregonensis*), but we decided to use a single account for this species with photos of both subspecies.

It is not possible in a publication of this scope to show all variations within each species, nor of course have we observed all such variations. To do so would require many more years of research and rearing; however, in those instances where we have observed significant variation, we have pointed this out in the accounts, in the hope that it will be instructive to future researchers. The majority of species covered in the book were each reared more than once. It is true to say that the extent of variation we encountered for many species in multiple rearings exceeded our expectations. Thus, our illustrations show the typical forms we saw in our rearings and simply provide guidance to identification. The reader should not necessarily expect to see exact replicas of all possible variations in our photographs. There are some species that are very consistent in appearance and change little with different climate, host plants, rearing conditions, etc., but these are generally in the minority.

Some examples of species in which we observed tantalizing variations that have been inadequately explained are given below. It is hoped that your curiosity and interest will be piqued.

Celastrina echo (Echo Blue): There is great variation in the boldness and coloration of dorsal plates on the larvae, particularly in the final instar. The significance of this variation for separating the two subspecies, *C. e. echo* and *C. e. nigrescens*, is poorly understood.

Polygonia oreas silenus (Oreas Comma): Adults and larvae from the west side of the Cascades (Nooksack Co., Wa.) are different from their counterparts east of the Cascades (e.g., Kittitas Co., Wa.), even though both areas are mapped to the same subspecies.

Pyrgus ruralis (Two-banded Checkered Skipper): High-elevation adults and larvae from western Yakima Co., Washington, are different from midlevel populations from Kittitas and Nooksack Cos., Washington.

Butterfly Habitats in Cascadia

Robert M. Pyle in his authoritative *The Butterflies of Cascadia* usefully defines the biogeographic region of Cascadia, and we encourage readers to refer to this account. Pyle selected a "conservative" definition of Cascadia for his book that comprises all of Washington and Oregon, the bordering sections of British Columbia, western Idaho, and northernmost California and Nevada. Our definition is similar except that we do not include the southern half of Oregon.

As Pyle notes, "Cascadia consists of a mosaic of landscapes, each with its own particular face, and influences in terms of butterflies." He describes 15 ecogeographic provinces in Cascadia with characteristic geology, climate, and ecology, resulting in distinct and identifiable butterfly faunas. Again, we refer readers to Pyle's book for further details on these ecogeographic provinces.

Our book concerns the immature stages of Cascadia's butterflies, so our consideration of habitats will center on their impact and control over life histories. Immature stages are relatively immobile (compared with winged adults, which at least have the potential to "escape") and are usually confined to a habitat for a long period of time, sometimes for 10–11 months of the year. The key adaptations needed to survive in specialized habitats, including dormancy, defense, and host plant utilization, are detailed elsewhere. Here, we take a more general and geographic approach, highlighting some important, characteristic, and butterfly-rich habitats in Cascadia and the adaptations needed by endemic species to survive. Pyle's fifteen Cascadian ecogeographic provinces fall equally and conveniently into three basic geographical units: coastal plains, highlands, and basin lands.

Shrub-heath habitat of western Cascadia Grapeview Loop, Mason Co., Wa.

Coastal Cascadia: The immediate coastal regions of Cascadia, with predominant cloudiness and cool temperatures for most of the year, provide a generally poor habitat for butterflies. Abundance and diversity are usually limited, particularly in immediate coastal areas of Washington and southern British Columbia. For example, Grays Harbor Co. in central coastal Washington records 30 species of butterflies. By contrast, Okanogan Co. in north central Washington, with components of highland and basin land fauna, records 100 more species (133). Species with short flight periods risk encountering prolonged unfavorable conditions on the coast, especially during spring, limiting activity, nourishment, and reproductive success. Cool temperatures and the lack of sunshine also restrain development of immature stages, prolonging exposure to disease, predation, and parasitism. Fungal and bacterial diseases promoted by moist, cool conditions are probably important regulators of butterfly populations in these coastal areas.

Inland from the coast but still west of the Cascades the climate and habitat improve significantly. The Olympic Mountains provide excellent butterfly habitat, particularly on the east (rain shadow) side, and because of the isolation of this "island range" from other mountainous areas, a number of unique subspecies have developed there. The southeast flanks of the Olympics provide the best known remaining habitat for the scarce Johnson's Hairstreak. The Sequim area of Clallam Co. receives rainfall nearly as low as that of the Columbia Basin as a result of the Olympic rain shadow effect, in the process creating a unique prairie habitat surrounded by wet maritime climate. Likewise in Oregon, the east flanks of the coastal mountains provide much improved habitat over the wet maritime coastal strip.

The Puget Prairies of Washington and the Willamette Valley of Oregon are well known for their unique habitats and butterfly populations. These prairie and valley habitats historically dominated large areas in Pierce and Thurston Cos., Washington, as well as the Willamette drainage of western Oregon, and in many areas were maintained prehistorically by Native American burning. Today the prairies are severely threatened, with only a few remnant patches still intact, and some of the butterflies found in these habitats are listed or proposed for listing. The prairies have long faced severe deterioration by human conversion to development, agriculture, and other uses. Less obvious but also severely damaging is the encroachment of nonprairie flora such as fir forest and nonnative shrubs, particularly Scotch Broom, into the prairie remnants. Many larval host plants are low growing or creeping and easily crowded out, leaving the butterfly larvae unable to survive.

In western Washington another unique habitat, with butterflies partially similar to those found in the prairies, is the shrub-heath plant community. This habitat is particularly well represented in Mason Co., from Shelton north to the lower Hood Canal area. This habitat is dominated by heaths, shrubs, and low or creeping plants, many of which are important hosts for larvae of a number of butterfly species. Typical plant communities in the shrub-heath include *Gaultheria shallon* (Salal), *Arctostaphylos uva-ursi* (Kinnikinnick), *Lotus crassifolius* (Big Deervetch), *Vaccinium ovatum* (Evergreen Huckleberry), and *Arctostaphylos* spp. (manzanita). Early-spring butterflies are typified by the hairstreaks *Callophrys augustinus* (Brown Elfin), *C. polios* (Hoary Elfin), *C. perplexa* (Bramble Hairstreak), and *C. eryphon* (Western Pine Elfin). Later, *Epargyreus clarus* (Silver-spotted Skipper) and a variety of swallowtails, greater fritillaries, and other species can be abundant. This fascinating habitat is not among our most scenic; indeed, the best shrub-heath habitat is found in recovering logging clear-cuts and power line rights-of-way. Following clear-cutting, the shrub-heath habitat typically begins supporting butterfly communities after about two years. One of the most important larval hosts is *L. crassifolius*, with at least five butterfly species wholly or largely dependent on it; this plant is one of the earliest to colonize following a clear-cut. The quality of the post–clear-cut habitat improves steadily for several years, peaking after roughly 5–7 years, at which time the ericaceous plants are firmly established, providing reliable larval host resources to a variety of butterflies. Then the encroachment of large shrubs and replanted evergreen trees begins to degrade the habitat, crowding out the smaller host plants, and after perhaps 10 years the butterflies have moved on to another recovering clearing. Because logging operations in this region are active and widespread, we can expect excellent shrub-heath habitat to continue to be available for the foreseeable future, although both the butterflies and the lepidopterist need to seek out the best new habitats on an almost yearly basis. We do not know to what extent populations move within the shrub-heath, or whether they wax and wane, with populations expanding and diminishing as the habitat ages. Whatever the means of adaptation, this unique and interesting suite of butterflies manages to adjust and thrive in this unusual and changing environment.

High mountain habitat of Cascades, Bear Creek Mt, Yakima Co., Wa.

Mountainous Cascadia: High-elevation areas, from modest "hills" and escarpments (1,000–3,000 ft) to snow-capped mountains (3,000–14,000 ft) are a dominating physical feature of the Cascadian landscape. The Cascade Mountains run the length of Cascadia, with Arctic/alpine habitats common in the north, becoming more subalpine in southern Washington and Oregon. The eastern slopes and canyons of the Cascades support many highly productive butterfly habitats. Other important highland areas include the Selkirk Mountains in the northeast, the coastal Olympic Mountains in the northwest, and the Blue-Wallowa Mountains. The butterfly fauna of low–high-elevation Cascadian habitats is exceptional in its diversity and sometimes abundance. Some of the best Cascadian butterfly habitats are found in the Cascade, Blue, and Wallowa mountains. Many high-elevation areas remain largely unexplored by lepidopterists (e.g., the Pasayten Wilderness in the northern Cascades) and likely hold many butterfly surprises.

While high-elevation areas with their ample rainfall usually abound with host plants in good condition, opportunities to develop, fly, feed, and mate are strongly limited by cool–cold and cloudy conditions. Consistent favorable conditions are characteristic of only 3–4 months annually, and most butterfly development and activity is concentrated in this short period. High-elevation species have evolved a number of adaptations that allow them to take maximum advantage of any favorable conditions. Most commonly, adults and larvae use dark coloration to absorb radiant energy and increase body temperatures, enabling activity and/or development. Many mountain butterflies have dark areas close to the wing bases to enable rapid transmission of heat from wings to the body. They also display "heliothermic" behavior, orienting when resting to achieve optimal exposure to sunshine and body heating. The temperature threshold for adult activity is also lower in most mountain butterflies than in lowland species. Some species (e.g., some satyrs) can even fly in overcast, rainy weather. Less research has been conducted on the adaptations of mountain butterfly larvae to maximize heating and development, but similar morphological and behavioral strategies likely occur. Larvae of fritillaries (*Argynnis* and *Boloria* spp.) are invariably dark colored and probably spend much of the day resting in warm microhabitats under sun-exposed rocks, vegetation, and soil. Larval development of some of these species is remarkably rapid in spring, taking as little as 4–6 weeks despite cool conditions. Similarly, development of black *Parnassius smintheus* (Mountain Parnassian) larvae at 3,000–5,000 ft in late spring–early summer is also rapid.

Sometimes conditions at high elevations are suboptimal for extended periods and timely development of some alpine butterfly species is impossible. Many species have adapted to this uncertainty by having the capability to postpone development until the following year, should conditions dictate. Others routinely take two years to develop from egg to adult. The immature stages of mountain butterflies must also have the ability to survive extended periods (up to 10 months at a time) of subzero temperatures and no food.

The telescoping of development and adult activity into a short period of time each year makes visiting of prime butterfly habitats in Cascadian mountains a memorable experience. A day in the mountains during June–August will often yield 50–60 species of butterflies, with some species (e.g., fritillaries, checkerspots, blues) highly abundant. The similar and profuse response of mountain flora to ephemeral optimal conditions, as well as being instrumental in butterfly abundance, simply adds to the spectacle. The abundance, diversity, and activity of butterflies also mean that opportunities for locating eggs and larvae abound in these habitats at this time. Whether you follow ovipositing females or search for eggs and larvae on host plants, it is usually easy to come up with at least some immature stages for rearing. Inspect *Sedum* (stonecrop) patches on the highest peaks or look for the giveaway bright yellow spots that adorn the black larvae of Mountain Parnassians. Look for ants crawling over lupine plants; invariably they will be tending larvae of Silvery, Arrowhead, or other blue species. Female skippers can often be seen ovipositing eggs at the bases of grasses, and swallowtail eggs may be found on leaves of host bushes along trails. Look at *Ceanothus* bushes for signs of *Nymphalis californica* (California Tortoiseshell) larvae, which in some years defoliate these plants. Bear in mind, though, that many of these mountain species will require a cold winter for their overwintering stage.

Many great butterfly habitats in the mountains of Cascadia are well known by lepidopterists, but many more remain to be discovered. Some notable locations include Slate Peak, north of Winthrop in northern Washington, Bird Creek Meadows near Mount Adams in Washington, Apex Mountain in southern British Columbia, Chumstick Mountain near Wenatchee in central Washington, and Diamond Peak in the Blue Mountains of southeast Washington. The Blue and Wallowa mountains of northeast Oregon also offer many great butterfly habitats, as do the peaks and ranges in the Goat Rocks wilderness area near Rimrock Lake, west of Yakima, Washington.

Shrub-steppe habitat of Cascadia Basin Lands, Waterworks Canyon, Yakima Co., Wa.

Basin Lands of Cascadia: The arid basin lands (shrub-steppe) of eastern Washington and Oregon occupy the greatest land area of Cascadia and are characterized climatically by hot, dry summers and cold, dry winters. Sagebrush is the dominant plant in the region, which is also known as the sage-steppe. Rainfall is sparse (8–15 inches annually), and precipitation rarely occurs in the summer months. At first sight these desert lands appear quite inhospitable for butterflies, and for much of the year they are; however, exceptions do occur, for example, in riparian areas adjacent to major waterways like the Columbia and Snake rivers as well as along smaller rivers and creeks. Irrigated agriculture and urban development in basin lands has removed native habitats and displaced some butterfly species but has also benefited some adaptable species. Moist areas near agricultural fields, orchards, vineyards, and gardens have allowed introduced mustard species (like Tumble Mustard) to remain green until midsummer, which in turn may allow an additional generation of some pierid species (e.g., *Pontia beckerii*, Becker's White). Some skippers such as *Atalopedes campestris* (Sachem), *Polites sabuleti* (Sandhill Skipper), *Hesperia colorado* (Western Branded Skipper), and *Ochlodes sylvanoides* (Woodland Skipper) thrive in waste, grassy areas adjacent to agricultural fields, and may expand populations as agricultural chemical inputs lessen in the future.

The best time to look for butterflies in shrub-steppe country is during March–June, when the majority of native plants bloom and flourish before browning off and becoming summer-dormant. The earliest butterflies eclose from overwintering pupae in late February and March (e.g., *Anthocharis sara*, Sara Orangetip; *Callophrys mossii*, Moss' Elfin) or appear from adult overwintering sites (e.g., *Nymphalis antiopa*, Mourning Cloak; *Polygonia* spp., anglewings). Others appear a little later following development of overwintered larvae, for example, Boisduval Blues (*Plebejus icarioides*) and Juba Skippers (*Hesperia juba*). The immature stages of shrub-steppe spring butterflies that oversummer and overwinter as pupae are characteristically fast developing because they are in a race against time. The pleasant, green, flowery shrub-steppe environment does not last long. In some years, optimal temperatures, sunshine, and host plants may last for only 4–6 weeks before excessive heat and lack of moisture cause plant senescence and curtail larval development. The pressure to develop rapidly has resulted in the ability of the larvae of some of these spring shrub-steppe butterflies to pupate at a smaller size than normal, which in turn results in undersized adults.

The majority of spring-flying shrub-steppe butterflies are dormant in one stage or another through the summer into autumn and winter. A few have evolved to exploit another season of the year with near optimal temperatures and host plant resources: autumn. Larvae of some shrub-steppe grass skippers like *Hesperia juba* (Juba Skipper) and *Amblyscirtes vialis* (Roadside Skipper) develop rapidly in spring to the final instar, then enter short-term dormancy during the hottest months, delaying pupation until the end of summer. Adults eclose, mate, and lay eggs during autumn, avoiding high summer, when host grasses are brown and unsuitable for larval development. Interestingly, some shrub-steppe species "ignore" spring, delaying flight activity until autumn. *Apodemia mormo* (Mormon Metalmark) is a good example, with prolonged larval development during spring and summer.

There are many shrub-steppe habitat sites in Cascadia that are well worth visiting in spring and autumn. Because of their seasonally ephemeral nature as good butterfly sites, many have yet to be discovered and should be sought on the first sunny days of spring. The flowery canyons and coulees in April near Yakima, Ellensburg, Wenatchee, and the Tri-Cities in eastern Washington are productive habitats for shrub-steppe butterflies, as are sites along the Columbia and Snake River valleys and their tributaries. The line of hills to the northwest of the Tri-Cities (Red, Candy, and Badger mountains) support good populations of swallowtails, checkerspots, and blues. Larvae of *Papilio zelicaon* (Anise Swallowtail) and *Chlosyne acastus* (Sagebrush Checkerspot) are readily found on *Lomatium* spp. (desert parsleys) and *Chrysothamnus viscidiflorus* (Green Rabbitbrush), respectively, on Red Mountain during April. Gray Rabbitbrush (*Chrysothamnus nauseosus*) flowers during late summer and autumn and is abundant in the shrub-steppe, attracting many butterflies to nectar, including skippers, fritillaries, sulphurs, and blues.

The life histories of Cascadia butterflies are molded by the three geographic/climatic habitat zones summarized above. Within these zones, up to five subzones add their influences to life history biology and ecology, creating characteristic butterfly faunas. Some of these faunas are imperfectly known and may contain further ecological subzones with "new" races, subspecies, or even species. Even in highly human-populated areas like the San Francisco Bay and Sacramento Valley, butterfly surveys/studies are lacking for some botanical/ecological zones (Shapiro and Manolis, 2007, p. 8). The thinly populated vastness of Cascadia is inevitably home to a good many butterfly surprises that await discovery. We strongly encourage the reader to explore undersurveyed areas for butterflies. Get off the beaten butterfly track. Many areas are pristine roadless wilderness and difficult to access, but some are very accessible but deemed nonattractive to butterflies as well as to people. Butterfly discoveries can still be made in wasteland and developed habitats.

Fieldwork

Finding butterfly eggs, larvae, or pupae in the wild can be exciting, enjoyable, and uniquely rewarding. Poets, historians, and scientists have written for centuries of the pleasure and fulfillment of watching butterflies glide and flit among wildflowers in beautiful summer meadows; we believe that the discovery and study of their immature stages takes us a step deeper into the appreciation of these interesting and beautiful insects. But finding eggs, larvae, and pupae is less a matter of chance or luck than of careful planning and research, together with solid basic knowledge. Whether the goal is to find eggs or larvae or to capture a gravid female to produce eggs, one's success is greatly increased by knowing when and where to search, how to recognize prime habitat, and how to recognize the sought-after species. Being well prepared does not diminish the enjoyment of the search or the excitement of the find, and in fact makes the eventual discovery even more rewarding. In contrast, searching blindly for immature stages is often so unrewarding that some will find it difficult to maintain interest. We hope this chapter will encourage field observers to take a little time to plan ahead in ways that will increase not only success but also the enjoyment of this delightful natural experience.

Preparation for Fieldwork: Preparation can greatly reduce wasted or unproductive field time. Searching for tiger swallowtail larvae in the early spring, or greater fritillary larvae in the summer, will be predictably unsuccessful, because these stages do not occur at such times. Tiger swallowtails overwinter as pupae, so their eggs are typically laid in late spring and their larvae develop in the summer. Greater fritillaries overwinter as unfed first instars, feeding and developing in the spring. Butterfly life histories are complex, further complicated by the fact that some species are double or triple brooded, repeating their life cycle multiple times each season. Choosing the best time to seek immature stages in the field is not as simple as consulting a cookbook of data. The best timing varies from year to year and with latitude and elevation, and populations of species often wax and wane in a cyclic manner. So preparation for fieldwork is an art as well as a science, a combination of judgment, experience, and knowledge, pieced together with various fragments of information like a challenging puzzle.

Probably the single most important piece of knowledge for Cascadia is the overwintering stage of each species. In more southerly, warmer regions the overwintering stage is less important, but in the Pacific Northwest winter is a great obstacle that must be overcome by every butterfly species. During winter there is no nectar and no edible larval host plant, and the cold weather does not permit flight. To survive, all butterflies must have an overwintering strategy, or they must leave the region by migrating south. Some butterflies overwinter as eggs, others as young, partly grown, or mature larvae, some as pupae, and some as adults. The window of opportunity for finding immature stages of any particular species is dependent on that species' overwintering stage. The overwintering stage of each species is described in our accounts and in some instances differs from the stages described from other geographic areas.

The flight period of each species is also important; this, together with the overwintering stage, provides important clues to when immature stages will be present. The adult flight period varies in different years; for example, an early spring usually results in early flight periods for many species. Published flight periods typically encompass an entire area, for example, all of Cascadia, but elevation exerts a great influence within that area: a species will fly early in the flight period at low elevations and later at higher elevations. Within a population, females tend to fly later than males and typically oviposit late in the flight period.

The rate at which eggs hatch and larvae grow also affects the timing of when they can be found. Information of this kind is generally unavailable in the literature, but we add to this knowledge base in our accounts. Generally, butterflies that feed on low-nutrition plants, such as grasses, sedges, or evergreen needles, grow slowly, whereas those using highly nutritious plants and parts of plants, such as buds and flowers, will develop faster. Temperature is the primary regulator of insect development, and all insects, including butterflies, develop much faster under warm than under cool conditions. Caterpillars developing in spring grow at a slower rate than those developing in midsummer; however, excessively warm conditions (i.e., above 35–40 °C) tend to inhibit insect development.

Identification: Many species of insects will be encountered while searching for the immature stages of butterflies. The eggs and larvae of moths, sawflies, and other insects are frequently found, often more frequently than those of butterflies. In addition, multiple species of butterflies may be found on the same host plant, so identifying a species when it is found is important if one is to avoid wasting a great deal of time rearing the wrong species. One of the purposes of this book is to present photographs of all stages of all our butterfly fauna, to help identify eggs, larvae, and pupae when they are found. In some species an accurate identification can be made at any stage, while in others only certain stages are unique. In others, typically in closely related species groups (e.g., the Northern, Anna's, and Melissa blues), accurate identification may not be possible at all until the adult ecloses, although other factors such as habitat or host plant may provide important clues. For those wanting to obtain eggs from gravid females, it is important that adults can be identified and that males can be separated from females. Many hours have been wasted by researchers waiting for males to oviposit in captivity!

Host Plants: The natural histories of butterflies are inextricably linked to their host plants. A few species of butterflies, such as the Painted Lady and Gray Hairstreak, are catholic in their tastes, utilizing a large variety of plants, but most use only a few related plants from a single family, and some feed on only a single plant species. A good knowledge of host plant species is imperative if one is to be successful in fieldwork. It is not necessary to be a botanist to master butterfly host plant identification, as there are many families of plants that are not used at all, but some study will be necessary to learn the most commonly used groups of plants. The amateur naturalist is likely to find this learning experience as interesting and rewarding as the study of butterflies themselves. When one has mastered the basics of butterfly host plant identification, a trip into the field will be a much more rewarding experience; in a glance one can determine which butterfly species are likely to be found nearby. Perhaps more importantly, one is able to tell which butterflies will not be present, thus avoiding wasting time searching inappropriate habitats. Plants, like butterflies, are seasonal, predictably developing and flowering at about the same time each year. *Arabis*, an important mustard genus that hosts a number of pierid butterflies, typically grows early in the spring at lower elevations and is virtually absent later in the season; consequently, the pierid butterflies that feed on *Arabis* must occur as larvae early in the season. Many butterfly larvae do not feed on all parts of a plant, so are present only when the appropriate part is available and in prime condition. For example, the Western Pine Elfin feeds on buds and tender new growth at the tips of pine tree branches, so larvae are present in spring before the tips mature and harden. Host plant information is available in many butterfly publications, including this book, but much remains to be learned and amateur naturalists can add greatly to our knowledge.

Planning your trip to search for butterfly immature stages can pay dividends. For greatest success we suggest that trips should concentrate on one or two target species, chosen in advance. All available information on these species should be reviewed to determine the best time and location for the search, and the literature should be consulted to refresh one's memory on identification of the host plant and the butterfly stages sought.

Finding Eggs: Butterfly eggs are tiny and easily overlooked. Small butterfly species such as blues have the smallest eggs, sometimes only 0.5 mm in diameter, while eggs of the largest species may approach 2 mm. Although tiny, eggs are not always difficult to find. They often contrast with host plant leaves or buds and are often predictably placed on a certain part of the host plant, greatly reducing the search area. Some species lay eggs in large masses that are more easily seen. When searching for eggs, identification of the host plant is essential. This is not always easily accomplished, as eggs are usually laid when the plant is very young, and easily recognizable parts of the plant such as flowers may not yet have developed. Eggs that overwinter may be laid indiscriminately, sometimes not on the host plant at all. Those that hatch and develop quickly are invariably placed on the most nutritious part of the plant needed by the larvae. Information on preferred parts of host plants is sketchy and scattered in the literature, but we have added to this knowledge in most of our species accounts. If eggs are laid at the bases of flowers, as with the Western Tailed Blue on vetches, only fresh-blooming plants need be searched, avoiding distractions on leaves and stems. Compton Tortoiseshell eggs are laid on the tips of small birch twigs, so it is unnecessary to search the leaves. If the part of the host plant used for oviposition is unreported, the difficulty of finding eggs is greatly increased. Eggs of species that utilize host plants dominating a habitat are particularly difficult to find. For example, searching for eggs of the Pine White in coniferous forests is like finding the proverbial needle in a haystack. Searching for eggs of grass-feeding satyrs in meadows is an equally daunting task. In contrast, locating eggs of species that utilize host plants that are limited in abundance in specific habitats is more profitable. Good examples include the Monarch, swallowtails, and whites. Once found, eggs should be kept cool and moist (not wet) in transit as they are vulnerable to desiccation. Host plant cuttings should also be collected and treated similarly. Searching for butterfly eggs requires patience and confidence.

A slightly more proactive method of locating butterfly eggs is to find and follow a gravid female exhibiting oviposition behavior. A female preparing for oviposition will typically have a fat abdomen and will flit from host plant to host plant, landing in places where it is apparent that she is not imbibing nectar. When preparing to lay eggs a female will often tap the host plant with her antennae and "taste" the plant with her tarsi (feet). When ready to oviposit, she will bend her abdomen down and touch the tip to the plant. Most species will quickly lay a single egg then move to another site, but some species, including checkerspots, tortoiseshells, and some other species, lay masses, strings, or stacks of eggs during a session that can last several minutes. When oviposition is observed, careful notes should be taken, as this may be new or poorly reported information. Females laying single eggs may cruise corridors or borders, such as roads, trails, foliage hedges, or even cliff faces, searching for oviposition sites. When oviposition is observed, the site should be checked quickly as it may be difficult to find again later. The location where oviposition occurred should be carefully noted and similar nearby sites searched.

Finding Larvae: As with eggs, locating larvae depends on finding and identifying host plants. An area where adults are abundant in season should be selected, and searching should be timed for when host plants are present in prime condition and larvae expected to be present. In most species larvae feed on specific parts of the plant, resulting in predictable and identifiable damage to the host. Once these characters are recognized, the search pattern can be focused and larvae more easily found. Other important factors to consider are whether the species is nocturnal or diurnal, whether it is gregarious or solitary, and whether it builds nests or feeds openly. If a nest-builder, the nest type (rolled leaf, folded leaf, silk shelter, inside hollowed fruit, etc.), and the part of the plant should be considered. The more information assembled before fieldwork, the better the search image and less time spent searching ineffectively.

A variety of search techniques may be used. Simple observation for nests or feeding damage can be highly effective for some species. The characteristic leaf-flap tying behavior of Propertius Duskywing larvae can often be spotted on trailside oaks. Mourning Cloak larvae are obvious when they defoliate a medium-size willow, Sara Orangetip larvae can be spotted by their leaf damage to mustard hosts, blackened Milbert's Tortoiseshell nests on nettle are conspicuous, and silken nests of some checkerspots can be easily spotted at a distance. In years of abundance it is easy to find larvae of Painted Ladies on thistles, and nettle patches are always worth searching for Red Admiral and Satyr Comma larvae. The brightly colored larvae of Anise Swallowtails are readily found on the green foliage of *Lomatium* spp. (desert parsley) in shrub-steppe country during May–June. The yellow-spotted black larvae of Mountain Parnassians are also relatively easy to find on *Sedum*-covered mountains in early summer. Searching the relatively bare branches of Green Rabbitbrush in late March/early April will often yield numbers of overwintered Sagebrush Checkerspot larvae sunning themselves. Other species are better found by shaking or beating a host plant, either into a net or over a white sheet; this is effective for Echo Blues and Brown Elfins on *Ceanothus*, Behr's Hairstreaks on bitterbrush, and *Euphilotes* spp. on buckwheats. Careful direct searching of host plants can be effective when larvae are fairly common and host numbers limited, always concentrating on the preferred parts of the host plant.

Finding Pupae: Finding pupae is more difficult than finding other immature stages for a number of reasons. They are less abundant; only a tiny percentage of eggs and a small percentage of larvae reach the pupal stage. Pupae are typically very well camouflaged and often located away from the host plant. Many species pupate in duff and debris or under rocks, and sometimes in holes below the ground surface. While pupae can occasionally be found in numbers, such as with California Tortoiseshells on *Ceanothus* during mass outbreaks, most pupae are found accidentally and only occasionally. Another exception is the Cabbage White in vegetable gardens, where pupae can be found on nearby structures like fences and buildings. Searching for pupae is usually not a productive use of field time; for most species, pupae are best obtained by rearing.

General Field Suggestions: Much has yet to be learned about the life histories of our butterfly fauna, so accurate and detailed records of findings are important. When immature stages are found, the date, locality, host plant, and elevation should be recorded, with a description of what caught the observer's attention. The position of the site, whether along an obvious flight corridor, in thick plant growth, in an open rocky area or elsewhere, should be recorded. Removing immature stages will result in fewer adults flying later, so it is best to take only a few individuals. If you are able to rear adults from the eggs, larvae, or pupae you find, we recommend making the effort to return them to where they were collected. The two larvae you take may be the only two destined to reach adulthood from a single female.

Identifying Females: This book is about immature stages, but to obtain eggs from butterflies in captivity, one must be able to recognize adult females. There is seldom any need to try to get butterflies to mate in captivity as the majority of females of most species seen or caught will have already been mated. Mating generally occurs within the first day or two of adult life, so simply obtaining females will usually provide a gravid individual. In many species the sexes can be identified by wing pattern, and a guide to adult butterflies in Cascadia (e.g., Guppy and Shepard, 2001; Pyle, 2002) should be consulted. For those species in which sexual dimorphism is minimal, females can be detected in several ways. Females tend to have an inflated abdomen (full of eggs), which in many species is conical in shape, like a pointed bullet with a round hole at the tip. Males have thinner abdomens, typically with an angled, truncated tip with a vertical slit. Very gentle squeezing of the male abdomen can often force out claspers and the genitalia, revealing their angular pointed processes and confirming the sex without harming the butterfly. Equally important and instructive is observing the butterfly's habits. Males typically perch conspicuously, darting out to challenge all passersby, or endlessly cruise along corridors, while females tend more often to stay out of sight. Females visit flowers for nectar more frequently than males, moving into cover when not nectaring. In some species, females are best found by "walking them up," disturbing the habitat near where males are perching in order to flush the females from their cover. In addition, males of some species habitually "hill-top"; while females visit such sites occasionally to mate, most individuals on hilltops are usually males. Butterflies that gather in aggregations at wet mud to imbibe liquids are nearly always males.

Rearing Techniques

For a field lepidopterist with unlimited time and energy, the ultimate challenge would be to find every stage of a butterfly in nature; however, for most of us, the only realistic way to observe all the life history stages of a butterfly is through captive rearing. In nature a variety of factors trigger and control development, and many of these factors are unknown or "invisible" to humans. The challenge of rearing is to emulate natural conditions well enough that the larvae develop, survive, and preferably thrive. In this chapter we describe some techniques we have found to be successful.

Obtaining Eggs: Unless you plan to search for eggs in the wild, the first ingredient for obtaining eggs in captivity is a gravid female butterfly. Females usually mate during the first day or two of adult life; thus most fresh-looking females will be fertile and able to lay eggs. Exceptions do occur, especially during spring, when conditions may be overcast and cool and opportunities for mating are limited. Worn females also usually contain viable eggs, but numbers may be limited. Butterflies overwintering as adults in reproductive diapause (e.g., anglewings and some other brushfoots), may be more than 8 months old before mating in the spring, so it is often difficult to determine the right time to capture a mated, gravid female. A number of species, particularly greater fritillaries and some satyrs, enter summer reproductive dormancy (aestivation) after eclosion and do not mature ovaries for 6–8 weeks. Greater fritillaries generally do not oviposit until August. Species that lay eggs in large clusters may oviposit only once or twice in their lifetime. The timing for these species is critically important and often difficult to achieve.

Handling gravid females should be minimized to prevent injury and disturbance. Females should be kept cool and in the dark during transit to prevent damage or exhaustion. Freshly captured adults may respond wildly for the first day or two but become tamer with time. Females need to be fed regularly, at least daily. In some instances they become so acclimated to feeding, they will land on the tender's hand when the cage door is opened. Feeding techniques vary; hand feeding with a cotton swab saturated with sugar water or commercially available butterfly nectar is time consuming but effective. Adults can be induced to feed by touching a nectar-saturated cotton swab to their antennae or tarsi. If the butterfly backs off or rotates away, wait a few seconds and try again. If the butterfly probes the swab with its antennae, it is about to feed. With larger butterflies a needle can be used to carefully extend the proboscis onto a nectar source. Some rearers have reported success with a brightly colored golf tee, placing a drop of nectar on the concave top. Saturated cotton balls can be left in cages for feeding, or live nectar plants placed with butterflies in larger cages. Adults should be placed in cages suitable to their size; a large cage is needed for swallowtails, and small enclosures are adequate for lycaenids. Crowding can be beneficial, especially if the female is forced into direct contact with the host plant. Many different types of cages can be used for oviposition, from small plastic boxes with mesh or muslin on one side for air exchange, to large walk-in netting-covered enclosures. We have obtained oviposition from blues and hairstreaks in plastic tubes (7 × 5 cm) with mesh lids. Greater fritillaries will oviposit in closed boxes and even paper bags, particularly when violet odors are present.

Confinement of females with fresh larval host plants is generally required for oviposition, although there are exceptions. For example, species that overwinter as eggs or unfed first instars will lay eggs on desiccated hosts and inert surfaces. Some species oviposit more readily when provided with a potted host plant rather than cut foliage in water; however, for the majority of species, the latter works well. Conditions that discourage oviposition include dim lighting, cool conditions, a poorly fed female, a wilted host plant, and lack of other butterfly activity in close proximity. The opposite conditions will encourage oviposition, particularly direct bright sunlight for at least a brief period. Placing oviposition cages outdoors under sunny conditions during spring or autumn provides optimal conditions and usually stimulates egg laying; however, provide partial shading, as continuous exposure to direct sunlight from May to September in most parts of Cascadia is lethal to most species. Oviposition is often best obtained indoors, where temperature and sun exposure can be controlled. Good exposure to sunlight appears to be most important in the first half-hour or so after feeding. Other species ovipositing in the same cage or nearby will sometimes create a group phenomenon in which all the females do likewise. Oviposition may occur quickly or may take several days or even a few weeks (some brushfoots). Sometimes a female will lay numerous eggs just prior to death, but these may or may not be fertile. If more than one species is kept in a cage, they should be sufficiently different so their eggs can be easily identified.

Care of Eggs: Eggs should be collected within a few days of oviposition. Eggs may fall, be lost, be attacked by predators, or hatch with larvae dying before being found. Eggs tightly attached to plant substrates (leaves, stems, petals, etc.) should be removed with the substrate. Eggs laid in batches are best moved intact by clipping the leaf or twig substrate. Lightly fixed or loose eggs can be carefully moved with a moistened camel-hair brush. Eggs may be stored in a Petri dish with a lightly moistened paper towel liner. Eggs often change color (darken) or develop visible embryos when they near hatching. At this stage they should be moved to a piece of fresh food plant so the larvae can begin feeding immediately. Overwintering eggs should be free of plant material to help prevent the growth of mold. Nonviable collapsed eggs will also grow mold and should be removed before overwintering.

Rearing Larvae: First instars (L1) are the most fragile larval stage, with the greatest losses. A 25–50% survival rate to L2 is generally considered acceptable. As with eggs, L1 should be handled carefully with a small moist camel-hair brush to avoid injuries. Small larvae are especially susceptible to predators like spiders and predatory bugs, so the containers should be inspected carefully, particularly when fresh food plant is introduced. Maintaining L1 can be tedious and time consuming, so patience is needed. The host plant must be changed (daily or on alternate days) and kept fresh; for best survival, the excrement (frass) should be cleaned from the container daily. Cut host plants should be kept fresh and cool under refrigeration in zippered plastic bags or in water, or potted plants can be kept for a ready supply. As the larvae grow, they can be moved to larger containers with host-plant cuttings or live plants. Cuttings can be placed with the larvae by inserting them in a small glass of water sealed at the top with kitchen plastic wrap, the plants extending through a hole of minimal size to prevent the larvae from drowning.

Some host plants can be difficult to obtain, particularly when rearing butterflies from different climate zones. Sometimes substitute host plants can be used; the species accounts in this book and elsewhere can be consulted for suggestions. If an alternate is used, it is advisable to start the larvae on the substitute plant from the beginning; switching host plants in midgrowth is often problematic. As the larvae grow, they require far greater quantities of food plant, particularly in the final instar, consumption reaching its fastest rate just prior to pupation. When wild plants are gathered, be wary of possible pesticide spraying, especially along roadsides or in agricultural fields. Larvae feeding on plants sprayed with insecticides will die quickly, while herbicide-sprayed plants tend to reduce feeding, lengthen development, and cause increased mortality.

Temperature, humidity, and photoperiod are important factors affecting development of larvae, sometimes in complex and less than intuitive ways. Some species prefer high humidity while others thrive in drier conditions. Some prefer heat, others, cool conditions. Generally, though, most species develop faster under warmer temperatures. Development at 25–30 °C is optimal for many species. Some Cascadia species are strongly influenced by photoperiod or, more precisely, whether days are lengthening or shortening. Change in daylength is the most important environmental cue for immature stages of butterflies in Cascadia to continue developing optimally or delay development to avoid unfavorable (too hot, too dry, too cold) conditions. There is still much to be learned about optimal rearing conditions for individual species, but when in doubt, one should imitate the natural climate/habitat/environment as closely as possible.

A variety of enclosures can be used for rearing, from Petri dishes to large walk-in screened enclosures. The size of the container should be matched to the larvae and to the required host plants. Early instars should be reared in small containers like Petri dishes, but different species may be reared optimally in different types of containers. You are encouraged to experiment. We developed a new rearing technique for early instars (L1–L3) of fritillaries that minimizes mortality and makes the best use of a limited supply of host plants. Platforms of small pieces of plastic (approximately 7 × 6 cm) are laid on saturated cotton wool in a Petri dish with small leaves of hosts (violets) resting on the plastic, with stems inserted into the wet cotton wool. Leaves remain fresh for about 7 days, and larvae feed and survive well in groups of up to 50 individuals. Nearly air-tight plastic boxes can be used for many species, but they must be kept out of direct sunlight and opened at least daily for moisture and oxygen control. Screened enclosures also work well, but cut plants tend to desiccate more quickly with ventilation. To prevent escapes, screening and ventilation openings must be smaller than the smallest larvae kept inside. If the larvae contract a disease, all containers and tools must be thoroughly washed in disinfectant. Some larvae, particularly gregarious species, thrive under crowded conditions, but others do not. Some larvae, particularly the lycaenids (coppers, blues, hairstreaks) are cannibalistic in crowded conditions. If a species is double brooded, it is easier to rear the spring brood than the fall brood, so overwintering can be avoided. The overwintered larvae of some species can be found relatively easily in spring (e.g., checkerspots), thus avoiding overwintering.

Overwintering: All Cascadia butterflies must diapause during winter or migrate to milder climes. Rearing butterflies that diapause as immature stages can pose additional challenges, for example, how to detect when diapause has begun, what to do at that time, how to maintain and protect the immature stages during winter, and how to "revive" them into activity the following spring.

Eggs entering diapause will fail to hatch within 2–3 weeks and often darken. Dead eggs usually buckle and collapse and can usually, but not always, be distinguished. Many species that overwinter as eggs develop mature embryos that are visible inside. Care must be taken not to refrigerate these eggs before the embryos are fully developed (typically 2–3 weeks). Larvae entering diapause typically darken, shrink, cease feeding, and wander under or off the host plant, seeking cover. Some species (e.g., greater fritillaries) overwinter as unfed L1, eating only the eggshell before seeking cover, often in aggregations. Diapause is genetically and hormonally controlled and obligatory for most single-brooded species. Multiple-brooded species are more opportunistic, entering diapause in response to environmental conditions such as photoperiod, temperature, and availability of fresh host plant. Manipulation of environmental conditions can sometimes postpone diapause. Information on diapause of individual species is given in species accounts.

Immature stages can be overwintered by placing them in dark refrigerated conditions, typically about 4–5 °C, avoiding freezing. Temperature and relative humidity requirements vary with species and are poorly understood. In particular, relative humidity can have a significant effect on overwintering survival. First instars of most species of greater fritillary overwinter well at relative humidities of 80–90%. A few species do better at 90% and greater, but all survive poorly at 50–60% or lower. If better information is not available, we recommend providing moisture to keep the humidity high, but wet conditions on or near eggs or larvae should be avoided to prevent mold growth, which can cause 100% mortality. A covered container with water in the bottom and eggs/larvae in Petri dishes on a rack above the water works reasonably well. Alternatively, immature stages can be overwintered outdoors. Complete mortality of overwintering Pine White eggs under various relative humidities prompted us to try overwintering eggs in a muslin bag attached to a pine tree branch 4 m from the ground. The majority of eggs in these batches survived and hatched in spring.

Breaking Diapause: This is accomplished when the overwintered larvae are brought out of the cold overwintering environment and begin feeding. Many are lost at this stage, often entire broods, when the larvae refuse to feed and then die. The conditions necessary to successfully end diapause are poorly understood for most species. A minimum period of "cold" or darkness before development competency is regained appears to be characteristic of some butterfly eggs and larvae. Temperature and relative humidity factors may also be involved. Successfully breaking diapause in immature stages of Cascadia butterfly species is often a matter of chance or experimentation; however, in our accounts we report considerable new information on our findings for specific species, including instances of breaking diapause during midwinter. However, much remains to be learned. In the absence of other information, our advice is to emulate natural conditions as closely as possible. One hedge against failure is to stagger the removal of individuals from cold storage over a period of time. Successful strategies should be recorded. There are many opportunities for observant amateur lepidopterists to contribute to our knowledge on this important aspect of butterfly biology.

Rearing sometimes produces outcomes quite different from those in the wild. The life history of a species may be different here than in warmer or colder climates; a species that is single brooded here may be multiple brooded in southern California, and the same species may be biennial in northern British Columbia. Similarly, because rearing conditions are invariably different from those in nature, the results are sometimes unpredictable. Single-brooded skippers often produce second broods in captivity, and some species may pupate after fewer instars than in nature. Other species may repeatedly reenter diapause in captivity without apparent reason, and others may develop through extra instars. Such are the vagaries and challenges of rearing. As our knowledge of natural triggers and environmental cues improves, many of these mysterious responses will become better understood. All who undertake rearing and keep careful notes on the outcomes can contribute significantly to this understanding.

Rearing large numbers of larvae is unnecessary and a great deal of extra work; it is usually preferable to rear a few. In a number of instances, we successfully reared species for this book from a single egg! However, it is more "normal" to expect a mortality rate of 50–75% during rearing. If you end up with too much livestock, share the excess with other interested rearers, including children and local schools; they will enjoy the experience and may provide a backup in case your brood fails.

Photographic Techniques

Documenting the immature stages of butterflies can be done by killing and preserving them in a liquid preservative such as alcohol, or by photography. There are pros and cons to each method. Preserved specimens in glass tubes and flasks are bulky and need to be maintained, and specimens tend to lose their natural colors; however, the finest structural details are preserved for later study. Larvae must be killed at each stage to record the entire life history, and this can be a problem when only a few specimens are available. Photography records natural colors faithfully and aesthetically, often in natural surroundings, without killing the subject, permitting all stages to be documented using a single individual (with a bit of luck!). However, each photograph records only one view or perspective, and minute details are typically not visible in photographs. This chapter describes some of the photographic techniques available for documenting the immature stages of butterflies.

Macro Photography Equipment: Photography of butterfly immature stages often involves some very small subjects, down to 0.5 mm. To obtain quality images of such tiny subjects, a digital single lens reflex (SLR) camera and a special macro lens are required, together with other equipment for stabilizing the camera and lighting the subject. While larger specimens can sometimes be photographed in nature, most "super macro" photography must be done indoors under carefully controlled conditions. Most of the photographs in this book were taken by the authors using a variety of Canon digital SLR cameras and a 1–5x macro zoom MP-E 65 mm Canon lens, resulting in images up to 8x natural size (equivalent to 8:1 in film cameras). Super macro photography is greatly affected by vibration; at such high magnifications, even the tiniest vibrations can blur an image. Equipment needed for high magnification work includes a good solid tripod and an electronic remote shutter release to reduce camera shake. Additionally the camera should be capable of "mirror up," also known as delayed shutter release; this allows the SLR camera to pause for a moment after the mirror flips up but before the shutter releases to further reduce vibration. Lighting can be accomplished with dual adjustable light sources, or with dual flashes. Another important accessory is a rack-and-pinion focusing platform that mounts between the tripod and the camera. This apparatus is very useful for centering and fine focusing a tiny subject, and it is mandatory if digital stacking is used.

Depth of field is always a problem in macro photography and a major problem with super macro work. When depth of field is poor, a thin slice of the subject will be in focus while closer and more distant parts are blurred. Depth of field can be improved by adjusting the camera to a small aperture (larger f-stop number) but this requires a longer exposure time to compensate for the light reduction, and there are limitations on how much the depth of field can be improved in this manner. Another technique is to use digital stacking, a method in which a series of photographs is taken, focusing first on the very front of the object and in tiny increments ever deeper into the subject until the back is reached. A very stable tripod and a rack-and-pinion focusing rack are needed, as well as a very still, cooperative subject. Special computer software is used to process the stack of digital photos to create a single composite picture that assembles all the best-focused parts of the individual photos. Some of the images in this book are the product of digital stacking. The technique is fairly time consuming, but the results can be excellent.

Techniques: Most other suggestions are much the same as in normal photography. A still subject is always important. Larvae are generally most "cooperative" under cool conditions (10–15 °C). If a larva is "squirmy," it can be cooled gently in the refrigerator, but usually it is best to leave it alone for a while and check back later; even the most uncooperative subjects will eventually stabilize and become stationary. Lighting can be hot and uncomfortable for larvae, causing excessive movement. Adjustable intensity lighting can be helpful; lights turned to a dim setting and very gradually brightened to higher intensity are often tolerated better than a sudden bright light. It is always advisable to be very familiar with your equipment so you can work rapidly when opportunity avails, to take several photos in quick succession before the subject begins moving.

Eggs, larvae, and pupae should be photographed at a minimum from the side and dorsally for each stage. Very early or very late instars should be avoided because these tend to be atypical. For example, late-instar larvae are usually duller in color and "bloated" at the head end as the integument tightens prior to splitting. Similarly, newly formed pupae are also usually very different in color and texture from aged pupae. Images of head capsules can be useful for separating species in some families. One should pay attention to aesthetics; when possible, place the larva on its natural host plant and move other plants into the background for natural colors. Artificial surfaces cannot be avoided in some cases (e.g., when pupae attach to a metal cage). White balance (adjusting the camera to correctly recognize pure white) is important with indoor photography, particularly when incandescent lighting is used. Colors can be computer adjusted later, but a good original photo will always produce the best final results. The size of the specimen should be measured carefully and recorded in a notebook, together with the stage, date, original location of the specimen, and species identification. All photographs should likewise be labeled with all the important data.

How to Use This Book

Names: The Latin names used for butterflies in this book follow Pelham (2008), with the exception of the greater fritillaries, for which we have elected to use the Old World genus *Argynnis*, reducing *Speyeria* to the level of subgenus in accordance with Simonsen (2006) and Simonsen et al. (2006). Our English names for butterflies generally follow the North American Butterfly Association (NABA) checklist, with some exceptions; we retain a few locally favored names such as Western Meadow Fritillary (instead of Pacific Fritillary), and we use widely accepted English names for a few taxa that NABA does not recognize as distinct species (e.g., Snowberry and Anicia checkerspots). Some taxa have been split since the NABA list was last updated (Halfmoon Hairstreak, Echo Blue, Lucia Blue, Northwestern Fritillary); in these cases we attempt to use names already published or otherwise gaining wide acceptance. We also suggest tentative names for a few recently recognized taxa that have no previously published English names (e.g., Summit Blue, Cascadia Blue). English names for plants are less standardized, often varying regionally. We attempted to choose English plant names most frequently used in Cascadia, but there is no accepted standard list of these names; we used Hitchcock and Cronquist (1973) or local field guides when possible, otherwise resorting to Internet sources for suggestions. Capitalization of English names follows the NABA protocol for butterflies, which is similar to that used by the American Ornithologists' Union (AOU) for birds and by the Dragonfly Society of the Americas (DSA) for dragonflies. In this protocol all proper names are capitalized but nonspecific names are lowercased; for example, fritillary is lowercased but Hydaspe Fritillary is capitalized. For consistency we extended this protocol to plant names as well.

Units of Measure: The reader will note that we use both metric and American units to some extent. For precise measurements of specimens and for captions of their images, we use the finely divided metric scale, with measurements reported in millimeters (mm). This is a widely accepted scientific standard and is much preferable to reporting small measurements in tiny fractions of an inch; however, for distance and elevation measurements we deviate from the all-metric standard, using miles and feet respectively. We do so because most Americans still "think" in these units, and we feel it more important to provide a user-friendly product than to strictly conform to an all-metric protocol.

In our species accounts readers may also note that, in the captions for the smallest specimens, we show sizes sometimes as integers and other times as decimal fractions of millimeters. Our measurement data were developed over a period of nearly fifteen years; during that time our techniques changed and improved. Our earlier measurements were recorded to the nearest integer millimeter, but as our techniques improved we made more precise measurements. Thus our data are a mix of new and old, integer and decimal fractions of millimeters. We are consistent in reporting the original data as recorded; we believe that retaining both the new and old data provides the reader with the best possible information, even though it gives the appearance of inconsistency.

Quick Photo Guides: Immediately following this section are three double pages of "quick photo guides," the first one a guide to eggs, the second to mature larvae, and the third to pupae. Each guide shows 24 sample photographs of representative species. In a book of this scope in which all immature stages are depicted, it would be complex and outside the scope of this work to develop a complete binomial key to the identification of all stages. The quick photo guides are a visual substitute for a key, certainly far from complete or perfect, but it is hoped they will be useful for rapid pre-identification of a related group of species. By visual comparison the user should be able to determine to which general group an egg, larva, or pupa belongs, and then to refer directly to that group without first having to peruse all the accounts in the book. Each quick photo guide image bears a caption indicating the species shown, also in most cases the general group to which that species belongs. Once an approximate visual match has been made, the user can refer to the table of contents at the front of the book to find the depicted group, or use the index to locate the exact species seen in the image. Once the source species account has been found, the user can explore other closely related species located nearby to find the best match for identification. The user is reminded, however, that not every stage of every butterfly can be visually identified to the species level. This is particularly true for those groups with many closely related species, such as the blues and hairstreaks.

The Species Accounts: The detailed species accounts follow the quick photo guides, these accounts covering 153 species plus 5 subspecies of Cascadia butterflies. Within the fauna of Washington State, only the species *Oeneis melissa* (Melissa Arctic) is entirely omitted, because of our inability to secure and rear this species. We also do not depict *Papilio canadensis* (Canada Tiger Swallowtail) as it is believed that most, perhaps all, of the *P. canadensis* swallowtails in Washington and southernmost British Columbia are actually hybrids with *Papilio rutulus*.

The species accounts for each family or significant subfamily are preceded by a one-page introduction to the characters common to all members of that group, together with a facing one-page photo montage of typical representatives. For artistic reasons the photo montages do not have identifying captions; however, in the Photo Credits and

Data section at the back of the book, the user will find complete identification data for all the montage photos.

Each species account shows photographs of all the stages (egg, all larval instars, pupa, and adult) for each butterfly species. Each image is captioned with the stage and size at the time the photograph was taken. Additional information for each photograph, including location and photographer, can be found in the Photo Credits and Data chapter at the back of the book. The text of each species account is arranged in four standardized sections: adult biology, immature stage biology, description of immature stages, and discussion.

The species are arranged in a commonly used biological order, beginning with the true butterflies and finishing with the skippers. The order of accounts is as follows:

1. Parnassians and Swallowtails
2. Whites and Sulphurs
3. Coppers
4. Hairstreaks
5. Blues and Metalmark
6. Fritillaries
7. Checkerspots and Crescents
8. Other Brushfoots, Admirals, and Monarch
9. Satyrs
10. Spreadwing Skippers
11. Monocot (Grass) Skippers

We hope you enjoy this book, and, more importantly, we hope you will be inspired to travel into the field to observe both adult and immature butterflies for yourself.

QUICK PHOTO GUIDES

Parnassians
Parnassius clodius

Swallowtails
Papilio eurymedon

Pine White
Neophasia menapia

Whites and Sulphurs
Colias alexandra

Coppers
Lycaena nivalis

Hairstreaks (some)
Callophrys mossii

Hairstreaks and Blues
Satyrium californica

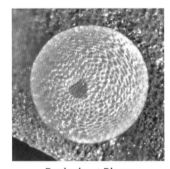

Buckwheat Blues
Euphilotes on *E. heracleoides*

Metalmark
Apodemia mormo

Fritillaries
Argynnis hesperis

Checkerspots and Crescents
Phyciodes pallida

Brushfoots (many)
Nymphalis l-album

Admirals and Sisters
Limenitis lorquini

Wood Nymphs
Cercyonis pegala

Alpines and Ringlets
Erebia epipsodea

Arctics
Oeneis chryxus

Cloudywings
Thorybes pylades

Duskywings
Erynnis pacuvius

Checkered Skippers
Pyrgus communis

White-Skipper
Heliopetes ericetorum

Sootywing
Pholisora catullus

European Skipperling
Thymelicus lineola

Grass Skippers (most)
Hesperia colorado

Dun Skipper
Euphyes vestris

QUICK PHOTO GUIDES

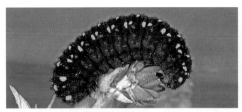

Parnassians
Parnassius smintheus

Parsley Swallowtails
Papilio machaon

Tiger Swallowtails
Papilio multicaudata

Whites (some)
Pontia beckerii

Sulphurs, some Whites
Colias interior

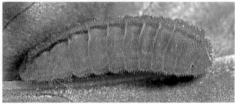

Coppers and Blues
Lycaena editha

California Hairstreak
Satyrium californica

Mistletoe Hairstreaks
Callophrys spinetorum

Cedar Hairstreak
Callophrys gryneus

Elfins (some)
Callophrys mossii

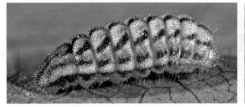

Buckwheat Blues
Euphilotes columbiae

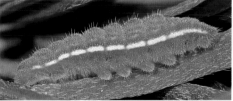

Blues (many)
Plebejus idas

Arctic Blue
Plebejus glandon

Metalmark
Apodemia mormo

Fritillaries
Argynnis atlantis

Checkerspots
Euphydryas colon

Brushfoots (typical)
Vanessa cardui

Admirals and Sisters
Limenitis archippus

Monarch
Danaus plexippus

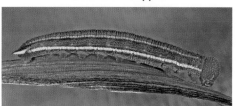

Satyrs
Cercyonis oetus

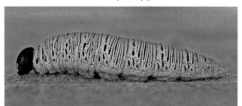

Silver-spotted Skipper
Epargyreus clarus

Duskywing Skippers
Erynnis propertius

Skipperlings
Thymelicus lineola

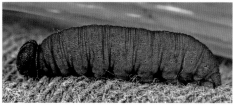

Skippers (many)
Hesperia comma

BUTTERFLY MATURE LARVAE

QUICK PHOTO GUIDES

Swallowtails
Papilio rutulus

Some *Pontia* (Whites)
Pontia beckerii

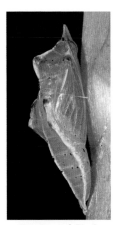

Pontia* and *Pieris
Pieris marginalis

Marbles, Orangetip
Anthocharis sara

Sulphurs
Colias alexandra

Greater Fritillaries
Argynnis zerene

Lesser Fritillaries
Boloria bellona

Crescents, Checkerspots
Chlosyne palla

Checkerspots (part)
Euphydryas colon

Brushfoots (many)
Polygonia gracilis

Brushfoots (many)
Nymphalis californica

Admirals and Sisters
Limenitis lorquini

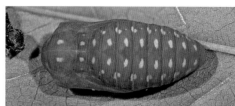

Parnassians
Parnassius smintheus

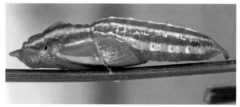

Monarch
Danaus plexippus

Satyrs (all)
Cercyonis pegala

Pine White
Neophasia menapia

Coppers, Hairstreaks, Blues (some)
Strymon melinus

Tailed Blues
Cupido comyntas

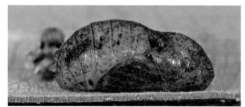

Coppers, Hairstreaks, Blues (some)
Celastrina echo

Arctic Blue
Plebejus glandon

Metalmark
Apodemia mormo

Skippers (most)
Pyrgus communis

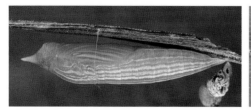

Skipperlings
Oarisma garita

Dun Skipper
Euphyes vestris

Family Papilionidae

PARNASSIANS and *SWALLOWTAILS*

The large and colorful swallowtails are among our best-known butterflies and found throughout the world. In Cascadia we have 7 species of swallowtails and 2 parnassians. Three Tiger swallowtails and the Old World and Anise swallowtails are widely distributed and common in Cascadia and have a high public profile. Their larvae are also well known and are often found by children in late summer when larvae wander, seeking sites for pupation.

At first glance the parnassians, with their large white wings, look like pierids (whites); however, their life histories have much in common with the more familiar swallowtails. Parnassians lay large spherical eggs with a textured or granular surface on or near the ground. The eggs overwinter, hatching in spring to yellow-marked, black velvety larvae. They have a dorsal unforked osmeterium (eversible scent gland), but it is rarely used. One of our 2 species uses bleeding hearts as its larval host, the other uses stonecrops. The larvae appear to mimic sympatric millipedes that secrete cyanide as a defense, but they may also be distasteful themselves. Pupation occurs in a rudimentary cocoon beneath host plants, adults eclosing in late spring–early summer.

Swallowtails lay large, smooth, spherical eggs that hatch into black larvae with white "saddles" that mimic bird droppings. Crypsis is replaced by warning coloration and markings in mature larvae. Most are green or black, some with "inflatable" bands or spots and yellow, blue, and black eyespots, which may mimic the head of a snake. All have a forked osmeterium concealed in a slit behind the head. When the caterpillar is disturbed, the bright-colored organ is everted, discharging a pungent odor. This has been shown to repel birds. Pupae overwinter and are formed nearly vertically and secured by a cremaster and girdle thread, usually attached to a small-diameter twig. Swallowtails take flight early in the season, and some species are still flying in fall. Greatest abundance occurs in the mountains during June–July. Larval hosts are umbellifers or *Artemisia* for black swallowtails and broad-leaved trees and shrubs for the tiger swallowtails. Some of our swallowtails are double brooded, but the parnassians are strictly single brooded.

PARNASSIANS and SWALLOWTAILS

Clodius Parnassian   *Parnassius clodius* Menetries

Adult Biology

Parnassius clodius is common in forested areas where it patrols roads and other corridors. Preferred habitats include forest openings, meadows, and roadsides, from sea level to 8,000 ft. It occurs on the coast and in the Cascades but is absent from most E areas except NE WA, the Snake River, and the highlands of SE WA and NE OR. Elsewhere it occurs from the AK panhandle S along the coast ranges and Sierras to central CA, and E to MT, WY, and UT. It flies in a single brood from early May to late Sep and both sexes visit flowers. *Eriogonum umbellatum* (Sulphur Buckwheat) was the primary nectar source used in WY (Auckland et al., 2004). Larval hosts include *Dicentra formosa* (Pacific Bleeding Heart), *D. uniflora* (Steer's Head), *D. cucullarea* (Dutchman's Britches) and *D. pauciflora* (Few-flowered Bleeding Heart) (Pyle, 2002; Warren, 2005). The average straight-line distance traveled per day by adults in a WY population was 202 m (Auckland et al., 2004). Males overpower females by grabbing them in midair and falling to the ground for mating (Scott, 1986a; Guppy and Shepard, 2001). The male secretes a waxy shield (sphragus) on the female's abdomen to prevent her from mating again. Eggs are laid singly, scattered haphazardly near the host plants.

Immature Stage Biology

We reared or partially reared this species several times from eggs. On July 29 and Aug 1 females from Bear Ck Mt, Yakima Co., WA produced ~100 eggs on each occasion which were held at 21–26 °C for 3–6 weeks before overwintering at 5 °C. After 117–147 days eggs were transferred to 22–25 °C under 13 or 24 hr light. Hatching occurred in 1–4 days and larvae were reared on cut *D. formosa* with L5 reached after 19 days. Due to a shortage of host plants, larvae were transferred to 17–20 °C with minimal food and 26 days were spent in L5. On Aug 17 females from Snoqualmie Pass, Kittitas Co., WA produced eggs which were reared to pupation on *D. formosa*. Some eggs removed in mid-April hatched quickly (1 hr), while others hatched up

to two weeks later; development from L1 to pupation took 18 days. On July 28, North Bend, King Co., WA females produced 65 eggs, 34 of which hatched during March–May. Some larvae were reared on cut *D. formosa*, others on live plants; some were kept at 18–20 °C, others in ambient outdoor conditions, but all refused to feed and failed to survive beyond early L2. Eggs mostly hatch in the evening and L1 leave eggshells uneaten. Larvae feed nocturnally on leaves, resting in curled leaves etc. on the ground during the day. Development is characterized by variable growth rates with multiple instars present at any one time. Larvae are sluggish, often resting in a partly coiled position. Feeding damage is seldom apparent as the host plant leaves are naturally deeply dissected. Late instars are cannibalistic when food is limited. Larvae refuse to feed on the Asian *D. spectabilis* but females will oviposit in its presence. In captivity pupation occurred at the plant base, often under paper toweling and within light silk webbing. Pupation in the wild occurs in a light silken cocoon near but above the ground (Scott, 1986a; Pyle, 2002). There are five instars, no nests are used and overwintering occurs as an egg. Protection is based on concealment (nocturnal behavior) and perhaps distastefulness (*Dicentra* host plants contain poisonous alkaloids). Scott (1986a) and Shapiro and Manolis (2007) suggest the larvae may mimic cyanide-containing millipedes. The larvae possess a small eversible tubercle (osmeterium) which may produce defense chemicals.

Description of Immature Stages

The egg is pinkish tan, hemispherical, and ornamented with intricate vermiform surface markings. The surface is often coated with a mustard-colored material. L1 is purplish brown to black with 12–14 tubercles per segment bearing multiple long dark setae. L2 is similar except the setae are shorter and there is a dorsolateral row of pale white–dull yellow spots, one spot per segment, most prominent in purplish larvae, indistinct in black larvae. The head is black. L3 is similar but setae are numerous, imparting a hairy appearance, and the pale lateral spots may be distinct to obscure. L4 may have additional yellow spots which form a nearly continuous, contrasting lateral stripe. Dark larvae retain a single more distinct spot per segment. A large black spot occurs above the yellow line (or yellow spot) on each segment. Lighter colored L4 have black markings dorsally that form a row of chevron-like patterns. Setae are numerous but very short. L5 is similar, with the yellow line more continuous and distinct in paler larvae, and the yellow spots in

Adult female

Egg @ 1.2 mm

Pupa @ 20 mm

L1 @ 2.5 mm

L2 @ 5 mm

L3 @ 9 mm

L4 @ 16 mm

L5 @ 36 mm

black larvae more vivid. The head, true legs and prolegs are black and increased numbers of setae give the larva a fuzzy appearance. Scott (1986a) reported that low elevation larvae are black and high elevation larvae gray-brown or pinkish gray. Our larvae were generally purplish-brown W of the Cascades (low elevation), and solidly black E of the Cascades (higher elevation). The pupa is robust and reddish brown; there are protuberances at the "shoulders" resulting in a squarish appearance anteriorly. Immature stages of *P. smintheus* are similar to *P. clodius* and could be confused in the early instars, although they are found on different host plants. Late-instar *P. clodius* have a single row of a few or many yellow spots while *P. smintheus* L4 and L5 have multiple bright yellow-gold spots in 4 rows. Mature larvae are pictured in Miller and Hammond (2003, 2007), Allen et al. (2005), and Neill (2007); Guppy and Shepard (2001) picture the egg and mature larva, and Woodward (2005) shows the mature larva and pupa. All images agree well with ours.

Discussion

Eggs and larvae are very difficult to find, but eggs are easily obtained from gravid females and are easily overwintered. Rearing larvae is difficult with early instars often refusing to feed and dying *en masse*. Larvae that do feed develop rapidly and reach adulthood within 6–8 weeks. The apparent occurrence of two larval color morphs is of interest and deserves further study to determine the causative factors. The immature stages are potentially toxic to natural enemies with alkaloids possibly sequestered from the host plants. Research is also needed on whether the osmeterium is functional. Scott (1986a) maintains that it is nonfunctional.

Mountain Parnassian *Parnassius smintheus* Doubleday

Adult Biology

Parnassius smintheus is a univoltine butterfly of the mountains, from upper foothills to the alpine zone on both sides of the Cascades of BC, WA and NE OR. Elsewhere it occurs in mountainous areas of western N America from AK to NM and central CA. Larval hosts are stonecrops including *Sedum lanceolatum* (Lanceleaf Stonecrop), *S. rosea* (Western Roseroot), *S. divergens* (Spreading Stonecrop), *S. oreganum* (Oregon Stonecrop) and *S. stenopetalum* (Wormleaf Stonecrop). Observations at Quartz Mt, Kittitas Co., WA, showed that males cruise in search of mates while females stay low. Mating lasts several hours (Scott, 1986a) and the male secretes a waxy structure (sphragis) on the female's abdomen to prevent further mating. We observed one female visiting a small stonecrop plant probing it several times with her abdomen. She then moved to a nearby blade of dead grass and laid a single egg. We found eight more eggs within an hour on small diameter (~ 1 mm) dead vegetation 1–2 inches above the ground, within 2–3 inches of stonecrop plants. Only one egg was found on a stonecrop. The flight period is late May to early Sep, depending on elevation. Adults visit flowers of Compositae and stonecrops (Scott, 1986a) and males select meadows on the basis of nectar and larval resources (Matter and Roland, 2002).

Immature Stage Biology

We partially reared this species many times during 15 years, but were unable to rear it uninterrupted from egg to adult. Gravid females were collected June 26 at Diamond Peak, Garfield Co., WA, July 17 and July 21 (different years), at Quartz Mt, Kittitas Co., on July 29 and Aug 16 at Apex Mt, BC. On all occasions numerous eggs were obtained by confining females with dried stonecrop stalks and foliage. Eggs were successfully overwintered, but no larvae survived past L3. L4 and L5 were collected six times between May 19 and July 18 at five localities in Kittitas, Chelan, Okanogan and Garfield Cos., WA. Larvae were found on stonecrops or basking nearby and in several instances were reared to adulthood. Eggs were held at ambient temperatures for ~1 month for embryonic development, then overwintered at 5 °C. After 83–221 days, eggs were exposed to 20–26 °C and hatched quickly, sometimes within 2 hours. L1 left shells uneaten and most eggs hatched within 2 days. Larvae sometimes fed poorly on *S. lanceolatum* and not at all on nursery *Sedum* spp. Mortality was high in L1 with some reaching L2 in 5–8 days (20–26 °C) and very few reaching L3, 13–14 days after egg-hatch. Similar mortality and developmental rates occurred on *S. spathulifolium* (Broadleaf Stonecrop). Mortality of early instars appeared in many cases to be due to disease. In the wild, larvae twitch violently when disturbed and drop to the ground, seeking cover. Larvae may bask openly or be concealed among *Sedum*. Larvae feed nocturnally, mostly on leaf tips, and feeding damage may be obvious. Mature larvae weave a loose cocoon and pupate within it; in captivity this occurs under cover. Guppy and Shepard (2001) reported that larval development takes 10–12 weeks and pupation occurs in a cocoon among loose debris on the ground. Adults eclose after ~14 days. There are five instars, no nests are made, and overwintering occurs as a mature egg, although Scott (1986a) suggests that some arctic populations are biennial, overwintering as older larvae or pupae in the second winter. Protection is based on chemical defense (eversible osmeterium), evasive reaction to disturbance, and perhaps distastefulness.

Description of Immature Stages

The egg is a white compressed sphere, flattened on the top. The micropyle is covered with about a dozen small dark purple beads. L1 is dark brown-black with a black head. Dorsal black tubercles each bear 3–4 long brown setae. Behind the head, this and all subsequent instars bear a yellow eversible osmeterium which likely emits defensive chemicals. L2 is dull black and setae are more numerous and shorter. Laterally, each segment has an orange-gold spot posteriorly. L3 is jet black with bristly setae and has 4 rows (2 dorsal, one on each side) of bright yellow-gold spots. The lateral row has two spots per segment (except segments 1–3 with a single spot/segment), a larger spot anteriorly and a smaller one posteriorly. Two shiny, flattened black tubercles border each dorsal gold spot front and back, and a row of similar tubercles is located just below the lateral gold spots. In L4 and L5 the pattern is similar.

Adult

Egg @ 1.6 mm

L1 @ 3.5 mm

Pupa @ 19.5 mm

L2 @ 6 mm

L3 @ 8 mm

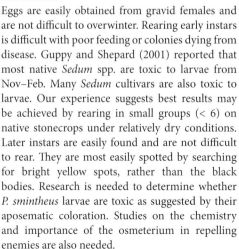

L4 @ 20 mm

L5 @ 27 mm

Mountain Parnassian Parnassius smintheus Doubleday

The yellow-gold spots contrast strongly against the black body and the larvae appear aposematic. Spots may be orange or white (Scott, 1986a). The smooth pupa is bright reddish brown, squarish anteriorly, tapering to a blunt point posteriorly. Dorsally there are four longitudinal rows of large contrasting off-white spots on the abdomen, and two spots on each thoracic segment. The immature stages are unique and should not be confused except possibly with *P. clodius*. A mature larva in Miller and Hammond (2003) is similar to ours, as are larvae and pupae in Scott (1986a); a CO larva in Allen et al. (2005) (as *P. phoebus*) differs in having smaller and fewer gold spots. Guppy and Shepard (2001) show electron micrographs and photos of eggs, mature larva, and pupa that agree closely with ours.

Discussion

Eggs are easily obtained from gravid females and are not difficult to overwinter. Rearing early instars is difficult with poor feeding or colonies dying from disease. Guppy and Shepard (2001) reported that most native *Sedum* spp. are toxic to larvae from Nov–Feb. Many *Sedum* cultivars are also toxic to larvae. Our experience suggests best results may be achieved by rearing in small groups (< 6) on native stonecrops under relatively dry conditions. Later instars are easily found and are not difficult to rear. They are most easily spotted by searching for bright yellow spots, rather than the black bodies. Research is needed to determine whether *P. smintheus* larvae are toxic as suggested by their aposematic coloration. Studies on the chemistry and importance of the osmeterium in repelling enemies are also needed.

Old World Swallowtail   *Papilio machaon* L.

Adult Biology

Papilio machaon is a holarctic species occurring across Eurasia, North Africa and North America. Numerous geographic forms and ssp. have been described and debate continues on the status of many of these including the Cascadian entity, which some authors (e.g., Pyle, 2002) consider to be a separate species, *Papilio oregonius* (Oregon Swallowtail). Old World Swallowtails occur in low-mid elevation habitats in E BC, WA, and OR including sagebrush canyons, along watercourses, and on nearby slopes and hilltops, particularly in the Snake and Columbia R drainages. There are 2–3 broods annually with adults flying from April to Oct. Males are avid hill-toppers and both sexes visit a range of flowers including thistles, to which they appear particularly attracted (Newcomer, 1964; Warren, 2005). The sole larval host is *Artemisia dracunculus* (Wild Tarragon). Many other *Artemisia* spp. occur in Cascadia but are not used by *P. machaon*, probably due to differences in chemistry (Murphy and Feeny, 2006). Larvae can be reared on umbelliferous plants like *Foeniculum vulgare* (Sweet Fennel), and may develop faster (Guppy and Shepard, 2001), but females do not choose these for oviposition. Females lay eggs singly and often choose isolated plants.

Immature Stage Biology

We reared this species several times from gravid females or wild larvae. Two females obtained from near the Snake R, N of Pasco, Benton Co., WA on April 30 and confined with cut foliage of *A. dracunculus* in a muslin-covered, steel-framed cage (55 x 30 x 20 cm), oviposited after a few hours in sunshine. Eighteen eggs were laid over 2 days and hatched after 6–7 days at 21–28 °C. Larvae were fed on *A. dracunculus* and developed from L1 to pupation in 21 days with 4, 3, 4, 3, and 7 days spent in L1–L5, respectively. The pupal stage lasted 12–14 days with adults eclosing during June 11–15. On July 15, L3 and L4 were collected in

Swakane Cyn, Chelan Co., WA and reared to adults during Aug 16–21. On Sept 9, in a side canyon of the Columbia R, N of Wenatchee, WA, 14 larvae were found on *A. dracunculus* representing all instars. Only one larva was found on any one plant. These larvae were at least second or possibly third brood, and the presence of all instars suggests that summer oviposition and development occurs over a period of 3–4 weeks. At "room temperature," Newcomer (1964) reported duration of larvae to be 30–35 days. Larvae fed openly at all times and no nests were made. L4 and L5 fed voraciously and attained a large size (up to 50mm). L5 wandered for 24–48 hrs before pupation, often on a host plant stem. Pupae from late summer larvae overwinter and may spend more than one winter in diapause (Acorn and Sheldon, 2006). Feeding is primarily on foliage, later instars consuming the elongate leaves from the terminal end. Defense is likely based on crypsis, young larvae sporting a "bird dropping" saddle and the boldly patterned older larvae blending in fairly well with Tarragon foliage. Birds reportedly reject larvae of other *P. machaon* ssp. after "tasting" them, so distastefulness may also be a defense (Scott, 1986a). Chemicals released from the osmeterium also repel ants and other invertebrate predators. Wild collected larvae may be parasitized by braconid wasps (Newcomer, 1964).

Description of Immature Stages

The egg is spherical, smooth, and flattened at the base. Creamy yellowish white when first laid, irregular, blotchy orange-brown markings appear after 2–3 days. L1 is black, becoming dark brown with maturity, with a dorsal white "saddle" on segments 6–7. Small, white-yellow spots also occur on most segments, particularly 1–3 and posterior segments. Six black bullae sporting 4–6 setae occur on each segment and the head is shiny black. L2 is similar but the bases of bullae are orange-brown. The white saddle is more prominent and there are a few white marks on the shiny black head. L3 is black with yellow-orange based black bullae, much reduced sublaterally, and the black head has two white stripes. White spots occur over most of the body, tending to transverse bands anteriorly. L4 is pale greenish white with numerous black and yellow spots and transverse intersegmental bands. The bullae are much reduced and the head is greenish white with black stripes and blotches. The true legs are greenish white with black tips and the prolegs each have a black dot. L5 is smooth with no bullae and is whitish green becoming green with transverse intersegmental black bands and alternating

Adult male

Egg @ 1.5 mm

Pupa @ 35 mm

L1 @ 3 mm

L2 @ 6 mm

L3 @ 14 mm

L4 @ 23 mm

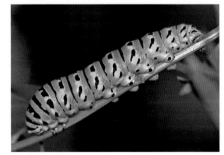

L5 @ 38 mm

Old World Swallowtail *Papilio machaon* L.

black and yellow spots arranged transversely on each segment to form bands. The green head is prominently marked with four black stripes and sometimes large spots. The pupa is brown or bluish green and has a moderately protuberant thorax, less pronounced than *P. zelicaon* pupae. There may be an indistinct dark lateral stripe and the green form is marked with yellow dorsally on the abdomen. Mature larvae are similar to and could be confused with *P. zelicaon*, but the latter does not feed on *A. dracunculus*. Images of larvae from Cascadia and AK appear in Guppy and Shepard (2001), Allen et al. (2005), and Neill (2007) and agree with ours.

Discussion

Obtaining eggs from gravid females and rearing larvae is relatively easy. Finding eggs and larvae on Tarragon in known habitats of *P. machaon* is not difficult, although the host is far more widespread and common than the butterfly. Larval coloration and markings do not appear to vary to the same extent as in the closely related *P. zelicaon*. The biology of this species is closely tied to its single host, which remains green and palatable during summer in arid environments, enabling two broods to be produced. It is unknown whether *Artemisia*-feeding larvae of *P. machaon* are distasteful to vertebrate predators, like some umbellifer-feeding races have been shown to be. Little is known concerning natural enemies except for the Newcomer (1964) report of parasitism.

Anise Swallowtail *Papilio zelicaon* Lucas

Adult Biology

Papilio zelicaon is familiar and well studied in W North America. It ranges from BC and SK to Baja CA, NM, CO, SD, and ND. It is found in most habitats including shrub-steppe, mountain meadows, suburban gardens, and along watercourses, but not in dense forests. In arid and high elevation habitats where it utilizes seasonally ephemeral larval hosts it is univoltine, but in coastal and wetter locations with season-long availability of hosts it may be double or triple brooded. The flight period is late March–late Sep, but in arid and high elevation areas it is a spring or high summer butterfly, respectively. A wide range of native and exotic umbelliferous hosts is used; Wehling and Thompson (1997) reported 69 hosts in 32 genera and Warren (2005) listed 39 species for OR. Hosts include *Lomatium* spp. including *L. dissectum* (Fern-leaf Parsley), *Angelica* spp. (angelica), *Cicuta* spp. (water hemlock), *Daucus* spp. (wild carrot), *Osmorhiza* spp. (sweet cicely), *Pteryxia petraea* (Turpentine Wavewing), *Foeniculum vulgare* (Sweet Fennel), *Glehnia littoralis* (Beach Carrot), *Carum carvi* (Caraway), *Heracleum maximum* (Cow Parsnip), *Ligusticum apiifolium* (Celery-leaf Lovage), *Sium suave* (Water Parsnip), *Zizia aptera* (Meadow Parsnip), and *Cymopterus terebinthinus* (Turpentine Cymopterus). Males patrol and perch on hilltops to await females near host plants. Both sexes visit flowers including lilac, lupine, mustards, balsamroot, desert parsley, yarrow, and thistles. Males are avid mud-puddlers. Females produce an average of 82 eggs (Wehling and Thompson, 1997) laid singly, preferring isolated and peripheral plants (Shapiro and Manolis, 2007).

Immature Stage Biology

We reared this species several times from gravid females or wild larvae. A female collected April 27 at Umtanum Cyn, Kittitas Co., WA, oviposited on *Lomatium dissectum*. Eggs hatched in 4 days, larvae were reared on *F. vulgare* at 20–22 °C, and pupation occurred 24 days post egg-hatch. Two females obtained from Bear Cyn, Yakima Co., WA on May 13 and confined with *Lomatium grayi* in a muslin-covered plastic bucket (28 cm dia, 30 cm tall) laid 15 eggs in two days. Eggs hatched in six days at 19–26 °C and larvae were reared on *F. vulgare*. Development was rapid with pupation 15 days post egg-hatch, and all pupae overwintered. Two–three days were spent in each instar except L1 and L5 which occupied 4 days each. Larvae fed openly at all times and no nests were made. L5 wandered for 24–48 hrs before selecting a site for pupation, often on a host plant stem. Despite their bold coloration, larvae blend well with umbelliferous foliage. Some authors consider swallowtail larvae like *P. zelicaon* to be warningly colored, mimicking an unpalatable model, e.g., *Danaus plexippus*. However, there is no evidence that swallowtail larvae are significantly distasteful or avoided by vertebrate predators (Brower and Sime, 1998). Intimidation by and release of defensive odors from the osmeterium is an additional line of protection. Pupae are capable of diapausing for at least two years (Sims, 1983).

Description of Immature Stages

The egg is spherical, smooth, and creamy yellowish-white. Irregular, blotchy, orange-brown markings appear in 2–3 days. L1 is black with a dorsal white "saddle" on segments 6–7. Small white spots occur on most segments, particularly 1–3. Six black bullae, each with 4–6 setae, occur on each segment. Small globules occur on setae tips. The head is shiny black. L2 is similar but the bases of bullae are orange-brown and the setae are shorter. L3 is black with orange-red based black bullae. The white saddle is more distinct; prominent white spots occur on anterior and posterior segments and above each true leg and proleg. L4 has reduced bullae and occurs in two forms, either black with orange and white spots or white with orange and black spots. The head in the black form is black with white markings and in the white form white with black markings. L5 is smooth, green or white with highly variable black and orange or yellow markings. In white larvae, the black transverse markings between and on segments present a banded appearance. The head has black stripes and yellow-orange blotches. The pupa is brown or green and angular with a protuberant thorax. Both forms have a dark lateral stripe; dorsally the brown form is whitish and the green form yellow. Mature larvae may be confused

Adult

Egg @ 1 mm

Egg @ 1.2 mm (mature)

L1 @ 5 mm

L2 @ 8 mm

Pupa @ 30 mm
(green form)

L3 @ 12 mm

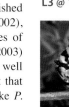

L4 @ 21 mm (white form)

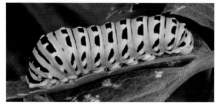

L5 @ 38 mm (white form)

L5 @ 35 mm (green form)

with *P. machaon* (Old World Swallowtail), but use different hosts. Early instars could be confused with *P. indra* (Indra Swallowtail) which occurs on the same hosts. Life stage images were published in Guppy and Shepard (2001), Pyle (2002), Woodward (2005), and Neill (2007). Images of mature larvae are in Miller and Hammond (2003) and Allen et al. (2005). All agree relatively well with ours, given the larval variation, except that of Allen et al. (2005), which looks more like *P. machaon*.

Discussion

Rearing larvae is not difficult as long as a good supply of host plant is available. Wild eggs can be found, their white color contrasting with the green host plant, and they can often be seen from above. Much still remains to be learned about the biology of this well-studied swallowtail, such as the genetic and environmental factors that control larval coloration. Over its range the Anise Swallowtail has multiple ecological races that differ in aspects of biology (e.g., voltinism, diapause, host plant use). The role of defenses like setal secretions (L1), concealment, possible aposematism and the osmeterium in survival need study. In N CA, up to 50% of larvae on *Lomatium californicum* were parasitized by a braconid wasp (Shapiro and Manolis, 2007).

Anise Swallowtail *Papilio zelicaon* **Lucas**

Indra Swallowtail   *Papilio indra* Reakirt

Adult Biology

Papilio indra is a charismatic small black swallowtail with reduced "tails" and a penchant for fast flight in wild, rocky places. It frequents hilltops, rocky slopes, and canyons but can also be found along rivers and roadsides and in open forest. It ranges from S BC to AZ and E to UT and CO. In Cascadia, populations are phenotypically similar and represent a single subspecies (*P. i. indra*). *Papilio indra* flies in Cascadia from mid-March to mid-Aug and is single brooded. Males are more commonly encountered than females, particularly early in the flight period. Females are best found 2–3 weeks after flight commencement on the upper reaches of rocky slopes near favored host plants, including *Lomatium grayi* (Gray's Desert Parsley), *Pteryxia petraea* (Turpentine Wavewing), and *Lomatium brandegeei* (Brandegee's Desert Parsley). Males often visit mud puddles, while females nectar on *Lomatium,* yellow asters, and brodiaeas. Females lay eggs singly on host plants, usually on outside bracts. Oviposition occurs in typical papilionid fashion with wings fluttering.

Immature Stage Biology

We reared this species twice from eggs obtained from Klickitat Co., WA females. On the first occasion Gary Pearson (Vancouver, WA) supplied us with 10 eggs on March 31. Most eggs died, but we were able to rear a single individual to adulthood. On the second occasion, 1 female caught near Klickitat in Klickitat Co. on May 1 was placed in a steel-framed muslin-covered cage (55 × 30 × 20 cm) with cut foliage of *L. grayi.* The female was held at 28 °C for 1 day and then placed outdoors in sunshine at 20–22 °C. Seven eggs were laid before the female died after 4 days. Eggs took 6–7 days to hatch at 20–28 °C. L1 consumed eggshells, and larval development was rapid, pupation occurring in 30 days at 20–23 °C and 22 days at 22–28 °C. Most time was spent in L5 (8–10 days), with 3–5

days in each earlier instar. Newcomer (1964) reported that larval development of *P. indra* took 18 days at "room temperature." Most growth occurred in L5, with larvae doubling in size from 23 to 50 mm. Final instars occurred in 2 color forms, banded or black. Of 6 larvae reared to L5, 4 were banded. Banded larvae tended to rest exposed on the host plants, whereas black larvae rested concealed within the foliage. The osmeterium was functional from L2 to L5. Pupation was on a cage side or branch. Pupae oversummer and overwinter, spending 11–12 months in this stage. Some pupae may remain dormant for 2 or more years (Shapiro and Manolis, 2007). Rapid larval development is critical to the survival of this species, as host plants senesce quickly. The black larval ground color aids in raising body temperature and hastening development. The bold banding of L5 may serve as a real or false advertisement of distastefulness. Protection may involve both concealment and intimidation.

Description of Immature Stages

The egg is spherical and smooth, flattened at the base. It is creamy white when first laid, and irregular, blotchy orangish markings appear in 2–3 days. L1 is black with irregular white spotting in 6 indistinct longitudinal lines, 2 dorsally and 2 laterally. Six black bullae with 4–6 setae occur on each segment. Small globules, which may be a defensive secretion, are present on the tips of setae. The head is shiny black, sparsely covered with short setae, becoming marked in L2 with 2 small lateral white spots and dense fine setae. The shiny black head and similar body patterning occurs in L2 through L4, with bullae bases becoming increasingly orange-brown and the setae proportionally shortened. The overall appearance is shinier from L2 to L4, and by L4 the larva appears orange-spotted. In L5 most individuals develop prominent transverse white-pink bands, 1 on each segment. The larva is smooth, and large yellow, pink, or white spots occur where bullae were earlier. The L5 head is yellow with 2 wide frontal black bands in an inverse V shape (a pattern noted in L4 by Newcomer, 1964), 2 lateral black dots, and a black area above the mouthparts. Unbanded L5 are black with a row of large yellow, white, or pink spots laterally, 1 per segment. Dorsally there are 2 rows of smaller orange spots, but overall the appearance is that of a jet black caterpillar. The pupa is relatively smooth, with few protuberances, and medium to dark brown with an olive cast in some individuals. Mature larvae of *P. indra* were illustrated by Pyle (2002), Miller and Hammond (2003), and Allen et al. (2005).

Adult

Egg @ 1.0 mm (fresh, *left*, mature, *right*)

L1 @3.0 mm

L2 @ 9 mm

L3 @ 14 mm

L4 @ 22 mm

L5 @ 30 mm (black form)

L5 @ 30 mm (banded form)

Pupa @ 26 mm

None of these is an accurate match for the larvae we reared, with the latter 2 publications showing more extensive banding and larger spots. An OR larva in Pyle (2002) is closest in appearance to ours, but the banding is yellow. Scott (1986a) described great variation in banding color of larvae.

Discussion

Obtaining gravid females in the favored rocky slope habitat can be difficult. Oviposition is not difficult to obtain when females are confined with fresh, unwilted hosts in sunshine. Rearing larvae is not difficult as long as a good supply of host plant is available. Larval coloration may help in identifying subspecies of *P. indra* (Emmel and Emmel, 1998). Additional study is needed to determine the genetic and environmental factors controlling larval coloration, which varies even in the progeny of a single female. The apparent difference in basking behavior between banded and unbanded larvae deserves further study, as does the suggestion by Emmel and Griffin (1998) that under arid conditions larvae may pupate in L4, and early maturing larvae may produce a second brood (Emmel and Emmel, 1998; Emmel et al., 1998). Little is known of natural enemies or the role of defenses like setal secretions (L1), concealment, possible aposematism, and the osmeterium in survival.

Western Tiger Swallowtail   *Papilio rutulus* Lucas

Adult Biology

Papilio rutulus is widespread and one of the commonest large butterflies in Cascadia, ranging from S BC to N Mexico and E to the Great Plains. Its large size and striking bright yellow-and-black tiger stripes catch the attention of even casual observers. It can be found in most habitats from sea level to ~7,000 ft, including gardens, alpine meadows, and riparian zones along watercourses. The single brood flies from late April to Aug, and eggs and larvae are found from mid-June to late Sep. In some years there may be a partial second brood. *Papilio canadensis* is rare in NE WA but common in S BC. Most individuals found near the border are likely hybrids with *P. rutulus*. The larval hosts include many species of common landscape trees, including *Salix* (willow), *Populus* (poplar, aspen), *Alnus* (alder), *Betula* (birch), *Acer* (maple), *Malus* (apple), *Fraxinus* (ash), and *Prunus* (cherry) (Warren, 2005). Males hill-top, perch, or patrol relentlessly, searching for females along corridors such as roads and foliage barriers. Both sexes visit a wide variety of flowers (Pyle, 2002; Shapiro and Manolis, 2007), and males commonly mud-puddle, often with other swallowtails. After mating, a female flies from plant to plant, searching for an oviposition site before depositing a single egg on the upper side of a leaf. She then flies to a different plant, often a considerable distance away, to search for the next site. Preferred oviposition sites are 5–8 ft above ground, sometimes overhanging inaccessible areas. Oviposition over a pond was observed in King Co., WA, and over a near vertical talus slope in Yakima Co., WA.

Immature Stage Biology

We reared or partially reared *P. rutulus* >10 times from eggs or larvae. A female laid a single egg on *Alnus* at Bethel Ridge, Yakima Co. on July 18 that hatched after 6 days at 22–26 °C. After consuming the eggshell, the L1 immediately fed on *Salix hookeriana* (Hooker's Willow) leaves, molting to L2 after 6 days. L2 and later instars select a leaf as a base, laying down a thin mat of silk on the upper surface. When the larva is not feeding, it returns to this silked leaf to rest. The silk mat draws the leaf into a gently concave shape opening upward, a characteristic shape and good search image for finding larvae. Ten and 14 days were spent in L2 and L3, respectively, with pupation occurring at the end of L4 (Sep 6), 43 days after egg-hatch. Guppy and Shepard (2001) reported 5 instars, and it is likely that instar number varies as a response to food quality, temperature, or photoperiod or other environmental factors. Mature larvae wander and turn dark brown prior to pupation. Feeding evidence typically includes jagged eaten leaf tips and edges; with later instars damage can be considerable. Larvae are solitary, and protection is derived from camouflage and the osmeterium, a Y-shaped eversible soft orange organ behind the head that emits a disagreeable odor to repel predators, mostly birds. Overwintering occurs as a pupa.

Description of Immature Stages

The eggs are ovoid and green with no discernible ribbing or markings, darkening with maturity. L1 and L2 are dark brown with a contrasting white saddle across the middle of the back; L3 is similar but green and brown. The larva undergoes a dramatic change in appearance in midgrowth, usually when it molts to L4; the "bird dropping" larva becomes smooth, green, and hairless with contrasting yellow false eyespots and a yellow collar behind the head. The pupa varies from dark brown to hoary gray with a longitudinal dark brown side stripe, well camouflaged against wood or dead leaves. Larvae are similar to those of *P. eurymedon*, *P. canadensis*, and *P. multicaudata*. The differences we have found consistently are limited: *P. multicaudata* is distinguished by its larger size, by fine black lines around the false eyespots (much heavier in *P. rutulus*), and by a yellow ring surrounding the blue false eyespot (*P. rutulus* has a heavy black ring). The larvae of *P. eurymedon* are more difficult to distinguish from *P. rutulus*, but the satellite false eyespot is usually disconnected from the main eyespot in *P. eurymedon* but attached to it by a blue band in *P. rutulus*. *Papilio rutulus* is bright yellow-green in the last instar while *P. eurymedon* tends to be more bluish green. Common larval hosts for *P. rutulus*, not used by *P. eurymedon*, are *Salix* spp., while *Ceanothus*, one of the hosts frequently used by *P. eurymedon*, is not used by *P. rutulus*. We have not reared *P. canadensis*, but immature stages are likely indistinguishable from *P. rutulus*, as shown by the images of this species in

Adult

Egg @ 1.75 mm

L5 @ 30 mm (head)

L1 @ 6 mm

Pupa @ 30 mm

L2 @ 9 mm

L3 @ 14 mm

L4 @ 22 mm

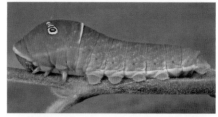

L4 @ 35 mm

L5 @ 30 mm

Western Tiger Swallowtail *Papilio rutulus* Lucas

Guppy and Shepard (2001). The immature stages of *P. rutulus* are well-known and the larvae have been illustrated frequently (e.g., Layberry et al, 1998; Guppy and Shepard, 2001; Pyle, 2002; Miller and Hammond, 2003; Allen et al., 2005). These images agree well with ours.

Discussion

P. rutulus adults require a large flight cage, but it is difficult to obtain eggs from gravid females. Obtaining livestock is best accomplished by finding eggs or larvae. Observing female behavior and flight patterns along a flight corridor, selecting host plants, which protrude, and searching for contrasting eggs or concave silken nest leaves 5–8 ft high may prove fruitful. Larvae and pupae are sometimes parasitized by ichneumonid wasps. Host plants for this species are readily available, and the larvae feed and grow well in captivity. The number of instars varies from 4 to 5; research is needed to determine the reasons why and the environmental cues involved. Similarly, studies are needed on the cues mediating pupal diapause and whether diapause can last for >1 winter as is the case with some other *Papilio* spp.

Two-tailed Swallowtail *Papilio multicaudata* W. F. Kirby

Adult Biology

Papilio multicaudata, the largest butterfly in Cascadia, flies from mid-April to late Aug and occurs E of the Cascades in S BC, WA, and OR. Elsewhere it is found throughout W US, S to Guatemala. Larval hosts include *Prunus* spp. (cherry, plum), *Fraxinus* spp. (ash), *Ptelea trifoliata* (Common Hoptree), and *Amelanchier alnifolia* (Serviceberry), but the most often used in Cascadia is *Prunus virginiana* (Chokecherry). Preferred habitats include canyon bottoms and roadsides in open pine forest, and shrub-steppe riparian areas. Both sexes nectar on flowers, including thistles, *Ceanothus*, and dogbane. Males often mud-puddle with other swallowtails and patrol roads and foliage barriers in search of mates. Females cruise along host-plant corridors in search of oviposition sites. In mid-May near Spokane, WA, we observed females patrolling the top of a cliff, visiting Chokecherry shrubs growing near the edge, alighting frequently, occasionally laying a single egg per plant on the upper side of the middle of a leaf. Females favored small shaded plants, and eggs were often placed on damaged leaves. A large, isolated Chokecherry at Naches, Yakima Co., WA attracted ovipositing females and patrolling males. Adults are long-lived, and this species is generally single brooded; Warren (2005) reported a partial second brood in OR. It is double or multiple brooded in N CA, CO, and farther S (Scott, 1986a; Shapiro and Manolis, 2007).

Immature Stage Biology

We reared this species 3 times from 4 eggs collected May 14 in Spokane Co., and 3 eggs (June 6, July 23) and an L4 (Aug 23) collected in Yakima Co. Three of the Spokane eggs hatched, and larvae were reared at 20–22 °C on potted *P. virginiana*; all pupated and overwintered, and 2 adults eclosed during June 8–10.

Adult

Development from L1 to pupation took 33 days. In the Yakima egg cohort, 2 larvae died in early instars and 1 reached L4. The collected L4 pupated. L1 consumed eggshells before feeding on terminal leaves. L2 spun a thin silk mat on the upper side of a leaf to which the larva returned to rest when not feeding. Older larvae produced similar mats, curving them into an upwardly convex shape; such "nest" leaves were not eaten. L2 larvae possess an eversible osmeterium behind the head that emits defensive chemicals when the larva is threatened. Older larvae eat large gaps in leaves or entire leaves, and sometimes rest on the underside of a leaf. Larvae are always attached to the plant with a strong silk tether to prevent falls or easy removal by predators, and often rest with the head humped up. Older larvae feed voraciously; 3 defoliated 2 Chokecherry plants, each 5 ft tall, and grew to 60 mm. The Yakima Co. larva, perhaps second brood, pupated at the end of L4 at 45 mm. Pupation occurs at the base of the host plant on the trunk or on stems. Protection is based on camouflage (green L4/L5 on green leaves, L1–L4 "bird-dropping" markings, pupa cryptic brown) and defensive chemicals (L1 setal droplets, L2–L5 osmeterium). Overwintering occurs as a pupa. There are 4 or 5 instars.

Description of Immature Stages

The yellowish green spherical egg turns orange with darker mottling near the base. L1 is black becoming dusky gray with irregular black patches dorsally and has a shiny black head. There are 6 protruding bullae on each of the first 3 segments, each bearing multiple white setae with droplets. Other segments have 2 dorsal bullae, larger posteriorly, each bearing setae. There are 2 protruding horns dorsally behind the head, each also bearing multiple setae. There is a dorsal yellowish saddle and there are additional small bullae laterally on all segments. L2 is similar, shiny black/brown, with horns and bullae proportionally reduced to rounded bumps, virtually absent in midbody. The saddle is white and setae are reduced to small stubble. The head and body of L3 are dark shiny brown and in some individuals the white saddle has 2 brown spots anteriorly. In the 3 segments behind the saddle and 2 anteriorly, the dorsal swellings bear contrasting bluish white spots. L4 is initially similar to L3, but during this instar there is a dramatic color change to bright green with 2–6 small blue spots on segments 3, 4, and 7–10. Between segments 3 and 4 there is a black transverse band anteriorly edged in white. The white part of the band often appears

Egg @ 2 mm

L5 @ 60 mm (head)

Pupa @ 37 mm

L1 @ 7 mm

L2 @ 10 mm

L3 @ 19 mm

L4 @ 30 mm

L5 @ 60 mm

Two-tailed Swallowtail *Papilio multicaudata* W. F. Kirby

wider than the black, which is largely obscured when the larva is resting; the black band expands dramatically when the larva is agitated. There are complex false eyespots dorsally on segment 3, mostly yellow narrowly outlined in black, with a blue dot center. L5 is similar except brighter green. Prepupal larvae become yellowish then brown prior to pupation. The brown or green pupa is cryptic with a roughened surface, sometimes with a midlateral dark brown stripe on the abdomen and dark brown wing cases and head. The immature stages of *P. multicaudata* are similar to those of *P. rutulus*, *P. eurymedon*, and *P. canadensis*. The false eyespots differ; in *P. multicaudata* they are less pronounced, with a finer black line around the blue ocellus and a smaller blue spot between the 2 halves of each "false eye." Allen et al. (2005) pictured a mature AZ larva similar to ours, and Neill (2007) showed a "half grown" larva similar to our L3.

Discussion

While this species is easily reared, it is difficult to obtain eggs in captivity. *Papilio multicaudata* needs a large cage for oviposition and, like other tiger swallowtails, oviposits poorly in confinement. Finding eggs or larvae is the easiest way to obtain immature stages. Eggs contrast with host leaves and are easy to see, and may be found on small plants along edges of flight corridors. Watching for ovipositing females is the best way to locate eggs. Research is needed on environmental cues, such as photoperiod or host quality, that may control instar number and diapause induction. Little is known of the natural enemies and other factors controlling abundance and distribution of this species.

Pale Swallowtail *Papilio eurymedon* Lucas

Adult Biology

Papilio eurymedon is common in Cascadia, often found in numbers in large mixed-species mudding congregations. It is found throughout much of Cascadia, absent only from the driest parts of the Columbia Basin, flying from late April to mid-Sep. Guppy and Shepard (2001) report the species is single brooded in BC but multivoltine farther S. We have not observed a second brood in WA; Warren (2005) reported the species is single brooded in OR, and Shapiro and Manolis (2007) described it as obligately single brooded in N CA. *Papilio eurymedon* utilizes a variety of larval hosts, including *Ceanothus velutinus* (Mountain Balm), *Ceanothus sanguineus* (Red-stem Ceanothus), *Prunus* (cherry, plum, bittercherry), *Crataegus* (hawthorn), *Ribes* (currant), *Alnus rubra* (Red Alder), *Holodiscus discolor* (Ocean Spray), *Rhamnus* (buckthorn, cascara), *Betula* (birch), and *Amelanchier alnifolia* (Serviceberry). *Malus* spp. (apple), *Spiraea douglasii* (Douglas Spiraea), and *Ceanothus integerrimus* (Deerbrush) are reported elsewhere. Guppy and Shepard (2001) reported hill-topping, and males often patrol forest roads and riparian corridors in search of mates. We have observed females ovipositing on *C. velutinus*, where they visit multiple leaves before choosing a site. A single egg is laid quickly, usually on the upper side, basal half, of a medium-sized new-growth leaf.

Immature Stage Biology

We reared or partially reared this species several times; on June 12 and 15 we collected 11 eggs but none hatched. On June 26 we collected 4 eggs and all developed to L5, 1 pupating. On June 8 we collected 3 eggs and an L1; the eggs hatched, but all larvae failed to survive past L2, dying from a virus. On July 24 we collected an L3 and reared it to pupation. In all cases the eggs and larvae were found on *C. velutinus.* Eggs hatched in ~4 days; early development was rapid but slowed in later instars. L2, L3, L4, and L5 were reached at 6, 11, 22, and 31 days post oviposition, respectively, and pupation occurred 9 days later at 20–30 °C. One egg was found on June 30 on *C. velutinus* at high elevation (5,200 ft) and was left in situ; development was followed to L3, when it was collected for rearing. This individual pupated at least 55 days post oviposition, considerably slower than the larvae reared in captivity, presumably as a result of lower temperatures. Larvae attain a large size, up to ~50 mm. In captivity, larvae were started on domestic prune leaves, then switched to *C. velutinus* on which they fed more readily. Only leaves were consumed, the larvae feeding from the leaf edges, with fresh young leaves preferred. Larvae do not make nests but spin a light pad of silk on a concave upper leaf surface where they rest. There are 5 instars, and the pupa overwinters. The survival strategy of *P. eurymedon* is a combination of camouflage and intimidation. The early instars have "bird dropping" markings for concealment from predators; later instars are green to blend with foliage and have distinct false eyespots to confuse and intimidate. An eversible osmeterium emits pungent, defensive chemicals if the larva is disturbed.

Description of Immature Stages

The egg is smooth, spherical, and pale green. L1 is dark brown with a contrasting white "bird dropping" saddle. There are 2–4 knobby dorsal horns on each segment, more and larger anteriorly. Between the largest horns the dorsal area is pale. L2 is reddish brown and smoother, the horns proportionally shorter. L3 is green and smooth, with reduced posterior horns. False eyespots appear on the second segment, the saddle is less distinct, and the body is covered with indistinct pale spots. L4 is green with no horns and numerous small lighter green spots; each of the central segments bears 3 round purple spots in a row down each side. A distinct yellow line between segments 3 and 4 is bordered with black posteriorly (often not visible). False eyespots are present on the third segment, each with 2 separate orange parts outlined with a thin black line and separated by a pale blue patch. The larger (outer) part of the eyespot includes an ocellus with a pink center and thick black outline, and a separate vertical black mark. L5 is similar, except the small green body spots become obscure. Prepupal larvae are dark reddish brown (Pyle, 2002). The pupa is marked in streaked browns and creams. In all stages *P. eurymedon* is similar to *P. rutulus,* and both use

Adult female

Egg @ 1.4 mm (fresh)

Egg @ 1.7 mm (mature)

L1 @ 5 mm

Pupa @ 40mm

L2 @ 7 mm

L3 @ 16 mm

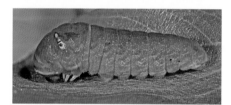

L4 @ 26 mm

many of the same hosts. Normal variation obscures interspecific differences, and we suggest they are not always reliably separated in the immature stages. The immature stages of *P. multicaudata* are also similar but usually grow larger, and the false eyespots are blue in the center surrounded with yellow. Miller and Hammond (2003) show a mature larva, Woodward (2005) shows a mature larva and a pupa, and Guppy and Shepard (2001) show the egg, mature larva, and pupa. All images agree with ours, except the black collar in Miller and Hammond (2003) is heavier.

Discussion

It is difficult to obtain eggs from caged females as they generally need a large flight cage. Eggs are relatively easy to find on *C. velutinus* by searching young leaves where adults fly. Larvae can also be sought, and Miller and Hammond (2007) suggest they can be obtained using a beating sheet. Larvae are generally easy to rear. Virus killed some of our early instars, and it is likely that many larvae are parasitized. Research on the natural enemies of *P. eurymedon* is needed to determine the extent of natural mortality levels in Cascadia.

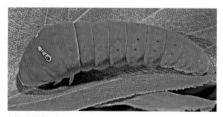

L5 @ 32 mm

L5 @ 48 mm (head)

Family Pieridae

WHITES and *SULPHURS*

The family Pieridae includes the whites, marbles, orangetips, and sulphurs and is represented by 18 species in Cascadia. Butterflies in this family generally have white or yellow wings dorsally, some with black, green, or orange markings.

Females lay single spindle-shaped eggs on the buds, flowers, or leaves of their hosts. The Pine White is an exception, laying rows of eggs on conifer needles. Eggs are typically pale blue or green maturing to orange, with the head of the larva showing through prior to hatching. Larvae of most species are cryptic, smooth, slender, and elongated, but densely covered with downy short setae that in early instars often carry sticky, oily droplets. These have been shown in some species (e.g., Cabbage White) to repel predators like ants.

Our most ornate pierid larvae are species of *Pontia*, including Becker's, Spring, Checkered, and Western whites. Most other species are uniformly green, although many have a strong lateral white stripe, sometimes brightened with red spots or dashes. Larvae of some species are sensitive to disturbance, wriggling violently or dropping to the ground on a fine thread of silk. Pupae are attached by a button of silk at the tail and a girdle around the middle. Some are strongly pointed at both ends, others more rounded, and coloration is influenced by the substrate.

Overwintering occurs as pupae among whites, marbles, and orangetips (except the Pine White, which overwinters as an egg), and as partially grown larvae in sulphurs. Many whites, marbles, and orangetips fly in early spring, utilizing the ephemeral growth of their mustard hosts to produce next year's generation. During some springs, almost every *Arabis* plant in the shrub-steppe of E WA seemingly has a pierid egg attached to it. Sulphurs generally fly in summer and use legume, bilberry, or willow hosts. Pierids occupy all habitats in Cascadia, from mountaintops to deserts to cities, and some species undergo periodic population explosions, e.g., Pine White and Orange Sulphur. The Cabbage White is a serious economic pest of crucifers worldwide.

Pine White   *Neophasia menapia* (Felder & Felder)

Adult Biology

The Pine White is a primitive white, with hosts, biology, and life history unique in our pierid fauna. It occurs from S BC to central CA, NM, and AZ and E to SD and NE. In Cascadia it flies in forested areas from sea level to mid-elevations except for wettest coastal areas in WA. Outbreaks reaching pest status have occurred (Pyle, 2002). A single brood flies late June–early Oct, peaking Aug–Sep. Adults, particularly females, spend much time resting in the upper canopies of conifers but glide down to ground level for nectaring on yarrow, thistles, asters, and Pearly Everlasting. Pyle (2002) reports nectaring morning and evening, but during late Sep in Yakima Co., WA, most nectaring occurred around midday. Males patrol canopies for females making short oviposition forays. Of 120 adults sampled during 4 days in Sep, only 20 were females. *Neophasia menapia* uses a wide range of native and introduced conifers as larval hosts, including *Pinus ponderosa* (Ponderosa Pine), *P. contorta* (Lodgepole Pine), *P. sylvestris* (Scotch Pine), *P. edulis* (Pinyon Pine), *P. strobus* (White Pine), *Pseudotsuga menziesii* (Douglas Fir), *Tsuga heterophylla* (Western Hemlock), and *Abies grandis* (Grand Fir) (Scott, 1986a; Guppy and Shepard, 2001; Pyle 2002). Eggs are laid in angled rows of 3–25 or more (Shapiro and Manolis, 2007), along conifer needles.

Immature Stage Biology

We obtained oviposition several times by gravid females from Yakima and Kittitas Cos., WA (2, 11, 32, and 37 eggs from 1, 10, 9, and 20 females, respectively, during Aug 9–Sep 20) but were successful in rearing to adulthood only once. Failures resulted from egg infertility or death during overwintering. Ovipositing females were fragile, living for only a few days. Oviposition was obtained on potted or cut *P. menziesii* in a variety of small and large cages. In the successful rearing, 37 eggs were laid in 5 batches (7, 3, 6, 10,

6) during Sep 18–20 on needles or cage sides, and were held at 15–21 °C until transfer to overwintering conditions on Oct 27. All batches except 1 were held at 5 °C/~80% RH. One batch was secured on an outdoor branch of *Pinus nigra* in Yakima. After 112–127 days, eggs held at 5 °C were transferred to 20–26 °C; outdoor eggs were transferred after 161 days (April 7). L1 hatched in 14–16 days from all or some eggs in each batch except 1. L1 did not consume eggshells, but most cannibalized 1–3 neighboring eggs. Development from L1 to pupation took 63 days, with L1–L5 occupying 13, 16, 10, 10, and 14 days, respectively. L1–L2 fed gregariously, typically 4–6 on a needle, most with heads toward the tip. Some fed from the tip, others from the sides. Feeding produced roughly notched needles, and most were consumed except for the basal stump. L3–L5 generally feed alone and are superbly camouflaged. L1 flicked frass away, but later instars simply dropped it. L3 possesses a bilobed gland ventrally on segment 1 that is everted when the larva is disturbed. In L4 and L5 this gland is smaller. Disturbed L1 may drop, suspended by silk; older larvae produce regurgitant and wave heads in unison. L1 do not possess droplet-bearing setae. Prepupal L5 wandered in captivity and pupated on twigs or needles; wild larvae are reported to lower themselves by silk for pupation on branches near the ground (Guppy and Shepard, 2001). Immature stages of *N. menapia* are attacked by at least 24 natural enemies (Scott, 1986a), but details are lacking. Protection is based on camouflage; there are 5 instars, no nests are made, and eggs overwinter.

Description of Immature Stages

The egg is flask-shaped, bluish to emerald green, with 20–22 vertical pale ribs. Around the narrow micropyle there are 9–11 bright white beadlike bumps. L1 is greenish yellow with a shiny black head. Each segment has 12–18 small dark spots, each bearing a short dark seta. The first segment is sclerotized dorsally. L2 is yellowish at first, becoming greener, and the head is medium-dark brown. Setae are more numerous, shorter and inconspicuous, and each segment has 4 transverse ridges posteriorly. L3 is yellow-green, darkening with maturity, with inconspicuous small white spots, each bearing a very short seta. Laterally, there is a spiracular yellowish white stripe, and the head is olive green with white spots. A thin inconspicuous yellowish stripe is present dorsolaterally. The true legs are black, tipped in orange, and the prolegs are green, each with a large black spot. L4 is dark conifer green with numerous small white spots and distinct

Adult male

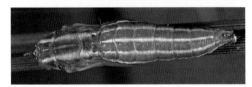

Pupa @ 20 mm (lateral, *above*, dorsal, *below*)

Eggs @ 1.5 mm

L1 @ 3 mm

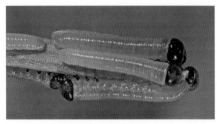

L2 @ 4.5 mm (top and bottom)

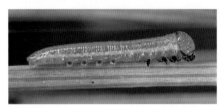

L3 @ 9 mm

L4 @ 15 mm

L5 @ 27 mm

lateral (thick) and dorsolateral (thin) yellowish stripes. There are 2 very short tail-like projections on the posterior segment. L5 is similar, the head green with yellowish spots, and the stripes may be yellow or white. The pupa is green with 2 white stripes dorsally and 3 on the thorax and abdomen. Laterally, there is a single broad white stripe on the abdomen and there are white markings on the wings. There is a black dot above the eye and the head has a brown conical protuberance. Scott (1986a) states that male pupae are green/yellow while females are dark brown. Immature stages of *N. menapia* are unlikely to be confused with other Cascadia butterflies but could be confused with some moths. The only other conifer-feeding butterfly, *Callophrys eryphon* (Western Pine Elfin), is a lycaenid. Published images include those in Pyle (2002) (eggs and larva) and mature larva in Miller and Hammond (2003) and Allen et al. (2005); all agree with ours.

Discussion

This is a difficult species to rear, and its immature stages are difficult to locate in the wild. Females are often egg-depleted or virgin, short-lived in captivity and reluctant to oviposit. Eggs are difficult to overwinter, succumbing to mold or dying from insufficient humidity. When eggs hatch, the larvae should be transferred immediately to prevent cannibalism of other eggs. Larvae develop slowly and suffer high mortality despite care. The ability of populations to sometimes "explode" is remarkable given the species' apparent fragility. Research on the population and environmental parameters leading to outbreaks is needed. The function of the ventral gland seen in L3 and apparent degeneration in later instars deserves study.

Pine White *Neophasia menapia* (Felder & Felder)

Becker's White   *Pontia beckerii* (W. H. Edwards)

Adult Biology

Pontia beckerii is a species of the harsher, drier environments of the arid interior, occurring from S BC to S CA, extending E to MT, WY, CO, and NM. It is common and widespread E of the Cascades in disturbed and undisturbed arid habitats, primarily shrub-steppe, deserts, canyons, and along watercourses. There are 2 broods in most of Cascadia, adults flying during March–May and Aug–Sep; a third generation may occur in S OR, and Guppy and Shepard (2001) suggest at least 3 generations occur in S BC. Availability of larval hosts in good condition may determine the number of broods in any specific habitat. Becker's White is an avid flower visitor, feeding on mustards, thistles, Alfalfa, sweet clover, and asters. Males patrol canyons in search of females. In mid-July 2 males were observed locating and attempting to mate with a resting, newly eclosed female in Benton Co., WA. After a prolonged male-male interaction, 1 male successfully mated. Larval hosts are various introduced and native mustards (Scott, 1986a). In Cascadia, *Sisymbrium altissimum* (Tumble Mustard), *Brassica nigra* (Black Mustard), *Stanleya pinnata* (Prince's Plume), *Schoenocrambe linifolia* (Plains Mustard), *Lepidium perfoliatum* (Field Peppergrass), *Descurainia* spp. (Tansy Mustard), *Arabis* spp. (rockcress), and *Thelypodium* spp. have been reported as hosts (Guppy and Shepard, 2001; Pyle, 2002; Warren, 2005). Eggs are laid singly, often on inflorescences and saliques.

Immature Stage Biology

We reared *P. beckerii* several times. Mature larvae collected from senescing *S. altissimum* in Benton Co. on July 3 pupated 3 days later. The pupae oversummered, then overwintered, and adults eclosed on April 20–22. On July 9 near Sun Lakes, Grant Co., WA, eggs, L3, L4, and L5 larvae were collected from *Thelypodium integrifolium* (Whole-leaf Mustard). The older instars pupated shortly after collection,

and adults eclosed a few days later. One female was collected on July 4 in E Kittitas Co., WA, where the only apparent host plant was *S. altissimum*. Several eggs were produced and reared to L2 on this plant, but they died when the plants senesced. One female obtained from near the Grande Ronde R, Wallowa Co., OR on May 18 and caged with cut *S. altissimum* in a plastic cylinder cage (16 cm dia × 8 cm tall) with muslin ends laid ~100 eggs during 4 days. Eggs were laid on all parts of the plant and inert surfaces. Eggs hatched after 3 days at 21–26 °C, with L1 partially eating shells, then feeding on *S. altissimum*. Larval development was rapid, with only 2–3 days spent in each instar and pupation beginning 14 days post egg-hatch. Adults eclosed in early–mid-June after 6–7 days, egg–adult development occupying little more than 3 weeks. Males eclosed 2–3 days before females. Larvae fed on all parts of *S. altissimum*, but L1–L2 preferred inflorescences. No nests were made, and despite crowding, larvae tolerated each other well. Survival is based on camouflage, and L1–L4 are likely chemically defended by droplets at the tips of setae, as is the case for *Pieris rapae* (Cabbage White) (Smedley et al., 2002). Prepupal larvae wandered, and pupae were formed on and off the host plant. Oversummering pupae were found on dried branches of *S. altissimum* in Benton Co. in July–Aug. Overwintering occurs as pupae, and there are 5 instars.

Description of Immature Stages

The spindle-shaped egg is pale yellow, turning orange within 24 hrs. There are ~12 ribs, each crossed by about 30 finer lateral grooves. L1 is initially brown, becoming yellow-orange with a shiny black head. On the first segment there are 2 small brown-black spots dorsally. There are 6 dorsolateral rows of brown bullae, each sporting a short seta with a droplet, a condition that persists until L4. Late L1 is creamy yellow with a speckling of tiny brown spots and an indistinct middorsal white stripe. L2 is light yellowish brown with black spots, becoming yellowish green. On each segment there are 20–24 large black raised bullae, each carrying a single short seta. Longer, pale setae occur ventrolaterally and on the gray head. L3 is bright yellowish green with distinct intersegmental yellow bands devoid of bullae. The raised black bullae are closer together, and each one carries a long, white seta, giving the larva a "hairy" appearance. The head is pale greenish yellow with small dark spots. L4 is very similar, with the development of tiny, black spots on the greenish segments and yellow intersegmental areas, and bearing longer white setae. L5 has a banded appearance with whitish segments, each dotted with

Adult

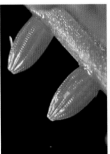

Eggs @ 1.0 mm

Pupa @ 15 mm

L1 @ 2.0 mm

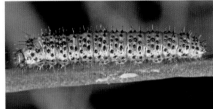

L2 @ 7 mm

L3 @ 11 mm

L4 @ 20 mm

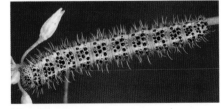

L5 @ 24 mm

Becker's White *Pontia beckerii* (W. H. Edwards)

20–24 black bullae sporting long white setae and separated by bright yellow intersegmental areas. There is a very indistinct pale middorsal stripe. The head is white and yellow with black spots. Prepupal L5 is dark pinkish brown. The pupa, attached by a silk girdle usually to a twig, resembles a bird dropping. The wing cases and much of the abdomen are creamy white, with the thorax and head gray-brown. There is a row of 4 black dots between the thorax and abdomen. Images of mature larvae appear in Pyle (2002), Allen et al. (2005), and Neill (2007) and agree with ours. The pupa pictured in Guppy and Shepard (2001) is similar to ours.

Discussion

This is an easily reared species with eggs readily obtained from gravid females. Larvae are also readily found on mustards in areas where the adults are common. Development is rapid, likely an adaptation to the ephemeral nature of hosts in arid habitats. Voltinism is variable in this species and may be related to host-plant quality. Larval populations on senescing *S. altissimum* in late spring produced oversummering and overwintering pupae, while larvae reared on good quality *S. altissimum* produced a second brood. Research is needed to elucidate the environmental cues involved in determining whether pupae diapause or not. It is possible that some pupae may undergo summer diapause only, producing autumn adults. Prolonged pupal diapause for more than a year may occur in this species and should also be investigated.

Spring White   *Pontia sisymbrii* (Boisduval)

Adult Biology

Pontia sisymbrii is a W North American species ranging from NW Canada to S CA. It flies E of the Cascades in WA as well as in the Olympic Mts and throughout most of OR. The flight period extends from late March in the low elevation shrub-steppe of E WA and OR to mid-Aug in mountainous areas. Males hill-top but also patrol roadways and open hillsides aggressively searching for mates. Females oviposit in late April to early May at lower elevations, with eggs laid singly, often on the plant stem. Warren (2005) summarized host plants utilized in Cascadia, including *Sisymbrium altissimum* (Tumble Mustard), *Arabis glabra* (Tower Mustard), *A. furcata* (Haired Rockcress), *A. sparsiflora* (Sicklepod Rockcress), and *A. holboellii* (Holboell's Rockcress), as well as numerous other mustards utilized in CA and CO, including *Streptanthus* spp. (streptanthus) and *Descurania* spp. (tansy mustards). Scott (1986a) reported that eggs are laid anywhere on the host plant and that females lay fewer eggs on plants that already have orange eggs. Some host plants (e.g., *Streptanthus*) have red-orange growths on the upper leaves resembling mature pierid eggs which deter oviposition by this and other pierid species (Shapiro, 1981). The Spring White is single brooded, and larval hosts include a variety of native and introduced mustards, particularly purple flowering *Arabis* spp.

Immature Stage Biology

We reared this species several times. Two cohorts were reared to pupae, while another 2 survived to L5; 3 cohorts were obtained from captive females, 1 from collected eggs, and all were obtained in Kittitas Co., WA between March 29 and April 29. A Yakima Co., WA cohort obtained from females caught on April 27 was reared on *S. altissimum*, while another cohort of eggs found on *Arabis* stems from the same county was reared on *Arabis* sp. Eggs hatch within 2–3 days at ~25 °C and L1 commence feeding immediately, sometimes

cannibalizing unhatched eggs. Eggshells are usually left uneaten and can be found in the wild, providing a clue that larvae may be present. Larval growth is rapid, successive instars molting approximately every 3 days, with L5 pupating after 5 days; development from egg-hatch to pupation occupies ~20 days at 22–27 °C. Leaves are favored, but flowers and saliques (seedpods) may also be consumed. Feeding typically begins near the top of the plant with the smallest terminal leaves, progressing downward, with the largest leaves used only if necessary. Consumed leaves are a good field mark for finding larvae. The larvae feed ravenously, and a single individual is capable of consuming an entire *Arabis* plant. The larvae are well camouflaged in the early instars but develop bright warning (aposematic) colors in the later stages, suggesting that they are distasteful. Alternatively, they may be mimicking a toxic model like the similarly marked larvae of *Danaus plexippus*. The larvae normally do not construct nests; however, Scott (1992) found 2 late instars that had silked together pods of *Arabis glabra* and were resting on silk mats. The larvae feed and rest positioned parallel to the stems, saliques, or leaf veins. *Pontia sisymbrii* overwinters as a pupa. In CA, adult eclosion extends over a long period, and multiyear pupal diapause (up to 4 years) has been reported (Shapiro 1981; Shapiro and Manolis, 2007). These adaptations to a variable spring environment also likely occur in Cascadia populations.

Description of Immature Stages

The pale eggs are spindle shaped, becoming bright orange after ~24 hrs. There are ~10 broad longitudinal ridges separated by slightly wider grooves; each ridge is crossed by numerous small lateral grooves. L1 varies from dull green to brown to orange and bears numerous black bullae, the largest arranged in 2 dorsolateral rows. Each bulla has a single black seta with a droplet at the tip, likely a defensive secretion as described for *Pieris rapae* (Smedley et al., 2002). L2 is brown–dark brown with vague lateral yellow banding forming distinct yellow blotches. The black head is speckled with white dots, which persists throughout development. In L3 yellow side blotches are distinct and yellow bands appear dorsally separated by gray-white areas. The middle of each segment in L4 is white, bordered by a black line on either side formed by coalescence of the black bullae, each bearing a short seta with a droplet. The intersegmental areas are bright yellow-green. L5 is beautifully patterned in aposematic colors, with each segment banded in bright yellow, porcelain white, and black. Prepupal

Adult

Egg @ 1.5 mm (fresh)   **Egg @ 1 mm (mature)**

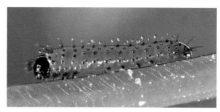

L1 @ 2 mm

Pupa @ 17 mm

L2 @ 4 mm

L3 @ 7 mm

larvae are dark and wrinkled and appear diseased. The pupa is greenish brown becoming black with numerous small pale spots. Contrary to Pyle (2002), the pupa is secured in the typical pierid way with silk girdle and cremaster attachments. In early instars, *P. sisymbrii* can be confused with several other pierids which may be found on the same plants; however by L4 the brightly banded yellow, black, and white larva is unmistakable. Guppy and Shepard (2001) illustrate an L4, and Allen et al. (2005) show an L5, both agreeing closely with our images.

Discussion

Eggs can be found relatively easily on host plants, particularly on *A. sparsiflora* in the shrub-steppe of WA, and larvae are easily reared. Other pierids, including *Anthocharis sara* and *Pontia beckerii*, often use the same host plants and are not readily separated. Larvae can be reared on standing stalks of *Arabis*, either live or cut in water. Rearing attempts using the commercial *Brassica rapa* (Field Mustard) resulted in fairly normal growth through L3 or L4, but in 2 separate attempts the larvae died before reaching L5. Larvae are prone to disease, thus attention to cage hygiene is important. The incidence of multiyear pupal diapause needs study as well as the extent of protracted emergence in spring. Research is also needed on whether the aposematic larvae are unpalatable to predators, or whether they are mimicking a toxic model. The secretions from the setae of early instars and their possible defensive function should be investigated, as well as the influence of host plants and their relative toxicity on larval survival and coloration.

L4 @ 15 mm

L5 @ 30 mm

Checkered White   *Pontia protodice* (Boisduval & Le Conte)

Adult Biology

Pontia protodice is rare in Cascadia, unlike most of the US, where in some places it can be abundant. It is highly vagile, expanding populations N during spring and summer, populating most states by autumn. In Cascadia it is infrequent, and records from WA are few (Pyle, 2002). It is most often seen in the SE deserts of OR (Warren, 2005). Its occurrence is dependent on migration from the S and it does not overwinter here. According to Shapiro and Manolis (2007), *P. protodice* is less abundant in N CA now than it was in the 1970s. This is a species of grassland, steppe, and disturbed areas, and it uses a large variety of brassicas as larval hosts. Scott (1986a) records more than 30; likely hosts in Cascadia include Tumble Mustard (*Sisymbrium altissimum*), rockcress (*Arabis* spp.), and peppergrass (*Lepidium* spp.). It can be very difficult to distinguish adult *P. protodice* from *P. occidentalis*, and it is likely that some individuals are overlooked. In N CA, *P. protodice* is multivoltine (Shapiro and Manolis, 2007), and Warren (2005) reported 2–3 broods may occur in OR. Checkered whites may occur in Cascadia from June to Oct and nectar on a wide variety of flowers, especially mustards, asters, and thistles. Eggs are laid singly, and females tend to avoid laying on plants that already host pierid eggs (Shapiro, 1981).

Immature Stage Biology

We reared this species twice, from females collected in S and N CA. A single female obtained in Los Angeles Co. on May 31 was held in a clear plastic cylinder with muslin lid (16 cm dia × 8 cm tall) with *S. altissimum*. An estimated 50 eggs were laid in 3 days at 18–25 °C. The eggs hatched after 3 days, but L1 began dying immediately. Two larvae molted to L2 3 days after egg-hatch, and 1 molted to L3 but then died, possibly as a result of pesticide contamination. The second cohort was from 2 females from the Feather R, Plumas Co., on Aug 26, which were caged with *S. altissimum* in a container (28 × 14 × 7 cm) with 1 muslin side on

Aug 31 and held at 21–28 °C. Eggs were laid within a few hours, and ~200 were laid during 3 days. Larvae began hatching on Sep 2 and were provided with *S. altissimum, Brassica nigra* (Black Mustard), and *Sisymbrium irio* (London Rocket). Feeding occurred on all hosts. Larvae molted to L2, L3, L4, and L5, 3, 6, 10, and 12 days post egg-hatch, respectively, at temperatures of 21–28 °C. The first pupation was 15 days after egg-hatch, and adults began emerging 28 days after oviposition. Males emerged 1–3 days before females. Larvae tolerated each other well when reared in large numbers (~50 per cage). Of an estimated 100 larvae reaching L5, 1 individual was melanic. Pupae were mostly brown-gray, but a few green pupae (~5%) also occurred. Larvae consumed all parts of host plants, but L1–L2 preferred inflorescences. Little is known of the natural enemies and defense strategies of this species, although small droplets on larval setae likely repel antagonists. It has been speculated that wasp parasitoids introduced to aid biological control of *Pieris rapae* may also attack *P. protodice* and be responsible for the observed declines in NE US (Wagner, 2005; Douglas and Douglas, 2005).

Description of Immature Stages

The pale yellow ribbed egg turns orange within 24 hrs. L1 is yellow-orange with a shiny black head with sparse short setae. There are 6 dorsolateral rows of black bullae, each sporting a single short seta with a droplet. As L1 matures the black bullae become less prominent and the ground color more yellow. Ground color in L2 is greenish gray with a granulated appearance, yellow-green bullae, and 2 reddish parallel lines dorsally. Early L3 has a granulated appearance with raised green bullae and a yellowish ground color; with maturity the bullae become black and 4 distinct, vivid yellow dorsolateral lines appear. The body is gray-green and the head has sparse black spotting. Similar patterns persist in L4 and L5 with the gray-green becoming dark bluish gray and the yellow lines more distinct. Setae carry droplets from L1 to L4. The anterior part of each proleg is yellow. The head in L5 is gray lightly spotted with black and has a small yellow patch on each side. The pupa is whitish gray to tan and covered with fine black spots and lines. More prominent spotting appears on the wing cases (along venation) and along the dorsal midline. Dorsal angular protuberances are tipped with orange-red and there is an orange line down each side of the abdomen. A few pupae (~10%) are green. The thorax is angular and there is a red dot at the base of the wing case. The immature stages of *P. protodice* are similar

Adult female (late brood)

Egg @ 0.9 mm

L5 @ 28 mm (head)

L1 @ 2.0 mm

Pupa @ 16 mm (green and tan forms)

L2 @ 4.0 mm

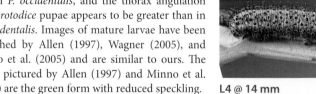

L3 @ 9 mm (pale) and L3 @ 10 mm (melanic)

L4 @ 14 mm

L5 @ 28 mm

Checkered White *Pontia protodice* (Boisduval & Le Conte)

to those of *P. occidentalis*. Shapiro and Manolis (2007) suggested that *P. protodice* larvae are more slender and their ground color bluer than larvae of *P. occidentalis*. The head of L5 *P. protodice* has fewer black markings and smaller yellow patches than that of *P. occidentalis*, and the thorax angulation in *P. protodice* pupae appears to be greater than in *P. occidentalis*. Images of mature larvae have been published by Allen (1997), Wagner (2005), and Minno et al. (2005) and are similar to ours. The pupae pictured by Allen (1997) and Minno et al. (2005) are the green form with reduced speckling.

Discussion

Obtaining eggs and rearing larvae is relatively easy, although care in providing clean plants and limiting the number of larvae per cage is recommended. This species has received considerable study, particularly on aspects of taxonomy and environmental influences on seasonal forms (e.g., Shapiro, 1976). In Cascadia and N CA, research is needed to determine why populations are declining. A study of the natural mortality factors affecting *P. protodice*, including native and introduced wasp parasitoids, would be instructive. The chemical nature and role of the setal secretion in L1–L4 also needs researching.

Western White *Pontia occidentalis* (Reakirt)

Adult Biology

Pontia occidentalis occurs in W North America from AK and BC to MB S to central CA and NM. In Cascadia, it is one of our most frequently seen high-elevation species found in open alpine habitats. It also occurs in lowland areas between the Cascades and Rockies, in fields, roadsides, and disturbed sites. Western Whites are common in much of E Cascadia, rare on the coast, with 1–2 generations (July–Sep) at high elevations. Two to 3 broods are reported in OR (Warren, 2005). Males patrol and "hill-top" in pursuit of females. Both sexes visit a variety of flowers, including thistles, asters, Alfalfa, mustards, and milkweeds, and males visit mud. Larval hosts include various native and exotic crucifers, with >20 species recorded (Scott, 1986a; Warren, 2005). In WA, *Lepidium* spp. (peppergrasses), *Anelsonia eurycarpa* (Daggerpod), and *Thelypodium saggitatum* (Arrow Thelypody) are known hosts (Pyle, 2002; Warren, 2005), as is *Arabis* spp. (our observations), and other mustards are likely used as well. Shapiro (1975) records the native crucifer genera *Streptanthus* and *Arabis* as the most commonly used hosts in undisturbed habitats in the CA Sierras. Eggs are laid singly on leaves, stems, and inflorescences, with females tending to avoid oviposition on plants already harboring pierid eggs (Shapiro, 1981); Scott (1992) observed that oviposition was often on a upper leaf surface.

Immature Stage Biology

We reared or partially reared *P. occidentalis* several times. On July 8, an L5 was collected from *Thelypodium* sp. in Grant Co., WA, and reared to adult. On July 5 (different year) an egg on *Arabis* and a gravid female were obtained from Lion Rock, Kittitas Co., WA; the egg hatched and developed to L4 in 16 days. The female laid 24 eggs on the wire mesh of a cage, and 16 larvae hatched 8 days later. The larvae were reared at 18–20 °C, feeding on the flowers and saliques of *Arabis* sp., and grew rapidly. Two pupated 16 days post oviposition, and adults eclosed after 2 weeks. Two females collected Aug 29 from Steens Mt in S OR were confined with cuttings of *Sisymbrium altissimum* (Tumble Mustard) in a plastic cylinder (16 cm dia × 8 cm tall) with muslin ends. Oviposition occurred within hours and ~150 eggs were laid over 7 days. Eggs hatched after 3 days at 21–27 °C, L1 partially eating shells. Larvae were fed on *Brassica nigra* (Black Mustard) and development was rapid, with 3 days spent in each of 5 instars. Adults eclosed after 10 days, 28 days after oviposition, males 1–3 days before females. One female from Bear Creek Mt, Yakima Co., WA on Aug 31 laid eggs on *Sisymbrium irio* (London Rocket), developing through only 4 instars and pupating 10 days after oviposition. Food in this rearing was limited and quality was poor. Larvae tolerated each other well when reared in large numbers (~50 per cage). All parts of host plants were consumed, including soft stems, but early instars preferred inflorescences. No nests were made, and defense is likely based on camouflage. Small droplets on the setae of L1–L4 likely repel natural enemies, as has been shown for *Pieris rapae* (Cabbage White) (Smedley et al., 2002). Pupation may occur on or off the host plant, and the pupa overwinters.

Description of Immature Stages

The spindle-shaped egg is pale yellow, turning orange-red within 24 hrs. L1 is yellow-orange with a shiny black head. On the first segment there are 2 brown-black spots dorsally. There are 6 dorsolateral rows of dark brown bullae each sporting a short seta with a droplet, a condition that persists until L4. L2 is yellowish gray becoming dark gray, with dorsal and lateral yellow stripes. The overall appearance is granulated, and there is an indistinct, narrow middorsal white stripe. L3 is similar with the gray and yellow stripes better defined. Setae are short and arise from raised black bullae. The head is black with small dorsolateral yellow patches. L4 is dark gray with the width of yellow stripes reduced. The head of L4 and L5 is gray with black spots and patches and a pair of dorsolateral yellow markings. L5 is similar, bluish gray with vivid yellow stripes and many short black setae.

Adult female

Eggs @ 0.9 mm **L5 @ 28 mm (head)**

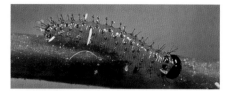

L1 @ 2.0 mm

L2 @ 5 mm

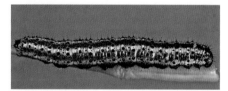

L3 @ 11 mm

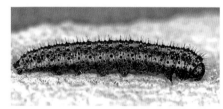

L4@ 14 mm

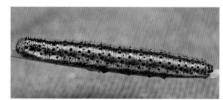

L5 @ 28 mm

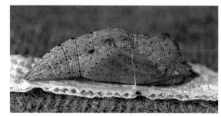

Pupa @ 16 mm

Raised black bullae give a spotted appearance. The pupa is bluish gray with irregular fine black spots. More prominent spotting follows venation on the wing cases. Dorsal angular protuberances are tipped with orange-red, and there is an orange line down each side of the abdomen. The thorax midline is relatively rounded. The immature stages of *P. occidentalis* are very similar to those of *P. protodice*. In our rearings the dark markings on *P. occidentalis* larvae were more extensive than those on *P. protodice*. The head of L3 is mostly black in *P. occidentalis*, gray-green in *P. protodice*. The head of *P. occidentalis* L5 has more extensive black markings and larger yellow patches than that of *P. protodice*. The pupal thorax is more rounded than in *P. protodice*. Images of eggs, a mature larva, and pupa were published in Guppy and Shepard (2001), and a mature larva image appears in Allen et al. (2005). All images agree with ours.

Discussion

This is a familiar species to mountain hikers but can also turn up in just about any locality E of the Cascades. It is easily reared from gravid females on native and weedy crucifers. Eggs may be found by searching mustards in habitats where the species is flying, although eggs may be confused with those of other pierid species. Development is very rapid, presumably an adaptation to the unpredictable and ephemeral climate of higher elevations. The ability to pupate after only 4 instars is also presumably an adaptive trait to ensure development is completed under suboptimal conditions. Pupae may diapause for >1 winter, a trait seen in other pierid species inhabiting unpredictable environments. Nothing is known of natural mortality factors, including natural enemies. Identification of the chemical droplets carried on the setae of L1–L4 and their role in defense deserves study.

Margined White   *Pieris marginalis* Scudder

Adult Biology

The Margined White is part of the circumpolar *Pieris napi* complex and occurs in wet forests in montane and coastal regions of W North America, particularly moist riparian zones along forest streams. In Cascadia it occurs locally in moist, wooded, cooler habitats from high-elevation (6,000–7,000 ft) damp subalpine meadows to the edges and glades of coastal coniferous forests. There are 2 broods in most mid-elevation habitats, the first flight March–June, the second flight July–Sep. In coastal areas it is multivoltine, with 3–4 generations, and at the highest elevations may be univoltine (Guppy and Shepard, 2001; Warren, 2005). Males patrol trails, roads, and shaded watercourses for females, and both sexes visit flowers, including dandelion, daisies, asters, mustards, cinquefoils, and geraniums. In Cascadia, larval hosts are native crucifers, including *Cardamine angulata* (Angled Bittercress), *C. breweri* (Brewer's Bittercress), *C. nuttallii* (Oaks Toothwort), *Nasturtium officinale* (White Watercress), and *Arabis* spp. (native rockcress) (Guppy and Shepard, 2001; Pyle, 2002; Warren, 2005). It may utilize some introduced crucifers like *Sisymbrium officinale* (Hedge Mustard) (Shapiro and Manolis, 2007), but these are usually uncommon in habitats used by *P. marginalis*. Eggs are laid singly, although often in fairly close proximity, usually on the undersides of leaves, sometimes on stems. Scott (1992) reported finding eggs in CO on the undersurfaces of *Cardamine* leaves.

Immature Stage Biology

We reared this species several times. On April 25 near Fall City, King Co., WA, females were observed landing on partially submerged *N. officinale* in a roadside ditch. One female produced 20+ eggs in captivity that hatched in 5 days. Larvae were reared to L3 on *N. officinale*, then moved to live garden cabbage plants, which they readily consumed. Development was rapid, the larvae pupating 18 days post egg-hatch and adults eclosing 8 days later. Young larvae grazed host-leaf surfaces halfway through, and later instars consumed entire leaves. Two females were obtained from near the Tucannon R, Columbia Co., WA on June 13 and confined with cut *Sisymbrium altissimum* (Tumble Mustard) in a plastic box (30 × 23 × 8 cm) with a muslin lid. Held at 22–26 °C, the females laid 22 eggs in 24 hrs and died shortly after. Eggs hatched in 4 days, and L1 partially consumed shells before feeding on *S. altissimum*. Pupation occurred 21 days post egg-hatch, with 4, 5, 6, 3, and 3 days spent in L1–L5, respectively. Adults eclosed after 9 days, 34 days after oviposition, and males eclosed before females. Larvae fed on all parts of the host and no nests were made. Survival is based on camouflage, the evenly green larvae blending well with foliage. Droplets, presumably containing defensive chemicals like those of the closely related *Pieris rapae* (Cabbage White) (Smedley et al., 2002), are produced on the tips of setae during L1–L5. Minute Pirate Bug nymphs killed many larvae in one rearing, hatching from unseen eggs on the provided host plants. Pupae were formed on and off the host plant. There are 5 instars and the pupa overwinters.

Description of Immature Stages

The white spindle-shaped egg yellows with maturity. There are 12 vertical ribs, each crossed by ~35 finer lateral grooves. L1 is tan turning light green after feeding with an indistinct middorsal white stripe. On each segment there are 10 small white bullae each bearing a short, pale seta with a droplet; these persist throughout development. The head is pale green with indistinct white spots. L2 is similar, light green with an indistinct lateral dark stripe and increased numbers of small white bullae ventrolaterally. Numerous small green bullae are present dorsally, each sporting a small pale seta. L3 is medium green dorsally with an indistinct dark stripe. Small dark green spots speckle the body, and large white bullae, 6 per segment, bear dark setae. Many small white bullae occur below the spiracles with pale medium-length setae. L4 and L5 are similar, medium-dark green dorsally, whitish green ventrally, bearing profuse pale setae. The head, true legs, and prolegs are green. Pupae vary from greenish yellow to tan, probably influenced by their surroundings as in *P. rapae*. Most of our pupae were green or yellow, sparsely and irregularly dotted with black. The thorax and dorsal ridges are tipped with reddish brown, and other parts of the ridges are whitish yellow. Adults of the spring and summer broods of *P. marginalis* are distinguishable by their

Adult

Egg @ 1.3 mm

Pupa @ 18 mm

L1 @ 5 mm

L2 @ 7 mm

L3 @ 15 mm

L4 @ 16 mm

L5 @ 26 mm

Margined White

Pieris marginalis Scudder

ventral wing markings, but in our rearings larvae from the different broods were not distinguishable. This species is most likely to be confused with *P. rapae*, but the 2 species tend to occupy different habitats in Cascadia. Both have larvae that are predominantly green; however, mature larvae of *P. rapae* are distinguished by having yellow dashes laterally. Images of mature larvae appear in Guppy and Shepard (2001) and Allen et al. (2005) and agree with ours. Images of tan-colored pupae are also shown by Guppy and Shepard (2001).

Discussion

This delicate butterfly is easily reared from gravid females, but adults tend not to live long in captivity. Larvae will feed on some weedy mustards (e.g., *S. altissimum, S. officinale*) but are reported not to fare well on some introduced species. The range of crucifers used by *P. marginalis* in Cascadia needs study as does comparative development and survival on different native and exotic species. Yellow violet and lawn daisy were reported as larval hosts by Pyle (2002), but this seems unlikely. Guppy and Shepard (2001) suggest that some pupae in each generation enter diapause, with adults emerging the following spring. The environmental and genetic influences determining the proportion of diapause to nondiapause pupae deserve study. The temperature and humidity tolerances of immature stages of this cool-habitat butterfly would also make an interesting study. Droplets were present on setae in L5, unlike most other pierid species which have droplets up to L4. The chemistry of the putative defensive secretions remains to be elucidated.

Cabbage White   *Pieris rapae* (L.)

Adult Biology

The Cabbage White is notorious the world over as one of the very few pestiferous butterfly species. It is one of the major insect pests of cruciferous crops throughout the temperate world. It has received much study both from the point of view of management and as a laboratory animal (Lokkers and Jones, 1999). A native of Europe, northern Asia, and Africa, this species was transported around the globe, presumably in shipments of vegetables. It spread throughout North America after its unintentional introduction to Quebec in 1860. Adults are highly mobile and occur in practically every open habitat, but they are generally more common in disturbed than in native habitats. In Cascadia, *P. rapae* flies from early March until Nov, producing 3 or 4 broods. Adults visit many types of flowers but prefer yellow and white species (Lazri and Barrows, 1984). Females generally mate during the first day of adult life and rarely more than once, copulation lasting about 85 min. Host plants include Cruciferae, Tropaolaceae, Resedaceae, and Capparidaceae. Commercial plants attacked include cabbage, cauliflower, brussels sprouts, radish, and turnip. Noncommercial hosts include *Arabis, Brassica, Sisymbrium, Nasturtium, Barbarea, Lepidium*, and *Cleome* (Warren, 2005). Eggs are laid singly on the undersides of leaves. Lifetime fecundity is 100–750 eggs.

Immature Stage Biology

We reared *P. rapae* several times from gravid females and field-collected eggs. Three females collected Sep 22 in Grant Co., WA, produced ~100 eggs. Spring brood females from King Co., WA (May 17) produced 12 eggs, all developing to adults. Eggs of both cohorts hatched in 4–5 days, and L1 consumed eggshells. The larvae developed from egg-hatch to pupation in 21 and 14 days, respectively, most pupae forming horizontally under a flat surface. Eclosion began 12 days later. Four females obtained from Horn Rapids, Benton Co., WA on Sep 27 produced >200 eggs over 3 days on broccoli flower heads. Eggs hatched in 3–4 days, and larvae were reared at 18–22 °C under naturally declining daylengths. Pupation occurred 15 days post egg-hatch and pupae entered diapause and overwintered. A subcohort of larvae was transferred to 28 °C and continuous light at the beginning of L5. These produced nondiapause pupae that eclosed after 7 days. A combination of temperature and photoperiod determines production of diapause or nondiapause pupae. Temperature above 24 °C prevents diapause regardless of daylength. At cooler temperatures, daylength must be less than 13 hrs to induce diapause (Lokkers and Jones, 1999). L1 move little, feeding from the undersides of leaves. Later instars are more mobile and prefer feeding on young growth; 85% of total diet eaten by larvae occurs in L5 (Slansky, 1973). Young larvae make holes in leaves but also eat leaf edges and mine into stems. Larvae are cannibalistic on eggs (Jones et al., 1987). Predatory and parasitic insects and spiders attack mostly L1–L3, whereas birds take larger larvae and pupae (Lokkers and Jones, 1999). Parasitic wasps are important regulators of *P. rapae* populations. Survival is based on camouflage and secretion of droplets from dorsal setae. Takabayashi et al. (2000) and Smedley et al. (2002) showed that these droplets are primarily unsaturated lipids, deterrent to ants. Shiojiri and Takabayashi (2005) showed the secretion does not repel a specialist parasitoid and may instead act as host-searching cue and oviposition stimulant for it. There are 5 instars.

Description of Immature Stages

The white spindle-shaped egg turns yellow-orange. There are 10 vertical ribs, each crossed by ~35 finer lateral grooves. L1 is tan, turning light green after feeding. On each segment there are ~10 white bullae, each bearing a short, pale seta with a droplet; these persist during development. The head and body of L2 are dull orange with green infusion dorsally and with many small setae. L3 is green, peppered with tiny black setae with an indistinct yellow middorsal line. Longer setae (with droplets) arise from ~10 white bullae per segment. A yellow dash occurs on each segment near the spiracle. L4 and L5 are similar to L3 except for increased density of ventrolateral setae, and the lateral yellow dashes and middorsal yellow stripe are distinct. In L5 there are 2 yellow dashes per segment. The head is green with many small setae. The head, true legs, and prolegs are green. Pupae formed on green surfaces are green, those formed on dark

Adult female

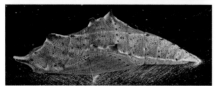

**Pupa @ 20 mm
(green and brown forms)**

Egg @ 1.1 mm (fresh, *left*, mature, *right*)

L1 @ 2.0 mm

L2 @ 5 mm

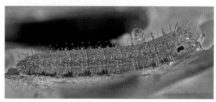

L3 @ 7 mm

L4 @ 17 mm

L5 @ 27 mm

surfaces are gray-brown. Both forms are speckled with tiny black spots, particularly on the wing cases (following venation) and dorsal head-thorax area. The middorsal ridge is tipped with black, white, and orange or yellow on the abdomen and brown-red on the thorax–head. The flared subdorsal ridges are white tipped with brown-red and the head cone is yellow-orange dorsally. This species is most likely to be confused with *P. marginalis*, but the 2 species tend to occupy different habitats. Both larvae are mostly green; however, mature larvae of *P. rapae* are distinguished by yellow dashes laterally and a middorsal yellow stripe. There are numerous published images of mature larvae, including Pyle (2002), Wagner (2005), Allen et al. (2005), and Neill (2007), all agreeing with ours.

Discussion

This is a highly successful species adapted to disturbed habitats and cultivated crucifers around the world. It is easily reared from gravid females, and immature stages are readily found in vegetable gardens. The species shows great variability in behavioral and physiological processes between separated populations. Varying selective pressures in different regions have produced rapid divergence between populations, contributing to its success; however, little is known concerning its adaptation to native environments in Cascadia. In undisturbed areas, *P. rapae* is not common and may have greater constraints on development and survival. An understanding of these constraints might provide some clues for enhancing management of the species in crop ecosystems.

Cabbage White *Pieris rapae* (L.)

Large Marble *Euchloe ausonides* (Lucas)

Adult Biology

The Large Marble ranges from AK to the Great Lakes in the E and S to CA and NM. In Cascadia it occurs E of the Cascades from BC to OR. An endangered subspecies, *E. a. insulana*, occurs on San Juan Is, WA. A wide range of habitats from sea level to almost 9,000 ft are occupied, including montane slopes, meadows, canyons, shrub-steppe, marine beaches, and desert washes. This species is univoltine, flying from late March to late July depending on elevation and seasonal conditions. Males patrol hilltops and disperse widely, flying low and often in a zigzag pattern looking for females. Both sexes visit flowers, particularly white and yellow species, including mustards and fiddlenecks. Larval hosts are crucifers; Scott (1986a) lists >20 native and introduced species. Guppy and Shepard (2001), Pyle (2002), and Warren (2005) report *Arabis* spp. (rockcress), *Descurainia* spp. (tansy mustards), *Schoenocrambe linifolia* (Plains Mustard), and *Sisymbrium altissimum* (Tumble Mustard) as hosts in Cascadia. Eggs are laid singly on buds and inflorescences and rarely on leaves. Opler (1974) stated that *E. ausonides* rarely places eggs anywhere but on unopened buds. Females tend to avoid ovipositing on plants already hosting eggs.

Immature Stage Biology

We reared or partially reared this species several times. On May 20, June 3, and June 4, eggs were collected from 3 locations in Kittitas Co., WA; all were laid on unopened buds of *Arabis* spp., and the resulting larvae were reared to pupae that overwintered. Development from egg to pupa took 19–22 days at 20–22 °C. The larvae fed on *Arabis* buds and flowers, then from L3 onward on saliques (seedpods); leaves and stems were not consumed. The saliques were eaten from the tips and the sides. Three females from Benton Co., WA (May 6) confined with *S. altissimum* in a plastic box with a muslin lid (36 × 23 × 12 cm) laid 6 eggs during 4 days. Eggs hatched in 4 days at 18–25 °C, and shells were partially eaten. Larvae fed on *S. altissimum* and development was rapid, pupation reached 18 days post egg-hatch. Pupae overwintered, with adults eclosing the following April. Larvae fed on all parts of the host, although L1 and L2 preferred terminal buds, inflorescences, and stems. Opler (1974) reported that early instars station themselves vertically among flower clusters and cover the portion of the plant on which they rest with loosely spun silk. L4 and L5 fed more on stems and saliques than on leaves. No nests were made, and survival appears to be based on camouflage and defensive chemicals produced by setae, as in many other pierids, including *Pieris rapae* (Cabbage White) (Smedley et al., 2002). Larvae are solitary, rarely with >1 per host plant, although they tolerate each other well in captivity. There are 5 instars, and pupation occurs mostly on host stems or other twigs. Prepupal larvae are purplish and wander prior to pupation (Opler, 1974).

Description of Immature Stages

The bluish white spindle-shaped egg turns orange after 24 hrs. There are ~16 vertical ribs, each crossed by ~30 finer lateral grooves. L1 is yellowish orange with green infusions behind the head and around the abdominal tip. There are large black bullae, ~8 per segment. Each bulla sports a short black seta carrying a droplet, this character persisting until L4. L2 is dull yellow with broad purple-gray stripes dorsally and laterally. There are 10 prominent black bullae per segment, each sporting a short seta with droplet. The head is tan with darker brown spots and a thin black collar. L3 is similar but has increased numbers of black bullae (~20 per segment), giving the larva a distinctly spotted appearance. The head is yellow-tan with black spots, and the ventral area below the spiracles is dull white. L4 is marked similarly but the bright yellow and blue-gray stripes are vivid. The head is dull yellow with black spots and the true legs and prolegs are pale green. The white ventrolateral stripe is bold and bordered below with a broken yellow line. In L5 the blue-gray stripes expand and are increasingly bold. Black bullae are more numerous (~28 per segment) and the setae are numerous and short and do not carry droplets. The head is gray with black spots and the legs are gray. Prepupal L5 are pinkish brown; Guppy and Shepard (2001) report prepupal L5 are brown in BC. The pupa is streamlined, with a long beak, and whitish with gray lateral and dorsal stripes, or it may be cinnamon brown in color, in either case blending with senescing vegetation. Larvae of this species are superficially similar to the yellow and gray striped larvae of *Pontia occidentalis* (Western White)

Adult

Eggs @ 1 mm

Pupa @ 30 mm

L1 @ 2.5 mm

L2 @ 6 mm

L3 @ 9 mm

L4 @ 15 mm

L5 @ 35 mm

and *Pontia protodice* (Checkered White), but they are readily separable on close examination. The larvae of the closely related *Euchloe lotta* (Desert Marble) are strikingly different, with no blue-gray or yellow striping. Images of L5 appear in Pyle (1981), Guppy and Shepard (2001), and Allen et al. (2005) and agree with ours. Pyle (1981) and Guppy and Shepard (2001) also show pupae, again similar to ours.

Discussion

This species is easily reared from gravid females, which will oviposit and feed on weedy crucifers like *S. altissimum*. Eggs may be found on host plants like *Arabis* but are difficult to separate from other pierid eggs. Larvae tolerate each other well in captivity, although care needs to be taken to prevent disease by keeping cages and host plants free from frass. Development is rapid, likely because of the ephemeral nature of many hosts in late spring. Natural mortality factors are unknown, but predators like Minute Pirate Bugs likely kill many larvae. Parasitic wasps are also natural enemies of late instars and pupae. Pupae may diapause for >1 winter to spread the risk of emergence in poor spring conditions. The crucifers used by this species in Cascadia need research, as does the chemistry of setal droplets.

Large Marble *Euchloe ausonides* (Lucas)

Desert Marble *Euchloe lotta* Beutenmuller

Adult Biology

The Desert Marble ranges from S BC to N Mexico, occurring W of the Rocky Mts but largely absent in coastal areas. In Cascadia it occurs E of the Cascades from BC to OR in shrub-steppe, desert, flat, gully, and hillside habitats. It is univoltine, flying early in the season, late March–late June, and is sometimes common in the Columbia Basin and adjacent areas. Males emerge before females and patrol gullies and hillsides for females. Hill-topping was observed in Benton and Kittitas Cos., WA during April. Females are usually found flying near host plants but are elusive and easily disturbed. Both sexes visit flowers, including phlox, mustards, and fiddlenecks. Larval hosts are many species of native and introduced crucifers. Opler (1974), Pyle (2002), and Warren (2005) recorded *Sisymbrium altissimum* (Tumble Mustard), *Descurainia pinnata* (W. Tansy Mustard), *Arabis furcata* (Columbia Gorge Rockcress), *A. sparsiflora* (Sicklepod Rockcress), *A. holboelli* (Holboell's Rockcress), *Halimolobos whitedii* (Whited's Fissurewort), and *Lepidium* sp. (peppergrass) as hosts in Cascadia. Eggs are laid singly on buds, inflorescences, stems, sometimes leaves (Opler, 1974).

Immature Stage Biology

We reared this species twice from gravid females obtained April 28 on Red Mt, Benton Co., and April 27, Schnebley Coulee, Kittitas Co., WA. The Red Mt female was caged in a plastic cylinder with muslin ends (13 cm × 8 cm) with *S. altissimum* and laid ~50 eggs over 4 days at 18–25 °C. Eggs were laid mostly on buds, flowers, and stems and hatched in 4–5 days. Larvae fed on *S. altissimum*, and pupation occurred 17 days post egg-hatch, with 3, 3, 2, 3, and 6 days spent in L1–L5, respectively. At Schnebley Coulee, females were observed frequenting *D. pinnata*, and 1 female was collected. Five eggs were produced and hatched quickly; the larvae fed modestly on flowers of *Descurainia*, then were switched to *Arabis*, which they relished. Pupation occurred 18 days post hatch. Pupae overwintered, with adults eclosing the following April. Larvae fed on all parts of the host, although L1 and L2 preferred buds and inflorescences. L3–L5 fed additionally on stems and leaves. Opler (1974) considered the *E. hyantis* group (including *E. lotta*) less specialized bud/inflorescence feeders than the Large Marble. No nests were made; survival is based on camouflage and defensive chemicals produced by setae, as in many other pierids, including *Pieris rapae* (Cabbage White) (Smedley et al., 2002). There are 5 instars, and pupation occurs mostly on host stems.

Description of Immature Stages

The greenish white spindle-shaped egg turns dark orange after 24–48 hrs. There are ~16 vertical ribs, each crossed by ~35 finer lateral grooves. L1 is yellow-orange with large black bullae, 8–10 per segment. Each bulla sports a short black seta carrying a droplet, a condition that persists until L4. L2 is greenish yellow with indistinct middorsal and midlateral dark stripes and a brownish black head. Each segment carries 10–12 prominent black bullae each with a short black seta. L3 is pale green with a greenish gray head, black collar, and yellowish lateral stripe bordered with black above. Each segment has ~40 black bullae, mostly very small, with tiny black setae. L4 is green with a distinct, broad, spiracular white stripe bordered below in bright green and above with peppered black spots. L5 is similar, with a more vivid lateral white stripe, sometimes slightly edged above in darker purple, and there are increased numbers of small black bullae and short setae. The head is green with black spots, and there is an indistinct middorsal pale stripe edged in black. Purple prepupal L5s wander before pupation. The pupa is streamlined with a long beak, whitish gray with small black spots, a lateral dusky stripe, and indistinct dark striping along wing case veins. After a few weeks pupae turn dark brown, presumably to blend with senescing vegetation. In the later instars the only larva likely to be confused with this species is that of *Anthocharis sara* (Sara Orangetip), which is also green with a prominent lateral white stripe; however, no purple edging occurs above the white stripe in this species, and unlike *E. lotta* the stripe continues onto the head. Images of mature *E. lotta* larvae appear in Pyle (2002) and Neill (2007), and an image of a pupa appears in Pyle (2002); all are similar to ours.

Adult

Pupa @ 22 mm (light and dark forms)

Egg @ .09 mm (fresh, *left*, mature, *right*)

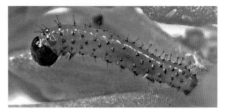

L1 @ 2 mm

L2 @ 6 mm

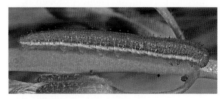

L3 @ 12 mm

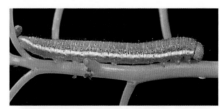

L4 @ 15 mm

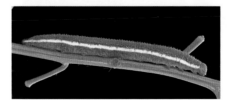

L5 @ 27 mm

Discussion

This species is easily reared from gravid females, which will oviposit and feed on weedy crucifers like *S. altissimum*. Obtaining gravid females is more difficult unless they are nectaring, in which case they are easily approached. Larvae tolerate each other well, although care should be taken to prevent disease and plant mold by keeping cages and host plants free from frass as much as possible. Development is rapid, likely because of the ephemeral nature of many of the mustard hosts in arid areas in late spring. Natural mortality factors are unknown, but small predators common on mustards, like Minute Pirate Bugs, probably account for many larvae. Parasitic wasps are also natural enemies of late instars and pupae. As with all Cascadia pierids, the chemistry of presumed defensive droplets produced by setae in L1–L4 is unknown. The importance of these droplets in deterring natural enemies also needs study. Pupal diapause may extend beyond 1 season as in some other desert-inhabiting pierids. Most of our reared larvae, and those pictured in Pyle (2002) and Neill (2007), showed very little evidence of purple edging of the lateral white stripes, a feature prominent in *E. hyantis* larvae (e.g., Allen et al., 2005). *Euchloe lotta* is closely related to *E. hyantis* (*E. lotta* was formerly considered to be a subspecies of *E. hyantis*), and the presence of distinct purple edging above the lateral white stripe might be a diagnostic character.

Sara Orangetip   *Anthocharis sara* Lucas

Adult Biology

Anthocharis sara is one of our most beautiful and earliest butterflies, a true harbinger of spring, flying in WA from mid-March to early Sep depending on seasonal conditions and elevation. It is our most easily recognized pierid and one of the most widespread in W North America, ranging from NW Canada to Mexico. In Cascadia, it occurs almost everywhere that mustards grow and a nectar source is available. It is less common in arid basin areas and more common in riparian canyons of foothills and lower mountains, but also occurs as high as alpine areas. Larval host plants are crucifers, including many species of *Arabis* (rockcress), *Barbera* (wintercress), *Thysanocarpus* (fringepod), and *Sisymbrium* (tumble mustard). *Arabis* is among the most favored, and on a sunny spring day any *Arabis* field is likely to have orangetips flying nearby. It is an avid flower visitor, seeking nectar from spring flowering plants like fiddlenecks, mustards, and phlox. Females lay their eggs singly, usually only 1 per plant unless overcrowding occurs or food plants are scarce. The slender eggs are laid vertically on various parts of the plant. Although *A. sara* enjoys a long flight period, it is single brooded in Cascadia. In cool, moist sites in CA it is partially double brooded (Shapiro and Manolis, 2007).

Immature Stage Biology

We reared this species numerous times from larvae and eggs found on *Arabis* spp. and from eggs obtained from gravid females. Eggs and larvae are found on all parts of the plant, including flowers, saliques, stems, and leaves. On June 4 we found scores of mature larvae in a field of *Arabis glabra* in Kittitas Co., WA. Many of these plants were virtually denuded, attesting to the appetite of these larvae and illustrating why females usually place only 1 egg per plant. One female from Umatilla NF, Garfield Co., WA on May 28 laid ~100 eggs on *Sisymbrium altissimum* (Tumble Mustard) over 2 days, and the eggs hatched 4 days later at ~25 °C. One female from Bear Canyon, Yakima Co., confined with *Arabis lyalli* (Lyall's Rockcress) laid 12 eggs that hatched in 3 days at ~27 °C. Development at 22–28 °C occupies 16–20 days, with ~3 days in each instar. Larvae feed on flower bases and saliques, then leaves and stems. They systematically consume hosts, working downward to finally eating the large basal leaves. The degree and progression of plant damage can reveal which plants are hosting larvae and their maturity. The larvae often rest on saliques or stems, where their slender green bodies blend well with the narrow green plant parts. L5 wander before pupating on a vertical stick away from the host plant; pupae are formed projecting upward at an angle of about 30 degrees. No nests are made; the pupa oversummers and overwinters and is cryptic, blending with green then dried vegetation by becoming more straw colored as the season progresses.

Description of Immature Stages

The pale yellow spindle-shaped eggs turn orange then reddish orange within 24–48 hrs. There are 13–14 bold, flat-topped vertical ribs crossed laterally by numerous narrow grooves. L1 is greenish amber with a shiny black setaceous head. Each segment has a lateral row of ~8 black tubercles, each with a single black seta. The setae produce small droplets of a presumed defensive secretion at their tips, a feature that persists throughout development. L2 is light gray with dull yellow spots laterally. L3 is pastel green with tiny black spots blending downward to white at and just below the spiracles, and below this line the color changes to dark green. L4 and L5 are 3-toned: light green dorsally, a white stripe laterally, and dark green ventrally. The pupa is green and narrowly pointed at each end, with an angular bulge in the middle and a longitudinal white stripe laterally; the upper pointed end curves gently away from the stick. After a few weeks the pupa turns straw colored. *Euchloe lotta* is similar to *A. sara* in all immature stages, but can always be distinguished by the lateral white stripe. In *A. sara* this stripe extends onto the head, but in *E. lotta* it does not. L5 of *E. lotta* is medium yellow-green above *and* below the lateral white stripe however, whereas *A. sara* is pale green above and very dark green below this stripe. Images of immature stages in the literature include an egg, mature larva, and pupa in Guppy and Shepard (2001), mature larvae in Allen et al. (2005), Miller and

Adult

Egg @ 1.5 mm

Pupa @ 19 mm

L1 @ 1.6 mm

L2 @ 6.5 mm

L3 @ 9 mm

L4 @ 16 mm

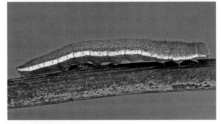

L5 @ 26 mm

Hammond (2003), Neill (2001), and Woodward (2005); our images agree closely with these.

Discussion

Eggs can be obtained fairly easily from gravid females by caging them with a host such as *Arabis* or *Sisymbrium*; however, the earliest females flying may be unmated. Eggs are easily found by examining *Arabis* spp. in areas where *A. sara* flies, but recently eclosed larvae are rarely discovered, suggesting that they immediately bore into flowers to feed. Eggs of this species are the most commonly found of Cascadia's pierids. Many pierid eggs are similar, however, rendering identification difficult, and up to 5 pierid species may oviposit on a single species of *Arabis* in the spring. Larvae of *A. sara* may be found from early May to early June depending on elevation, most numerous where adults were common a week or two earlier. Rearing *A. sara* is relatively easy as the species is quite hardy; cut *Arabis* will remain fresh in a refrigerator for up to 2 weeks, and the widespread introduced weed, *S. altissimum,* is an alternative host. Attention to cage hygiene is important to prevent disease. Commercial cultivars of *Arabis* are available, but none we have tested appears acceptable to *A. sara* or other pierid species. Stick supports should be provided for pupating larvae, and pupae can be stored outside in a ventilated enclosure. Some pupae may take 2–3 years before producing adults (Shapiro and Manolis, 2007). This behavior, commonly shown by pierids like *A. sara* and *Pontia beckerii* in the Columbia Basin (about 10% of reared individuals), is thought to be a "bet-hedging" strategy ensuring survival of at least some of the population should a "catastrophe" occur, for example, nonemergence of host plants because of drought.

Clouded Sulphur   *Colias philodice* Godart

Adult Biology

Colias philodice ranges over all of subarctic, temperate, and subtropical North America except FL and W CA. In Cascadia, it is resident and common throughout the Columbia Basin and E montane areas, but less common W of the Cascades. It is found in a variety of native, agricultural, and disturbed habitats, from Alfalfa fields to mountain meadows. There are 2–3 broods in most low–mid-elevation habitats, with adults flying early April–early Nov. Males are notorious mud-puddlers, and both sexes visit many flowers, including thistle, aster, Alfalfa, dandelion, and clover. Males patrol in a low, erratic, and rapid flight through fields in search of females. Mating generally takes place in mid–late morning, and oviposition occurs from midmorning to late afternoon. Courting males hover over females diffusing a pheromone that entices the female to extend her abdomen for copulation (Scott, 1986a). Under cool conditions adults bask laterally, exposing the ventral surface of the wings perpendicularly to the sun. Many native and introduced leguminous host plants (Scott, 1986a, records 24 species) are used, including species of *Medicago* (Alfalfa), *Trifolium* (clover), *Baptisia* (false indigo), *Vicia* (vetch), *Astragalus* (locoweed), *Lathyrus* (wild pea), *Melilotus* (sweet clover), *Hedysarum* (sweet vetch), *Lotus* (trefoil), *Lupinus* (lupine), and *Thermopsis* (goldenbanner). Range expansion and increased abundance of *C. philodice* in W US is due largely to increased acreage of Alfalfa and apparent preference for this crop, in which it is sometimes a pest (Tabashnik, 1983). Eggs are laid singly on host-plant leaflets but in high-density populations are laid in large numbers anywhere on the plant (Douglas and Douglas, 2005).

Immature Stage Biology

We reared *C. philodice* several times from gravid females. Three females collected April 18 from nr Wanapum Dam, Kittitas Co., WA produced 20 eggs

on *Astragalus* sp.; the larvae were reared on Alfalfa and developed from oviposition to pupae in 19 days at 20–22 °C. Two females obtained from Juniper Dunes, Franklin Co., WA on April 6 and confined in a steel-framed muslin cage (55 × 30 × 20 cm) with cut *M. sativa* laid ~100 eggs over 6 days. Three females obtained from Apex Mt, BC on July 29 laid ~100 eggs on cut *Vicia vilosa* (Hairy Vetch). Eggs hatched in 3 days at 25–28 °C and 7 days at 18–25 °C. Development from L1 to pupation took 20 days on *M. sativa* under the warmer temperatures and 25 days on *Trifolium medium* (Zigzag Clover) in the cooler conditions, and adults eclosed after 7–10 days. Shigeru (1958) reported much faster development, 10 days from egg-hatch to pupation on *Vicia* sp. at 27 °C. Three larvae in the BC cohort suspended development in L3 and entered diapause. Scott (1986a), Guppy and Shepard (2001), and Pyle (2002) report L3 or L4 as the overwintering stage; however Allen (1997) and Douglas and Douglas (2005) state overwintering occurs as a pupa in WV and the Great Lakes region. L1 consume most of their eggshell and eat holes in leaves, usually on either side of the midrib. Early instars usually rest on midribs and tolerate high densities well in captivity. Later instars consume entire leaves from the edge. Defense is based on camouflage, larvae blending extremely well with host foliage and stems. Mature larvae vary greatly in size depending on host-plant type and availability. Parasitic wasps, predatory bugs, spiders, and birds are likely important in regulating populations.

Description of Immature Stages

The white spindle-shaped egg turns yellow then orange-red. There are 20–22 vertical ribs, each crossed by ~50 finer lateral grooves. L1 is tan-brown turning dull yellowish green. Each segment has 5–6 transverse folds, and there are 10 rows of short setae along the body, each bearing a droplet. L2 is dark green becoming light green, with a vague middorsal dark stripe. Above the spiracles is an indistinct pale stripe, and each segment has ~80 small black bullae, each with a short pale seta. The bifurcated head is pale green-tan with black bullae and short white setae. L3 is green with a lateral supra-spiracular white stripe that becomes more prominent with maturity. Setae are reduced and the head is green peppered with black spots; the true legs and prolegs are green. L4 and L5 are similar, dark to medium green with a thick spiracular white stripe. There is a vague dark stripe middorsally and a vague yellow dorsolateral stripe in some individuals. The terminal part of the posterior segment is identically colored to the head, light green

Adult

Egg @ 1.1 mm
(fresh)

Eggs @ 1.1 mm
(mature)

Pupa @ 17 mm

L1 @ 2 mm

L2 @ 5 mm

L3 @ 9 mm

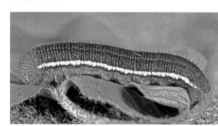

L4 @ 15 mm

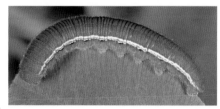

L5 @ 25 mm

Clouded Sulphur *Colias philodice* Godart

with black dots. In L5 the white stripe is punctuated with a broken orange-red line, reduced in some individuals to small dashes or entirely absent. The pupa is green with a thick lateral yellowish stripe on the abdomen, extending along the wing case margin. A red-purplish stripe occurs below the yellow stripe on 3 abdominal segments, along with a small black spiracle on each segment. The larvae of all *Colias* spp. are very similar and easily confused. Clouded Sulphurs occur in the same habitats and on the same hosts as the Orange Sulphur and are most likely to be confused with this species. Images of immature stages in the literature include Allen (1997), Guppy and Shepard (2001), and Wagner (2005). All are similar to ours.

Discussion

Colias philodice and the closely related *C. eurytheme* are favorite research animals for insect biologists and ecologists. Consequently there is a substantial amount of information available on population biology, host-plant relationships, courtship, mating, nectar use, development, thermal ecology of larvae, and genetics of these species. Clouded Sulphurs are easily reared on a wide range of hosts, and an artificial diet has been developed (Taylor et al., 1981). They can withstand moderate crowding, but care is needed to prevent disease outbreaks to which they are prone. The apparent differences in overwintering stages used by E and W US populations deserve research, as do the environmental cues determining diapause. The setae of L1 carry droplets like *Pieris* spp., but unlike *Pieris* spp. not in the later instars. The defense strategies of *C. philodice* have not been researched and would make a rewarding study.

Orange Sulphur   *Colias eurytheme* Boisduval

Adult Biology

Colias eurytheme occurs throughout the US and Mexico and is one of our most abundant butterflies. It occurs in all open habitats from sea level to mountaintops and sometimes swarms by the millions in Alfalfa fields. In Cascadia it occurs everywhere as a seasonal migrant except in far NW WA. It is unable to survive winters in Cascadia because of the absence of a diapausing stage (unlike *C. philodice* which diapauses as L3–L4), although there are a few reports of spring individuals (Pyle, 2002; Warren, 2005). Immigrants arrive in late spring–early summer, producing 2–4 generations, with adults still flying in Nov–Dec. Orange Sulphurs are commonly found in disturbed habitats, Alfalfa fields, pastures, dry meadows, and lawns. Males are avid mud-puddlers; flight is erratic, rapid, and low. Males patrol, challenging other males in aerial battles, and chase down females. Successful mating is mediated by UV reflection signals and pheromones (Silberglied and Taylor, 1978). Flowers are visited by both sexes, including milkweed, Alfalfa, clover, asters, and sunflowers. Many native and introduced leguminous host plants are used; Scott (1986a) records 44 species, including *Medicago* (Alfalfa), *Trifolium* (clover), *Baptisia* (false indigo), *Coronilla* (crown vetch), *Vicia* (vetch), *Astragalus* (locoweed), *Lathyrus* (wild sweet pea), *Melilotus* (sweet clover), *Lotus* (trefoil), *Lupinus* (lupine), *Phaseolus* (bean), *Pisum* (pea), and *Thermopsis* (goldenbanner). The great abundance of *C. eurytheme* in the W US is due largely to increased acreage of Alfalfa and an apparent host preference shift to this crop, in which it is sometimes a pest (Hoffman, 1978). Females produce 700–1,000 eggs (Guppy and Shepard, 2001; Douglas and Douglas, 2005), laying them singly on leaves. Fertility of eggs ranges from 96–100% (Shigeru, 1958).

Adult male

Immature Stage Biology

We reared *C. eurytheme* several times from gravid females. Three females obtained Sep 22 in Alfalfa fields (Grant Co., WA) produced numerous eggs that were reared to adults at 20–22 °C on Alfalfa. Development from egg-hatch to pupation took 27 days. A female obtained July 27 from the Umatilla NF, Garfield Co., WA and confined with *Trifolium repens* (White Clover) laid ~100 eggs over 3 days. Eggs hatched in 4–5 days at 22–28 °C; larvae exited by chewing open the top of the egg and consumed most of the shell. Larvae were fed on *T. repens*; L1 skeletonized leaves by feeding between veins on both sides of the midrib, creating a characteristic feeding pattern. Development was rapid, pupation reached after 15 days with ~3 days spent in each of the 5 instars. Shigeru (1958) reported faster development, 10 days from egg-hatch to pupation on *Vicia* sp. at 27 °C. According to Shigeru (1958), overwintering occurs as a larva, with reduced growth under cool temperatures with no diapause. Allen (1997) reported overwintering occurs as a pupa in WV. Early instars rest on midribs and tolerate high densities well in captivity. Later instars consume entire leaves from the edge or tip. Defense is based on camouflage, larvae blending extremely well with host foliage and stems. Little is known concerning natural enemies, but parasitic wasps, predatory bugs, spiders, and birds are likely important in regulating populations.

Description of Immature Stages

The white spindle-shaped egg turns yellow then orange-red. There are 18–20 vertical ribs, each crossed by ~40 finer lateral grooves. L1 is tan-brown turning dull yellowish green after feeding. Each segment has 5–6 transverse folds, and there are 10 rows of short setae along the body, each bearing a droplet. L2 is dark green becoming light green, with a vague middorsal dark stripe. Above the spiracles there is a distinct white stripe, and each segment has ~100 small black bullae, each bearing a short pale seta. The bifurcated head is pale green-tan with dark bullae and short white setae. L3 is green–dark green with an indistinct middorsal dark stripe and a lateral supra-spiracular white stripe. Setae are reduced and the head is green peppered with tiny black dots. L4 and L5 are similar, dark green with a thick spiracular white stripe. There are increased numbers of tiny black dots/short pale setae and there is a vague dark stripe middorsally and a yellow dorsolateral stripe in some individuals. The head is light green peppered with tiny black dots. In L5 the white stripe has an intermittent rusty red

Egg @ 1.3 mm

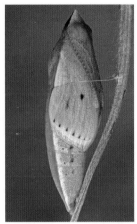

Pupa @ 22 mm (pharate)

L1 @ 4 mm

L2 @ 8 mm

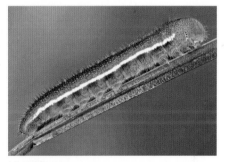

L3 @ 12 mm

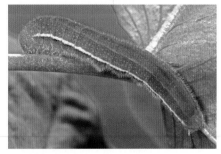

L4 @ 20 mm

L5 @ 32 mm (prepupal)

stripe running through it. In some individuals this is reduced to small dashes or is entirely absent. According to Shigeru (1957), L5 reared in 10 hr photoperiod have black patches on the white stripe. The pupa is green with a thick lateral whitish yellow stripe on the abdomen. The antennae are magenta, a single small black dot occurs on each abdominal segment, and each wing vein ends in a black dash. The larvae of all our *Colias* spp. are very similar and easily confused. Orange Sulphurs occur in the same habitats and on the same hosts as the equally common *Colias philodice* (Clouded Sulphur) and are most likely confused with this species. The black dashes on wing veins in *C. eurytheme* pupae do not appear to occur on *C. philodice* pupae. Images of immature stages appear frequently in the literature; recent examples include Allen (1997), Allen et al. (2005), and Minno et al. (2005) and are similar to ours.

Discussion

Orange Sulphurs are easily reared on a wide range of hosts, especially Alfalfa and clover, and an artificial diet has been developed (Taylor et al., 1981). Eggs and larvae are readily found in the wild, particularly around Alfalfa fields. Larvae can withstand moderate crowding in cages, but care is needed to prevent disease outbreaks to which they are prone. The minimum temperatures necessary for survival and development have not been determined. This information would enable a better understanding of the geographical limits of overwintering and the possibility of *C. eurytheme* becoming a permanent Cascadia resident if the climate warms. The setae of L1 carry droplets like the setae of *Pieris* spp., but not in later instars. The defense strategies of *C. eurytheme* have not been researched and need study.

Western Sulphur   *Colias occidentalis* Scudder

Adult Biology

The Western Sulphur is virtually a Cascadian endemic, ranging from S BC to NW CA eastward to W ID. Its distribution within Cascadia is patchy, occurring on Vancouver Is, the W slopes of the Cascades in BC, NW WA, the Cascade Mts from C WA to C OR, and the Ochoco and Siskiyou ranges. It is found in meadows, along roadsides, and on slopes in forested areas, from sea level to higher elevations (6,000 ft). It is univoltine, flying from May to late Sep, with most mid-elevation populations peaking in June–July. Flight is fast and erratic, particularly when alarmed. Males emerge ~1 week before females and patrol streams, meadow edges, trails, and roads. Males mud-puddle, and both sexes visit a variety of flowers, including thistles, dogbane, and milkweed. Larval hosts are mostly native and some exotic legumes, including *Lathyrus lanszwertii* (Mountain Pea), *L. nevadensis* (Nevada Deervetch), *L. pauciflorus* (Steppe Sweetpea), *Lupinus albicaulis* (Sickle-keeled Lupine), *L. arcticus* (Subalpine Lupine), *L. sericeus* (Silky Lupine), *Melilotus alba* (White Sweet Clover), *Vicia augustifolia* (a cultivar vetch), and *Vicia sativa* (Common Vetch) (Scott, 1986a; Pyle, 1981, 2002; Warren, 2005). We observed females ovipositing in early July in Kittitas Co., WA, visiting multiple vetch plants, placing single eggs on the middle of an upper leaf surface, then quickly moving on.

Immature Stage Biology

We reared or partially reared this species several times from gravid females. Several eggs collected July 1 at Taneum Cr, Kittitas Co., were reared to diapausing L2 on *Vicia* at 20–22 °C; they did not survive the winter. One female obtained from Bear Canyon, Yakima Co., WA on June 29 confined with *Trifolium medium* (Zigzag Clover) in a plastic box with a muslin lid (32 × 20 × 9 cm) and held outdoors in sunshine laid ~50 eggs over 2 days. One female from Umtanum Ridge (June 3), and 5 females from Satus Pass (June 23),

Yakima Co., WA, caged and confined with *M. alba* did not oviposit; however, ~100 eggs were laid over 4 days by the Satus Pass females when *M. alba* was replaced with *T. medium*. Eggs hatched after 6–7 days at 20–30 °C; larvae exited by chewing open the top of the egg and consumed most of the shell. Larvae were reared on *T. medium*; L1 and L2 skeletonized leaves by feeding between veins on both sides of the midrib, creating a characteristic feeding pattern. L2 was reached after 7–10 days. After 7 days most (>90%) larvae changed from green to tan and sought out curled/dried leaves in which they spun a silk pad and rested. The remaining larvae molted to L3 after ~10 days and immediately became brown, sought refuge, and entered dormancy. All larvae in both cohorts were dormant by late July–early Aug and the Satus Pass cohort (*n* = ~50) was overwintered at 5 °C from Aug 13. The Bear Canyon cohort (*n* = ~30) was held indoors (21–26 °C, ~30% RH) until Oct, when examination revealed all were dead. After 10 weeks, the Satus Pass cohort was removed from overwintering; examination revealed only 1 larva, an L3, was still alive. Held at 27 °C under continuous lighting, it commenced feeding on *T. medium* after 48 hrs; L4 and L5 were attained after 10 and 16 days, respectively, and pupation occurred 20 days after removal from overwintering. Held at 15–20 °C, a male eclosed after 7 weeks. Early instars rest on midribs and tolerate high densities in captivity. Later instars consume entire leaves from the edge or the tip. Defense is based on camouflage, the larvae blending extremely well with host foliage and stems. In L1 the tips of setae carry droplets, which may be defensive. Little is known concerning natural enemies, but parasitic wasps, predatory bugs, spiders, and birds are likely important in regulating populations. Significant mortality in L1–L3 was caused by Minute Pirate Bugs (*Orius tristicolor*) in our rearings.

Description of Immature Stages

The white spindle-shaped egg turns orange-red. Eggs near hatching have white tips and bases. There are 16–18 vertical ribs, each crossed by ~40 finer lateral grooves. L1 is tan-brown turning dull yellowish green after feeding. Each segment has 5–6 transverse folds, and there are 10 rows of short white T-shaped setae along the body, each bearing a droplet. L2 is green, lighter with maturity, with dorsal and dorsolateral dark stripes. There is a dorsal dark green stripe, and each segment has ~100 small dark bullae each sporting a short pale seta with a droplet. The bifurcated head is olive green with short pale setae. L3 is similar with more setae and has a more prominent

Adult

Egg @ 0.8 mm

Pupa @ 20 mm

L1 @ 2.8 mm

L2 @ 4.5 mm

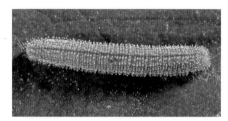

L3 @ 8 mm

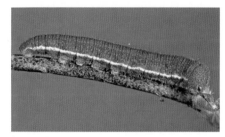

L4 @ 15 mm

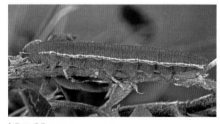

L5 @ 25 mm

dorsal dark green stripe. Overwintering L2 and L3 are straw-tan colored. L4 is bright green with a middorsal dark stripe and reduced setae. The body is peppered with tiny black dots, and a bold lateral white stripe is present with slight yellow markings on the lower edge. The head is green with tiny black dots. L5 is slightly darker, bluish green with a middorsal dark stripe and distinctive red blotches in the lateral white stripe. The pupa is green with an indistinct middorsal black stripe. There is a thick lateral whitish yellow stripe on the abdomen and a red-purple dot on each of the 3 abdominal segments closest to the wing case. These dots may form a line. The larvae of all *Colias* spp. are easily confused, and larvae of legume-feeding species need to be reared to adulthood to be sure of identity. Images of mature larvae appear in Allen et al. (2005) and Neill (2007), who also illustrates a pupa. The latter are similar to ours, but the larva in Allen et al. (2005) lacks red markings in the lateral stripe.

Discussion

Western Sulphurs may be reared on clover (recorded here as a host for the first time) but in our experience refuse to oviposit on Yellow Sweet Clover, which has been recorded as a host. Lupines may be an acceptable alternative host for rearing. Overwintering of larvae is difficult. Rearing L1–L2 under increasing daylength may avoid overwintering. Droplets on the tips of setae were present in L1–L2 and may have a defensive function (as in *Pieris* spp.), but this needs research. Ecological factors restricting the distribution of this species need study. A relatively small host range and an inability to use Alfalfa as a host may be part of the reason.

Queen Alexandra's Sulphur   *Colias alexandra* W. H. Edwards

Adult Biology

Colias alexandra is found throughout Cascadia E of the Cascades except for a narrow E–W belt through central WA. Beyond Cascadia the range extends from central AK to AZ and from the Cascade-Sierra Mts E into the W Great Plains. The single brood flies late April–early Sep; however, bivoltine populations have been reported in several other areas. Scott (1986a) reported that males average 6–12 days lifespan, females less. Males patrol open corridors in search of females, often seen flitting between flowers. The preferred habitat includes prairies, foothill grasslands, and wet woodland meadows (Allen et al., 2005). The reported larval hosts are a variety of legumes, including *Astragalus filipes* (Threadleaf Milkvetch), *A. canadensis* (Canada Milkvetch), *A. miser* (Weedy Milkvetch), *A. purshii* (Woollypod Milkvetch), *A. lentiginosus* (Freckled Milkvetch), *Oxytropis* spp. (locoweed), *Thermopsis montana* (Mountain Thermopsis), *Lathyrus* spp. (wild pea), *Hedysarum* spp. (sweet vetch), *Lupinus* spp. (lupine), *Trifolium* spp. (clover), and *Medicago sativa* (Alfalfa) as well as numerous additional hosts beyond Cascadia. Females lay up to 600 eggs (Guppy and Shepard, 2001), mostly on the uppersides of leaves; however, Scott (1992) observed oviposition on both leaf surfaces in CO.

Immature Stage Biology

We reared this species once and partially reared it 3 times. On July 16, gravid females from Pend Oreille Co., WA were confined with cut *M. sativa* and *Melilotus alba* (Yellow Sweet Clover) in a plastic box (30 × 23 × 12 cm) with a muslin lid. Oviposition occurred on both hosts within hours, and ~50 eggs were laid over 5 days. Eggs hatched after 4 days at 24–27 °C and natural daylengths. L1 refused to feed on *M. sativa* but accepted *Trifolium repens* (White Clover)

Adult

and *Lupinus lepidus* (Pacific Lupine). Larvae reached L3 and diapaused Aug 12. One female from Penticton, BC, on Aug 10 laid ~40 eggs on *A. miser*. Eggs hatched in 6 days (21–26 °C and natural daylengths), and L1 fed on *T. repens* and *Oxytropis monticola* (False Locoweed). L2 was reached in 5 days, diapausing in late Aug. Larvae in both cohorts were overwintered at 5 °C. Groups of larvae were exposed to warm temperature (28 °C) and continuous lighting during Dec–March. Larvae exposed to warm conditions after 77 days did not break diapause and died; however, feeding started in 16, 11, or 8 days after 92, 103, or 107 days overwintering, respectively. Some larvae molted to L4 18 days after first feeding (*T. repens*) but reentered diapause after 7–10 days in L4. On Aug 12, females from Okanogan Co., WA produced 73 eggs. Reared on *Lotus pedunculatus* (Greater Bird's-foot Trefoil), larvae reached L2 in ~2 weeks. Larvae were overwintered at 4 °C from mid-Sep. Most of the cohort was exposed to 20–22 °C and natural daylengths in mid-March but failed to feed and died. One larva removed from overwintering on April 9 broke diapause quickly, fed on *L. pedunculatus*, and molted to L3 after 6 days, pupating 22 days later. The adult eclosed after 11 days. Guppy and Shepard (2001) reported that larvae overwinter in L3. In our rearings, overwintering occurred in L2 or L3 depending on the instar reached by late Aug, when diapause appears to be induced. L1–L2 fed on the upper surface of leaves, producing pale "window pane" patches, then holes. Postdiapause larvae ate leaves from the edges. All instars rested on upper surface midribs when not feeding. L4 fed on stems when leaves were consumed. Guppy and Shepard (2001) reported larvae are solitary; Hayes (1981) found they may be cannibalistic. Protection is based on camouflage and head "thrashing" when disturbed. Frass is forcibly expelled, probably to divert predators. Nests are not used, and there are 5 instars.

Description of Immature Stages

The pale yellow egg is spindle shaped and turns bright orange-red in <2 days. There are ~12 longitudinal sharp-crested ridges separated by broad concave spaces, and ~35 fine lateral cross ridges. L1 is brown becoming green after feeding; each segment has 4 folds posteriorly and bears ~10 white setae with bulbous, T-shaped tips, usually with a droplet. The black head is granular, bearing white setae. L2 is forest green, strongly textured with numerous small black spots (~60 per segment) and a bulbous-tipped white seta arising from each. Prediapause L3 is similar to

Egg @ 1.6 mm

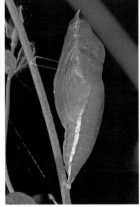

Pupa @ 15 mm

L1 @ 1.5 mm

L2 @ 6 mm

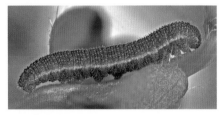

L3 @ 7 mm

L4 @ 15 mm

L5 @ ~30 mm

L2. Postdiapause L3 is medium-dark green with a dark green head. There are many neat lateral rows of small, lighter green bullae with a dark spot and white seta at the apex of each. The head bears small black spots with white setae, and there is a bold contrasting light green lateral stripe. L4 has a bold lateral white stripe and is dark green below the stripe, lighter green above. The body and head are covered with numerous short setae, giving a slightly shaggy appearance. The head, true legs, and prolegs are green. L5 is similar to L4, bright forest green with a contrasting white lateral spiracular stripe with small orange markings anteriorly. The pupa is irregularly elongated, pointed at both ends, shiny forest green with a white lateral stripe posteriorly. The larvae of legume-feeding *Colias* spp. (*C. alexandra, C. occidentalis, C. eurytheme,* and *C. philodice*) are very similar and variable. While body striping and red markings in L5 may differ, these species are probably not reliably separated (Allen et al., 2005). There are few published images of immature stages of *C. alexandra*; Allen et al. (2005) illustrate a mature larva which is very similar to ours except with more extensive red markings.

Discussion

It is relatively easy to obtain eggs from gravid females; however, mortality can be high in L1–L2 and overwintering is difficult. The minimum period for overwintering appears to be ~3 months, although 5–6 months is likely optimal for rapid postdiapause development. Short but increasing daylengths for postdiapause larvae may also be important. Larvae reenter dormancy under continuous lighting. If postdiapause development is delayed in spring, long daylengths in midsummer may result in late larvae overwintering a second time. Larvae are extremely well camouflaged and difficult to find. Little is known of natural enemies and other factors affecting population dynamics.

Queen Alexandra's Sulphur *Colias alexandra* W. H. Edwards

Labrador Sulphur   *Colias nastes* Boisduval

Adult Biology

Colias nastes is circumpolar, occurring across most of Arctic North America. In W North America, *C. n. streckeri* is found in BC and AB, S to extreme N WA and MT. It is confined to alpine tundra, favoring windswept scree slopes and summits. An apparent association with *Phyllodoce empetriformis* (Pink Mountain-heather) has been noted (Pyle, 2002). Our observations support this; it is unlikely this is a larval host, although adults may nectar on it. A single brood flies mid-Jun–early Aug in WA, June–Sep farther N, peaking Jul–Aug. It is univoltine in BC (Guppy and Shepard, 2001) and presumably WA, but may be univoltine or biennial in AK (Harry, 2009) and other N parts of its range (Scott, 1986a). Adults fly only in sunshine at temperatures >7 °C (Roland, 2006); flight is rapid, low, and distinctive. Darker individuals are more active and move greater distances because of ability to maintain higher body temperatures in sunshine (Roland, 2006). Both sexes visit asters, locoweeds, and fleabanes, and females hover over host plants. Larval hosts are a variety of legumes, including *Astragalus alpinus* (Alpine Milkvetch), *Hedysarum alpinum* (Alpine Sweetvetch), *Oxytropis campestris* (Slender Crazyweed), *O. splendens* (Showy Locoweed), and *O. borealis* (Sticky Crazyweed) (Pyle, 2002; Scott, 1986a; Guppy and Shepard, 2001; Harry, 2009). Eggs are laid singly on flowers and leaves.

Immature Stage Biology

One female obtained from Apex Mt, BC on July 27, held in sunshine (22–28 °C), laid ~20 eggs on *O. campestris*; however, the female was unmated and the eggs collapsed after 24 hrs. Images and discussion of larvae and pupae are courtesy of Nicky Davis (UT), who reared *C. n. nastes* from eggs obtained from Baffin Is in July by Jack Harry. Eggs were laid on *A.*

Colias nastes habitat, Apex Mt, BC

alpinus and hatched July 27–29. Larvae were reared at ~25 °C under continuous light on *Astragalus cicer* (Chickpea Milkvetch). This host appeared to be unfavorable for many larvae, with ~70% mortality during L1. L2 was reached after 6 days, and most (~88%) of the survivors entered diapause in this instar or L3. Diapausing larvae remained green to yellow-green. Two larvae continued development, pupating ~35 days after egg-hatch. Adults eclosed after 7–8 days. Harry (2009) presented life history information for *C. n. aliaska* obtained by in-field rearing in AK. Eggs were obtained from females caged with living *O. borealis*. Larval development was completed in either 1 or 2 years, with L1 overwintering in the first year and mature L4 in the second year. Guppy and Shepard (2001) show diapausing larvae that appear to be L4. Every instar except L5 has been recorded as an overwintering stage for *C. nastes*. Once L5 is attained, completion of development is assured (Harry, 2009). Harry (2009) reported that *A. cicer* was not eaten by *C. nastes* larvae in AK. It also appeared to be an unfavorable host for early-instar *C. n. nastes*, but ~30% of larvae were able to develop on it. Larvae skeletonize leaves in early instars, feeding from leaf edges later. Protection is based on camouflage, with larvae resting on the midrib of upper leaf surfaces. There are 5 instars, and nests are not used.

Description of Immature Stages

The pale yellow egg turns orange; it is spindle shaped with ~18 longitudinal ridges and ~50 lateral cross ridges. L1 is pale green with a dark head (Guppy and Shepard, 2001). L2 is dull green with dorso- and ventrolateral white stripes and a dark setaceous head. There are ~50 small black dots on each segment, each bearing a short black seta. L3 is olive green with prominent white dorso- and ventrolateral stripes and an olive green head. The dorsolateral stripes are bordered below by a row of 10 elongated black dots or

Adult female

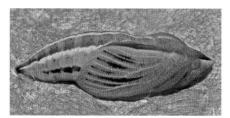

Pupae (*lower*, pharate) (photos N. Davis)

Egg @ 1.1 mm

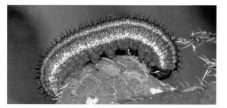

L2 (photo N. Davis)

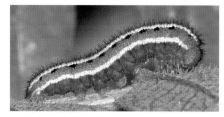

L3 (photo N. Davis)

L4 (photo N. Davis)

L5 (photo N. Davis)

L5 (photo N. Davis)

dashes. There is a middorsal dark line and the body and head are covered with short black setae. The green prolegs and true legs are tipped in orange-brown. L4 is similar, with increased density of black setae and pink spiracles. L5 is bright green with a light green head and proportionately shorter setae. The white stripes and middorsal black line are prominent. The pupa is green with black lines on the wing cases between veins. Dorsally there is a pair of white stripes with a dark middorsal line. There is a whitish yellow lateral stripe on the abdomen with a shorter black and red stripe below. The larvae of all *Colias* spp. are similar; however *C. nastes* is the only sulphur in Cascadia with both dorsolateral and ventrolateral white stripes. Images of larvae in Guppy and Shepard (2001), Miller and Hammond (2007), and Harry (2009) are similar to ours.

Discussion

It is challenging to obtain gravid females of *C. nastes* in Cascadia. The summits of S BC and N WA offer the best opportunities but the butterflies are elusive. There appears to be a narrow window of time when gravid females are present; males dominate the active population during much of the flight period. Rearing larvae under long daylengths or continuous light may allow complete development, but most will enter diapause. The flexibility of the diapause stage and incidence of annual or biennial populations deserve further study. Miller and Hammond (2007) considered *C. nastes* to be sensitive to climate change and vulnerable to elimination in Cascadia if temperatures increase. Comprehensive studies on all factors regulating populations of this species are urgently required.

Pelidne Sulphur   *Colias pelidne* Boisduval & Le Conte

Adult Biology

The Pelidne Sulphur occurs in 3 disjunct areas, the W High Arctic (YT and AK), E High Arctic (NF, LB, Baffin Is), and down the Rockies from AB to MT, WY, ID, and OR. The E and W populations may represent different species (Warren, 2005). In Cascadia, *C. pelidne* is a rare species found only in SE BC, the Wallowa and Steens Mts of E OR, and the high peaks of W ID. *Colias pelidne* flies in a single generation from late June to Sep in high-elevation (6,000–9,000 ft) habitats, tundra, cirques, and flowery slopes. Of ~30 individuals seen near the summit of Steens Mt on Aug 19, only 3 were females. Males patrol all day, low to the ground, searching for females, which spend much of their time resting. Pelidne Sulphurs are erratic and rapid fliers, difficult to capture except when nectaring or when females fly over dwarf huckleberries seeking oviposition sites. Both sexes visit flowers, particularly thistles and asters. Males visit mud and moist soil. The larval hosts are various low-growing *Vaccinium*, although the species used are generally unknown except for *V. caespitosum*, reported by Miller and Hammond (2007). *Gaultheria hemifusa* (Creeping Wintergreen) has also been recorded as a host (Scott, 1986a; Guppy and Shepard, 2001; Pyle, 2002; Warren, 2005). Oviposition in captivity on *Vaccinium uliginosum* (Bog Blueberry) was reported by Harry (2009), who presumably also reared it on this host. Eggs are laid singly, usually on leaves.

Immature Stage Biology

We partially reared this species once to L3 from a gravid female obtained from Steens Mt, OR on Aug 19. Our images of L5 and pupa are courtesy of Nicky Davis, who reared the species from a North Slope Alaska population. Confined in a plastic cylinder (13 cm dia × 8 cm tall) at 22–26 °C with cut foliage of *Vaccinium scoparium* (Grouseberry), our female began laying eggs within 30 min. During 4 days, ~40 eggs were laid. Eggs hatched in 6 days at 22–26 °C and L1 immediately commenced feeding on *V. scoparium*,

developing to L2 in 5 days. Feeding by L1 and L2 causes characteristic skeletonizing of leaves, beginning as "window paning." Larvae rest on upper leaf midribs when not feeding. Feeding and development slowed in L2, with some individuals resting in dried, curled leaves after 3 days. Twelve days after molting to L2 all larvae were dormant, resting in refugia, thus were transferred to 5 °C for overwintering. Small cohorts (1–4 larvae) were exposed to 19–27 °C under continuous lighting or 12–14 hr light, after 53, 80, 87, 101, or 141 days and were provided new growth of *Vaccinium parviflorum* (Red Huckleberry). Minor feeding was observed in some cohorts, but all larvae died after 7–25 days without developing. A cohort of 2 larvae exposed to 19–27 °C, ~14 hr light after 167 days and provided young foliage of a commercial blueberry cultivar, fed avidly within 24 hrs, creating large holes in leaves, and 1 molted to L3 after 6 days; however, after exposure to continuous lighting for 48 hrs, the larva ceased feeding and reentered dormancy. A final cohort of 14 larvae removed from overwintering after 180 days and exposed to 19–27 °C, ~14 hr light with foliage of another commercial blueberry cultivar, largely refused to feed, and most died within 7 days. Two larvae did feed and molted to L3 after 11 days. Feeding and development was limited in L3; 1 larva died, the other reentered diapause after 7 days in L3. Failure to obtain continued postdiapause development in warm conditions under ~14 or 24 hr daylengths suggests that *C. pelidne* may be a facultative or obligate biennial species on Steens Mt. A potential larval host, *Gaultheria shallon* (Salal), was rejected by postdiapause larvae. Postdiapause development (as L4?) on an optimal host is likely to be rapid, enabling adult eclosion by late June or early July in the high-elevation habitats occupied by *C. pelidne*. Larval defense is based on camouflage, the green larvae blending well with the foliage of *Vaccinium*. One larva in our rearing was parasitized with a tachinid fly that emerged postdiapause. There are 5 instars.

Description of Immature Stages

The egg is white turning yellow then orange-red within 24–30 hrs. There are 16–18 vertical ribs and 50–60 lateral cross ridges. L1 is yellow-tan with a black head and is sparsely covered with tiny white setae, 8–10 per segment. The posterior segment is dark dorsally. L2 is dull green tending yellowish anteriorly and posteriorly, covered with tiny white setae, 40–50 per segment, and has a yellowish setaceous head. L3 is green, densely covered in tiny white setae arising from dark bases, with spiracular white stripes. The head is green peppered with tiny black spots. L5 is bright

Adult male

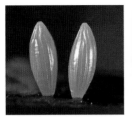

Eggs @ 1.3 mm (fresh, *left*, mature, *right*)

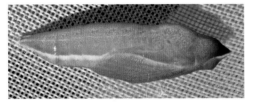

Pupa (photo N. Davis)

L1 @ 3.0 mm

L2 @ 4 mm

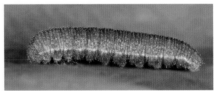

L3 @ 8 mm

L5 (photo N. Davis)

L5 (photo N. Davis)

bluish green with prominent white spiracular stripes edged in orange. Two yellow stripes are positioned dorsolaterally and the head is green with indistinct black spotting. The pupa is green with indistinct vermiform black markings. There is a thick lateral whitish stripe on the abdomen, also a shorter, narrower red stripe. The larvae of all *Colias* spp. are very similar, but *C. pelidne* and *C. interior* are the only sulphurs in Cascadia that feed on *Vaccinium* spp. Early instars of *C. pelidne* and *C. interior* are similar, but *C. interior* lacks the dorsolateral yellow stripes of *C. pelidne* in L5. Neill (2007) presented images of an egg and prewintering (L2?) and postwintering (L4?) larvae, which appear similar to ours. A mature larva was illustrated in Harry (2009) that also appears similar to ours.

Discussion

Dornfield (1980), Guppy and Shepard (2001), Allen et al. (2005), and Miller and Hammond (2007) all state that the immature stages of *C. pelidne* are unknown. Our account provides the first detailed observations on the immature stages of this rarely encountered species; however, much remains to be learned about the life history of this alpine sulphur. Larvae from Steens Mt entered diapause in L2, contrary to Harry (2009), who found L4 to be the overwintering stage in a cohort obtained from AK. Our experience with overwintered larvae entering a second dormancy in L3 suggests that the Steens Mt population may be biennial. Further rearing is required to determine whether Cascadia populations of *C. pelidne* are biennal or univoltine and what the environmental cues are. The host range of *C. pelidne* in Cascadia needs study, as does the role of biotic and abiotic factors in regulating population densities.

Pelidne Sulphur *Colias pelidne* Boisduval & Le Conte

Pink-edged Sulphur   *Colias interior* Scudder

Adult Biology

The Pink-edged Sulphur is a N species ranging from BC to NF S to the Appalachians, the Great Lakes, and MT, ID, and OR in the W. It occurs in the Cascades in S BC and N WA, the Blue/Wallowa Mts, and the OR Cascades S to Crater Lake. It is found in mid–high elevation forests (3,000–7,000 ft) in grassy clearings, meadows, and marshes and along roads and trails. It is univoltine, flying from June to Sep, peaking in July. It is relatively sedentary and a weak flier. Both sexes spend much time nectaring and resting on low vegetation. Red, blue, and yellow flowers appear to be favored, including red clover, Self-heal, dandelion, and trefoils. Males, especially newly eclosed individuals, mud-puddle (Douglas and Douglas, 2005) and patrol slowly to seek females (Scott, 1986a). Larval hosts are various species of *Vaccinium*; in Cascadia, *V. caespitosum* (Dwarf Bilberry), *V. uliginosum* (Bog Blueberry), and *V. myrtilloides* (Velvetleaf Huckleberry) have been recorded (Guppy and Shepard, 2001; Pyle, 2002; Warren, 2005). Females oviposit throughout most of the day (Allen, 1997), and eggs are laid singly, usually on the upper surface of leaves.

Immature Stage Biology

We reared this species once from a gravid female obtained on July 16 at Tiger Meadows, Pend Oreille Co., WA. Using a small plastic box with a muslin lid (28 × 16 × 7 cm), we confined the female on July 17 with a small potted *Andromeda polifolia* (Bog Rosemary), the only ericaceous plant available. No eggs were laid. Two days later, cut foliage of *Vaccinium scoparium* (Grouse Whortleberry) was provided, and oviposition occurred immediately. An estimated 60 eggs were laid over 2 days, mostly in groups of 10–15

Adult

on foliage tips. Hatching began after 4 days at 22–28 °C. Development was rapid, L1 occupying 3–4 days and L2, 8–10 days. Newly molted L3 stopped feeding Aug 10–15, sought curled leaves, and formed a silk mat on which they rested. Dormant larvae remained green. Most huckleberry and blueberry hosts of *C. interior* are deciduous; thus overwintering must occur on the ground. Foliage and dormant larvae were transferred to 12 °C on Aug 18, then to 5 °C on Sep 12. After 90 days, 2 L3 were transferred to *V. scoparium* leaves on wet cotton wool in a Petri dish and held at 27 °C and continuous light. After 72 hrs, 1 larva died, but the other started wandering, feeding, and "flinging" frass. After 10 days the larva molted to L4, increasing its size by only 1 mm during L3. Eleven and 10 days were spent as L4 and L5, respectively, the larva pupating 31 days after removal from overwintering. Adult eclosion occurred after 17 days at 20–25 °C. Termination of overwintering varied, some larvae feeding and molting, then returning to dormancy, suggesting the possibility of a 2-year life cycle in some individuals. Feeding damage by L1–L2 is distinctive, with entire leaves skeletonized. L3–L5 consume leaves from the edges or tip. Cannibalism was observed, with an L5 feeding on a newly formed pupa. Larvae rest on the upper surfaces of leaves on the midrib. Pupation occurs on host-plant twigs. Defense is based on camouflage, larvae blending well with host foliage and stems. Little is known concerning natural enemies, but parasitic wasps, predatory bugs, and spiders are likely to be important in regulating populations.

Description of Immature Stages

The white spindle-shaped egg turns orange-red. There are 18–20 vertical ribs, each crossed by ~80 finer lateral grooves. L1 is brown becoming orangish yellow after feeding. There are 10 rows of short, pale setae along the body, each bearing a droplet. The slightly bifurcated head is black with short setae. L2 is green and textured, becoming lighter. The body is covered with tiny green and black bullae, each bearing a short pale seta with a droplet. The head is olive green with black dots and short setae. L3 is similar, with a vague middorsal dark stripe. L4 is dark green and textured, becoming smoother and lighter with reduced bullae and setae. An indistinct spiracular white stripe becomes bolder and infused with red on the lower half with maturity. L5 is bluish green peppered with many tiny black dots and short pale setae. The spiracular white stripe is bold with an orangish red line through the center; some individuals lack the red line. The head, true legs, and prolegs are lighter green

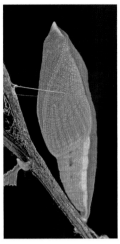

Pupa @ 16.5 mm

Egg @ 1.8 mm

Eggs @ 1.8 mm (mature)

L1@ 3 mm

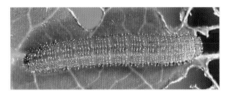

L2 @ 4 mm

L3 @ 6 mm

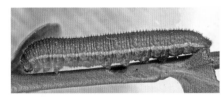

L4 @ 15 mm

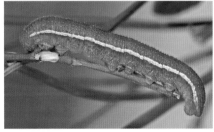

L5 @ 28 mm

with black dots. The spiracles are white encircled with black. The pupa is green with a thick lateral whitish yellow stripe on the abdomen and a red-purple dot on each of the 3 abdominal segments closest to the wing case. These dots may form a line. Douglas and Douglas (2005) report some pupae are brown. The larvae of all *Colias* spp. are very similar and easily confused; however, this species and *Colias pelidne* (Pelidne Sulphur) are the only sulphurs in Cascadia that use *Vaccinium* as hosts. Images of mature larvae appear in Allen (1997) and Allen et al. (2005). Both are similar to ours, although lighter in color. Allen (1997) illustrates a pupa that lacks red-purple dots.

Discussion

Pink-edged Sulphurs may be reared on Grouse Whortleberry and probably other *Vaccinium* spp. The full host range of *C. interior* in Cascadia needs study. The distinctive skeletonized damage to *Vaccinium* leaves caused by L1–L2 creates a distinctive field mark that may help in locating larvae during July–Aug; however, adults are easily caught and females oviposit readily in captivity. Droplets on the tips of setae were present in L1–L3 and may have a defensive function (as with *Pieris* spp.), but research is needed. Overwintering of our larvae occurred in L3 as also reported by Allen (1997), Pyle (2002), and Allen et al. (2005). Klots (1951) and Douglas and Douglas (2005) suggested overwintering may occur in L1, and Lyman (1897) reported L2 as the overwintering stage. The possibility of overwintering occurring in these earlier instars should be investigated. The possibility of a 2-year life cycle occurring in some populations also needs study.

Dainty Sulphur   *Nathalis iole* Boisduval

Adult Biology

The Dainty Sulphur is a rare visitor to Cascadia, with R. M. Pyle's report of 7 fresh males in SE WA in 1975 and an earlier record from ID the only records so far of this tiny sulphur in the Pacific Northwest (Pyle, 2002). It is resident in most of S US and N Mexico, migrating N in spring, and is found in summer in most of central US, sometimes extending into S Canada by autumn. A warming climate may increase the chances of finding this species more often in Cascadia. The annual N spread of Dainty Sulphurs from S deserts is facilitated by a broad host-plant range, rapid development, and exploitation of watercourses and highways as travel routes. Migration may be wind assisted, and each successive generation moves farther N. Weedy, dry, open, and disturbed habitats are preferred, including roadsides, grazed arid lands, riparian areas, and canyons. *Nathalis iole* is multivoltine, with many generations in the S and 2–3 generations in the N parts of its range. Breeding continues throughout the year in the S with no diapause stage. It is most likely to be seen in Cascadia during Aug–Oct. Adults are erratic fliers, usually staying close to the ground and basking laterally when conditions are cool. Many types of flowers are visited, often small, yellow composites, and males mud-puddle. Scott (1986a) provided notes on courtship behavior. Larval hosts are weedy Asteraceae, including *Dyssodia papposa* (Fetid Marigold), *Dyssodia pentachaeta* (Golden Dyssodia), *Bidens pilosa* (Hairy Beggarticks), *B. aurea* (Arizona Beggarticks), *B. ferulaefolia*, *Helenium autumnale* (Common Sneezeweed), *Cosmos* spp., *Tagetes* spp. (marigolds), *Palafoxia linearis* (Desert Palafox), *Thelesperma* spp. (greenthreads), *Hymenothrix wislizenii* (Thimblehead), and *Hymenothrix wrightii* (Wright's Thimblehead) (Scott, 1986a; J. Brock, pers. comm.). Eggs are laid singly on young leaves, sometimes sepals (Scott, 1986a).

Immature Stage Biology

We reared this species once from livestock obtained from Pima Co., AZ (provided by Jim Brock). Eggs, L2 and L4 were collected on April 21 from *H. wrightii* and reared on *Tagetes* sp. at 19–21 °C. Eggs hatched within 24 hrs and L1 consumed eggshells before feeding on young leaves and petals of *Tagetes* sp. Larvae molted to L2, L3, and L4 after 5, 6, and 7 days, respectively. Pupation occurred after a further 5 days with <24 hrs spent as a prepupa. There were only 4 instars. Adults eclosed 15 days later. Development from egg-hatch to eclosion occupies ~38 days at 18–21 °C. Feeding by L1 causes "window paning" of leaves; later instars consume leaves and flower petals from edges. Larvae rest on stems or leaf midribs, feed mainly at night, and do not build nests. Pupation usually occurs on the host plant, with the pupa formed horizontally on the underside of a leaf or vertically on a stem. Defense is based on camouflage, setal droplets (particularly L1–L3), a ventral gland on segment 1 that likely emits defensive chemicals, and intimidatory behavior in the final instar (waving of anterior end when threatened). Coloration is highly variable, with larvae entirely green or strongly marked with red-magenta stripes. This appears to be dependent on host-plant type, with larvae feeding on red-marked hosts (e.g., beggarticks) more likely to have red markings than those feeding on hosts without red markings (e.g., marigolds) (Minno et al., 2005).

Description of Immature Stages

The yellow-orange egg is very finely sculpted with ~30 vertical ribs and does not change color with maturity as occurs in other sulphurs. L1 is dull green with a shiny black setaceous head. There are ~10 simple black setae on each segment arising from swollen green bullae, giving the larva a slightly textured appearance. Each seta carries a droplet, a condition that persists until pupation. L2 is medium green with a dark brown setaceous head. Red markings may appear in L2 and usually consist of a wide middorsal stripe. L3 is green with or without a red dorsal stripe and has 6–7 transverse folds on each segment. Green larvae have an indistinct spiracular pale stripe and an indistinct middorsal dark stripe, both of which may be red in red-marked individuals. Indistinct tiny black spots and setae cover the body, which is also sparsely clothed (~6 per segment) with medium-length black setae. The setaceous head is green. A pair of pink-red enlarged bullae immediately behind the head are present in all larvae. L4 is green with indistinct pale vermiform markings and middorsal and spiracular

Adult male

Egg @ 0.9 mm

L1 @ 2 mm

Pupa @ 11 mm

L2 @ 5 mm

L3 @ 9 mm

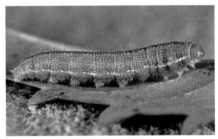

L4 @ 16 mm

L4 @ 16 mm

stripes that may be pale, red, or bold magenta. A pair of enlarged pink, red, or magenta bullae behind the head are present in all individuals. The head, true legs, and prolegs are green and the spiracles are white. Prepupal larvae do not change color. The pupa is bright green with tiny, indistinct white vermiform markings and white spiracles. A pair of small pink-red dots occur dorsally on the head, and the cremaster is also pink-red. The head end is rounded, lacking the projection characteristic of the pupae of other sulphur species. *Nathalis iole* larvae lacking red-magenta stripes are similar to larvae of *Colias* spp., but the latter do not feed on Asteraceae. Heavily marked *N. iole* larvae are distinctive and should not be confused with any other species (see Allen et al., 2005, p. 47). An image of a mature green form larva appears in Minno et al. (2005) and is similar to ours.

Discussion

This nomadic and unmistakable little sulphur may turn up in Cascadia more frequently in the future if the climate of W US warms. *Nathalis iole* is easily reared, accepting a number of weedy composites as larval hosts. Oviposition is likely to be easily obtained in captivity, and larvae develop rapidly even under cool conditions. The incidence of red-marked larvae and the role of host-plant species in determining coloration deserves study, as does the significance and function of the enlarged pair of pink-red bullae. The chemistry of setal droplets, which are present throughout larval development, and emissions from the ventral gland also deserve study. Very little is known of the biotic and abiotic factors enhancing or depressing population expansion and migration in *N. iole*.

Dainty Sulphur *Nathalis iole* Boisduval

Family Lycaenidae
Subfamily Lycaeninae
COPPERS

There are 7 copper species in Cascadia. Adults are dimorphic, with males coppery or blue dorsally and the females brown or gray. Flight periods are typically brief, and all species except the Purplish Copper (which is multibrooded) have a single annual generation. Coppers are sedentary, rarely moving far, and occupy habitats from the lowlands to alpine summits. Some require habitats with permanent moisture so are often found near watercourses and ponds.

Eggs with developed first instars overwinter in all species except 1 and are laid singly near or on senescing hosts. Lustrous Coppers overwinter as partially grown larvae and may be biennial. Eggs are partially compressed spheres with distinctive coarse honeycomb surface ornamentation. Larvae are similar to those of other lycaenids but with somewhat narrower, more elongated bodies on average. The head is small and hidden under the first segment. Mature larvae are generally green, downy, and well camouflaged and generally lack dorsal chevron markings. Many have a strong dorsal red-purple stripe at some stage. The minutely plumed mushroom-shaped setae of coppers are unique and without a microscope appear to be tiny white dots.

Copper caterpillars feed primarily on leaves and usually rest on the undersides of leaves. Some species create characteristic feeding marks that are good field signs for locating larvae. Larval hosts are docks, knotweeds, and sorrels for most species; others use buckwheats, cinquefoils, or huckleberries. There are usually 4 instars, but coppers that overwinter as larvae sometimes have 5 instars. Pupae are smooth and rounded, brown or green, sometimes matching substrate color, and are usually oriented horizontally with a silken girdle thread secured around the middle.

Larvae of about half our species are known to be attended by ants, despite the lack of a honeydew gland. There are other potential ant-attracting glands present that may mediate the observed interactions. The Ruddy Copper rests and pupates in underground ant chambers on occasion.

Lustrous Copper   *Lycaena cupreus* (W. H. Edwards)

Adult Biology

Lycaena cupreus is rare in WA, barely entering the N Cascades, the only known locality being at Slate Peak, Okanogan Co. This species occurs in broadly separated populations in the Rocky Mts, SW BC to N WA (several ranges, including the Kootenays and Selkirks), and in the S Cascade and Sierra Mts of S OR and N CA. Warren (2005) indicated that the Rocky Mt and OR populations (ssp. *L. c. cupreus*) differ from the SW BC and N WA ssp., *L. c. snowi*. The known flight period in WA is very short, late July–early Aug, but is likely underrepresented because of the rarity of the species. WA and BC populations are known to use only 1 host plant, *Oxyria digyna* (Mountain Sorrel), while the S subspecies utilizes *Rumex pauciflorus* (Alpine Sorrel) and various docks. The Lustrous Copper is a species of steep alpine scree slopes and barren ridgetops in Cascadia, but the S subspecies occurs at lower elevations. Males perch and patrol for females in hollows of open areas, crisscrossing steep slopes in alpine areas. The females are more sedentary, nectaring on asters, *Arnica*, and small daisies, also visiting *Oxyria* plants for oviposition. We have not observed oviposition; however, in captivity, females placed eggs singly on either the panicles (hanging flower parts) or the upper surfaces of leaves. Scott (1992) observed oviposition twice on rocks near *O. digyna* in CO, with the eggs placed just below a rock overhang, and on other occasions found hatched and unhatched eggs in similar positions. *Lycaena cupreus* is single brooded (possibly biennial in some locations) throughout its range.

Immature Stage Biology

We attempted to rear this species twice with eggs from captive females. Two females obtained July 31 at Slate Peak, Okanogan Co., laid 4 eggs on hanging petals of an *Oxyria* panicle; these eggs were very difficult to find, as the flowers hang in confused loose clusters and the eggs are extremely well camouflaged

against the rust-colored plant. The eggs hatched, but the larvae would not feed on a substitute host, *Rumex obtusifolius* (Bitter Dock). Two females collected Aug 6, same place, different year, produced 7 eggs, this time on the upper surface of *O. digyna* leaves, where they were easily found. Three eggs hatched after 5 days, but the remainder were sterile. The larvae developed well on *Oxyria digyna*, surviving to L2 or L3. L1 fed on leaves, resting prone in the middle of a leaf and eating a tiny hole halfway through. L2 was reached 8 days post oviposition and ate small round holes on the upper side, leaving a translucent leaf membrane on the opposite side. Feeding occurred anywhere on the leaf surface except the edges. At 13 days post oviposition, 1 larva reached L3 and, together with an L2, fed heavily before entering diapause in late Aug, about 3 weeks post oviposition. None of the larvae survived overwintering. We searched *O. digyna* plants exhaustively at Slate Peak without finding larvae; however, others have found L2 and L4 on undersides of leaves. There are 4 instars, siblings are tolerated well, and no nests are built. Protection appears to be based on camouflage. From our experience, the overwintering stage appears to be L2 or L3, which is unique among our coppers; other species of *Lycaena* overwinter as eggs. Ballmer and Pratt (1988) reared CA *L. cupreus* at 25–27 °C, and although some larvae completed development, most entered diapause in L3. Scott (1992) found a hatched pupa on the underside of a rock within several meters of *O. digyna* plants. In the same paper he speculated that *L. cupreus* may be biennial, overwintering twice in larval stages before reaching adulthood. Kiyoshi Hiruma (pers. comm., 2005) speculated similarly after finding L2 and L4 at the same time.

Description of Immature Stages

The eggs of *L. cupreus* are beautifully ornamented with a coarse latticework of bold white ridges forming irregular large polygons over the egg surface. The large spaces within the polygons have a granular texture and greenish white color, and the micropyle is small and indistinct. L1 is creamy tan with a purplish cast. It bears long black setae along the dorsum and head and a ventral fringe of shorter blond setae and has rows of distinct brown spots. L2 is similarly patterned except the dorsum is bright magenta and the body is green, much darker on the anterior two-thirds. In L3 there are increased numbers of proportionally shorter setae and a bright magenta stripe is present along the dorsum. L4 is uniformly green, and there are numerous short setae. We have not seen the pupa, but an image in Guppy and Shepard (2001), presumably

Adult

Eggs @ 0.8 mm

Oxyria digyna
(host plant)

L1 @ 1.6 mm

Lustrous Copper   *Lycaena cupreus* (W. H. Edwards)

from CA, shows it to be rather light brown with a number of small irregular black spots and blotches. Other Cascadian *Lycaena* spp. have similar green larvae but more finely sculpted eggs. The larvae of *Lycaena mariposa* are the most similar; however, the pupa is pale green and the larvae feed on *Vaccinium*. In *L. rubidus* L2 is reddish, the final instar occurs in 2 color forms, and the pupa is lighter. In *L. helloides* the larvae are green at all stages, the pupa is darker, and the food plants are various docks. *Lycaena nivalis* and *L. editha* have magenta dorsal stripes throughout development and both feed on docks and knotweeds. *Lycaena heteronea* has green larvae and pupae. Guppy and Shepard (2001) illustrated a mature *L. cupreus* larva closely resembling ours.

L2 @ 3.0 mm

Discussion
This species is difficult to find, difficult to rear, and difficult to provide with correct host plants. To date we have found no usable substitute host plant and the normal host is usually not available in commercial nurseries or within reasonable access. In our experience with captive females, eggs were laid on flowers in 1 instance and on leaves on another occasion, while others have recorded oviposition on rocks in the wild. More research is needed to determine oviposition preferences in our area. Additional research is needed to determine whether this species is univoltine or biennial, and whether it is capable of developing through 5 instars as are alpine populations of the closely related *Lycaena phlaes* (Ballmer and Pratt, 1989). Finding a substitute host plant would also be useful. *Lycaena cupreus* occupies a precarious alpine habitat which is vulnerable to climate warming. The impact of warming on *L. cupreus* populations in Cascadia warrants urgent research.

L3 @ 5 mm

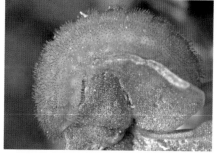
L4 (photo M. Peterson)

Edith's Copper   *Lycaena editha* (Mead)

Adult Biology

The range of *Lycaena editha* extends throughout the NW US from CA to MT, E to WY and CO. In Cascadia it is found mostly in S and NE OR extending into the Blue Mts of SE WA. A single adult was found near Spokane, WA in 2000 (Pyle, 2002), and in mid-July 2006 and 2009, it was found at 2 localities in Pend Oreille Co., NE WA. It should also be looked for in S BC. Its likely presence in AB in the 1920s was demonstrated by Anweiler and Schmidt (2003). *Lycaena editha* occurs in montane (>3,000 ft) habitats (meadows, roadsides) in a single brood from mid-June to early Aug (WA) and as late as early Sep on Steens Mt in SE OR. In some locations, *L. editha* is abundant, often occupying the same habitats as *L. nivalis* and *L. helloides*. It is strongly attracted to flowers, particularly yarrow, dogbane, and asters, and males aggressively guard low perches. Larval hosts are species in the Polygonaceae, primarily *Polygonum* (knotweeds and smartweeds) and *Rumex* (docks and sorrels) (Pratt et al., 1993; Scott, 1986a, 1992; Warren, 2005; Austin and Leary, 2008). Dornfield (1980), Pyle (1981), and Neill (2007) also recorded species of *Potentilla* and *Horkelia* (Rosaceae) as hosts, but these need verification. Scott (1992, 2006) observed CO females landing on *Rumex acetosella* (Sheep Sorrel), crawling down the plant, and ovipositing on litter, also ovipositing on *Polygonum douglasii* (Douglas' Knotweed) and *P. bistortoides* (American Bistort). He found a definite preference for *R. acetosella* over *P. douglasii*. Adults appear to be relatively long lived, at least in captivity, surviving for 2–3 weeks. Females lay eggs singly.

Immature Stage Biology

We reared this species twice. Six females obtained from the lower Blue Mts near the Tucannon R in Columbia Co., WA on July 27 laid 40 eggs on green leaves of potted *Rumex venosus* (Winged Dock) during 2 weeks. Eggs were overwintered in an unheated

Adult female

garden shed from Oct onward. Five females obtained from near the summit of Steens Mt in Haney Co., OR on August 29 laid ~100 eggs on dried foliage of *R. acetosella* during 2 weeks. Eggs were overwintered at 5 °C from Sep onward. Small numbers (7, 6, 14) of eggs in the WA cohort were removed from overwintering on Jan 20, Feb 5, and March 16. Jan eggs were held at 15–21 °C under daylengths of 12–13 hr and hatched after 10–14 days. Larvae fed on *Rumex crispus* (Curly Dock) but died during L1 and L2. Feb eggs were held at 25 °C and continuous lighting and hatched within 48 hrs. Larvae fed on *R. acetosella* and took 22 days to reach pupation. Adults emerged after 20 days at 20 °C. Throughout development larvae fed only on leaves, grazing mostly on lower surfaces, eventually eating holes through them and later feeding on leaf edges. No nests were built, and the green larvae with prominent dorsal red line rely on camouflage for protection. Sheep Sorrel is an identical green color with red lines on the stems. Larvae tolerate each other well in captivity but are not communal. There are 4 instars and ant associations are important (Ballmer and Pratt, 1988). Ballmer and Pratt (1991) reported ant attendance by *Formica alipetens* and showed *L. editha* larvae to be the most attended species by *Formica pilicornis* among 59 lycaenid species in an experimental setup. The egg is the overwintering stage.

Description of Immature Stages

The egg is greenish white, ornamented with a coarse latticework of bold white ridges that form large, irregular, and deep polygons. L1 is whitish tan becoming tan-orange with a black head, and the first and last segments have distinct black dorsal sclerotized plates. Each segment has ~12 black spots with the anterior dorsal pair bearing long backward-swept dark setae. Shorter pale setae arise from lateral spots. Late in L1 a distinct middorsal red line develops. L2 is green with a prominent middorsal red-magenta stripe and black head. A pair of medium-length black setae arises from dorsal black spots on each segment. L3 is similar, with a green head and with all setae shorter, especially laterally, giving the appearance of very fine black peppering. L4 is darker green with a dense covering of short orange-brown setae on the body and the green head. Small white spots pepper the body, and the spiracles are pink-orange encircled in brown. Prior to pupation the magenta stripe fades. The pupa is ovate, orangish tan marked with numerous small dark brown-black spots and dashes. Dorsally the pupa is sometimes lighter with fewer spots and an indistinct dark stripe. Among Copper species, the larvae of *L.*

Egg @ 0.7 mm

L1 @ 2 mm

Pupa @ 12 mm

L1 @ 3 mm (late)

Edith's Copper *Lycaena editha* (Mead)

rubidus most closely resemble those of *L. editha* but do not possess the prominent paired dark setae seen in L2 and L3 and they also lack a dorsal stripe in L4. L4 of *L. nivalis* have a red ventrolateral stripe as well as a dorsal one. Our images appear to be the first published of *L. editha*. *Lycaena dione* (Gray Copper), a fairly similar but larger species usually found at lower elevations, has been reported in the ID panhandle very close to WA (Pyle, 2002). We know of no published images of *L. dione* immature stages, but in a brief description in Guppy and Shepard (2001), they seem similar to *L. editha*.

L2 @ 5 mm

Discussion

This species oviposits readily in captivity as do all our coppers and is easily reared on various *Rumex* and *Polygonum* spp. The range of hosts used in Cascadia is uncertain and deserves study. Of particular interest is the possible utilization of *Polygonum douglasii* (reported by Scott, 1992, in CO) and the rosaceous plants *Potentilla* and *Horkelia* spp. reported by some authors to be hosts. This species appears to be undergoing a range expansion N and should be looked for in NE WA, N ID, and S BC. Whether this expansion is due to climate warming or other environmental factors or is an artifact of inadequate sampling, clarification studies are needed. Research on the impact of natural enemies on the population biology of *L. editha*, as well as the identity and role of ants in its ecology in Cascadia, would be rewarding. The possible presence of *L. dione* in far NE WA warrants investigation.

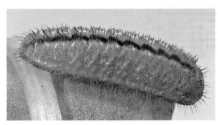

L3 @ 10 mm

L4 @ 20 mm

Ruddy Copper   *Lycaena rubidus* (Behr)

Adult Biology

Lycaena rubidus flies in arid areas in the Columbia Basin and E OR. Beyond Cascadia it ranges throughout NW America into S AB. The flight period is early May–late July, and it is single brooded. Generally scarce throughout the region, *L. rubidus* can be locally common in prime habitat; we once observed hundreds in flight on May 23 at Juniper Dunes, Franklin Co., WA. Males perch conspicuously in search of mates while females flit through lower vegetation seeking nectar. Adults are sexually dimorphic; however, some females are orange like the males, which may allow them to avoid males and feed and oviposit more freely than their gray counterparts. Larval hosts include *Rumex venosus* (Winged Dock), *Rumex salicifolius* (Willow Dock), and *Oxyria digyna* (Alpine Sorrel); Scott (1986a) lists 6 additional *Rumex* hosts from CO. Females lay eggs singly on debris and inert surfaces near the base of the host plant. Scott (1992) reported finding numbers of eggs in CO laid on litter under several *Rumex* spp., as well as a few near *Polygonum douglasii*.

Immature Stage Biology

We reared this species twice from gravid females and once from field-collected larvae. On April 18 we collected 17 L3 and L4 at Juniper Dunes and reared them on *R. venosus* to adulthood. Plants hosting larvae were easily found by characteristic leaf damage. Larvae were found in dense grass or under debris at the bases of *R. venosus*, all within ~8 cm of the plants. Most larvae were concealed at ground level, but a few were on the plants or in voids and holes under the plants; 2 larvae were attended by ants. In sandy soils, larvae can be found with tending ants in holes in the loose sand (J. P. Pelham, pers. comm.). Several gravid females were obtained from Juniper Dunes in each of 2 years (May 23 and 25) and confined with *R. venosus*

and *Rumex acetosella* (Sheep Sorrel), laying 200 and 22 eggs, respectively. Eggs were cemented to the substrate and sticks, with a preference for crevices and cracks. The low number of eggs obtained from the May 25 females may have occurred because *R. acetosella* (not recorded as a host) was used. The eggs were allowed to mature and were overwintered after 2 weeks or left under ambient conditions until mid-Sep, when transferred to 5 °C. Six and 2 adults were reared on *Rumex obtusifolius* (Bitter Dock) and *Rumex crispus* (Curly Dock), respectively. Development from egg-hatch to pupation took 19–22 days at 20–22 °C, nearly half of this time spent in L4, and adults eclosed 10–17 days later. Sixteen eggs from the May 25 cohort were removed from 5 °C on Dec 7, and hatching occurred within 4–6 days at 27 °C. Larvae were provided with *R. acetosella*, but only limited feeding occurred and all died within a few days, suggesting this host may be unsuitable for development. Six eggs were placed on *R. crispus* on Feb 5, and larvae fed well, developing to pupation in 22 days at 25 °C. Larvae show a strong preference for the basal parts of *Rumex* leaves, feeding mostly at night. Larvae eat large holes through the leaves, or feed from the leaf margins. The larvae occur as green or brown forms. Pupation occurred on hosts or inert surfaces, some attached by a weak cremaster to *Rumex* flower heads, others unattached on leaves or hidden on the cage bottom. Protection depends on concealment, camouflage, and tending by ants. The larvae are solitary but tolerant of one another. No nests are constructed, but refugia are sometimes used. Ballmer and Pratt (1991) showed that this species is strongly attended by the ant *Formica pilicornis*.

Description of Immature Stages

The greenish white egg is covered with a latticework of oval indentations. L1 has a black head and is pale yellow-brown with dark brown rounded sclerotized plates on the first and last segments. Light brown tubercles sport long brownish setae along the ventral margin, behind the head, and in a double row dorsally. Late in the instar the color is purplish with lighter longitudinal stripes. L2 is longitudinally striped in yellow and brown (white and gray early in the instar), with a broad dorsal magenta stripe and a small, dark sclerotized round dorsal plate on the first segment; setae are as in L1. L3 is densely covered in shorter setae, brown dorsally and pale along the ventral margins. The body is green with yellow wavy stripes and a magenta dorsal stripe. L4 is green (or reddish brown in the alternate color form), densely covered with short orangish setae. The dorsal stripe

Adult male

Ruddy Copper *Lycaena rubidus* (Behr)

Egg @ 0.8 mm

L1 @ 2.7 mm

Pupa @ 14 mm

L1 @ 2.3 mm and L2 @ 5 mm

L3 @ 7 mm

L4 @ 16 mm and 18 mm (two color forms)

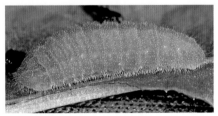

L4 @ 22 mm

persists but is less distinct. Numerous white spots cover the body, and the spiracles are pale, ringed by a thin brown outline. The pupa is ovate, light brown with numerous contrasting irregular dark brown spots. The immature stages of *L. rubidus* are similar to those of other coppers. *Lycaena heteronea* and *L. helloides* larvae lack the purple dorsal stripe. *Lycaena cupreus* is found in restricted alpine habitats and *L. nivalis* larvae have a purple ventrolateral stripe; *L. mariposa* is similar except for the pupa, which is pale green. *Lycaena editha* is similar in all stages except the eggs, which are more coarsely latticed. Pratt et al. (1993) showed that *L. rubidus* and *L. editha* are closely related. Published images of *L. rubidus* are few; Allen et al. (2005) show a mature AZ larva that is green like ours but with dark spiracles.

Discussion

Larvae are not difficult to find in habitats where the species is common and at the right season, and eggs are readily obtained from gravid females confined with *R. venosus*. Eggs and larvae are not difficult to rear provided fresh food plants are available. Little is known of the natural enemies or the abiotic factors influencing the patchy distribution and local abundance of *L. rubidus* populations. The incidence, ecology and competitiveness of orange females deserves study, as does the ecology of the 2 larval color forms. Ant attendance appears to be important in the defense ecology of this species, although Funk (1975) reported ant collection of *L. rubidus* eggs which may have been fed to the colony or perhaps overwintered in the ant nest. Clearly the relationship between *L. rubidus* and ants deserves further study.

Blue Copper   *Lycaena heteronea* Boisduval

Adult Biology

The Blue Copper ranges throughout the North American W from mid-BC to mid-AZ and E to the plains. In Cascadia it is found in and E of the Cascades in a wide range of habitats and elevations, from shrub-steppe to open pine forest to the subalpine. *Lycaena heteronea* flies in a single generation from early May to early Sep with a peak in most places in July. This large copper can be abundant in localized areas but is moderately common to uncommon over much of its range. Buckwheats (*Eriogonum* spp.) host this butterfly, providing nourishment for adults and larvae. Scott (1986a) lists 9 *Eriogonum* hosts in CO, and at least 7 species are reported in Cascadia, including 4 of our most widespread species, *E. sphaerocephalum* (Round-headed Buckwheat), *E. compositum* (Northern Buckwheat), *E. strictum* (Strict Buckwheat), and *E. heracleoides* (Parsley Desert Buckwheat); of these, *E. compositum* appears to be the most frequently used. Males perch or search for mates, flying near buckwheats even after the prime bloom has passed; males also visit mud. In early Aug we observed a female in extended oviposition behavior on *E. heracleoides*, touching her abdomen repeatedly to the senesced brown flower heads and flower stalks immediately below, although no eggs were found. Scott (1992) reported finding numbers of *L. heteronea* eggs laid on the "umbel subtending bracts" just under the flower heads of several species of *Eriogonum*, adding that females do not crawl down the host-plant stems to oviposit at the base as some other lycaenids do.

Immature Stage Biology

We reared this species 3 times from females collected in mid-July to early Aug at mid-elevation in Kittitas

Co., WA, near the Tucannon R in Columbia Co., WA, and at high elevation (7,500 ft) on Steens Mt, OR. Approximately 25 (Kittitas), 100 (Columbia), and 50 (Steens) eggs were laid on loose, dried *Eriogonum compositum* and *E. heracleoides* leaves and stems on the cage bottom within 1–12 days. After a few weeks at ambient temperatures, eggs were overwintered at 4–5 °C for 100+ days. After removal from overwintering, eggs hatched within 24–48 hrs of exposure to 20–25 °C. L1 did not eat their eggshells but began feeding on leaves of *E. compositum* or *E. niveum* (Snow Buckwheat). The latter was an acceptable host up to L2, but *E. compositum* was found to be necessary for further development. An additional acceptable host for later instars was *Eriogonum elatum* (Tall Buckwheat). Larvae do not build nests. Throughout development larvae eat round holes halfway through host leaves, often from the upper side of the leaf, using their extendable necks to hollow out the tissue from the inside in a circular area around the hole. Adjacent leaf tissue turns pale yellow, contrasting with the dark green leaves, a search pattern useful for finding larvae. Scott (1986a) reported that young larvae feed on undersides of leaves. Ant attendance has been reported. The Kittitas larvae pupated quickly, 17–21 days after egg-hatch, although the Columbia Co. cohort took longer, 35 days at 25 °C. Some larvae pupated on the underside of *E. compositum* leaves; a pupa was also found attached in this manner in the wild. Adults eclosed 13–14 days after pupation, developing from L1 to adult in 30–48 days.

Description of Immature Stages

The eggs are white, flattened spherical, and ornamented with a distinct lattice pattern of 4-sided irregular polygons. Eggs from Columbia Co. females were marked with yellow-orange streaks. L1 is brownish green with a black head. Each segment has 22 distinct round black spots, each giving rise to a single, moderately long pale seta. L2 is light milky green with slightly shorter setae and pale patches laterally. L3 is darker forest green with alternating light green and forest green longitudinal stripes. L4 is uniform bright forest green; the black spots are limited to the ends or absent with white spots replacing them. Short setae, mostly white but some black in a dorsal strip, give a slightly shaggy appearance. Columbia Co. larvae had a distinct pale yellow-white lateral stripe in L4. Ballmer and Pratt (1988) reported a pale yellow or white lateral line in some *L. heteronea* larvae. The pupa is light green, mottled with many tiny darker green splotches. The mature larva was illustrated by Allen et al. (2005), and Guppy and Shepard (2001) illustrated the mature

Adult male

Egg @ 0.8 mm (color variations)

Pupa @ 11.5 mm

Pupa @ 10 mm (pharate)

L1 @ 1.5 mm

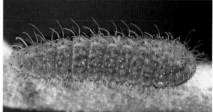

L2 @ 3 mm

L3 @ 8 mm

L4 @ 13 mm

L4 @ 18 mm

larva and pupa; all images agree closely with ours. The mature larvae of *L. heteronea* are similar to those of other coppers but should be separable by host plant. *Lycaena mariposa* is found on *Vaccinium* spp., and *L. rubidus*, *L. helloides*, and *L. nivalis* are found on docks. The larva of *L. nivalis* can be distinguished by its distinct purple dorsal stripe.

Discussion

Gravid females near the end of their flight period generally lay eggs readily on dried foliage and stems of *E. compositum* and *E. heracleoides* and probably other *Eriogonum* spp. Larvae are relatively hardy and easy to rear on *E. compositum* or *E. elatum*. *Eriogonum niveum* is a suboptimal host, apparently not enabling full development. In areas where the species is common, immature stages can be sought by searching *Eriogonum* plants. Rapid egg-hatch following overwintering indicates dormant eggs contain fully developed embryonic larvae; thus it is important to allow prediapause development to occur under ambient temperatures during late summer–fall before overwintering. We overwintered eggs successfully in Petri dishes stored over water in a domestic refrigerator or in an unheated garage.

Purplish Copper   *Lycaena helloides* (Boisduval)

Adult Biology

Lycaena helloides is common and widespread in Cascadia and the only copper found in all parts of our region. It is found throughout much of W North America and Canada, including AK. The flight period is very long, early May–early Oct, and many habitats are utilized, including backyards, disturbed areas, coasts, urban weedy fields, agricultural fields, streamsides, and open woodlands from sea level to 10,000 ft. Larval hosts include a wide variety of species in the *Polygonaceae*, including *Polygonum* (knotweeds and smartweeds) and *Rumex* (docks and sorrels); additionally *Potentilla* (cinquefoil) is sometimes utilized (Scott, 1986a; Pyle, 2002; Shapiro and Manolis, 2007) but requires confirmation. Adults nectar readily on a wide range of flowers, including asters, thistles, and clovers. Males defend low perches, flying out aggressively at intruders. Females lay eggs singly on and near their host plants, particularly on the leaves and complex flower heads of docks. In our area this species is at least double, sometimes triple brooded and continues breeding as long as the season permits.

Immature Stage Biology

We reared or partially reared this species several times. On June 11, a single spring-brood female collected near Ellensburg, Kittitas Co., WA produced 40 eggs on *R. obtusifolius* (Bitter Dock) leaves; eggs hatched in 6 days, and larvae were reared on this plant at 20–22 °C. Five larvae pupated 17 days after egg-hatch and eclosed to adults 7 days later. In early June, females from Okanogan Co., WA produced ~70 eggs that hatched in 5 days, larvae pupating after 22–24 days at 20–22 °C. In late Aug, a female from high elevation in Kittitas Co. laid ~85 eggs on *R. obtusifolius*. Most eggs were tucked deep into the seedy flower head; some were placed on leaves and stems near leaf attachments. Only 10% of the eggs hatched, and larvae were reared on *R. obtusifolius* to

adulthood in late Sep, the remaining eggs diapausing through winter. Five females obtained from Conrad Meadows, S of Rimrock Lk, Yakima Co., WA, on Aug 6 laid 40 eggs on dried foliage of *Rumex crispus* (Curly Dock). Most eggs overwintered with fully developed embryos, but ~25% hatched after 8–10 days. Overwintered eggs hatched readily when exposed to springlike conditions after 3–5 months. The Yakima Co. cohort was reared on *Rumex acetosella* (Sheep Sorrel) and hatched after 9 days at 15–21 °C under 9 hr daylengths. Development from L1 to pupation took 45 days, and adults eclosed after a further 11 days. The much faster development of spring cohorts under similar temperatures may have been due to longer (natural) daylengths. On hatching, larvae chewed a neat circular exit hole at the top of the egg but did not consume the shell. Throughout development the larvae fed on leaves, eating holes randomly and leaving only a transparent membrane. Later instars fed on leaf edges and also consumed *Rumex* seeds when available. Larvae are protectively camouflaged; they are tolerant of each other but not communal. There are 4 instars and no nests are built; ant associations have not been reported. Dornfield (1980) stated that pupae hibernate, but this is contrary to our experience and other reports of eggs overwintering (Scott, 1986a). In CO, Scott (1992) found eggs in Aug–Sep on dead basal leaves of *Rumex* and *Polygonum* spp.; in an area where seasonal inundation occurred, he found eggs on upper parts of the plants.

Description of Immature Stages

The white egg has an irregular pattern of deep polygons (often triangles) formed by intersecting surface ridges. L1 is pale olive green dorsally and yellowish ventrally. There are ~12 small dark spots per segment, and the anterior dorsal pair give rise to long black swept-back setae. L2 is pale green, lightly striped dorsally, and bears numerous long blond setae, the dorsal pairs the longest. L3 is pale green with indistinct wavy yellow dorsal lines, turning bright grass green; the paired dorsal setae are dark and contrasting. Numerous small white and dark spots give the larva a salt-and-pepper appearance. In L4 the dark spots disappear and the white spots are prominent and numerous. L4 is uniformly yellow-green, with a darker green dorsal stripe and a contrasting yellow-green, sometimes pinkish, ventrolateral line. The pupa is black, brown, or greenish brown, matching the background. Lighter pupae have darker wing cases and broad dark stripes dorsolaterally. Numerous small white spots are noticeable on black pupae. All

Adult male

Eggs @ 0.8 mm

Pupa @ 9 mm

L1 @ 1.5 mm

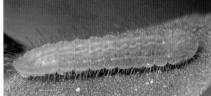

L2 @ 6 mm

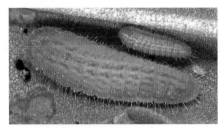

L3 @ 9 mm (larger larva)

L4 @ 17 mm

L4 @ 18 mm

Purplish Copper *Lycaena helloides* (Boisduval)

coppers have similar larvae, although in 5 species they bear a distinct purple dorsal stripe which only *L. helloides* and *L. heteronea* (Blue Copper) lack; *L. heteronea* is distinguished by its buckwheat hosts and pale mottled green pupa. Surprisingly for such a common butterfly, there are few published photographs of *L. helloides* immature stages. Guppy and Shepard (2001) illustrate the egg and mature larva of *L. helloides*, and Allen et al. (2005) show a mature larva; these images agree well with ours.

Discussion

This species oviposits readily in captivity and is easily reared. The eggs overwinter but must be allowed time for embryonic development. We noted classic "bet hedging" in mid–late-season broods, some eggs hatching to attempt one last generation while the majority hibernated. This aggressive strategy assures that the species exploits resources for as long as they are available. Elucidation of the environmental cues determining diapause or nondiapause warrants study. The impact of daylength on the rate of immature development also deserves study. Pupal coloration appears to vary considerably (see Scott, 1986a), and determination of the genetic and environmental causes would make an interesting study. Little is known of the impact of natural enemies and climate. Such knowledge may help in understanding the recent decline of this species in N CA where it is now scarce and largely confined to marshes and damp grasslands (Shapiro and Manolis, 2007). No decline appears to have occurred in Cascadia, and this pretty species can occur nearly anywhere.

Lilac-bordered Copper *Lycaena nivalis* (Boisduval)

Adult Biology

Lycaena nivalis flies from mid-May to late Aug in Cascadia, frequenting the E flanks of the Cascade Mts as well as other mid- and high-elevation areas (up to 10,000 ft) in S BC, WA, and OR. Except for an isolated population in the Olympic Mts, the species is absent near the Pacific coast. This species is seldom seen in numbers, but adults of both sexes come readily to nectar, especially at flowering buckwheats, spireas, asters, and yarrow. The territorial males guard perches on low vegetation and rocks, and females are usually seen in the vicinity of larval hosts. This species ranges from CA to BC, E to WY and CO, but there is only 1 known larval host plant, *Polygonum douglasii* (Douglas Knotweed), an inconspicuous plant of spotty distribution. When the adults are on the wing, the host plant has usually senesced and appears to be absent, so females oviposit in the vicinity where the plants grew earlier. In captivity, females oviposit readily on *P. douglasii* and *Rumex acetosella* (Sheep Sorrel), always placing the eggs on the underside of plant debris on the cage bottom. *Lycaena nivalis* is single brooded and its appearance often dependent on the timing of snowmelt.

Immature Stage Biology

We reared or partially reared this species 4 times; on 2 occasions, eggs laid on dried foliage of *R. acetosella* failed to survive overwintering at 5 °C and 50–60% RH. On another occasion a small number of eggs hatched but were lost from the host plant in L1. Seven females were collected on July 5 in Kittitas Co., WA, and caged for 15 days with cuttings of *R. acetosella* and *Rumex obtusifolius* (Bitter Dock) on

the cage bottom. About 200 eggs were produced, 85% glued to the undersides of *R. acetosella* cuttings and 15% on nearby inert surfaces; no eggs were found on *R. obtusifolius*. The eggs were held until embryos developed, then overwintered. On March 1 ~20 eggs were removed and began hatching 3 days later at 20–22 °C. Larvae left their eggshells uneaten save for the escape hole and fed on *R. obtusifolius*, eating holes halfway through leaves, the resulting holes closely matching the length and width dimensions of the larvae. The larvae developed quickly, with L2, L3, L4, and pupa reached 7, 11, 14, and 25 days post egg-hatch; the adults began eclosing 11 days later. During L1–L3 holes were eaten halfway through leaves, always in the middle of a leaf, but L4 fed from leaf edges, consuming the leaf in irregular outlines. Variation in developmental rates led to considerable staggering of adult eclosures. Populations in high-elevation habitats spend 8–10 months as dormant eggs (Newcomer, 1963), and hatching may not occur until mid–late May. The larvae were tolerant of each other in captivity, fed only on host-plant leaves, and did not build nests. Larval camouflage is excellent and they are rarely found in the wild. We have not observed ant attendance; Ballmer and Pratt (1991) showed this species to be poorly attended by the ant *Formica pilicornis*. While *R. acetosella* has not been reported as a host, the females readily oviposited on this plant, so it may be used. Larvae readily accepted *R. obtusifolius* as a host although females did not oviposit on it.

Description of Immature Stages

The egg is a white compressed sphere, the surface covered with distinct rounded perforations through the white outer wall. The micropyle is medium sized and distinct. L1 is pale orange with a bluish infusion near the head, also to a lesser extent on the posterior segments. Small dark tubercles carry long paired dark dorsal setae, 2 per segment. L2 is fuchsia pink with purplish sides and an indistinct white dorsolateral stripe. The setae are numerous, pale, and of medium length. L3 coloration is green laterally, with broad purple dorsal and ventrolateral stripes. The first segment is infused with purple, and there are numerous blond setae throughout the body. L4 is bright green with numerous small white spots. There is a contrasting reddish dorsal stripe bordered with white on each side, and a paler reddish ventrolateral stripe bordered with white below. The pupa is secured with a silk girdle, and is broadly ovate and yellowish with sooty black smudges on the wing cases. The abdomen has rows of black and white spots, and

Adult

Lilac-bordered Copper *Lycaena nivalis* (Boisduval)

Egg @ 0.8 mm

L1 @ 2 mm

L2 @ 5 mm

L3 @ 9 mm

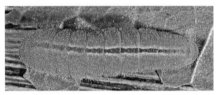

L4 @ 18 mm

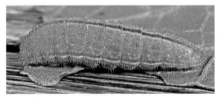

L4 @ 18 mm

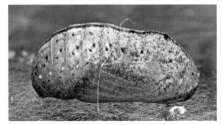

Pupa @ 10 mm

there is a strong purple and black middorsal line. Under magnification tiny trumpet-shaped setae are visible (reported earlier by Newcomer, 1963, and Scott, 1986a). The immature stages of all coppers are relatively similar; however, the larvae of *L. heteronea* and *L. helloides* lack purple dorsal stripes, and in L4 so does *L. mariposa*. *Lycaena cupreus* flies in restricted alpine habitats where *L. nivalis* is absent. The larvae of *L. editha* and *L. rubidus* are fairly similar, but the pupae of both species differ from *L. nivalis* in coloration and markings. Ballmer and Pratt (1988) show a CA L4 which is mostly golden with green infusion, a faint dorsal purple stripe, and indistinct ventrolateral yellowish stripe. Another Pratt photo appears in Guppy and Shepard (2001), this one green like ours but lacking a distinct ventrolateral line. Allen et al. (2005) show a mature CA larva, also green but with diminished dorsal and ventrolateral stripes.

Discussion

Little is known of the natural history of *L. nivalis* in Cascadia, although Newcomer (1963) provided some notes on biology. While adults are readily observed, we have not found immature stages in the wild. In areas where adults are common we have searched unsuccessfully for the food plant both before and after the flight period. Our females oviposited on *R. acetosella*, and larvae readily accepted *R. obtusifolius*. These observations suggest hosts other than *P. douglasii* may be utilized, but additional field and laboratory work is needed to clarify this. The behavior of larvae, particularly in terms of defense and feeding (nocturnal or diurnal), as well as possible intrapopulation differences in development rates, would also make interesting studies. Research is needed on the parasitoids, predators, and diseases that impact this species as well as the abiotic factors that influence distribution and abundance.

Mariposa Copper   *Lycaena mariposa* (Reakirt)

Adult Biology

The Mariposa Copper is a NW North American species ranging from the Yukon to WY, S to central CA. In Cascadia it occurs in most montane areas of BC, WA, ID, and OR excluding coastal ranges, and has been recorded up to 7,600 ft in OR (Warren, 2005). The single generation flies from early June to Sep, peaking in Aug, and is locally common. Favored habitats include bogs, wet meadows, and roadsides in forested mountains as well as a few lowland coastal bogs. Adults are frequent flower visitors, feeding on Pearly Everlasting, yarrow, asters, sedum, and spirea. Males perch on low-growing plants, darting out to intercept females and rival males. Larval hosts are various ericaceous plants, including *Vaccinium caespitosum* (Dwarf Bilberry) and likely *V. scoparium* (Grouseberry) in the mountains and *V. oxycoccos* (Wild Cranberry), *V. uligonosum* (Bog Blueberry), and maybe *Andromeda polifolia* (Bog Rosemary) in lowland bogs in W BC (Guppy and Shepard, 2001). Reports of *Polygonum* (knotweed) as a host appear to be in error. In captivity, females place their eggs on host-plant stems or leaves at the plant base.

Immature Stage Biology

We reared this species 3 times with eggs obtained from females collected from Okanogan, Yakima, and Pend Oreille Cos., WA. The Okanogan Co. females (2) were caged with *V. caespitosum* clippings, while females from Yakima (6) and Pend Oreille (8) were caged with *V. scoparium*. Within 3–4 days, 15–200 eggs were laid on old dry cuttings of the hosts. The Pend Oreille cohort was initially supplied with dried stems and leaves of *A. polifolia*, but no oviposition occurred until this potential host was replaced with *V. scoparium*. Limited oviposition was obtained by Yakima Co. females when supplied with dead foliage of a commercial heather (*Phyllodoce* sp.) but increased greatly when *V. scoparium* was provided.

No eggs were laid on fresh clippings of any host. The eggs overwinter, remaining dormant for 7–10 months under natural conditions. Eggs laid in late July, held at seasonal temperatures until mid-Sep, then transferred to 5 °C, hatched after 6 days when placed at 26 °C on Dec 7. Overwintered eggs hatched within 3 days when removed after 7–8 months. Minor feeding occurred on fresh clippings of *V. caespitosum* leaves (Okanogan Co.), but larvae died after <7 days. L1 provided with *V. caespitosum* flowers fed immediately, eating through the bases of the flowers and adjacent stem and shallowly into the undeveloped fruit ovary. Throughout development this cohort fed only on small green bilberry fruits. Several larvae disappeared, apparently victims of cannibalism, only 2 surviving to pupation, 23–28 days after egg-hatch. Adults eclosed 14 days after pupation. None of the eggs from Yakima Co. females hatched. A 95% hatch rate occurred in eggs from Pend Oreille Co. females, and L1 immediately fed on clipped *V. scoparium* leaves. Larvae were held at 25 °C under continuous lighting, pupating 31 days after egg-hatch. Although a few flower buds were offered and eaten, virtually all the larval feeding was on leaves. Pupae produced adults after 17 days under the same rearing conditions. *Vaccinium scoparium* and *V. caespitosum* are likely the preferred hosts in Cascadia but appear to differ in the acceptability of leaves. This species has 4 instars, and ant attendance has been reported elsewhere (Warren, 2005). The larvae do not construct nests at any stage, but their green coloration is very effective camouflage against the backdrop of green host-plant leaves.

Description of Immature Stages

The egg is a white, slightly compressed sphere with many close-packed round openings penetrating the surface. L1 is light purplish brown and has a shiny black head; each body segment has 4 black spots dorsally and 3 brown spots laterally, each spot bearing a single very long seta, black dorsally and blond laterally. Laterally, L2 is strongly striped longitudinally in wavy cream and brown bands; down the dorsum is a broad pink stripe and long blond setae. In Okanogan Co. L3, a bright pink dorsal stripe dominated larval appearance, while the sides were greenish with pale lateral stripes. Pend Oreille Co. L3 were bright green with a red dorsal stripe and pronounced lateral dark wavy stripes. L4 from both cohorts were bright green with a slightly darker green dorsal stripe, bordered on each side by a row of fairly short black setae. Elsewhere setae were numerous and blond, imparting a shaggy appearance. The fresh pupa is very pale green

Adult

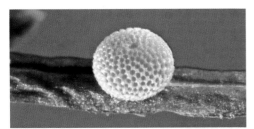

Egg @ 0.7 mm

L1 @ ~1.2 mm

Pupa @ 10 mm

L2 @ ~3 mm

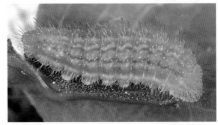

L3 @ 7 mm

L4 @ 8 mm (early)

L4 @ 13 mm

with a translucent appearance on the anterior half and is secured with triple silk girdle threads. In captivity the preferred pupation site appeared to be the upper surface of a leaf. The mature pupa is dark and has many small black spots and blotches and a dorsal brown stripe. Guppy and Shepard (2001) and Woodward (2005) published images of the mature larva and pupa very similar to ours. The mature larvae are similar to those of *L. heteronea, L. rubidus,* and *L. helloides*; however, neither these species nor any other Cascadia coppers use ericaceous larval host plants.

Discussion

In captivity it can be difficult to obtain viable eggs from females, which tend to be fragile, feed poorly, and die quickly. We tried unsuccessfully for several seasons to obtain viable eggs using incorrect hosts (*Polygonum* spp.). Once gravid females were placed with dried foliage and stems of *V. caespitosum* or *V. scoparium*, they oviposited reasonably well if provided with nectar and kept at warm (~25 °C) temperatures out of direct sunlight. Okanogan Co. larvae were apparently cannibalistic, but this was not observed among the Pend Oreille Co. cohort. It is very difficult to find larvae as the *Vaccinium* host plants can be exceedingly abundant. Also, throughout development the larvae probably stay on the flowers and fruits, most of which are hidden under the low dense canopy of leaves. Nothing is known about the biotic and abiotic factors affecting the abundance and distribution of this species.

Mariposa Copper *Lycaena mariposa* **(Reakirt)**

Family Lycaenidae
Subfamily Theclinae

HAIRSTREAKS

There are 18 species of hairstreaks in Cascadia. Hairstreaks are named for the tail-like extensions on the wings of some species; however, the majority of ours lack tails. The *Satyrium* hairstreaks are handsome and nectar lavishly on milkweeds and buckwheats. The genus *Callophrys* includes our only 3 green butterflies, while the remaining species are brown, handsomely marked with white lines and sometimes red or orange spots.

Some hairstreaks, particularly those whose larvae feed on shrubs or trees (i.e., *Habrodais*, *Satyrium*) overwinter as eggs; their caterpillars develop to pupation during spring and eclose as adults in summer. All other Cascadia species overwinter as pupae, producing spring-flying adults. Most of our hairstreaks are single brooded, but the Gray Hairstreak has multiple broods, and 2 others have limited second broods. Caterpillars of nearly all species have a honeydew gland, and about 1 in 4 of our species are known to be ant attended.

Eggs are uniform compressed spheres with fairly fine ornamentation, most white or light green. Larvae are similar to those of coppers and blues and are fairly similar in size and shape, but colors and patterns vary widely. Some species vary in color depending on the part of the host eaten; thus, Gray Hairstreak larvae feeding on flowers of a plant differ in color from those feeding on leaves of the same plant. Pupae are similar in most species, typically brown, stubby, and rounded on both ends, the abdomen swollen.

Hairstreak and elfin caterpillars use a very wide variety of hosts. Some species prefer to feed on flowers and fruits; others are leaf feeders. The Golden Hairstreak feeds on chinquapin leaves, and most *Satyrium* species use a variety of shrubs (one using lupines). The *Callophrys* hairstreaks and elfins feed on vetches, buckwheats, stonecrops, and Kinnikinnick; 2 species feed on evergreen needles, and 2 use dwarf mistletoe that grows on evergreens. The Gray Hairstreak has one of the widest ranges of hosts of any North American butterfly, and caterpillars may be found on many plants.

Golden Hairstreak   *Habrodais grunus* (Boisduval)

Adult Biology

Habrodais grunus is one of the rarest butterflies in WA, flying only where its rare host *Chrysolepis chrysophylla* (Giant Chinquapin) grows. The only known locality for *H. grunus* in WA is in Skamania Co. near the Columbia R Gorge, the N extent of its range. It is common to the S in the Cascades and coast ranges of OR, also CA and N Mexico. In S OR and CA, several species of live oak (*Quercus*) are also utilized as larval hosts. In Cascadia the single brood appears to fly mid-Aug–early Sep, although unsubstantiated reports suggest adults may occur in June in WA. Warren (2005) reports the flight is late July–late Sep for OR. Shapiro and Manolis (2007) state that *H. grunus* in N CA ecloses in late spring–early summer, then enters aestivation until Sep. Aestivating adults sit quietly in bramble thickets or in leaf litter, on cool, moist forest floors, but are active in late afternoon and even fly in twilight. In Sep they emerge from the forest to lay eggs. Marked individuals indicate an adult life span of at least 3.5 months (Shapiro and Manolis, 2007). It is likely that adults also emerge in late spring–early summer in Cascadia and aestivate during the summer, but research is needed to confirm this. Scott (1986a) reported that females oviposit on host-plant twigs; we found that eggs are laid on leaves, as did Pyle (2002). Warren (2005) reported adults spend most of the day resting in the shade until about 3:00 PM, when they become active, similar to the findings of Shapiro and Manolis (2007). Miller and Hammond (2003) reported *H. grunus* to be diurnal, but this may refer to non-aestivating populations. Scott (1986a) and Shapiro and Manolis (2007) reported that adults do not appear to visit flowers, although Pyle (2002) reported nectaring on late-season composites. Males patrol the host-tree canopies seeking females (Scott,

Adult

1986a). Overwintering eggs may be found on the undersides of Giant Chinquapin leaves, 0.5–1.5 mm from the outer leaf margin.

Immature Stage Biology

We partially reared this species twice. On June 30, we collected an L3 that pupated 20 days later but died as a pupa. On the same day we found 6 overwintered eggs on leaves; however, all were parasitized by wasps. On May 5 (different year), we found 2 overwintered eggs that hatched 9 days later (20–22 °C); however, Giant Chinquapin buds had not yet opened and the larvae refused to feed on *Quercus garryana* (Garry Oak) leaves. On the same day we also found 2 recently hatched eggs, indicating that larvae may hatch before chinquapin buds have opened. Our L3 developed fairly rapidly, feeding heavily on terminal chinquapin leaves. Larvae attain a large size for hairstreaks before prepupal shrinking and pupating. Pupation occurred on the underside of a Giant Chinquapin leaf near the stem in captivity; a wild hatched pupa was found in the same position. The larva and pupa are well camouflaged against the yellow underside of chinquapin leaves, so camouflage is probably the prime survival strategy of this species. Ballmer and Pratt (1991) showed that larvae of *H. grunus* in CA are relatively poorly attended by *Formica pilicornis* ants, but field observations on the importance of myrmecophily are lacking. The larvae do not construct nests, there are 4 instars, and the egg overwinters.

Description of Immature Stages

The pinkish white egg is a compressed sphere, heavily ornamented with numerous tiny bulbous projections; the micropyle is broad and pale blue. L1 is dull orange-brown and has a shiny black head, a shiny black sclerotized dorsal plate behind the head, and a smaller plate on the posterior segment. Small clustered black spots dorsally, laterally, and ventrolaterally give rise to rather long pale setae. L3 is yellow-green and has a shaggy appearance owing to numerous yellowish setae of varying length. There are an indistinct white ventrolateral stripe and indistinct vertical white markings on the sides, and dorsally the segments are vaulted into humps. L4 is similar in color and general appearance, but the setae are more numerous and proportionally shorter and the body has numerous small brown spots. With maturity L4 becomes a glossy grayish gold, the dorsal setae are greatly reduced, and the brown spotting retreats mostly to the 2 ends. Longer white setae persist along the ventral margin. The pupa is gray on the anterior half, the abdomen is brown dorsally and pale gold

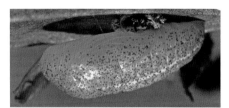

Pupa @ 12 mm

Egg @ 0.8 mm

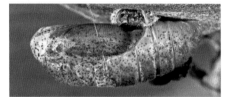

Pharate pupa @ 12 mm

L1 @ 1.6 mm

on the sides, the wing cases are brown with black tips, and there are numerous small dark brown spots over the surface. Published images of mature larvae appear in Ballmer and Pratt (1988), Miller (1995), Miller and Hammond (2003), Allen et al. (2005), and Neill (2007). These images show similar morphology but considerable variation in color; the 2 Miller images, apparently from OR, are dull gold (like ours) and green (as is the Neill image); the other images, apparently from CA, are almost white with a pale bluish cast. Neill (2007) also shows the egg. We know of no other species whose immature stages would be easily confused with *H. grunus*, and no other lycaenid is known to feed on Giant Chinquapin in Cascadia.

L3 @ 8.5 mm

Discussion

The life history of this species needs clarification, and there are many opportunities for research. Our observations are few and fragmentary, and several aspects of the life history are unclear for Cascadia populations. In particular, studies are needed to determine when adults emerge and whether they enter aestivation as they do in CA (Shapiro and Manolis, 2007). The phenology of oviposition and development of immature stages also needs study. The Giant Chinquapin hosts in WA are currently protected from cutting by the Forest Service, but many are located in dense Douglas Fir plantations; the maturing firs are crowding them and competing for sunlight, and some chinquapins are dying. Modified forest practices should be considered to clear a buffer zone around the chinquapins to allow them to survive and flourish. A tiny black wasp parasitizes many overwintering eggs and the impact of parasitoids and other natural enemies on *H. grunus* deserves study.

L4 @ 12 mm

L4@ 16 mm

Golden Hairstreak *Habrodais grunus* (Boisduval)

Coral Hairstreak *Satyrium titus* (Fabricius)

Adult Biology

Satyrium titus is relatively scarce in Cascadia, seldom common even locally, although the species can be found in the same habitat year after year. Adults are often found in the higher elevation fringes of the shrub-steppe near the lower timber line, areas where Chokecherry and milkweeds occur together, but they also occur in hotter, drier parts of the Columbia Basin. The flight period in WA extends from early June to mid-Aug. *Satyrium titus* is a fairly widespread North American species, occurring across Canada from BC to the Atlantic, and S through much of the US. Within WA, OR, and BC, *S. titus* is entirely restricted to the E side of the Cascades. The preferred larval host is usually *Prunus virginiana* (Common Chokecherry), but *Rosa woodsii* (Wood's Rose) and *Amelanchier alnifolia* (Western Serviceberry) have also been reported. Scott (1992) reported *Prunus americana* (Wild Plum) is used in CO. When we caged females with *R. woodsii* and *P. virginiana* they oviposited only on the Chokecherry. Both sexes are strongly drawn to flower nectar, particularly Showy Milkweed. Allen (1997) reported that males perch on trees or shrubs awaiting females. Eggs are oviposited in mid to late July in small clusters or in lines. They are strongly glued in small crevices, on dead or living leaves, on protected woody stems, or on ground debris near host plants. Scott (1992, 2006) observed oviposition in CO; he recorded females landing on *P. americana*, walking down the stems, and ovipositing small clusters of eggs on the ground or litter at the base of the host. Allen (1997) reported that eggs are usually deposited in the rough bark at the base of small host trees within an inch of the ground. He also reported *Prunus serotina* (Black Cherry) as the preferred host in WV. *Satyrium titus* is strictly single brooded.

Adult

Immature Stage Biology

We reared this species once from 5 gravid females collected July 16 in E Kittitas Co., WA. Twenty-six eggs were glued tightly into rivet holes in a metal rearing-cage frame. The eggs were allowed 2 weeks for embryonic development, then refrigerated at 3–5 °C to overwinter until the following spring. Nine eggs were removed on March 26, and hatching began within 4 days; 4 developed to adults. Larvae were reared on Chokecherry, and development from L1 to pupation took 23 days at 20–22 °C; adults began eclosing 22 days later. The larvae fed on young leaves, eating holes midleaf or from the edges. Scott (1986a) reported that CO larvae eat leaves and fruits at night and rest at the plant base during the day; Allen (1997) reported the same in WV, and Nielsen (1999) provided an image of 3 larvae resting at the base of a tree. Larvae tolerate crowding and often rest in close proximity; we saw no evidence of aggression or cannibalism. The larvae did not make shelters; the survival strategy is likely based on camouflage and nocturnal activity. Larvae pupated under debris on the cage bottom. Scott (1986a) reported pupation under debris at the base of the host plants. He also reported that the larvae are attended by ants, and Ballmer and Pratt (1991) reported tending by the ant *Formica pilicornis* in an experimental setup. Douglas and Douglas (2005) reported in the Great Lakes region that ants carry larvae from plant to plant as well as transport larvae from the base of the plant to the flowers and fruits at night. There are 4 instars.

Description of Immature Stages

The white eggs are covered with numerous rounded mounds. L1 is reddish brown and has a shiny black head; dorsal and ventrolateral tubercules are pale and contrasting, giving rise to long, backward-curving blond setae. Shiny black dorsal sclerotized plates occur on the anterior and posterior segments. L2 is similar but more reddish in ground color; the setae are proportionally shorter and the sides are a diffused pale greenish color. L3 is dull green with chartreuse highlights in the middle and bright wine-red anteriorly and posteriorly. Dorsally a narrow shaft of wine-red coloration partially bisects the body from both ends. Body setae are numerous, fairly short, stiff, and light brown. L4 is brighter green with wine-red markings, resulting in a unique and handsome larva. Setae are very numerous but reduced to stubble length. The spiracles are pale tan and somewhat contrasting against the green sides. The pupa is fairly dark brown with many small black spots posteriorly, coalescing into heavy black vermiculae anteriorly,

Eggs @ 1.0 mm

L1 @ 2.4 mm

Pupa @ 13 mm

L2 @ 6 mm

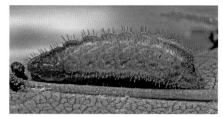

L3 @ 11 mm

and the surface is covered with numerous short blond setae. The immature stages of *S. titus* are unique and unlikely to be confused with any other Cascadian species except perhaps in early instars or the pupa. Published images of mature larvae include those in Ballmer and Pratt (1988), Allen (1997), Guppy and Shepard (2001), Wagner (2005), Minno et al. (2005), and Allen et al. (2005). While these images are from both coasts of the US, they are all very similar.

Discussion

While gravid females can be scarce, they oviposit fairly readily in captivity, eggs overwinter well, and the larvae are easily reared if fresh host plant is available. Care must be taken to allow embryonic development before overwintering the eggs. *Satyrium titus* is far less common than its larval hosts, so other factors control distribution and abundance and warrant investigation. We have searched chokecherry unsuccessfully for larvae, but as *H. titus* is nocturnally active it may be necessary to search under host plants showing feeding damage. As 3–4 weeks is spent as a pupa, mature larvae should be sought well before the flight period, probably around early–mid-May. Chokecherry plants can be large, and the role of ants in carrying larvae from the ground to flowers and fruit deserves study. Very little is known of diseases, predators, or parasitoids affecting this species. Why certain host plants are readily used in some areas but are avoided elsewhere (e.g., *Rosa woodsii*) is not known and deserves study. The bright wine-red larval coloration may be aposematic.

L4 @ 19 mm

L4 @ 19 mm

Coral Hairstreak *Satyrium titus* (Fabricius)

Behr's Hairstreak   *Satyrium behrii* (Edwards)

Adult Biology

Behr's Hairstreak is a W US species ranging from S BC to S CA and E to CO and NM. Single brooded, it flies from mid-May to mid-Aug in WA. In Cascadia it is found only E of the Cascades in shrub-steppe habitat and adjacent pine forest edges where the larval host plant *Purshia tridentata* (Bitterbrush) grows. Warren (2005) reports this species occurs up to 7,000 ft in OR, although most records are between 2,000 and 5,000 ft. Adults are sometimes common; however, populations fluctuate considerably and may be relatively scarce for several years at a time. Adults readily nectar, favoring plants like milkweed, thistles, *Ceanothus*, and *Eriogonum* (buckwheats) where they grow in close proximity to *Purshia*. After mating, typically in mid-July, females lay their eggs singly on twigs and sometimes leaves of the host plant.

Immature Stage Biology

We reared this species from eggs twice, and from half-grown field-collected larvae on a number of occasions. In mid-July we obtained gravid females in the shrub-steppe of Kittitas and Yakima Cos., WA, caging them with Bitterbrush. Within a few days >90 eggs were oviposited, most directly on *Purshia* stems, where they contrasted sharply. The eggs were not hidden or tucked into crevices, rather were placed on exposed stem surfaces. In early Aug, the eggs were transferred to overwintering conditions (4–5 °C). In mid-April, after Bitterbrush had begun sprouting, the eggs were transferred to 20–22 °C. Hatching began after 3 days, indicating that the embryos had developed the previous season prior to overwintering. L1 eclosed through a hole chewed through the top of the egg, with the remainder of the eggshell left uneaten. Initially the larvae fed only on the buds of *P. tridentata*, hollowing

Adult

out the contents completely with their extendable necks. Larvae stayed on the *Purshia* plants, wandering only if the food became stale. There was no evidence of feeding on leaves in the early instars; however, when the larvae were half-grown they switched from buds to eating leaves. The mature larvae wandered from the host plant and found cover, where they pupated 24 days after egg-hatch; the adults eclosed 16 days later. Ballmer and Pratt (1991) reported that this species is facultatively myrmecophilous, but we have not observed ant associations. There are 4 instars, and larvae do not construct nests. Larvae blend very well with the pastel green host plant, providing excellent camouflage and protection from enemies.

Description of Immature Stages

The eggs are pure white compressed spheres, sometimes with a yellow stain, and covered with a pattern of well-rounded nodes; between the nodes are irregular small perforations. L1 is cinnamon brown with a black head, each segment bearing several small black spots, each spot with a single long blond seta. L2 is bluish green, bearing many small black spots, each with a proportionately fairly short blond seta. Ventrolaterally there is a greenish white line, and the sides bear faint dark green diagonal marks. L3 is brighter green, highlighted with a single white dorsal line and diagonal white dorsal lines on each segment outlining indistinct dorsal plates. The larva is covered with numerous tiny black spots from which short bristly setae arise, and there is a prominent white line along the ventrolateral margin. L4 is dark forest green highlighted with white marks. The longitudinal white line along the ventrolateral margin is prominent, bordered above and below with thin dark green lines. Each segment has a distinct dorsolateral diagonal white line, bordered below with dark green and above with lighter green. The pupa is greenish brown with small dark brown spots and a line of small pinkish spots laterally. The larvae of *S. behrii* are distinctive, and while other lycaenids may be similar in shape, size, and even color, the patterns of white lines are unique. Several images of the immature stages have been published, including a mature larva and pupa in Guppy and Shepard (2001) and mature larvae in Miller (1995), Miller and Hammond (2003), and Allen et al. (2005); their images agree closely with ours.

Discussion

Eggs can be obtained by caging gravid females with cuttings of Bitterbrush hosts. The eggs should be left

Egg @ 1.0 mm

Hatched eggs

Pupa @ 11 mm

L1 @ 1.5 mm

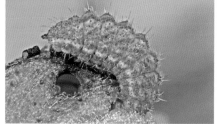

L2 @ 3.3 mm

L3 @ 7 mm

L4 @ 11 mm

L4 @ 16 mm

to mature for a couple of weeks prior to placing them in an overwintering environment as the embryos need to mature prior to hibernation. Larval survival may be poor in the early instars, perhaps because of the volatiles produced by cut *Purshia* food plants. Allowing eggs to hatch in a Petri dish without the food plant may minimize plant volatile–related mortality. It is very difficult to find larvae on *Purshia* as they blend extremely well with the plant; however, it is quite easy to shake them from the plant. For best results one should wait until *P. tridentata* is in prime bloom, then, in a locality where *S. behrii* is known to be common, shake the blossomed branches vigorously into a butterfly net or onto a sheet spread under the plant. The contents can be spread on a white sheet and searched for larvae. We have obtained larvae in this manner on many occasions, typically during the first 2 weeks of May, sometimes as late as mid-June along higher shrub-steppe hills and ridges.

Behr's Hairstreak   *Satyrium behrii* (Edwards)

Halfmoon Hairstreak   *Satyrium semiluna* Klots

Adult Biology

Satyrium semiluna, previously a subspecies of *S. fuliginosa* (Sooty Hairstreak), was recently elevated to the species level (Warren, 2005). All populations in WA and BC are included in *S. semiluna*. The range is imperfectly defined but includes WA and OR, E to WY and S to NV and CA, and probably to CO. The single brood flies from mid-May to mid-June at low elevations, to mid-Aug–early Sep in the mountains. In May–June *S. semiluna* is found in shrub-steppe localities in E WA and OR. In eastern mid-elevation areas, adults are often found on windswept ridges and hillsides. In WA, in Jul–Aug the species flies in mountain locations such as Bear Creek Mt in Yakima Co. Adults of *S. semiluna* are closely associated with their lupine host plants, *Lupinus lepidus* (Prairie Lupine) and *L. sericeus* (Silky Lupine). Females lay their eggs early in the flight period near host plants, often on inert surfaces at the plant base. Low–mid-elevation adults are strongly marked whereas high-elevation butterflies have relatively faint maculation.

Immature Stage Biology

We reared or partially reared this species 6 times: 5 cohorts from shrub-steppe populations and 1 from a high elevation (~6,000 ft). Wenas Lake, Yakima Co., WA (shrub-steppe): adults collected in mid-June, confined with cuttings of *L. sericeus,* produced 30 eggs in 5 days, mostly in small clusters tucked into crevices and cracks. Eggs are laid singly or in small clusters. On April 7, 13 eggs were removed from overwintering and placed with *L. lepidus*. One larva reached L2 10 days post hatch before dying. Another 27 eggs were removed on April 18, and 7 days later the larvae were L2, feeding lightly, preferring cut stems and stalks, and boring deeply into the cut ends; however, all L2 were dead by 20 days post hatch. Four L3 and L4 were collected from Yakima Co. shrub-steppe in mid-April at the base of *L. lepidus* plants. Two were tended by

Low-elevation adult, Yakima Co., WA

ants. The larvae fed on leaves, leaving brown spots on the upper leaf surface and a typical lycaenid hole below. They survived well on lupine leaves but later preferred leaf stems, cutting the stems from the plants and quickly defoliating them. The larvae were sensitive to light, moving under cover, suggesting that natural feeding is probably nocturnal. Three larvae pupated on May 9, producing adults on May 23. A pupa was also found in Yakima Co. under an *Astragalus purshii* plant. Although it may have simply wandered there from a nearby lupine, the possibility that this plant is a new host cannot be excluded. Bear Creek Mt (high elevation): females collected Aug 13 laid ~100 eggs on *L. lepidus* stalks or dried leaves within 5 days. After overwintering at 5° C, 24 eggs were placed in summerlike conditions (25 °C, 24 hr light) on March 29 on a commercial lupine cultivar, *Lupinus* var. Russells. Hatching occurred within 48 hrs, and L1 displayed photonegative behavior, dispersing and seeking shelter. No feeding occurred and all larvae died. Remaining eggs were used to repeat the process on April 7. L1 transferred readily to *L. lepidus*, and feeding commenced immediately on leaves and flowers. Molting to L2 occurred 2–3 days post hatch, and some L3s appeared 4 days post hatch. Feeding was intense, with larvae aggregating on unopened buds and flowers. Much excavating of stems also occurred. L4 was reached 9 days post hatch, and pupation occurred after 19–20 days. Adults eclosed 13 days later, 33 days post egg-hatch. Eggs overwinter, there are 4 instars, and no nests are constructed at any stage.

Description of Immature Stages

The egg is light green (mountain) or golden brown (shrub-steppe), ornamented with intersecting ridges. The micropyle is large, distinct, and contrasting dark green. The eggs from shrub-steppe females were covered with a thick shiny layer of transparent cement. L1 varies from bright pink (shrub-steppe) to creamy pale pink (mountain) with a shiny black head. Rows of black spots give rise to simple long setae. Mountain L1 have numerous red spots on the body. The first segment bears a shiny black dorsal shield. L2 has pink to light olive green lateral and dorsal broken stripes and 2 parallel light stripes down the dorsum. Shrub-steppe L3 is light gray to pale green with 2 diagonal white stripes per segment; the mountain L3 is similar to L2. In both forms numerous small black spots cover the body, each with a single dark seta of moderate length. Mountain L4 occurs in 2 color forms, reddish pink or green. Both have

Eggs @ 0.8 mm

L1 @ 2 mm (early)

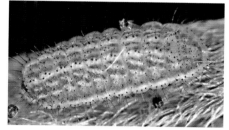

L2 @ 3 mm

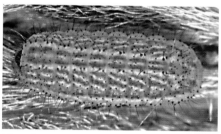

L3 @ 6 mm

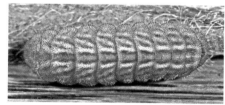

L4 @ 15 mm

Pupa @ 9 mm

white dorsal and ventrolateral markings similar to the shrub-steppe population. Shrub-steppe L4 is similar to L3. The pupa is brown and shiny and bears numerous inconspicuous stubbly blond setae; laterally there is a row of obscure reddish spots. *Satyrium semiluna* larvae differ from most other *Satyrium* spp. *Satyrium sylvinus sylvinus, S. s. nootka*, and *S. saepium* larvae are most similar but are always more green in color; both have dark pupae, and they use different host plants. Allen et al. (2005) stated that the larvae of *S. fuliginosa* (which presumably are similar to *S. semiluna*) are more easily confused with larvae of blues than other hairstreaks, and that *S. fuliginosa* cannot reliably be separated from *Plebejus icarioides, Glaucopsyche lygdamus*, and *G. piasus*. While these species, all of which use lupines, are fairly similar to *S. semiluna*, we found significant differences. Mature *S. semiluna* larvae have 2 distinct white dorsal stripes, but *P. icarioides* has no dorsal stripe, and both *Glaucopsyche* spp. have a single dark dorsal line. The larvae of *S. semiluna* have a dark dorsal shield, which is lacking in all 3 blues. The eggs of the blues are white and lack the heavy glue covering seen in some *S. semiluna* populations, and the pupae of *Glaucopsyche* spp. are darker. Guppy and Shepard (2001) illustrated L4 and pupa (as *S. fuliginosum*); their images agree closely with ours except their pupa is lighter. We know of no published images of *S. fuliginosa* larvae to compare with *S. semiluna*.

Discussion

Eggs may be obtained from gravid females by caging them with dried cuttings of their lupine hosts. Larvae can be found in April/May at low-elevation shrub-steppe localities by searching under host lupines. Early instars appear fragile but later instars are easily reared. Further research is needed to determine whether there are differences between the immature stages of *S. semiluna* and *S. fuliginosa*. Rearing of *S. semiluna* cohorts from different elevations is needed to determine whether the differences in egg coloration and in L1 and L3 coloration, and presence/absence of cement described here are consistent.

California Hairstreak *Satyrium californica* (W. H. Edwards)

Adult Biology

Satyrium californica occurs throughout the intermountain west from BC to Baja. A single generation flies in midsummer, early June–early Aug, in low and mid-elevation shrub-steppe habitats of ID, E WA, and OR along the flanks of the Cascade Mts, plus a few outlier localities. It is scarce throughout its range but sometimes common locally, often found with other *Satyrium* hairstreaks. The most commonly used host plant in Cascadia is *Purshia tridentata* (Bitterbrush); *Ceanothus velutinus* (Mountain Balm) is also used. Beyond Cascadia, *Amalanchier* (serviceberry), *Prunus virginiana* (Chokecherry), and *Salix* (willow) are reported (Pyle, 2002), and Scott (1992) adds *Cercocarpus* (mountain mahogany) in CO. Both sexes visit buckwheats, milkweeds, and mints for nectar. After mating, the females lay eggs in small clusters, averaging 3.7 eggs per cluster (Scott, 1992). The eggs are sometimes stacked like partially offset pancakes, typically in crevices, under loose bark, or in a narrow crotch on the woody parts of the host plant. Shiny, clear glue securely anchors the eggs in place and helps camouflage and protect them. Scott (1992) described oviposition: "Female crawled on and probed *P. tridentata* branches, outer branches first, and finally laid 6 eggs in 80 sec on branch under strip of loose bark 7cm above ground." In captivity, females caged with Bitterbrush, Mountain Balm, and chokecherry oviposited only on Bitterbrush or inert surfaces.

Immature Stage Biology

We partially reared *S. californica* from WA several times. Females collected July 24 (Okanogan Co., WA) and July 29 (Kittitas Co., WA) produced infertile eggs. Females collected Aug 6 at Chumstick Mt, Chelan Co., WA produced 10 eggs that overwintered and were reared on *P. tridentata* at 20–22 °C, natural daylength; 1 reached L5 but did not pupate. On April 19, May 27, and June 2 (different years), single L2, L3, and L4

were collected from *P. tridentata, C. velutinus*, and *P. tridentata*, respectively, all in Kittitas Co. The April 19 larva died, but the others were reared to adulthood at 20–22 °C on their respective hosts. Overwintered eggs appear soiled and darkened. Emerging larvae chew a small exit hole but leave the eggshells uneaten and still encased in the thick glue binding them together. Eggs hatched 5 days after transfer from overwintering to 20–22 °C, indicating that the embryonic larvae had developed prior to diapause. L1–L4 and pupa occupied 3, 3, 4, 4, and 12 days, respectively. On Bitterbrush, larvae fed on leaf buds early, switching to young terminal leaves later; L2 collected in April on *P. tridentata* was found adjacent to several leaf buds with feeding holes. The larva found on *Ceanothus* was early L3 when discovered on tight unopened flower buds in the early season; it fed strictly on flower buds throughout development, leaving them only to pupate. Larvae produced considerable frass, the timing of which indicated diurnal feeding at least in L3 and L4. The larvae do not construct nests and are cryptic, matching the buds and woody parts of the host, and clinging tightly. Protection appears to be based on camouflage and concealment. Ballmer and Pratt (1991) showed strong attendance by ants (*Formica pilicornis*) in the laboratory; however, we have not observed ant attendance. The larvae are solitary; there are 4 instars, and the egg overwinters.

Description of Immature Stages

The greenish eggs are delicately sculptured with ropy ridges intersecting to form numerous triangular shapes on the surface. L1 is medium brown with a dark brown head and an equally dark thoracic shield. Each segment bears 3 or 4 small dark spots on each side, each spot giving rise to 1–2 pale setae of moderate length. There is a rounded dark dorsal plate on the terminal abdominal segment. L2 is reddish magenta; broken cream dorsolateral stripes border the dark central area and pairs of pale setae arise from small dark spots. L3 is similarly patterned but darker. Large dark purple dorsal spots dominate the dorsum and numerous small dark bullae sport setae, black dorsally and pale peripherally. Within the pale dorsolateral stripe each segment bears ~3–4 dark setae per side, and there are vague diagonal light streaks laterally. L4 is similar except the ground color is more brownish. White diagonal stripes are well developed laterally, and the purple dorsal spots have large bluish centers. The pupa is dark brown with medium brown splotches and unkempt pale setae protruding at all angles. The larvae of *S. californica* are unique and should not be confused with other Cascadia species; the magenta

Adult female

Eggs @ 0.8 mm

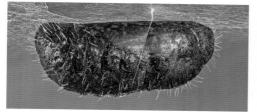

Pupa @ 11 mm

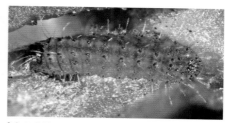

L1 @ 1.5 mm

L2 @ 2.7 mm

L3 @ 5 mm (early)

L4 @ 18 mm

California Hairstreak *Satyrium californica* (W. H. Edwards)

color and dark dorsal spots are distinctive. The same hosts are used by *S. behrii* and *S. saepium*; however, these species are green and easily differentiated. Guppy and Shepard (2001) and Allen et al. (2005) illustrate mature larvae that agree closely with ours. Miller and Hammond (2003) and Ballmer and Pratt (1988) show larvae that are generally darker and less magenta than ours. Emmel and Emmel (1973) describe mature larvae in CA as gray-brown with gray spots, and Ballmer and Pratt (1988) describe them as chocolate brown dorsally and white ventrally.

Discussion

Larvae are difficult to find because of their camouflage and general scarcity. Beating larvae from host plants such as *Purshia* can be productive for other species, but *S. californica* larvae cling so tightly that this technique is less successful. It is difficult to obtain oviposition in captivity; females that appear to be gravid are often either unfertilized or devoid of eggs, and the eggs obtained often fail to hatch. Eggs are difficult to detach from the woody substrate to which they are glued, but if left attached the wood may promote mold growth during overwintering. L1–L2 are fragile, and few survive past L3 in captivity. There is much we do not know about this species, including the full range of host plants, possible variability in larval coloration, the impact of parasitoids, predators, and ant attendance, and the reasons for periodic population swings. At one site near Ellensburg, WA, where the species was once common, a black blight has infected *P. tridentata*, and *S. californica* has disappeared locally.

Sylvan Hairstreak *Satyrium sylvinus sylvinus* (Boisduval)
(light form)

Adult Biology

Two subspecies of the Sylvan Hairstreak, one light ventrally with pale markings and the other much darker and more boldly marked, occur in Cascadia. The light form appears to be restricted to warm shrub-steppe riparian and canyon habitats in E and SE WA, while the dark form (*S. s. nootka*) occurs more widely in Cascadia and W US. The single generation of *S. s. sylvinus* flies from early June to late July and the larval host plant is primarily *Salix exigua* (Coyote Willow). Other *Salix* spp. are also likely used as hosts. *Satyrium sylvinus sylvinus* is not abundant, but small colonies occur reliably, typically in areas where Showy Milkweed grows together with stands of *S. exigua*, often along waterways or roadside ditches. Adults fly for ~2 weeks, with males and females often nectaring together on milkweed, Spreading Dogbane, and Pearly Everlasting. There is little information on oviposition behavior, but in our rearing, adults oviposited in mid-July, placing eggs in small clusters in protected areas, especially in very acute angled crotches of branchlets at diameters of 2–3 mm. The eggs were anchored with tough, water-resistant glossy glue from which the eggs are difficult to remove. Some eggs were placed on edge, wedged up against adjacent eggs.

Immature Stage Biology

We reared this subspecies once, obtaining 178 eggs from 4 females obtained near Ellensburg, Kittitas Co., WA, in mid-July. The following March and April, eggs were removed from overwintering in 3 separate cohorts of ~10–15 eggs each. First-cohort eggs began hatching March 19, 4 days after removal from overwintering. One larva survived to L4 but failed to pupate. Second-cohort eggs were removed on April 3 and placed on a *Ceanothus velutinus* (Mountain Balm) leaf as a hatching platform. L1 fed on *C. velutinus* but refused to switch to *S. exigua*; they developed slowly,

with 5 individuals reaching L4, but then developed black spots, and all but one died. One pupated after 35 days but failed to eclose. It is uncertain what caused this mortality of late instars, but it may be related to *C. velutinus* being a suboptimal or toxic host. A third cohort of eggs was removed from cold storage on April 18, and the larvae grew well on *S. exigua*, a number pupating and 6 eclosing to adults by May 24. Several larvae were initially provided *Prunus virginiana* (chokecherry) as a host, but they refused to feed and were moved to *S. exigua*, on which they fed well. L2, L3, L4, pupa, and adult stages of this cohort were reached 5, 8, 11, 20, and 32 days post egg-hatch, respectively. L1 chews an escape hole in the egg but leaves the eggshell uneaten, still covered with the tough glue that anchored and protected it during overwintering. L1 eat small oval holes through *S. exigua* leaves, leaving a thin transparent membrane on the opposite surface. By L2, the larvae ate holes through leaves randomly over the leaf surface, and often wandered off the plant. Holes were generally not at leaf edges and were usually oval or round and about the size of the larva. Larvae fed nocturnally and diurnally. Larvae confined together showed high mortality, suggesting cannibalism. The mature larvae pupated almost in unison among the *S. exigua* leaves, rather than wandering off the plant to pupate. The larvae did not construct nests, although they tethered themselves to the host plant with a single strand of silk. There are 4 instars, and the egg overwinters. The survival strategy of *S. s. sylvinus* appears to be based on camouflage in all immature stages. Ant attendance has not been observed in our area, although Ballmer and Pratt (1991) showed experimentally that CA *S. sylvinus* larvae were strongly attended by *Formica pilicornis*.

Description of Immature Stages

The eggs are dark purple with a bumpy surface and a deep, wide micropyle. Scott (1986a, 1992) described eggs of *S. sylvinus* from CO as greenish white or greenish tan, becoming pale tan-yellow, unlike ours and more similar to eggs of *Satyrium californica*. L1 is light brown, bearing numerous long white setae. The head is shiny black, and dorsally there are 2 rows of whitish spots. In L2 the body color is light green with numerous tiny dark green spots, and there are dorsolateral and ventrolateral broken white lines. In L3 the color remains light green and the white lines are distinct and continuous. Laterally the segments each bear 3 diagonal white lines and the setae remain numerous but are proportionally shorter. L4 is similar

Adult

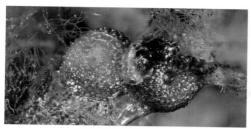

Eggs @ 0.8 mm

Pupa @ 12 mm

L1 @ 1.6 mm

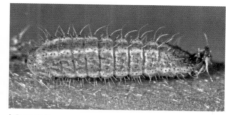

L2 @ 3.6 mm

L3 @ 7.5 mm

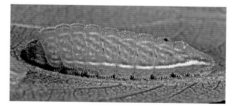

L4 @ 15 mm

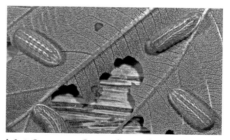

L4 @ 9 mm on *Ceanothus velutinus*

Sylvan Hairstreak (light form) *Satyrium sylvinus sylvinus* (Boisduval)

to L3, although the setae appear stubbly. The pupa is bright tan-brown and bears short, stubbly setae and a broken dark dorsal line. Emmel and Emmel (1973) and Scott (1986a) described the pupa as pale olive green or greenish brown mottled with dark green and brown. Guppy and Shepard (2001), Ballmer and Pratt (1988), and Allen et al. (2005) published images of mature larvae, all of which agree closely with ours. The closest relatives in Cascadia include other *Satyrium* hairstreaks. The larvae of *S. semiluna* are similar although they have a brownish cast on the sides and are found on lupines rather than willows. The larvae of *S. saepium* are also similar but are always found on *Ceanothus*. Other *Satyrium* spp. found beyond Cascadia may also have similar larvae.

Discussion

While we have used the name *S. s. sylvinus* to designate the light form of this species, it is believed to be an undescribed subspecies and a formal description is needed. *Salix exigua* is a very difficult host plant to examine for immature stages; the best way to obtain rearing stock is to capture gravid females and cage them with the host plant. Additional work is needed to determine if this subspecies can utilize any larval hosts other than *S. exigua*. Additional attempts at rearing this subspecies on *Ceanothus* might provide insight on plant toxicity and on its ability to adapt to new hosts. The range and impact of natural enemies affecting *S. s. sylvinus* are unknown, as is the degree of ant attendance in Cascadia.

Sylvan Hairstreak *Satyrium sylvinus nootka* M. Fisher
(dark form)

Adult Biology

Adults of *Satyrium sylvinus nootka* are very similar to *Satyrium californica* (California Hairstreak), suggesting that *S. s. nootka* might be more closely related to *S. californica* than to *S. s. sylvinus*. Adults of *S. s. nootka* are best separated from *S. californica* by minor differences in the markings on the ventral hindwing. As differences are relatively minor, comparison of immature stages may help further clarify their separation. The range of *S. s. nootka* in Cascadia includes the mountainous areas surrounding the Columbia Basin in E WA, including the Blue Mts, also the Olympic Peninsula and SW WA. It also occurs in S BC and throughout most of OR. Elsewhere, it ranges S to S CA and E to CO. Habitats include the upper edge of the shrub-steppe to elevations of 3,500 ft or more in timbered mountains, where it flies in flowery meadows and natural forest corridors and along roadsides. The larval hosts of *S. s. nootka* have not been confirmed but are almost certainly *Salix* spp. (willows). *Salix exigua* (Coyote Willow), the host plant of *S. s. sylvinus*, occurs sparsely near some of the lowest colonies of *S. s. nootka*, but is absent where this subspecies occurs in the mountains. Willows such as *Salix sitchensis* and *S. scouleriana* are likely hosts and have been used for rearing but have not been confirmed in the wild. An anomalous population of probable *S. s. nootka* flies in mid- June in the shrub-steppe at Quincy Lakes in Grant Co., WA, where it is associated with *Salix exigua,* but this population has not yet been adequately studied or reared. The single brood of *S. s. nootka* generally flies later in the season than *S. s. sylvinus*, from late June to early Sep. While *S. s. nootka* is widespread, it is never present in large numbers, and little is known of its courtship or oviposition behavior. Adults seek nectar on flowers such as buckwheats. One female was observed in apparent oviposition behavior on *Ceanothus velutinus* (Mountain Balm) in Okanogan Co., WA, but no eggs were found. Adults are sometimes found together with *S. californica* as well as with other *Satyrium* hairstreaks. This species is single brooded.

Immature Stage Biology

We reared this species once from 32 eggs laid by 2 females captured in a mountain meadow July 15 at 3,500 ft in Pend Oreille Co., WA. Ten eggs were removed from overwintering on March 26 and hatched 3–6 days later at 20–22 °C. The larvae were reared on *S. scouleriana*, which they readily accepted, 9 surviving to L4; however, apparently healthy late instars then began dying, the smallest first. While the *S. scouleriana* host may ultimately have proven toxic, it is more likely that the larval deaths were due to nocturnal cannibalism by the larger larvae. Only 1 individual survived to L4, pupated, and eclosed. L2, L3, L4, pupa, and adult stages appeared at 3, 7, 12, 26, and 40 days post egg-hatch at ~20–25 °C. Some eggs were laid singly, but most were in small clusters of 3–4 glued together with shiny, water-resistant glue. Eggs were stuffed into holes and crevices on host-plant twigs or inert surfaces and were well camouflaged. Slices that had been cut into host-plant twigs for oviposition sites were utilized, as were natural sites such as narrow crotches at the bases of leaf petioles. L1 left eggshells uneaten and fed well on *S. scouleriana* terminal leaves, each larva skeletonizing a small patch of leaf and leaving a transparent membrane. L2 rested in small congregations on the undersides of leaves. The larvae blended well with the green willow leaves, and the presence of morning frass indicated nocturnal feeding. L2 and later larvae consumed entire leaves from the sides with no remaining membranes and apparently preferred the distal halves of the leaves. Throughout development, survival appeared to be based on camouflage and concealment. There are 4 instars, the egg is the overwintering stage, and the larvae do not construct nests.

Description of Immature Stages

The eggs are dark purple with a prickly surface and numerous short conical spines at the intersections of low ropy ridges in mostly triangular lattices. The micropyle is wide, deep, and distinct. L1 is dark brown with long unpigmented setae and a black head. L2 is light greenish brown with pairs of unpigmented setae arising from brown spots dorsally and on the sides. L3 are green, the number of setae greatly increases, and distinct yellowish longitudinal stripes are found dorsolaterally and low on the sides. Each segment has 2 diagonal yellowish lines. L4 is pale green, with a

Adult

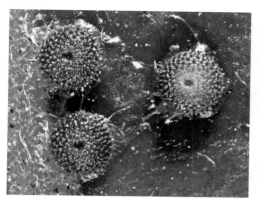

Eggs @ 0.7 mm

L1 @ 1.5 mm

L2 @ 2.8 mm

L3 @ 6 mm

L4 @ 15 mm

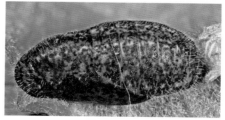

Pupa @ 8.5 mm

Sylvan Hairstreak (dark form)

Satyrium sylvinus nootka M. Fisher

broad darker green dorsal stripe bordered on each side with distinct white lines. The diagonal white side stripes and the white ventral line are obscure. The pupa is dark brown with irregular light brown splotches, a dark brown dorsal line, and numerous short, light brown setae. In all immature stages, *S. s. nootka* is similar to *S. s. sylvinus* and different from *S. californica*. *Satyrium sylvinus nootka* and *S. s. sylvinus,* based on adult and immature characters, are undoubtedly closely related and distinct from *S. californica*. Some distinctions between immature stages of *S. s. nootka* and *S. s. sylvinus* appear to exist. In *S. s. nootka*, L1 is darker and in L3 the setae are darker, but the greatest difference appears to be in the pupal stage. The pupa of *S. s. sylvinus* is light brown while in *S. s. nootka* it is darker. There are no known published photographs of the immature stages of *S. s. nootka* with which to compare our images; the mature larva pictured by Guppy and Shepard (2001) is apparently from CA, but the subspecies is not stated.

Discussion

Little is known about this subspecies and its biology. Subspeciation of *S. sylvinus* has also occurred elsewhere, for example CA (Emmel and Emmel, 1973), but little study has been done on differences in immature stage morphology and biology. A major need is to identify the natural host plants of *S. s. nootka* in Cascadia. Additional work is needed in comparison of adult collections of *S. s. nootka* to *S. s. sylvinus*; DNA analyses may prove useful in determining whether there are significant differences between these 2 subspecies. Little is known of the natural enemies of this species. The Quincy Lakes population deserves specific study to determine if it is indeed *S. s. nootka*, and to investigate why it occupies a shrub-steppe habitat, unlike other populations of this subspecies.

Hedgerow Hairstreak   *Satyrium saepium* (Boisduval)

Adult Biology

The Hedgerow Hairstreak is a common species in Cascadia, a summer resident of mountains and foothills where *Ceanothus* host plants grow. This species is found from BC S into N Mexico, and E to the Rocky Mts. In WA it flies E of the Cascades; in OR it is found in the Cascades, the coast range, and the NE mountains. The flight period of the single brood in WA is late June–mid-Sep, depending on elevation, earlier to the S and shorter in the Rocky Mts. Larval hosts include *Ceanothus velutinus* (Mountain Balm) and *C. sanguineus* (Redstem Ceanothus). Scott (1986a) lists 5 additional species, 2 of which, *Ceanothus integerrimus* (Deerbrush) and *C. cuneatus* (Common Buckbrush), occur in Cascadia and may be used. Pratt and Ballmer (1991) reared *S. saepium* to adulthood on *Lotus scoparius*, but there is no evidence that this host is used naturally. Both sexes readily visit flowers, with buckwheats commonly used. In captivity eggs are laid on the leaves of *Ceanothus*, mostly on the undersides with no preference for position. Newcomer (1973) reported eggs were laid on stems and undersides of leaves. In CO, Scott (1992, 2006) observed oviposition in July on *Ceanothus fendleri*, always on stems 1.5–3 mm in diameter. He also found an overwintered egg in April in a similar position.

Immature Stage Biology

We reared this species to adulthood from eggs obtained on July 13 from Chelan Co., WA females. Eggs were also obtained from females collected on Aug 10 in Klickitat Co., WA, with development to L3. Larvae were also found on numerous occasions from localities in Kittitas, Chelan, Okanogan, and Klickitat Cos., WA; several were reared to adults. These larvae were usually associated with *Ceanothus* leaves; however, on one occasion we found a half-grown

Adult

larva May 28 feeding on *Ceanothus* flower buds, and reared it to an adult on the same. Gravid females held at 20–26 °C confined with *C. velutinus* in plastic or aluminum screen cages laid 50–200 eggs on leaves and inert surfaces. Two weeks were allowed for the embryos to develop before the eggs were transferred to 3–5 °C for overwintering. Eggs were removed on March 26 or May 7 and began hatching after 3–4 days at 20–28 °C. Larval development was rapid on *C. velutinus* and *C. sanguineus*, with 4–5 days in each instar, pupation occurring 19 days after egg-hatch. Pupation occurred on the cage bottom beneath a cage liner. Adults began eclosing 17 days after pupation, 36 days after egg-hatch. L1 ate holes halfway through the middle of young *Ceanothus* leaves; later instars ate holes through the leaves. Scott (1986a) reported that larvae eat the upperside of leaves; however, larvae we have found have always been on undersides of *C. velutinus*, and Pyle (2002) reported that larvae feed on the undersides of *Ceanothus* leaves. Newcomer (1973) reported feeding occurred on flower buds as well as leaves. The survival strategy of *S. saepium* appears to be based on camouflage and concealment. Ballmer and Pratt (1991) showed strong attendance by ants (*Formica pilicornis*) to CA *S. saepium* larvae, but we have not observed myrmecophily in Cascadia. There are 4 instars, the larvae do not build shelters, and the egg overwinters.

Description of Immature Stages

The whitish green egg is covered with a latticework of intersecting white ridges forming numerous small oval openings. Numerous hairlike projections radiate outward from intersections of the white ridges. L1 is whitish olive with a darker dorsal stripe, pale dorsal paired tubercles, and dark circular spots laterally. Long swept-back pale setae arise from the dorsal tubercles, also along the ventrolateral margin and in a single lateral row. The head is shiny dark brown. L2 is pale green, laterally marked with diagonal white lines, 2 or 3 per segment. There is a strong dorsolateral white stripe, and another along the ventral margin. L3 is similarly patterned with a shiny black head; however, the white markings are smaller and less contrasting and the ground color is bright green. The setae are reduced to light brown stubble, and there is a light brown plate dorsally on the first segment. L4 maintains these patterns, with narrow white stripes dorsolaterally and around the ventral margin. The lateral diagonal white marks are reduced, and the green larva is densely covered in very short setae. The pupa is ovate with a posterior swelling, brown with numerous dark brown spots, especially

Egg @ 0.8 mm

Pupa @ 11 mm

L1 @ 1.2 mm

L2 @ 2.5 mm

L2 @ 4 mm

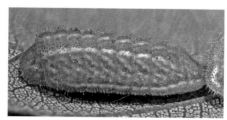

L3 @ 8 mm

L4 @ 15 mm

anteriorly. The larvae of *Satyrium sylvinus* are similar, although eggs and pupae are separable and the species does not use *Ceanothus*. Immature stages of *Glaucopsyche* and *Cupido* are similar; however, they are never found on *Ceanothus*. There are several published images of mature larvae, including Pyle (1981), Miller (1995), Miller and Hammond (2003), Allen et al. (2005), and Neill (2007). These are generally similar to ours except that diagonal side markings are less distinct. Guppy and Shepard (2005) published images of mature larvae and pupae similar to ours, although their prepupal larva is pink, a transition color we did not observe.

Discussion

Unlike most Cascadian hairstreaks, the larvae of *S. saepium* are relatively easy to find. Larvae can be recovered from *Ceanothus velutinus* by beating terminal branches onto a sheet or into a hand net, with mature larvae usually present when the flowers are in prime bloom. Alternatively, larvae can be found by hand-searching the undersides of new-growth terminal leaves. Hedgerow hairstreak larvae often occur on *Ceanothus* with *Celastrina echo* and *Callophrys augustinus* (both flower feeders), so care must be taken to correctly identify the larvae. All 3 species are similar in early instars, but distinct in L4. Very little is known of the parasitoids, predators, or diseases of *S. saepium* and their impact on population dynamics. The possibility of ant attendance also needs investigation. The favored habitat of *S. saepium* is open forested areas where *Ceanothus* grows in the understory. Increased incidence of wildfires may be impacting the habitat of this and many other species. These impacts may be positive or negative and need study.

Hedgerow Hairstreak *Satyrium saepium* (Boisduval)

Bramble Hairstreak   *Callophrys perplexa* W. Barnes & Benjamin

Adult Biology

Callophrys perplexa is 1 of 3 similar green hairstreaks found in WA and OR. All green *Callophrys* in our area that feed on *Eriogonum* (buckwheats) are *C. sheridanii* or *C. affinis*, while all *Lotus* feeders are *C. perplexa*. *Callophrys perplexa* is confined to the far W US from WA to S CA. There are 2 subspecies in Cascadia, *C. p. perplexa* and *C. p. oregonensis*, the former W of the Cascades, from Hood Canal in Mason Co., WA, S into W OR, and the latter on the E flanks of the Cascades from Yakima Co., WA, S to Deschutes Co., OR. The larval host of *C. p. perplexa* is *Lotus crassifolius* (Big Deervetch) and that of *C. p. oregonensis*, *Lotus nevadensis* ("Cowpie" Deervetch), although other species of *Lotus* may be used as well. Adults are inconspicuous, flying low near their hosts, both sexes readily visiting flowers for nectar. The preferred habitat is open heath-Salal areas, power line rights-of-way, roadsides, and clear-cuts, where males perch on Salal and huckleberry waiting for passing females and defending their territory against other males. Females lay eggs singly on unopened buds or newly opened terminal leaves. The flight period of this univoltine species in WA is mid-April–mid-June, slightly later in OR. In captivity, 1 reared female lived for 7 weeks at 17–20 °C.

Immature Stage Biology

We reared this species 3 times. On April 29 and May 3 (different years), 7 and 30 eggs were obtained from females of *C. p. perplexa* from Mason Co., WA. These were reared to 5 and 16 pupae on *L. crassifolius* and *L. pedunculatus,* respectively, and most eclosed to adults the following spring. On June 28 we obtained eggs from females of *C. p. oregonensis* from Satus Pass, Klickitat Co., WA. Larvae of *C. p. perplexa* reared at 20–22 °C developed from L1 to pupa in 23–27 days. Pupation occurred at the cage bottom under a liner. In the wild, pupation occurs in leaf litter under the host. Pupae were held for several weeks before overwintering at

3–5 °C. The pupae were removed in April, and adults eclosed 6 days later. Eggs of *C. p. oregonensis* hatched in 4 days, and larvae reared on *L. nevadensis* pupated 18 days later at 22–28 °C (mid–late July). Pupae were overwintered at 5 °C for 4.5 months and eclosed after 12 days at 18–20 °C. Larvae of both subspecies fed readily, only on leaves, stripping plants especially just prior to pupation. L1 and L2 grazed on the surface of leaves, creating pale areas; later instars fed from leaf edges. The larvae are solitary and do not construct nests. The green larvae are camouflaged well on host plants; there are 4 instars, and the pupa overwinters. This species may be ant attended but ranked fairly low in an experimental test of ant attendance in CA (Ballmer and Pratt, 1991).

Description of Immature Stages

The green egg is a compressed sphere with a white layer of raised intersecting ridges forming tiny surface polygons. L1 is light tan in *C. p. perplexa*, green in *C. p. oregonensis*, with 2 rows of small dorsolateral tubercles giving rise to 4 long pale swept-back setae per segment. L2 is covered with numerous shorter pale setae, producing a shaggy appearance. Dorsally there is indistinct pale longitudinal streaking. L2 is grayish green in *C. p. perplexa* and green in *C. p. oregonensis,* speckled with dark brown spots. In L3 *C. p. perplexa* develops from pale bluish green to green, and has a distinct white ventrolateral stripe, while in *C. p. oregonensis* it is pale green with paired broken reddish purple dorsal stripes and a bold dark reddish purple ventrolateral stripe above the white stripe. In L4 the subspecies are similar, bright green with a bold white ventrolateral stripe and a dense covering of tiny setae; however, the spiracles are dark red in *C. p. oregonensis*, pale whitish in *C. p. perplexa*. The pupa is ovate with a swollen abdomen, rusty brown with dark spots posteriorly, including 2 lines of larger spots, darker brown-black anteriorly and on the wing cases, and covered with numerous bristly short setae. The most similar species are *C. sheridanii* and *C. affinis*. The larvae are best separable in L4; *C. sheridanii* and *C. affinis* have yellow-gold highlights on the pale longitudinal stripes whereas in *C. perplexa* the lateral stripe is cold white. In *C. sheridanii* and *C. affinis*, segments have sculpted concave sides and sometimes paired rows of raised dorsal tubercles; in *C. perplexa* the sides are uniformly convex and there are no dorsal tubercles. The larvae of *C. sheridanii* and *C. affinis* are often brightly marked with red highlights, a character not seen in mature *C. perplexa* larvae, although L3 of *C. p. oregonensis* do have

Adult female (ssp. *perplexa*)

Bramble Hairstreak *Callophrys perplexa* W. Barnes & Benjamin

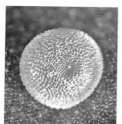

Egg @ 0.7 mm
ssp. *perplexa*

L1 @ 1.5 mm
ssp. *oregonensis*

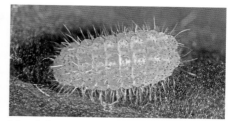

L2 @ 2.5 mm ssp. *perplexa*

L2 @ 4 mm ssp. *oregonensis*

L3 @ 7 mm ssp. *oregonensis*

Pupa @ 8 mm
ssp. *oregonensis*

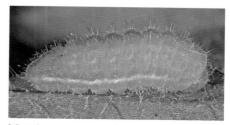

L3 @ 8 mm ssp. *perplexa*

reddish markings. The pupae of *C. sheridanii* and *C. affinis* are smooth and lack dark spotting; the pupa of *C. perplexa* is more roughly textured with heavy dark maculations. Allen et al. (2005) show an image of a "Bramble Hairstreak" and 2 of "Coastal Bramble Hairstreak," larvae from CA; however, all are attributed to *Callophrys dumetorum*, now considered a subspecies of *C. sheridanii* (Warren, 2005). The "Coastal" images appear similar to our *C. sheridanii*, while their "Bramble" image is fairly similar to our *C. perplexa*. Neill (2007) shows an image of a mature larva that agrees with ours.

Discussion

In addition to *C. perplexa*, a complex of other species, including *Glaucopsyche lygdamus*, *Strymon melinus*, *Erynnis persius*, *Epargyreus clarus*, *Thorybes pylades*, and perhaps *Cupido amyntula* use *Lotus*, so care should be taken to correctly identify field-collected eggs and larvae. Additional rearings are recommended to confirm the differences we found between *C. perplexa* and the other green hairstreaks. It would be interesting to determine whether *C. perplexa* can be reared on *Eriogonum* and if larvae would be marked differently. Obtaining eggs from gravid females is relatively easy. Other legumes should be tested for acceptance by ovipositing females and larvae.

L4 @ 15 mm ssp. *oregonensis*

Western Green Hairstreak *Callophrys affinis* (W. H. Edwards)

Adult Biology

Callophrys affinis is 1 of 3 similar green hairstreaks that occur in Cascadia. The Cascadian range of *C. affinis* includes a band along the E flanks of the Cascades in WA and BC, also in the Blue Mts, extending SW into central OR. Beyond Cascadia, *C. affinis* ranges through much of the west to CO and Baja, CA. The single generation flies from early April to mid-June. The preferred habitat includes shrub-steppe ridges at lower elevations and open lithosol slopes at higher elevations. Males perch on sticks or plants above the ground awaiting females. The similar *C. sheridanii* occurs in some of the same areas but flies earlier, with some overlap, and frequents the bottoms of canyons and gullies, usually resting on the ground or low plants. Larval hosts are buckwheats, including *Eriogonum sphaerocephalum* (Round-headed Buckwheat) and *E. elatum* (Tall Buckwheat); Pyle (2002) also reported *E. umbellatum* (Sulphur Flower) and *E. heracleoides* (Parsley Desert Buckwheat), and Scott (1986a) listed additional species in CO. Captive females lay eggs singly on leaves, but Pyle (2002) stated that oviposition is mostly on the flowers and buds of *Eriogonum*. We suspect that females oviposit on buckwheat leaves early in the season before flowers are available, and on buds and flowers later. Scott (1992) observed numerous ovipositions and reared *C. affinis* on *Ceanothus fendleri* and on *E. umbellatum* in AZ. *Callophrys affinis* is not known to use *Ceanothus* in Cascadia.

Immature Stage Biology

We reared this species twice from eggs obtained from captive females. Females collected in Kittitas Co., WA on May 25 were placed with *E. sphaerocephalum* cuttings, on which they oviposited sparingly. *Eriogonum elatum* was later provided, and substantial oviposition occurred on the leaves. Females from Umtanum Ridge, Yakima Co., WA on June 5, confined with *E. elatum* leaves and flower heads in a plastic cylinder with muslin ends (13 × 8 cm), laid 20 eggs in 4 days. Some eggs were laid on leaves, but most were placed on flowers. Eggs hatched in 5–6 days at 21–29 °C, and larvae were fed on *E. elatum* leaves at 20–22 (mild) or 23–29 (warm) °C. Development from L1 to pupation took 34 and 23 days under the mild and warm temperatures, respectively. Under warm conditions, 4, 5, 3, and 11 days were spent in L1–L4, respectively. The larvae are well camouflaged against the green host leaves. Early instars eat halfway through leaves, producing yellow, circular scars, whereas later instars eat holes through leaves and leaf edges. Flower-feeding larvae appear more likely to develop red coloration than leaf feeders, likely improving protection of larvae resting on flower heads. Pupation occurred under the cage liner in captivity; in nature it likely occurs in the ground litter under the host plant. Adults eclose in 12–23 days from overwintered pupae held at 5 °C for as little as 50 days, then exposed to 22–28 °C. Larvae are solitary and do not build nests or shelters; there are 4 instars, and the pupa overwinters.

Description of Immature Stages

The egg is pale bluish white with a broad depression and an obscure micropyle. The surface is ornamented with interconnected white ridges forming numerous small rounded polygons. L1 is pale green with obscure paired dorsal tubercles, each sporting a few long pale setae with dark bases. The dorsum and ends of the body are whitish gray. Pale dorso- and ventrolateral stripes develop during L1. L2 is yellow-green and has numerous spiny setae of varying lengths, producing a shaggy appearance. The paired dorsal tubercles are more conspicuous, each whitish in color. Between the rows of tubercles are 2 broken brown dorsal stripes. The body is covered with numerous small brown-black spots, one at the base of each seta. In L3 the longitudinal striping is more pronounced, and strong white lines are present low on each side and dorsolaterally. In some individuals there are rows of red dots associated with the white stripes. There is considerable color variation in L4. The body may be bright green or pale silvery green; the lateral and dorsolateral stripes may be white tinged in pink or white highlighted in gold, and bright red spot rows may or may not be present above the light lines. Setae are numerous and short, imparting a fuzzy appearance. The segments are indented laterally, and dorsal tubercles are present but not exaggerated. In red-marked larvae the spiracles are also red. The pupa is ovoid, posterior swollen, orange-medium brown, with darker wing cases. The abdomen has a

Adult

Eggs @ 0.5 mm

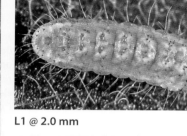

L1 @ 2.0 mm

Pupa @ 10 mm

L2 @ 4 mm

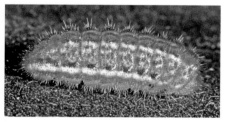

L3 @ 8 mm

L4 @ 15 mm

L4 @ 15 mm

black dorsal stripe, there are 2–4 indistinct rows of small brown spots, and numerous small bristly brown setae cover most of the pupa. The immature stages are very similar to those of *C. sheridanii* and the 2 species may not be reliably separable. Variability in both species is substantial, obscuring any consistent differences. Mature larvae of *C. perplexa* have a shaggier appearance and are green and white mostly lacking yellow and red markings. The *C. perplexa* pupa is darker and has numerous random dark spots. We know of no images of immature stages of *C. affinis* in the literature. Scott (1992) presented written descriptions of several variants of immature stages of *C. affinis* in CO, some of which sound similar to ours.

Discussion

This is an easily reared species, and the larvae withstand crowding well. Additional rearing is needed to further determine the amount of larval variability in this species. Experiments rearing larvae on the flowers or leaves of buckwheats may provide interesting insights into color variations. The possible use of *Ceanothus* as a host in our area needs investigation, as does the potential use of additional buckwheat species. *Callophrys affinis* occurs in association with *Eriogonum douglasii* (Douglas' Buckwheat) at some WA locations. The extent and impact of natural enemies affecting this species is unknown and deserves study. Captive rearing with *Ceanothus* may prove interesting, and further rearing may provide better clues for separating *C. affinis* from *C. sheridanii*.

Western Green Hairstreak   *Callophrys affinis* (W. H. Edwards)

Sheridan's Hairstreak   *Callophrys sheridanii* (W. H. Edwards)

Adult Biology

Callophrys sheridanii is common in Cascadia, occurring in low-elevation shrub-steppe to subalpine mountains. This species flies in much of WA E of the Cascades, and on both sides of the Cascades in OR. Beyond Cascadia it ranges throughout much of the W US, from S BC to central CA and NM. The single generation flies from late Feb to early Aug, with timing a function of season and elevation. Males perch on rocks and bare ground in canyon or ravine bottoms, challenging passersby in search of females. Females inspect several new buckwheat leaf sprouts before depositing a single egg on a bud or upper leaf surface. Oviposition is from mid-March to early April in low-elevation shrub-steppe habitats, to late July at high elevations. *Callophrys sheridanii* most often uses *Eriogonum compositum* and *E. elatum* as hosts but at times also *E. heracleoides* in WA. Pyle (2002) also reported *E. douglasii* and *E. umbellatum* as hosts, and Warren (2005) reported *E. nudum, E. sphaerocephalum,* and *E. ovalifolium* in OR.

Immature Stage Biology

We reared this species several times from field-collected eggs and larvae. On 2 occasions we obtained eggs from females confined with cut *E. elatum* leaves. One female obtained April 6 from Waterworks Canyon (1,700 ft), Yakima Co., WA, laid ~40 eggs in 48 hrs at 19–22 °C. Eggs hatched in 6 days at 24 °C, and larvae were reared on *E. elatum* and *E. compositum* at 25 °C. L1, L2, and L3 occupied 6 days each. L4 occupied 11 days, pupating 35 days post oviposition. Larvae from eggs found March 27 and May 9 (different years)

Adult

at Schnebley Coulee, Kittitas Co., WA (1,600 ft) developed faster, pupation occurring 28 and 31 days post oviposition, respectively. Mature larvae seek out refugia like curled, dead leaves for pupation. At a high-elevation locality (4,500 ft, Colockum Pass, Kittitas Co.), larval development occurred about 1 month later, and at still higher elevation (5,700 ft, Chumstick Mt, Chelan Co., WA), development occurred about 3 months later, with pupation on Sep 1. L1 feed on upper surfaces of new leaves, creating grooves and holes penetrating halfway through, leaving extensive round yellow spots. A significant feeding variation occurred at higher elevations where *C. sheridanii* larvae fed on *Eriogonum* flowers. Protection is based primarily on camouflage, the larvae blending well with their host plants. The larvae are solitary, have 4 instars, and do not build nests or shelters. Pupae remain dormant during the summer and winter. Hiruma et al. (1997) reported that some pupae in captivity eclosed as quickly as 30 min after exposure to warmth the following spring, while others had delayed eclosures. A sharp chill after pupation was shown to result in earlier eclosures, whereas a slow cooling at less severe temperatures resulted in prolonged eclosures the following spring.

Description of Immature Stages

The egg is pale greenish white, covered with tiny bumps. The micropyle is inconspicuous but located in a broadly concave indentation. L1 is pale yellow becoming greenish yellow, bearing long pale setae in 2 dorsal rows and around the ventral periphery. There is a broad green dorsal stripe bordered by indistinct white stripes. L2 is light green with numerous shorter setae and has pale white longitudinal stripes, 2 dorsally and 1 low on each side. L3 is densely covered with setae, imparting a shaggy appearance. The head is green and the spiracles are orange. Dorsolateral and ventrolateral stripes are yellowish and indistinct to white and bold, forming rough chevrons dorsally. L4 is green in low-elevation populations, with numerous short stubbly setae and a distinct yellow ventrolateral stripe below the spiracles, and there are 2 indistinct yellow stripes dorsally. Viewed from above, a broken central line of yellowish spots can be seen. In higher-elevation populations feeding on *Eriogonum* flowers, bright red markings characterize mature larvae, and a double row of dorsal bullae can be prominent. The brightest larva we collected was on *E. heracleoides* and at maturity was small, 11 mm compared to 16 mm in shrub-steppe populations. Warren (2005) reported that higher-elevation *C. sheridanii* on *E. heracleoides*

Egg @ 0.7 mm

L1 @ 2.2 mm

L2 @ 5 mm (late)

Pupa @ 10 mm

L3 (strong markings, early)

L3 @ 7.5 mm (low elevation, late)

L4 @ 10 mm (high elevation)

L4 @ 11 mm (high elevation)

L4 @ 16 mm (low elevation, mature)

in OR are smaller. Mature larvae become pink just before pupation. The pupa is reddish brown with numerous short setae. There is a faint dark dorsal stripe. *Callophrys perplexa* is a closely related *Lotus*-feeding species in which the mature larva has a prominent white lateral line and the pupa is darker brown, speckled with black spots. Mature larvae of *Callophrys affinis* occur in green and red-spotted forms. In the green form the dorsal stripes are broken into dashes and are brighter yellow. *Callophrys affinis* flies later than *C. sheridanii* with some overlap, preferring ridges rather than ravine bottoms. The pupa of *C. affinis* varies from orangish to dark brown and has a few blackish spots, which are lacking on *C. sheridanii*. Allen et al. (2005) published a high-elevation Yellowstone NP image of a brightly colored *C. sheridanii* larva similar to our brightest individual. Guppy and Shepard (2001) published images of eggs and L1, and Scott (1986a) pictured a green L4, all similar to our shrub-steppe form.

Discussion

Eggs can be found by searching *Eriogonum* spp. leaves shortly after they sprout in early spring, in ravine bottoms where *C. sheridanii* flies. Later, larvae can be found by searching the same plants, looking for yellow feeding marks on the leaves. Still later in the season, larvae can be found at high elevations by searching the flowers of food plants where this species flies. Further research is needed to determine why the high-elevation forms are brightly colored, and if this coloration is linked to flower feeding.

Johnson's Hairstreak   *Callophrys johnsoni* (Skinner)

Adult Biology

Callophrys johnsoni is a relatively rare species, owing to its highly specific habitat requirements. It occurs only in far NW North America, in a spotty distribution from S BC to N CA. In WA, *C. johnsoni* occurs W of the Cascades in areas where its only confirmed host plant, *Arceuthobium tsugense* (Western Dwarf Mistletoe), grows on old-growth Western Hemlock. As few old-growth forests remain in the Pacific Northwest, much of the original habitat has been lost, and *C. johnsoni* is under consideration for listing as a state endangered species. In the isolated areas where this hairstreak is still found, adults spend much of their time high in the trees where the host plant grows. Adults come to the ground occasionally to find nectar, bask, or visit mud. When disturbed, these butterflies tend to flush upward, flying high into the trees. In more S latitudes, *C. johnsoni* has been reported as double brooded, but it is likely only single brooded in Cascadia. The flight period is from mid-May to mid-June. Little is known about courtship, but gravid females captured in early June oviposited readily when caged with *Arceuthobium*. Eggs are laid singly, tucked deep into a crevice or a protective crotch of the host plant, or are placed in the concave tip of a mistletoe stalk. Sometimes multiple eggs are placed close to each other or on the hemlock immediately adjacent to the *Arceuthobium*. The stark white eggs contrast sharply with the yellow-green *Arceuthobium*.

Immature Stage Biology

We reared this species once, from eggs obtained from a female on June 10 from Mason Co., WA. The eggs hatched in ~7 days, and L2, L3, L4, and pupa were reached 8, 12, 19, and 31 days post egg-hatch, respectively. Larvae blend perfectly with the

Arceuthobium **host plant**

similarly colored host plant and cling tenaciously to it. The larvae feed on various parts of *Arceuthobium* but strongly prefer the pale blue terminal buds. The larva feeds by chewing a small round hole in the side of a bud and extending its head and long neck into the plant to hollow it out from the inside, leaving behind a distinctive and recognizable damage pattern. No larvae were observed to enter entirely into the host plant while feeding. Older larvae continue this feeding pattern, except that the final 2 instars completely consume the buds. The larvae do not construct nests but rely on superb camouflage for protection from predators. In captivity, about 4 weeks after hatching the mature larvae moved to areas of dense needles in or adjacent to *Arceuthobium*. There they pupated about a week later, anchoring themselves with a thin silk girdle thread. In captivity, 8 larvae wandered to inert surfaces (cage or paper liner) to pupate, while 14 others pupated in dense evergreen needles adjacent to *Arceuthobium*. There are 4 instars, and the pupa overwinters.

Description of Immature Stages

The white egg is a compressed sphere ornamented with numerous small divots. L1 is yellow-orange with a black head, and the body has long setae projecting backward; there are many brown and white speckles on the body. L2 is greenish with a gray head; setae are sparse and have an unkempt appearance. In L3 dorsal plates appear, yellow dorsolaterally and fringed at the front in white. The third segment is darkened dorsally, and a distinct white ventrolateral line contrasts with the yellow-green body. In L4, 2 color forms occur, green and dark brown. In both forms the dorsal plates contrast strongly with the body color, and the white dorsolateral processes on each plate are large and prominent bullae. The first and third dorsal plates are darker and more contrasting than the others, particularly the third segment, which is

Adult

Eggs @ 0.7 mm

L1 @ 2 mm

L2 @ 4 mm

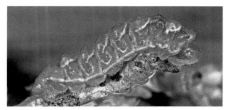

L3 @ 12 mm

L4 @ 19 mm (green form)

L4 @ 19 mm (brown form)

Pupa @ 11.5 mm

almost entirely dark brown. Each segment has a ventrolateral bulla tipped with brown. In the dark form the sides are infused with brown, and the ventral margins are fully dark brown. The pupa is dark brown, covered with numerous setae and secured by silk girdle threads. Larvae of this species are similar to *Callophrys spinetorum* (Thicket Hairstreak), the greatest differences being the usually darker first and third dorsal segments and the more prominent white edgings to dorsolateral bullae in *C. johnsoni*. Images of a late instar and pupa reared by D. McCorkle in OR appear in Guppy and Shepard (2001) and Pyle (2002); these images are similar to ours.

Discussion

Immature stages of *C. johnsoni* reside high in large trees, the larvae only occasionally found on lower branches, so the life history is best observed by obtaining eggs from gravid females and rearing larvae. Obtaining the *Arceuthobium* host and keeping it fresh is a challenge. *Arceuthobium tsugense* may be replaced with the more common *A. campylopodum*, found primarily on Ponderosa Pine; the larvae feed equally well on these dwarf mistletoes. *Arceuthobium* ages and decomposes rapidly when the host branch is cut, disintegrating into numerous small segments after 5 or 6 days. Dwarf mistletoes can be kept fresh longer by collecting sizable infested branches (e.g., ~3 ft in length) bearing evergreen needles, and keeping the stem in water in a cool place. With a good supply of fresh host plant, larvae are relatively easily reared; however, the unknown temperature and moisture requirements of overwintering pupae appear to be critical, and keeping pupae alive during winter is difficult. Captive pupae tend to eclose very early in the spring, long before *Arceuthobium* has sprouted, which could be problematic for any artificial reintroduction program. Little is known of the parasitoids, predators, or diseases of *C. johnsoni*, or of other factors controlling population densities.

Thicket Hairstreak   *Callophrys spinetorum* (Hewitson)

Adult Biology

Callophrys spinetorum is a handsome butterfly of forested areas, found from BC to Baja, and from the coastal mountains to the Rockies. In Cascadia it occurs E of the Cascades but is absent from unforested and arid areas. *Callophrys spinetorum* flies mid-April–late Aug near evergreen trees, particularly Ponderosa Pine, infested with *Arceuthobium campylopodum* (Western Dwarf Mistletoe), the larval host plant. In BC, *Arceuthobium americanum* (American Dwarf Mistletoe) is also used (Guppy and Shepard, 2001), and Scott (1986a) lists other mistletoes. In Cascadia, only 1 other species feeds on *Arceuthobium, Callophrys johnsoni* (Johnson's Hairstreak), a species whose range barely overlaps with that of *C. spinetorum*. The Thicket Hairstreak is relatively scarce in Cascadia; adults are most often seen basking, nectaring, or sipping mud along unpaved forest roads. Warren (2005) noted that adults appear to "drop out of the forest canopy"; thus they may spend much time in the treetops, contributing to their apparent scarcity. Females oviposit in June, placing eggs singly on *Arceuthobium*, often tucking them deep into crevices. In WA there is a partial second brood, and multiple broods occur in N CA (Shapiro and Manolis, 2007), but Warren (2005) reports the species is univoltine in OR.

Immature Stage Biology

We reared or partially reared this species several times from eggs produced by captive females. Two females collected from Mt Hull, Okanogan Co., WA on June 28 produced 3 eggs; 2 were reared to pupae at ~20–22 °C in 33 days. Two females obtained from Keremeos, BC on June 16 were caged with *A. campylopodum* on June 22 and laid ~200 eggs. Eggs hatched after 7 days at 22–27 °C, and larvae were reared on *A. campylopodum* from L1 to pupation in 29 days at 20 °C. Two pupae eclosed after 21 days (Aug 20),

Adult female

and a third overwintered. L1 were well camouflaged on the host; larvae are slow moving, clinging tightly to a bud or fruiting head and moving only after it has been consumed. Larvae fed on any part of *Arceuthobium* but showed a strong preference for the pale green terminal buds, which were consumed by eating a round entry hole and hollowing out the inside. The larvae did not fully enter buds while feeding, remaining on the surface and relying on very effective camouflage for protection. Prepupal larvae ceased feeding, became stationary in seclusion, and shrank prior to pupating 5–6 days later. Pupation occurred in late July–early Aug, the pupa anchored with a single silk girdle thread. In captivity, pupation occurred under cover at the cage bottom, but Pyle (2002) reported that pupae overwinter in the treetops. No nests are constructed, and the larvae are solitary, perhaps myrmecophilous (Ballmer and Pratt, 1991), with 4 instars. In one of our rearings, predatory mirid bugs (*Deraeocoris brevis*), present on mistletoe, killed many larvae. Overwintered pupae held outdoors eclosed in early April.

Description of Immature Stages

The egg is white, the surface patterned with fine rounded indentations, and the micropyle is deep and green. L1 is orange-brown, similar to the *Arceuthobium* host, has a black head, and bears fairly long pale swept-back setae dorsally and ventrolaterally. Black sclerotized plates occur dorsally on the first and last segments. L2 is similar, but the setae are proportionally shorter. Numerous brown speckles dot the body, and each segment bears 2 obscure dorsal tubercles. L3 is green speckled with tiny brown dots, punctuated with a reddish yellow ventral peripheral stripe. The paired dorsal tubercles are prominent and yellow, and on segments 1, 3, and 8 they also bear reddish brown highlights; a broad green dorsal stripe bisects the tubercles. L4 is green–dark green and bears a broad darker green dorsal stripe. The dorsal tubercles are enlarged, prominent, and mostly orange, each bearing a reddish brown spot; each is bordered with a diagonal white line anteriorly, and beyond the white line the segment is contrasting black. A ventrolateral row of spots is a combination of white, yellow, orange, and dark brown, and the head is dark brown. The pupa is ovate, enlarged posteriorly, and medium brown with black around the thorax; there are numerous black spots and the surface is covered with setae. In some individuals there is an indistinct middorsal black stripe on the abdomen. The larvae of *C. johnsoni* are very similar; however, in L4 the dorsal

Egg @ 0.5 mm

Arceuthobium campylopodum (host plant, eaten)

L1 @ 1.5 mm

L2 @ 3.7 mm

L3 @ 9 mm

L4 @ 17 mm (mature)

Pupa @ 10 mm

tubercles on the first and third segments are much darker than in most *C. spinetorum*, particularly when viewed from the side; also the diagonal white lines on the tubercles are heavier. No other Cascadian larvae are likely to be confused with *C. spinetorum* in L3 and L4; however, earlier instars are similar to those of other *Callophrys* spp. No species other than *C. spinetorum* and *C. johnsoni* are found on *Arceuthobium*. Several publications illustrate mature larvae, including Ballmer and Pratt (1988), Neill (2001, 2007), Miller and Hammond (2003), and Allen et al. (2005); Guppy and Shepard (2001) show L4 and pupa. Of these, both the images in Guppy and Shepard and Neill (probably from CA and OR, respectively) are much darker than ours, the Miller and Hammond larva is fuchsia (probably prepupal), and the Allen et al. image (from AZ) is dull orange, lacking green coloration altogether. The Ballmer and Pratt image is similar to ours.

Discussion

Rearing *Arceuthobium*-feeding species can be challenging as it is difficult to maintain a fresh supply of these plants for the larvae. The generally scarce nature of the adults renders it difficult to obtain eggs. Late instars can be found on *Arceuthobium* in mid-July; terminal buds with round feeding holes may provide a good search image, but the larvae are very well camouflaged and are difficult to detect. The timber industry endeavors to control parasitic *Arceuthobium*, and the impacts of control measures on *C. spinetorum* need study. In the few areas where *C. spinetorum* and *C. johnsoni* may occur together, it would be of interest to study any tendencies for the 2 species to hybridize, particularly since *C. johnsoni* may be a future candidate for state or federal endangered species listing.

Thicket Hairstreak *Callophrys spinetorum* (Hewitson)

Cedar Hairstreak *Callophrys gryneus plicataria* K. Johnson

Adult Biology

The cedar-feeding subspecies of *Callophrys gryneus, C. g. plicataria,* occurs in NW OR, W WA, and S Vancouver Is (Warren, 2005). *Callophrys gryneus plicataria* is single brooded in WA, flying from mid-April to early July. Adults are typically scarce, localized, and rather unpredictable, but population outbreaks occasionally occur. On May 26 we observed an "outbreak" along the upper Nooksack R (Whatcom Co., WA), when hundreds of adults were on the wing. Adults fly in open forested areas near large cedars, along roadsides, and in flowery clearings. Females lay eggs singly on new-growth needles at the branch tips of mature cedar trees; in captivity, eggs may be placed on either surface of the needles. Adults often fly to ground level for nectar, utilizing a wide variety of flowers, including dandelions, black caps, and forget-me-nots. Steep hillsides where the ground is at eye level with higher tree branches can be good locations to find them. Females are similar to males but can usually be identified by their larger abdomens, swollen with eggs. The primary larval food plant is *Thuja plicata* (Western Red Cedar).

Immature Stage Biology

We reared or partially reared this subspecies 3 times. In mid-May, eggs were obtained from Mason Co., WA females; the eggs hatched, but the larvae failed to survive the early instars. On July 1, a worn female from Pend Oreille Co., WA, laid several eggs, 1 of which produced a larva that survived to L4 before dying. Females obtained May 26 from near Nooksack Falls, Whatcom Co., WA, produced ~150 eggs, and the resulting larvae were reared at 20–24 °C on *T. plicata.* Survival was low, 6 reaching the prepupal stage and 2 pupating; our immature stage images are from this cohort. Females oviposit readily on fresh, new-growth cedar branch tips. Eggs hatch in 5–6 days, and larval development from egg-hatch to pupation occupies ~39 days. The larvae feed on new spring-growth terminal needles of cedar branches, chewing small round holes in the needles, usually near the tips, using their extendable necks to hollow out the insides. This species does not construct nests, relying on superb camouflage for protection. Pupation probably occurs high in the trees, with the overwintering pupa secured by a silk girdle. According to Ballmer and Pratt (1988), this and closely related species have 5 instars; however, we documented only 4. Larvae are solitary and possibly cannibalistic if crowded.

Description of Immature Stages

The partially flattened white–pale green eggs are covered with numerous small tubercules, contrasting with the dark green *Thuja* needles on which they are laid. In every instar the larvae bear numerous unbranched setae. The setae are very long in L1, becoming more numerous but proportionately shorter with successive instars, reduced to short bristles by L4. L1 is pale creamy amber, becoming olive green by L2 then darker green thereafter. L4 is a rich, glossy dark forest green. All instars after L1 have 4 parallel longitudinal yellow-white stripes, 2 dorsal and 1 ventrolateral on each side; these stripes become prominent but broken in L4, giving the larva a braided green and white-yellow appearance, closely resembling the "woven" pattern of *Thuja* needles. The dark brown pupa is ovate, larger at the posterior end and covered with short bristles. It is secured with a thin silk girdle. Larvae reared from widely separated populations in Mason, Whatcom, and Pend Oreille Cos., WA, showed little variation; however, larvae appear to differ from those of the sibling subspecies *C. g.* nr *chalcosiva* (Juniper Hairstreak). The white markings of the Juniper Hairstreak are larger, bolder, and more intricately patterned. *Callophrys g.* nr *chalcosiva* feeds on juniper in arid areas of E WA and OR. No other Cascadia butterfly larvae are likely to be confused with this species except perhaps in the early instars. Guppy and Shepard (2001) illustrated a mature larva and pupa of both *Mitoura rosneri* and *M. siva;* however, the localities are not stated, so it is unclear which subspecies they represent; in both cases the larvae appear intermediate between our *C. g.* nr *chalcosiva* and *C. g. plicataria.* Woodward (2005) illustrated a mature larva of *C. g. rosneri* fairly similar to ours, but the locality is not reported. Neill (2007) illustrated an adult and mature larva of "*Mitoura grynea*" that appears to be *C. g. plicataria.*

Adult

Egg @ 0.7 mm

L1 @ 1.7 mm

Pupa @ 10 mm

L2 @ 3.5 mm

L3 @ 5 mm

L4 @ 10 mm (early)

L4 @ 17 mm (mature)

Cedar Hairstreak *Callophrys gryneus plicataria* K. Johnson

Discussion

Eggs can be obtained by caging gravid females with fresh cuttings of *T. plicata,* a widespread and common plant in Cascadia. We found mortality in the early instars to be very high. Tending the early instars requires patience, as each time the food plant is changed it is necessary to search the old plant material for the well-camouflaged tiny larvae and to carefully transfer them with a moist camel-hair brush. The cedar branches are odorous when cut, perhaps a defense against insect foraging. These odorous chemicals may have contributed to the observed poor survival of early instars in our rearing. The taxonomy of *C. gryneus* has been the subject of much debate among lepidopterists; in the past as many as 5 species have been split from this taxon (see detailed discussion in Warren, 2005). One of the most persistent splits separated the cedar feeders from the juniper feeders. Our rearings suggest consistent differences in larval coloration and markings between *C. g. plicataria* and *C. g.* nr *chalcosiva*; however, Allen et al. (2005) considered cedar- and juniper-feeding larvae to be "indistinguishable." Further work, particularly on the biology and ecology of immature stages, is needed to clarify or refute the differences between *C. g. plicataria*, *C. g. rosneri*, and *C. g.* nr *chalcosiva.* Little is known of the biotic and abiotic factors controlling the distribution and abundance of this and related species.

Cedar (Juniper) Hairstreak   *Callophrys gryneus* nr *chalcosiva* Clench

Adult Biology

The juniper-feeding Cedar Hairstreaks of E OR and S central and SE WA are included in this taxon; adults are typified by their striking bright purple ventral wing coloration. The subspecies name *barryi*, widely applied to juniper feeders over the past 3 decades, was found to be used in error (Warren, 2005). Warren (2005) determined that our juniper-feeding populations most closely resemble the UT subspecies *chalcosiva*. *Callophrys g.* nr *chalcosiva* has a short flight period, usually between early April and late May in SE WA, and its appearance is strongly influenced by seasonal temperatures. In warm seasons, adults appear in early April, but in cool springs they do not fly until early May. Known larval hosts include *Juniperus occidentalis* (Western Juniper) and *J. scopulorum* (Rocky Mountain Juniper); *Calocedrus decurrens* (Incense Cedar) may be used in OR. Typical habitat is dry shrub-steppe areas with stands of juniper, where adults seek nectar from desert parsleys and buckwheats close to the junipers. Males perch conspicuously and defend territories, while females are more likely to be out of sight on juniper boughs. Eggs are laid singly, directly on juniper needles. This species is single brooded in Cascadia but double brooded farther S.

Immature Stage Biology

We reared this species twice. On April 18, several females were collected at Juniper Dunes, Franklin Co., WA. The females were caged with cuttings of *J. occidentalis* from the collection site, also with clippings of an ornamental cultivar juniper. Oviposition occurred on the cultivar or nearby inert surfaces. *Juniperus occidentalis* was avoided

Adult

by the females, perhaps because of profuse pitchy secretions. Approximately 30 eggs were laid over 10 days. Hatching commenced on April 29, and development to pupation took 30 days at 20–22 °C. One female obtained from Juniper Dunes on May 6 laid 15 eggs within 24 hrs on potted *J. occidentalis* and sprigs of *J. scopulorum*. A total of 40 eggs were laid during 2 weeks, mostly on the undersides of needles or on cage surfaces. Hatching commenced on May 14, and development to pupation took 50 days at 20–25 °C. The slower development of the second cohort is unexplained. The larvae feed entirely on juniper needles, excavating holes on upper surfaces in early instars and progressing to consumption of needles with maturity. Camouflage in this species is superb and surely the basis of the survival strategy. In captivity, the mature larvae leave the host plant and wander prior to pupation, which occurred at the bottom of the cage in folds of paper toweling. Some larvae become fuchsia pink during prepupal wandering. Larvae are solitary, and cannibalism may occur in crowded conditions. No nests are constructed. We recorded 4 instars, contrary to Ballmer and Pratt (1988), who reported 5; the pupa overwinters.

Description of Immature Stages

The egg is whitish green with irregular surface indentations and the micropyle is deep and green. L1 is orangish with slight gray infusion at both ends, becoming pale olive green. Each segment bears 4 long, pale brown, backward-curved setae dorsally, and a fringe of stiff pale setae around the ventral margin. The head is light brown. Later in the instar irregular reddish markings appear laterally and dorsally. L2 is dark olive green with an indistinct broad pale dorsolateral stripe and a strong ventrolateral pale stripe. Setae are numerous but proportionally shorter and stiff. The head is green. L3 is similar but becomes dark green with white markings, precursors to the bold markings of L4. The mature L4 has a green head and is bright, shiny, dark forest green punctuated by bold, highly contrasting white markings, thickly covered with short blond setae. There is a ventrolateral line of very bold white bars, 1 per segment, each bar enlarged anteriorly. A second bold white marking occurs dorsolaterally on each segment, shaped somewhat like the head of an open-end wrench opening posteriorly. Anterior segments also have a small white lateral spot, and there is an obscure white line along the ventral margin. These white markings are bolder and more intricate than corresponding marks on the sibling subspecies *C.*

Egg @ 0.6 mm

Pupae @ 11 mm (dark and light forms)

L1 @ 1.1 mm (early)

L2 @ 3 mm

L3 @ 9 mm (late)

L4 @ 15 mm

L4 @ 15 mm

g. plicataria. The pupa is broadly ovate, covered with numerous short bristly setae, and occurs in 2 color forms, dark "burnt" brown and black. The L4 markings of *C. g.* nr *chalcosiva* are similar to those found on juniper-feeding *C. gryneus* in CA and AZ. Miller (1995) illustrates a *C. gryneus barryi* larva (presumably *C. g.* nr *chalcosiva*); Ballmer and Pratt (1988) illustrate a mature CA larva of *C. (m.) siva*; and Allen et al. (2005, p. 63 pl. 3 only) illustrate an AZ *C. gryneus* larva, all of which are similar to our *C. g.* nr *chalcosiva*.

Discussion

In WA *C. g.* nr *chalcosiva* is listed (as *Mitoura grynea barryi*) by the Department of Fish and Wildlife as a Species of Concern and may be listed as a state endangered species. Juniper habitats in SE WA are limited, and the Juniper Dunes Wilderness is one of the few locations where this species may reliably be found. Degradation of this wilderness by livestock and motor vehicles may threaten the species, and it appears to be less common there now than formerly. Adults should be sought early in the season when they fly together with *Strymon melinus* (Gray Hairstreak). A great deal of comparative work has been done on the adults of *C. gryneus* and on analysis of blend zones where subspecies meet; nevertheless, dialogue continues as to whether this "super species" is truly a single species. Much less attention has been given to the comparative biology of the immature stages, and much work remains to be done in this regard. Little is known of the impact of natural enemies or environmental factors on the ecology and abundance of this subspecies. WA populations fly earlier than those reported in NE OR, where Warren (2005, p. 126) reported late May–late June. An investigation of these differences may prove interesting.

Cedar (Juniper) Hairstreak *Callophrys gryneus* nr *chalcosiva* Clench

Brown Elfin *Callophrys augustinus* (Westwood)

Adult Biology

The Brown Elfin flies mid-March–late July and is one of Cascadia's commonest and earliest hairstreaks. *Callophrys augustinus* is found throughout most of forested WA and OR; beyond Cascadia it occurs in most of Canada, extending down both coasts of the US. Typical habitat includes forest openings, power line rights-of-way, tree plantations, heath scrublands, chaparral, and forest roadsides. It is a generalist, using many larval hosts, including *Gaultheria shallon* (Salal), *Ceanothus* spp. (Mountain Balm, Redstem Ceanothus), *Arctostaphylos* spp. (Kinnikinnick, Manzanita), *Vaccinium* spp. (blueberry), *Arbutus menziesii* (madrone), *Pyrus malus* (apple), *Purshia tridentata* (Bitterbrush), *Berberis* spp. (Oregon Grape), *Amelanchier alnifolia* (Serviceberry), and others (Scott, 1986a). *Callophrys augustinus* is single brooded, with males flying earlier than females, staking out territories, and perching conspicuously. On May 18 in King Co., WA, we observed a female land on terminal clusters of flower buds of Salal and lay a single egg on the tip bud after ~2 min of walking and "inspecting" the raceme. Later, another female showing similar behavior failed to oviposit on a raceme with a hatched egg, suggesting its presence may have been inhibitory. Other eggs were found in similar positions on or at the base of unopened flower buds at the tips of the flower racemes. Captive females on Oregon Grape and Kinnikinnick strongly preferred the latter for oviposition.

Immature Stage Biology

We reared this species 3 times. One female obtained May 16 from Kittitas Co., WA, produced 20 eggs, which were reared on *Arctostaphylos uva-ursi* to 12

Adult

pupae. One female from Klickitat Co., WA on April 22 laid 20 eggs that were reared on apple (*Pyrus malus*) to 1 pupa, and on May 18 in King Co., 7 eggs were found and reared on *G. shallon* to 3 pupae. Larvae collected in June and July in Kittitas, Chelan, Okanogan, and Klickitat Cos., WA on *Ceanothus velutinus* (Mountain Balm) were also reared. Rate of growth was variable; the *Gaultheria* and *Pyrus* cohorts developed from L1 to pupa in 22–25 days (on *Pyrus*, 4, 3, 5, and 13 days in L1–L4, respectively), while on *Arctostaphylos*, development took 34 days, all at 20–22 °C. On *G. shallon*, larvae fed only on green, unopened flower buds by extending their heads and necks and eating a deep hole in the side of a bud. Sticky sap on the buds did not hinder them. Small larvae also entered *G. shallon* flowers to feed on the small fruits inside; 2 L4 left the host before pupating, whereas a third pupated on the plant. In nature, the larvae probably pupate and overwinter in the leaf litter near the base of the host. On *Pyrus*, all instars fed on leaves, making partial holes as L1 and L2, feeding from the edges thereafter. Buds and petals of Wallflower (*Erysimum cheiri*) were also eaten by this cohort. The larvae are well camouflaged and have 4 instars; they do not use nests or shelters, other than the hollowed-out fruits in which they partially reside at times. The larvae are solitary and can be cannibalistic in small cages. Ballmer and Pratt (1991) reported this species to be facultatively myrmecophilous, but we have not observed ant associations.

Description of Immature Stages

The egg is a light blue-green compressed sphere with a delicate white surface lattice of tiny rounded polygons. L1 is pale cream-tan, the head is shiny light brown, which persists to pupation. There are 4 rows of long pale setae dorsally and laterally. Dorsal sclerotized plates on the first and posterior segments are brown. L2 is whitish becoming pale green with indistinct white markings; numerous setae of varying length give the larva a shaggy appearance. There are very faint longitudinal green stripes on the sides, and dorsally the segments are vaulted into humps. L3 is pale green and bears an obscure pale dorsal stripe. Laterally, each segment bears a white diagonal mark above and a white splotch below and there is an indistinct ventrolateral white stripe. L4 is bright green accented with yellowish white markings, including a narrow lateral stripe and diagonal dorsolateral marks forming dorsal chevrons. The first segment has a bright fuchsia dorsal sclerotized plate, and the fourth segment may have 2 small bright red dorsolateral spots; in some individuals these spots

Egg @ 0.6 mm

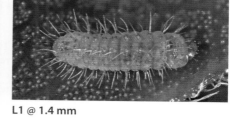

L1 @ 1.4 mm

Pupa @ 10 mm

L2 @ 2.5 mm

L3 @ 5 mm

L4 @ 13 mm

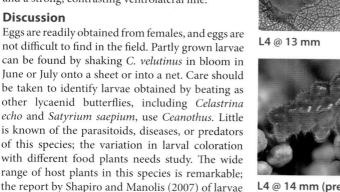

L4 @ 14 mm (prepupal)

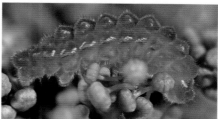

Brown Elfin *Callophrys augustinus* (Westwood)

occur on all segments. The pupa is medium brown with irregular black markings, darkening with age, covered in dense short setae. A number of Cascadian hairstreaks have green larvae with pale markings that are superficially similar to those of *C. augustinus*, but most should be separable by referring to our images. The well-developed yellow dorsal chevrons appear to be diagnostic, but variability in *Callophrys* spp. larvae is considerable. There are several published images including mature larvae in Ballmer and Pratt (1988), Allen (1997), Miller and Hammond (2003), and Allen et al. (2005). Guppy and Shepard (2001) picture an egg and L4, and Woodward (2005) shows an L4 and pupa. Most of these images are fairly similar to ours; however, the Ballmer and Pratt larva (CA) is dramatically marked with hoary dorsal tubercles and a strong, contrasting ventrolateral line.

Discussion

Eggs are readily obtained from females, and eggs are not difficult to find in the field. Partly grown larvae can be found by shaking *C. velutinus* in bloom in June or July onto a sheet or into a net. Care should be taken to identify larvae obtained by beating as other lycaenid butterflies, including *Celastrina echo* and *Satyrium saepium*, use *Ceanothus*. Little is known of the parasitoids, diseases, or predators of this species; the variation in larval coloration with different food plants needs study. The wide range of host plants in this species is remarkable; the report by Shapiro and Manolis (2007) of larvae feeding on *Cuscuta* (dodder) and our observations of feeding on *Erysimum* suggest many more hosts may be utilized.

Moss' Elfin   *Callophrys mossii* (Hy. Edwards)

Adult Biology

Callophrys mossii is a relatively scarce and localized species in Cascadia; however, within colonies it can sometimes be common. Scott (1986a) stated that adults move only 50 m during their 1-week life span. Moss' Elfin is a W North American species, occurring in WA, OR, N CA, ID, and S BC, with isolated populations in central WY and N CO. In WA and OR, *C. mossii* occurs on both sides of the Cascades in canyons, south-facing rocky slopes, and gullies below 5,000 ft with abundant *Sedum* (Warren, 2005). Both sexes visit mud, and both nectar on early spring flowers such as Spring Beauty, phlox, and Salt and Pepper Parsley. Of our 4 elfins, *C. mossii* flies the earliest, late Feb–late May, often near patches of snow and while host plants are barely emerging. Larval hosts include *Sedum lanceolatum* (Lance-leaf Stonecrop), *S. spathulifolium* (Spatula-leaf Stonecrop), *S. divergens* (Spreading Stonecrop), *S. oreganum* (Oregon Stonecrop), and *S. rosea* (King's Crown) in WA; Warren (2005) adds *S. obtusatum*, *S. laxum*, and *S. stenopetalum* for OR. Males perch on branch tips and fly out to challenge passersby. Females hide inside shrubs, venturing out to nectar or when flushed. Scott (1986a) stated that in courtship the male pursues the female; they land, flutter, and nudge slightly, and then they join. We found that captive females oviposit directly on stonecrops, laying eggs singly, often in a crotch where a leaf petiole joins a stem. *Callophrys mossii* is single brooded.

Immature Stage Biology

We reared or partially reared this species several times. Females collected in April or May during 3 different years from Reecer Canyon, Kittitas Co., WA, produced eggs that were reared to >30 pupae.

Adult

Two larvae were found at this locality on May 4, 1 reared to pupation. Eggs were obtained from a female collected at Waterworks Canyon, Yakima Co., WA, on April 20 but were infertile. Development from L1 to pupation took 25 days for an April cohort reared at 20–22 °C, and 30 days for a May cohort reared outdoors during cool spring weather. At 20–22 °C, each of the first 3 instars occupied about 5 days, and L4 took 10 days. Larvae fed well on *Sedum* leaves when flowers were not available. Early instars bored into the plant, especially at an apex, until only half of the larva was visible. Larvae fed on the tips of *Sedum*, then worked their way down, often severing a stem and allowing the tip to fall to the ground, sometimes with the larva clinging to it. Later instars are well concealed on the host plants or in soil and debris at the base of the plants. Scott (1986a) reported that L1 and L2 eat leaves, whereas older larvae prefer flowers and fruits. Warren (2005) found that larvae of early-flying populations of *C. mossii* in OR feed on *Sedum* leaves, whereas later-flying populations feed on flower heads, which are then available. Ballmer and Pratt (1991) reported larvae of *C. mossii* are facultatively myrmecophilous in CA, but we have not observed ant attendance. The larvae are solitary but seem to tolerate each other well in captivity. The larvae blend very well with host plants, so camouflage is likely an important component of their survival strategy. There are 4 instars, no nests are made, and the pupa overwinters.

Description of Immature Stages

The pinkish white egg is delicately ornamented with a pattern of numerous small oval openings. L1 is straw colored, white at both ends, and there are brown sclerotized plates dorsally at the front and rear. Small brown nodes give rise to pairs of long, pale, swept-back setae dorsally, laterally, and on the terminal segment. L2 is pale bluish laterally, tan-reddish dorsally, with numerous brown setae, dark brown nodes at their bases. L3 is pinkish orange, covered with numerous pale setae. L4 is variable, pale green with reddish tones dorsally or brightly marked with red dorsal stripes, dorsolateral chevrons, and lateral markings. Segments have strong dorsal humps, with red lateral marks edged in contrasting white posteriorly. Scott (1986a) reported that CA larvae are scarlet or pink, reddish with green tones, yellowish or greenish. Allen et al. (2005) described larvae as variable, orange, yellow, pink, or red, usually red with white chevrons. The pupa is ovoid, reddish brown, and covered with numerous short blond setae. There are 2 rows of dark brown spots on the abdomen. We suspect that larvae

Egg @ 0.6 mm

L1 @ 1.5 mm

Pupa @ 10.5 mm

L2 @ 2.7 mm

Moss' Elfin *Callophrys mossii* (Hy. Edwards)

of later-season populations feeding on flowers may be more strongly colored than earlier season leaf feeders. In the immature stages *C. polios* is most similar; however, it does not feed on *Sedum*. Published images of mature larvae include those by Guppy and Shepard (2001), a larva from OR similar to ours but with more red; Neill (2007), a mature larva with very little red coloration; and Allen et al. (2005), a dark maroon larva from CA.

Discussion

Immature stages are not easily found; however, searching stonecrops with feeding marks can produce an occasional larva. Shapiro and Manolis (2007) claim larvae are usually easier to find in N CA than adults. Eggs can be obtained from gravid females confined with cut *Sedum* in plastic vials as small as 7 × 5 cm. *Callophrys mossii* larvae have highly variable coloration, possibly related to the food plants consumed, but this is not fully understood. Rearing a cohort from a single source but on several different *Sedum* species might produce interesting results; also, feeding some larvae only leaves and others only flowers might provide insight into color variations. Why CA larvae are routinely more brightly colored than those from Cascadia is not known. Why *C. mossii* is far more restricted in its range than are its stonecrop host plants is not known and would provide a productive subject of study. The predators, parasitoids, and diseases impacting this species are little known and deserve study, as does the incidence and importance of myrmecophily.

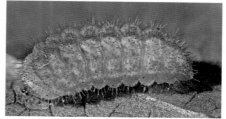

L3 @ 8 mm

L4 @ 13 mm (brightly colored)

L4 @ 15 mm

Hoary Elfin   *Callophrys polios* (Cook & Watson)

Adult Biology

Callophrys polios ranges throughout boreal North America from AK to NM and ME to WA. In Cascadia it occurs in disjunct populations in S BC, ID, N and W WA, and coastal and NE OR. It flies in a single brood from late March to early June and is a species of scrub-heath habitat, including clear-cuts, roadsides, power lines, and prairie remnants. Males fly earlier than females and establish territories to attract mates. Flight is very low and inconspicuous, and the sole known larval host is the low-growing, spreading *Arctostaphylos uva-ursi* (Kinnikinnick). Captive females laid about 90% of their eggs directly on new-growth terminal leaves, with no apparent preference for upper or lower surfaces, but showing a strong preference for leaves close to the ground. Scott (1986a) reported that CO females preferred flower stalks or leaf buds as oviposition sites. In WA, *Arctostaphylos* has typically just begun leafing out when oviposition occurs.

Immature Stage Biology

We reared *Callophrys polios* twice from females collected April 23 and 28 (different years) in Mason Co., WA and once on May 16 from females from Moses Meadows, Okanogan Co., WA. Oviposition occurred on budding leaves of *Arctostaphylos*. Eggs hatched in unison after 5–9 days, even though oviposition occurred over several days. L1 did not feed on eggshells but immediately began feeding on flowers and unopened buds, switching to new-growth leaves as soon as they were available. Larvae did not feed on old foliage. Scott (1992) also found that larvae prefer to eat young leaves. Molting to L2 occurred ~6 days after hatch, and L2 fed by "mining" a new-growth leaf, creating a translucent leaf membrane with crisscrossing vermiform feeding tracks. Some larvae also fed on leaf stems. L3 occurred ~11 days post egg-hatch, larvae changing from brown to bright green, both forms superbly camouflaged. Many L3 severed fresh leaves and fell with them to the ground, where they continued feeding. As the host is a low-growing, creeping plant, such falls should not be a threat to survival. It is unclear why leaves were severed unless it was to feed on the leaf stems, which they apparently relish. This pruning habit defoliates the host in captivity if multiple larvae are present. Albanese et al. (2007) described a host-plant girdling habit in *Callophrys irus* in MA, which may be analogous to our observations. L4 occurred ~16 days after egg-hatch and appeared to prefer feeding on new-growth stems. Mature larvae abruptly ceased feeding and wandered off the host plant, seeking shelter. About 25–29 days post egg-hatch, pupation occurred in protected locations. The pupa overwinters with the adult apparently nearly fully formed, enabling early eclosion in spring as soon as conditions permit. In captivity, the adults eclosed 3–6 days after pupae were removed from cold storage. The larvae do not build nests or shelters, and while they are not gregarious, they appear to tolerate each other well. There are 4 instars, and the pupa overwinters.

Description of Immature Stages

The egg surface is white, overlaying a grass green interior. A neat latticework of intersecting white ridges covers the egg in an irregular honeycomb; between the ridges the green interior is visible. The micropyle is large, recessed, and green. L1 is pale blond-brown with a chestnut brown head. A diamond-shaped brown thoracic shield and a similar mark on the distal end of the abdomen are present. There are numerous small brown tubercles on the body, each giving rise to 1 or 2 long blond setae. Late in L1 the larva develops wine-colored markings, solid at the posterior end but reduced to spaced streaks at the anterior end. In L2 the wine-colored markings are prominent on the posterior half, diffusing to amber on the front half. The setae are blond, more numerous, and proportionately shorter, and the thoracic shield is reduced and indistinct. L3 is bright grass green, with numerous fairly short setae originating from small pale bullae, and there is an indistinct pale yellow-green lateral stripe near the spiracles. L4 is similar to L3 except larger, darker green, and covered with a greatly increased number of short setae, and the yellow side stripe is more distinct. The pupa is dark brown with

Adult

Hoary Elfin   *Callophrys polios* (Cook & Watson)

Eggs @ 0.6 mm

L1 @ 1.4 mm

Pupa @ 10 mm

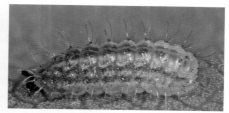

L1 @ 2.2 mm (late)

L2 @ 5 mm

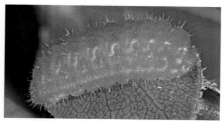

L3 @ 9 mm

L4 @ 15 mm

numerous setae; the abdomen has 2 rows of spots, 1 dark and the other pale. Scott (1986a) described the larva as having a green middorsal stripe and pale oblique dashes on the flanks; however, we found neither of these markings on our larvae. Layberry et al. (1998) illustrated a larva similar to ours but with a more prominent side stripe. Allen et al. (2005) and Wagner (2005) pictured mature larvae very similar to ours. Most similar species are other elfins in the subgenus *Callophrys* (*Incisalia*); of these, *C. augustinus* is the most similar, but larvae have double dorsal lines and pupae are darker. *Callophrys mossii* are more colorful, with distinct dorsal plates, and stonecrops are the only hosts. *Callophrys eryphon* larvae are dark green with 2 very strong white dorsal and lateral stripes.

Discussion

Callophrys polios is relatively easy to rear once eggs have been obtained from caged females. We found it difficult to find eggs or larvae on the low, spreading food plant; it was much easier to obtain gravid females and cage them with Kinnikinnick whereupon they oviposited readily. In the Pacific Northwest many nurseries stock the host plant, so it is readily available for rearing. Research should be conducted to determine whether the leaf-cutting behavior occurs in wild populations and, if so, to determine its function and benefit. One possibility is that the stem feeding observed after leaf cutting confers nutritional or defense benefits to the larvae. Little is known of the biotic and abiotic factors limiting the abundance and distribution of this species.

Western Pine Elfin *Callophrys eryphon* (Boisduval)

Adult Biology

The Western Pine Elfin is common in Cascadia, flying in a single brood mid-March–late July depending on elevation and seasonal conditions. *Callophrys eryphon* is found in most forested areas, but populations are sparse W of the Cascades. The greater range includes the S half of Canada and much of the American west. Larval hosts in Cascadia include *Pinus ponderosa* (Yellow Pine) and *Pinus contorta* (Lodgepole Pine); Pyle (2002) thought *Pinus monticola* (White Pine) was likely also used. Scott (1986a) and Austin and Leary (2008) listed 13 additional host trees found beyond Cascadia, including *Pinus banksiana, P. resinosa, P. rigida, P. virginiana, P. strobus, P. sylvestris, P. echinata, P. jeffreyi, P. lambertiana, P washoensis,* and *P. taeda* as well as *Picea mariana* (Black Spruce) and *Larix laricina* (Larch); according to Allen et al. (2005), the species is "also reported on firs." Pratt and Ballmer (1991) surprisingly reared *C. eryphon* to adulthood on the legume *Lotus scoparius* (Deervetch), but this is unlikely to serve as a natural host. The preferred habitat is open forest where pine trees, particularly *P. ponderosa*, grow near meadow openings with blooming nectar flowers. Many flowers are visited, including yarrows, desert parsleys, dandelions, and willow catkins. Males emerge 1–2 weeks before females, staking out perches near pine trees and returning repeatedly to the same spot; they usually perch <2 m above ground level (Warren, 2005). Females spend much of their time flying between nectar sources or perching discreetly on the host plant. Eggs are laid singly on the soft new-growth branch tips of the host trees. The eggs are typically tucked deep into a crevice close at the base of a slanted pine needle, on the protected underside.

Adult male

Immature Stage Biology

We reared this species 5 times. Two females obtained May 9 near Cle Elum, Kittitas Co., WA, produced 20 eggs on cuttings of *P. ponderosa*. Larvae were reared on *P. ponderosa* with 1 surviving to pupation. Females collected in Klickitat Co., WA, April 24 laid ~25 eggs on *P. ponderosa* and were reared to L2 before dying. Two eggs were collected from *P. contorta* on May 30 near Hood Canal, Mason Co., WA; larvae were reared on *P. contorta* and 1 pupated. On May 16 a female from Moses Meadows, Okanogan Co., WA, laid 5 eggs on Lodgepole Pine; the larvae were reared on this host and 1 survived to pupa. Four females from near the Tucannon R, Columbia Co., WA laid ~150 eggs over 5 days on *P. ponderosa*, with 27 reared to pupation. All pupae were overwintered at 3–5 °C for at least 114 days, and adults eclosed within 7–10 days of exposure to 18–21 °C. Eggs hatched after 5–10 days at 20–24 °C; development from L1 to pupation took 33–39 days at 20–22 °C, with each instar occupying ~7–8 days, except L4, which took 12 days. Faster development (23 days) occurred at 25–27 °C. Newcomer (1973) reported larval development took only 2 weeks when fed solely on catkins of *P. ponderosa*. The cryptic L1 feed on new-growth needles at the branch tips, mining the surface initially and consuming the needles later. The tiny larvae are difficult to find in crevices of new-growth needles. They do not wander, staying on the new growth, refusing to feed on older needles, and growing slowly. Yellow frass is produced in considerable quantities, providing a good clue to the whereabouts of larvae. When cut, pine branches and needles exude sticky pitch that can engulf and kill eggs and early instars. As the larvae feed, they cut the new-growth needles near the base, dropping them to the cage floor. The prepupal larva ceases feeding, shrinks, appears moribund, then pupates normally after 2–4 days either on the host or off (under cage liner). Pupae are not securely attached and are easily dislodged. The larvae are colored and striped identically to the host needles, producing superb camouflage. The larvae are solitary and do not make nests. There are 4 instars and the pupa overwinters. We have not observed ant associations.

Description of Immature Stages

The egg is greenish white and has a beaded surface. L1 is greenish cream darkening to brown. There are 2 widely separated dorsal rows of nodes, each producing a pair of long blond setae. Early L2 is cinnamon brown with longitudinal bold white stripes dorsolaterally and below the spiracles. Mature L2 is

Egg @ 0.8 mm

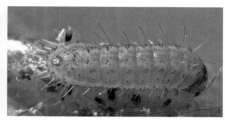

L1 @ 2 mm

Pupa @ 10 mm

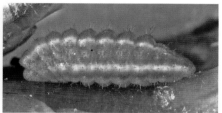

L2 @ 6 mm

L3 @ 9 mm

L4 @ 14 mm

L4 @ 15 mm

Western Pine Elfin *Callophrys eryphon* (Boisduval)

forest green, with setae and white stripes. L3 is darker green and the white stripes are narrower and bolder. L4 darkens again and the white dorsal and lateral stripes are prominent and contrasting. The pupa is ovate, enlarged at the posterior end and dark brown, covered with numerous short setae. There are several Cascadian lycaenids that have green larvae with white stripes. The most similar include *Callophrys gryneus* (both subspecies), *C. perplexa, Satyrium sylvinus* (both subspecies) and *S. saepium*. All can be separated by referring to our images. No other lycaenid in our area uses pines as hosts (although *C. gryneus* uses cedars and junipers). Images of mature larvae are found in Ballmer and Pratt (1988), Layberry et al. (1998), Guppy and Shepard (2001), and Allen et al. (2005). All are similar to ours, although the white stripes are tinted with yellow in all except Allen et al. (2005).

Discussion

Eggs and larvae can be found by searching small pine trees in areas frequented by adults by gently bending a new-growth branch tip and rotating it between the fingers so that the new needles spread outward and rotate, exposing all surfaces. Yellow frass provides a good clue to larval presence. Pitch on cut branches is troublesome during rearing; we avoided this by bending a single new-growth pine tip into a Petri dish, with the stem wrapped in moist paper towel, and rearing the larvae there. Alternatively, larvae may be reared on pine catkins (Newcomer, 1973) or *L. scoparius* and perhaps other *Lotus* spp. (Pratt and Ballmer, 1991). Little is known of parasitoids, predators, or diseases in this species; thus research is needed.

Gray Hairstreak *Strymon melinus* Hubner

Adult Biology

Strymon melinus is one of our commonest hairstreaks, found throughout the US and S Canada. It occurs everywhere in Cascadia except coastal WA. The flight period is very long, late March–early Oct. An extraordinary diversity of larval hosts are used, including *Trifolium* (clover), *Sedum* (stonecrop), *Quercus* (oak), *Fragaria* (strawberry), *Humulus* (hops), *Polygonum* (knotweed), *Rumex* (dock), *Lotus* (deervetch/trefoil), *Vicia* (vetch), *Salsola* (Russian Thistle), *Eriogonum* (buckwheat), rosaceous plants, corn, beans, and mallows. Scott (1986a) lists many others. *Strymon melinus* occurs in most habitats, from coastal scrub-heath to E shrub-steppe as well as foothill, mountain, and subalpine areas, avoiding only thick forests; however, in spite of its amazing adaptation, *S. melinus* is typically found in small numbers, abundant only in low-elevation shrub-steppe riparian habitats along major rivers. Males perch and wait for females, mating soon after females eclose. Multiple ovipositions were observed on April 20 along the Snake R, Whitman Co., WA, where pristine females flitted among *Vicia* plants in new leaf but not yet blooming. Eggs were laid singly on unopened coiled terminal flower buds or on nearby leaflets. This species is multibrooded, reproducing as host plants and weather permit, with 3 broods likely in Cascadia.

Immature Stage Biology

We collected and reared larvae of *S. melinus* many times. We also reared it from field-collected eggs twice, and twice from gravid females obtained March 23 in Benton Co., WA and April 10 from Franklin Co., WA. Mid-instar larvae were collected Sep 5 at mid-elevation (3,000 ft, Chelan Co., WA). Oviposition was obtained on commercial cultivars of lupine and *Clematis* sp. as well as *Amsinckia lycopsoides* (Tarweed Fiddleneck). Eggs hatch after ~3 days, and molts to L2, L3, and L4 occurred at 7, 13, and 24 days post oviposition, respectively, at 20–25 °C and natural daylengths. Pupation occurred after 27 days, and adults began eclosing 37 days post oviposition. Eggs are laid on flowers or leaves, well camouflaged or hidden in flower parts. Eggshells are uneaten except for an escape hole, and the larvae immediately seek flowers to feed on if they are present. If flowers are not available, the larvae will feed on leaves. Feeding on flowers is the primary behavior (Scott, 1986a; Guppy and Shepard, 2001; Allen et al., 2005; Pyle, 2002), but the ability to utilize leaves permits exploitation of multiple hosts and habitats. When flowers were withdrawn from larvae, they switched easily to leaves. Leaf-fed larvae were green throughout development, whereas those fed lupine flowers and leaves were tan with white markings, becoming pink and white at maturity. Camouflage provided by adopting the color of host plants appears to be an important survival strategy for *S. melinus*. Allen (1997) reported ant associations, and Ballmer and Pratt (1991) ranked it 10th among 59 lycaenid species for ant attendance. Pupation occurs in a sheltered location like a curled leaf. This species has 4 instars and is solitary but tolerates siblings. The larvae make no nests, and the pupa overwinters.

Description of Immature Stages

The egg is pale green with a fine latticework of polygonal cells. The micropyle depression is darker green, broad, and fairly deep. L1 is yellow-cream punctuated with small brown spots giving rise to long unpigmented setae in 2 dorsal rows and around the ventral periphery. The head is cinnamon brown with dark eyespots, and there is a light brown thoracic shield and a round dorsal abdominal shield. L2 is covered with setae and has 2 dorsal dark lines, and broad dorsal plates are present on each segment. All 3 characters persist until pupation. In L3 each segment has a dark lateral diagonal mark bordered in white above, and there is a white ventrolateral line below the spiracles. L4 is similar but with the characters more pronounced and with more numerous and shorter setae. The pupa is distinct, light tan to dark "burnt" brown, covered with short setae and bearing dark markings in a predictable pattern not found in other

Adult

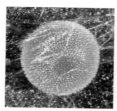

Egg @ 0.7 mm

L1 @ 1.2 mm

Pupa @ 11 mm

L2 @ 3 mm

lycaenids. There is a black patch on the central dorsal line where the abdomen and thorax join and a black triangular mark on each side touching the wing case. Larval coloration is extraordinarily variable. We show a few colors, including tan, gray, and green, but larvae may also be yellow, orange, pink, purple, or olive. Because of this variability and the numerous host plants used, *S. melinus* larvae may be confused with other lycaenids, but close attention to a few characters should eliminate most confusion. Of importance is the host plant, whether leaves, flowers, or pods/seeds are used, the patterns of setae, and the double dorsal line. Few other lycaenids attain the size of a mature *S. melinus* larva, up to 17 mm. There are many published images of *S. melinus* immature stages, with larvae in Tveten and Tveten (1996), Miller and Hammond (2003), and Wagner (2005), and larvae and pupae in Allen (1997) and Guppy and Shepard (2001). Accounting for normal color variability, all images agree well with ours.

L3 @ 6 mm

Discussion

Partly grown larvae can be found by beating host-plant flowers such as *Eriogonum elatum* into a hand net in July or August, or hand-searching flowers of known hosts. Obtaining eggs from females is sometimes difficult as females may be "choosy" concerning oviposition host; however, providing a "smorgasbord" of potential hosts will often result in eggs on at least 1 host. Observing female behavior may reveal the preferred host plant at that locality. The occasional impact of *S. melinus* on commercial beans has led to some studies, but we still know little of its biology and ecology. The abundance of and relative ease in rearing this species makes it well suited for studies on biology in the laboratory and ecology in the field (e.g., Alcock and O'Neill, 1986, 1987).

L3 @ 10 mm

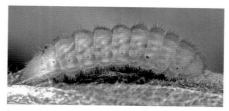

L4 @ 14 mm

L4 @ 17 mm

Gray Hairstreak *Strymon melinus* Hubner

Family Lycaenidae
Subfamilies Polyommatinae and Riodininae
BLUES and *METALMARK*

There are 19 species of blues in Cascadia. Males are typically blue dorsally, whereas females are usually brown. About half our blues fly in a single annual generation whereas the rest have 2 broods, at least on occasion. The blues have adapted to many habitats, from hot prairies to alpine summits and most environments in between. Blues are often abundant and can be seen in large mixed-species mudding parties that may number hundreds at a single site.

Blues lay finely ornamented, flattened, sphere-shaped eggs, usually on flower buds or new growth of hosts. Larvae are similar to coppers and hairstreaks, generally sluglike in appearance, cryptic in green, pink, or gray shades, often with dorsal or lateral stripes and diagonal side markings. There are usually 4 instars. The pupae are smooth, with the abdomen larger than the thorax and a silk girdle thread around the middle.

The closely related Anna-Melissa-Northern Blue group overwinters as eggs, while the other *Plebejus* spp. (Greenish, Boisduval's, Acmon, and Lupine blues) spend the winter as half-grown larvae. The tailed blues overwinter as mature larvae, and the remaining 10 species overwinter as pupae. With these varied life histories, blues can be found in all seasons from early spring to late fall.

Most blue larvae have a "honeydew gland" on the seventh abdominal segment. About half our species are myrmecophilous (ant attended). Blues use a wide variety of larval host plants, including vetches, peas, lupines, buckwheats, saxifrages, dogwoods, and the flowers of various other hosts.

Metalmarks (subfamily Riodininae) are sometimes grouped with the Lycaenidae, and we include our single species here. The Mormon Metalmark overwinters as an egg and grows very slowly on buckwheat hosts, reaching adulthood in late summer.

BLUES and METALMARK

Western Pygmy Blue *Brephidium exilis* (Boisduval)

Adult Biology

Brephidium exilis, arguably the smallest butterfly in the US, is a year-round resident of arid regions in S states. Populations expand N in spring and summer, reaching S OR most years. It occurs sporadically in SE ID and E OR to the SE edge of the Blue Mts (Pyle, 2002). Warren (2005) stated that *B. exilis* may be abundant some years in the SE deserts of OR from mid-July to Sep. It was recorded for the first time in WA on Sep 17, 2004 by Robert M. Pyle at the Umatilla NWR along the Columbia R. *Brephidium exilis* reproduces quickly in S US with many multiple broods (potentially 8–9 annually); a warming climate may increase the chances of this species occurring and breeding farther N. Preferred habitats are disturbed weedy areas, roadsides, and gullies, also alkali flats and salt marshes. Adults are inconspicuous in flight; males fly endlessly searching for females, which often remain in the interior of host plants, ovipositing or sipping nectar. Adults may roost communally on dead or leafless shrubs, with up to 8 or 9 in close proximity. Host plants are chenopods (goosefoots), including *Atriplex* spp. (saltbush), *Chenopodium* spp. (pigweed), *Salicornia* spp. (pickleweed), *Salsola tragus* (Russian Thistle), *Suaeda* (seepweed), *Sesuvium* (sea purslane), and *Halogeton glomeratus* (Halogeton) (Scott, 1986a; Shapiro and Manolis, 2007). Pratt and Ballmer (1992) showed that *B. exilis* can be reared on *Lotus scoparius*. Eggs are laid singly on all parts of the host plant, particularly undersides of leaves (Scott, 1986a).

Immature Stage Biology

We reared this species twice from gravid females. Seven females collected Nov 7 at Indio, S CA, laid ~90 eggs in hot sun on *Atriplex polycarpa* (Cattle Saltbush) and cage surfaces over 2 days. Eggs hatched after 3–4 days at ~20–25 °C, and L1 bored into *A. polycarpa* seeds, where they fed, producing frass around entry holes. Feeding also occurred on leaves, hollowing them out from the upper surface, but none of the leaf feeders survived. Mortality was high, with most dying in L1. L2 also fed inside seeds, but later instars fed from the exterior. The pupa was anchored by a thin silk girdle to a small stem within a seed cluster. Development was rapid, pupation occurring 26 days after oviposition and adults eclosing 10 days later. One female collected from Suisun Marsh, N CA, on Oct 11 laid ~25 eggs on leaves and flowers of *Honckenya peploides* (Sea Sandwort, host used at Suisun Marsh) and *S. tragus*. Eggs hatched in 4–5 days at 19–22 °C, and larvae were fed on *S. tragus*. Feeding was on leaves and flowers, larvae remaining hidden in the bracts. Mortality was low (<10%), and pupation occurred 26 days after oviposition, with adults eclosing 8–10 days later. Larvae do not construct nests, and there are 4 instars. Protection is based on camouflage, concealment, and attendance by ants (Ballmer and Pratt, 1991). Development continues during winter in S US; Shapiro and Manolis (2007) report that there is no diapause stage in N CA. This species is unable to survive winters in Cascadia. Scott (1986a) reported that "pupae hibernate" but did not state where this occurs.

Description of Immature Stages

The tiny white to pale green eggs are ornamented with tidy spiral rows of interconnected diamonds. L1 is greenish yellow with brown spots (~14 per segment). The head is setaceous and purplish black. L2 is light brown-green peppered with dark spots of variable size, concentrated laterally, each carrying a tiny, backward-curled white seta, this character persisting to pupation. There are 2 dorsal rows of low humps, sometimes separated by reddish spots. L3 is light green with a black shiny head, an indistinct ventrolateral white line, and numerous splotchy white markings laterally. In some individuals the ventrolateral line is suffused with reddish pink, and there is a broad dorsal reddish pink stripe. L4 is similar, with dorsal yellowish white markings forming 2 lines of small chevrons in the green form. In the red form, the dorsal stripe is dark red and intermittent and the ventrolateral stripe is reduced with red suffusion at the posterior end. Many red-form individuals lose their pigment during L4. The pupa occurs in at least 2 color forms according to background. Pupae formed on plants are light yellow-green with whitish markings on the abdomen. In some there are 2 dorsal rows of dark brown-black

Adult male

Egg @ 0.4 mm (side)

Egg @ 0.5 mm (top)

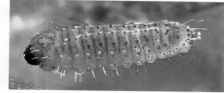

L1 @ 1.0 mm

Pupa @ 7 mm

L2 @ 2.5 mm

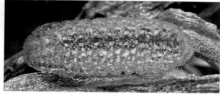

L3 @ 5 mm (red form)

L3 @ 5 mm (green form)

L4 @ 9 mm (green form)

L4 @ 11 mm (red form)

dots, 2 per segment, also a few on the wing cases. Pupae on nongreen surfaces are tan–light brown with darker brown spots and a middorsal dark line, most pronounced on the thorax. Larvae could be confused with other lycaenids but the strongly curled stubbly setae appear to be unique; also the chenopod hosts should distinguish *B. exilis* from most lycaenids. *Strymon melinus* (Gray Hairstreak) may feed on chenopods and could resemble *B. exilis* but grows to almost twice the size. Ballmer and Pratt (1988) illustrated a green-form L4 and pupa, and Allen et al. (2005) pictured a red-form L4.

Discussion

Obtaining eggs from gravid females is the best way to rear this species. Females oviposit well when exposed to sunshine and hot conditions. Rearing early instars can be frustrating as they are tiny and cryptic and mortality can be high, but later instars are hardier. Larvae are difficult to find, but the presence of attending ants may provide clues. The occurrence of red-form larvae in some populations deserves study. In our N CA cohort, red forms comprised 90% of larvae in L3, but this was reversed in L4. The incidence of red forms may be influenced by host plant. Camouflage of red-form larvae on red and green stems of *S. tragus* was clearly enhanced. Temperature thresholds for development and survival are unknown but would provide important clues to the ability of this species to invade and reside in more N areas.

Western Pygmy Blue   *Brephidium exilis* (Boisduval)

Western Tailed Blue   *Cupido amyntula* (Boisduval)

Adult Biology

Cupido amyntula is common in Cascadia, frequenting woodland clearings and roadsides, where it can be found in mixed aggregations of blues at mudding areas, or at flowers where both sexes seek nectar. It is widely distributed in forested areas of BC, WA, and OR; beyond Cascadia it occurs throughout W North America and central Canada. The flight period in Cascadia is late March–early Aug, with some records to early Oct. The larval hosts are in the pea family and include *Vicia americana* (American Vetch) and other vetches, *Lathyrus* spp. (wild pea), *Astragalus* spp. (rattlepod), and *Lotus* spp. (deervetch). Females lay eggs singly, typically placing each at the base of an opened pea flower, either on the flower sepal or on the stem immediately below. Up to 3 eggs may be placed on a single plant when resources are limited. *Cupido amyntula* is single brooded in BC but double brooded in OR and perhaps WA.

Immature Stage Biology

We reared this species on several occasions using *Vicia americana*, *V. villosa* (Woolly Vetch), and *Pisum sativum* (Garden Pea) as hosts. Eggs collected June 2 near Liberty, Kittitas Co., WA, were reared to adults, as were eggs obtained from females from Umtanum Canyon, Kittitas Co., WA on May 20. Eggs found June 2 and July 6 at Mineral Springs, Kittitas Co., and near Usk, Spokane Co., WA, respectively, were reared to L2, and 4 L4 collected July 22 at Hurley Creek, Kittitas Co., were reared to adults. Eggs hatch after ~4–5 days at 19–25 °C and development is rapid, taking 14–25 days from L1 to mature L4. In the Liberty cohort, 1 larva pupated June 23, 18 days post egg-hatch, and eclosed ~2 weeks later. Two larvae in the Hurley Creek cohort pupated; 1 eclosed on Sep

25, the other overwintered. All mature L4 in the Umtanum Canyon cohort turned reddish brown and sought refuge in curled leaves at the bottom of the cage. These larvae were transferred to overwintering conditions (5 °C) on Sep 14. After 103 days, survivors (60%) were exposed to 27 °C and continuous light. No feeding occurred and all pupated after 11 days, with adults eclosing 13 days later. Thus, it appears that *C. amyntula* in WA may be facultatively double brooded; sporadic late-summer reports of adults supports this notion. *Cupido amyntula* is multiple brooded farther S (Scott, 1986a). Larvae feed primarily on flowers and fruits (peas), but L1 and L2 may also feed on leaves, excavating holes and resting in them. After hatching, early-season L1 immediately cut an access hole through the base of a flower sepal, move into the basal internal flower, and feed on the flower or developing pea. When only peas are present, L1 feed directly on them. Garden peas in pods provide a suitable host for rearing larvae. Larvae on larger peas sometimes bore entirely into pods; some pupate within the pod. We have not observed larvae plugging entry holes behind them, as reported by Scott (1986a). Larvae do not build nests, although they may use pods as natural shelters. Ballmer and Pratt (1991) reported *C. amyntula* as facultatively myrmecophilous, attended by the ant *Formica obscuripes* in CA. Survival is based on camouflage, concealment, and ant attendance; there are 4 instars, and larvae are solitary and cannibalistic when crowded.

Description of Immature Stages

The egg of *C. amyntula* is a compressed sphere, white with a delicate lattice of rounded surface polygons; the micropyle is fairly large and green. L1 is light tan with a greenish flush and small brown spots, each giving rise to a long, pale, backward-curving seta. The head is shiny black. L2 varies from whitish with brown or magenta markings to magenta. There are 2–3 darker lines laterally and a narrow dark dorsal line. There are 2 rows of long pale dorsal setae, and numerous short dark setae cover the body. L3 is pale greenish to yellowish, with 3 diagonal olive or rufous side markings on each segment. Setae are numerous but shorter than L1 and L2. L4 is variable, pale greenish yellow to green or reddish, with a darker dorsal stripe. There are numerous small pale spots and a pale ventrolateral line; the sides have variable darker diagonal wavy markings, and the head is dark brown. The pupa is pale tan initially, aging to dark purplish brown, and bears numerous long pale setae. There are 2 dark spots on each side of the thorax, a single

Adult

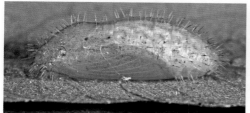

Pupa @ 9 mm (mature, *above*, fresh, *below*)

Egg @ 0.5 mm

L1 @ 1.0 mm

L2 @ 4 mm

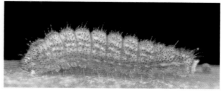

L3 @ 6 mm

L4 @ 10 mm

L4 @ 9 mm (feeding inside *Vicia* pod)

silk girdle thread, and a narrow brown dorsal line. Heavier dark markings and numerous dark spots in rough bands on the wing cases become apparent in aged pupae. Immature stages of *C. amyntula* are very similar to those of *C. comyntas*, from which they may not be separable, although larvae of the latter are often greener. The long pupal setae are unique to *Cupido*, so no other lycaenid pupae should be confused. Several lycaenid larvae are fairly similar, particularly *Plebejus melissa* and *P. anna*, both of which use plants in the pea family. Published images of *C. amyntula* include an L4 and pupa (CA) in Guppy and Shepard (2001) and an L4 (E Canada) in Allen et al. (2005); all are within the range of variation we found.

Discussion

The eggs of *Cupido amyntula* are easily found on the flowers of *V. americana*. Later, larvae can be found in the peas of *V. americana* by searching for entry holes. Early instars are difficult to rear and are easily lost in decomposing cut flowers. Little is known about second-brood adults, what larval hosts they use for oviposition, and whether larvae reach the overwintering stage. Similarly, the environmental factors determining the production of second-generation adults or diapausing L4 are unknown. Although temperature and daylength are likely important, host quality (e.g., flowers, pods, or leaves) may also be influential. Larval color variation is considerable and may be influenced by host type and whether feeding occurs on flowers, seedpods, or leaves. The parasitoids, diseases, and predators of *C. amyntula* are unknown and await study.

Western Tailed Blue *Cupido amyntula* (Boisduval)

Eastern Tailed Blue   *Cupido comyntas* (Godart)

Adult Biology

The Eastern Tailed Blue ranges throughout E US, with disjunct populations in the W. *Cupido comyntas* is generally uncommon in Cascadia, with populations in SE BC, NE WA, N ID, and W OR. Favored habitats include dry disturbed and weedy sites near canals, creeks, and fallow fields and along roadsides. Pyle (2002) reported an early April–early Aug flight period for Cascadia, including W OR, where it is double brooded. Guppy and Shepard (2001) reported a shorter flight of mid-June–mid-July in SE BC, where it is single brooded. The flight period in WA is poorly documented, but adults have been reported in late July at Spokane. Males fly low to the ground seeking females near hosts, and both sexes visit many types of low-growing flowers like clovers. Oviposition was observed on *Lotus unifoliatus* (Spanish Clover) in the Willamette Valley, OR on May 18. The female flew very low, with inconspicuous short hops of a few inches between plants, carefully probing terminal leaves with her abdomen. An egg was placed on the upper surface of a leaf near the stem, in the first cluster of leaflets below the terminal cluster. The female then flew 8–10 ft and began probing additional plants in short hops. She repeated the process 3 times, ovipositing twice. When a cloud obscured the sun, flight and oviposition ceased. Females continued ovipositing in captivity until May 24. Eggs were laid on both surfaces of the upper 3 sets of *Lotus* leaflets but not on *Trifolium* (clover). Many legumes have been reported as host plants elsewhere, including species in the genera *Lathyrus, Astragalus, Vicia, Lotus, Melilotus, Lupinus,* and *Trifolium*. In the Willamette Valley, *C. comyntas* was observed to reject 3 species of nonnative *Vicia* (*V. cracca, V. hirsuta, V. sativa*). A succession of host

Lotus unifoliatus (host plant)

plants is used during the season in N CA (Shapiro and Manolis, 2007).

Immature Stage Biology

We reared this species from Willamette Valley, OR females that oviposited on *L. unifoliatus*. Approximately 60 eggs were obtained from 10 females; 23 were reared to pupae, all eclosing to second-brood adults. Oviposition occurred for 1 week and the eggs began hatching when 5 days old. L1 fed on partially opened terminal leaf bundles of *L. unifoliatus*. L1 and L2 fed by eating halfway through the leaves in small patches from the underside. The larvae developed rapidly, molting to L4 10 days post egg-hatch. Pupation began 20 days post egg-hatch, and adults eclosed after a further 8 days. Larvae skeletonized leaves, each larva tethering itself to the host with a single silk thread. Some of the smaller larvae disappeared, suggesting cannibalism. Second-brood larvae may feed on flowers and/or peas available at that time, as do the related *C. amyntula*. L4 were transferred to a substitute nonnative plant, *Lotus pedunculatus*, which may have contributed to slower development in this instar, although larvae seemed to feed well. Scott (1986a) reported that the nearly mature larva overwinters, sometimes in host pods. In a laboratory experiment, Ballmer and Pratt (1991) ranked this species 10th out of 59 lyacaenid species in degree of attendance by the ant *Formica pilicornis*. Camouflage is likely an important factor in survival as the larvae are small and green and blend well with host plants. No nests are constructed.

Description of Immature Stages

The egg is white and has a golf ball–like surface texture of low ridges and divots; the micropyle is fairly large and grass green. L1 is pale tan with a purple-black head grading to cream at the back. Numerous long blond setae are present on the body and head. L2 is light green with numerous small brown spots, each with a long blond seta, and segment joints on

Adult female

the sides are darker green. L3 is darker green, bearing numerous small white or dark brown spots with setae of variable length, imparting a shaggy appearance. Most segments bear 2 or 3 indistinct darker green diagonal marks laterally, and there is a blurry white lateral stripe above the spiracles. L4 is bright grass green and bears numerous white spots and short setae. The lateral stripe is yellow but indistinct. L3 and L4 have a darker green, slightly depressed dorsal groove. The pupa is pale yellow-green, bearing numerous very long setae and has a few irregular dark brown contrasting spots anteriorly. Dornfield (1980) reported that the pupae possess stridulating organs between the abdominal segments and can produce sound. The closely related *C. amyntula* is similar except that the dark lateral stripes persist to the final instar. *Glaucopsyche lygdamus* often flies with this species, but its larvae grow larger, have a darker and tapered dorsal stripe, and have a pupa with large spots and lacking setae. *Plebejus icarioides* larvae are similar but darker and have a double dorsal line and shorter setae. Other *Plebejus* spp. larvae may be similar. The immature stages of *C. comyntas* have been described by many authors, including a meticulous description by Lawrence and Downey (1966). Guppy and Shepard (2001), Allen (1997), and Allen et al. (2005) illustrated mature larvae of *C. comyntas* similar to ours except for diagonal green streaks which we saw in L3, and Tveten and Tveten (1996) showed a larva with a white ventrolateral stripe.

Discussion

This species is best reared by obtaining gravid females and caging them with a live host plant. Determining the local host plant can be difficult but could be important, as the females may be selective, as we experienced with the OR population we reared. Care should be taken collecting wild eggs as several other species of blues also use similar legumes. First instars are inconspicuous and easily lost in captivity. Care should be taken obtaining stock from NE WA as *C. amyntula* populations there are heavily maculated on the ventral hindwing and easily mistaken for *C. comyntas*; habitat may provide a good clue in that area as *C. comyntas* prefers dry scrubby habitat while *C. amyntula* prefers wet areas. This species needs field study in WA to determine whether it is double brooded and to better define the flight period. Research on host-plant preferences of geographically isolated populations would be valuable and likely provide interesting information on historical distributions and dispersal.

Egg @ .05 mm

L1 @ 1.0 mm

L2 @ 2.5 mm

L3 @ 4 mm (late)

L4 @ 11 mm

Pupa @ 8.5 mm

Eastern Tailed Blue *Cupido comyntas* (Godart)

Echo Blue   *Celastrina echo echo* (W. H. Edwards)

Adult Biology

Previously considered a subspecies of *C. argiolus* (Spring Azure), *Celastrina echo* is now considered a full species (Pyle, 2002; Warren, 2005) occurring from BC to CA and E to TX, and *C. echo echo* is the nominate subspecies. It occurs throughout most of Cascadia and is the only *Celastrina* found W of the Cascades. Favored habitats include shrubby, riparian areas from sea level to high elevations. The flight period is very long, late Feb–early Oct, according to elevation and seasonal conditions, and the species is double brooded in many areas. Farther S (N CA), 3 or more broods may occur annually (Shapiro and Manolis, 2007). This common butterfly is one of the earliest species to fly in the spring, eclosing shortly after temperatures exceed ~15 °C. Males are common visitors to mud, ash, and dung, and both sexes utilize a wide variety of flowers. Adults spend much time over tall host shrubs, and males constantly patrol for females. *Celastrina echo* utilizes a remarkable variety of larval hosts (see Scott, 1986a, for list), feeding on new buds and fruits of *Cornus sericea* (Red Osier Dogwood) and *Ceanothus velutinus* (Mountain Balm), probably the 2 most commonly used plants in Cascadia, as well as *Spiraea*, *Arctostaphylos* (manzanita), *Vaccinium* (blueberry), *Viburnum*, *Holodiscus discolor* (Ocean Spray), *Sambucus* (elderberry), *Arbutus* (madrone), and others. A gravid female will visit several host-plant buds, probing with her antennae before selecting an oviposition site. Pushing her abdomen down between the clustered terminal buds, she places a single egg as far out of sight as possible before moving on to another site. Aphid-infested *C. sericea* appear to be avoided by ovipositing females.

Immature Stage Biology

We reared this species several times from gravid females and eggs and larvae obtained from *Cornus* and *Ceanothus*. Females obtained from Klickitat and Garfield Cos., WA during April 27–May 30 laid eggs on *C. sericea, C. velutinus*, or *H. discolor*. Eggs were laid mostly on the terminal buds of host plants. Eggs hatch in 48 hrs at 22–27 °C, and L1 chew round holes into buds, hollowing out the inside. Larvae do not enter the buds entirely but often have their heads and necks extended deeply. Larvae feed voraciously and develop rapidly, some taking as little as 12–14 days from L1 to pupation at 22–27 °C; however, there is much variability in development of individuals within cohorts. Some adults eclose 9–10 days later, but many enter summer then winter dormancy, eclosing the following spring. In 1 rearing, 66% of the spring pupae overwintered. After 120 days storage at 5 °C, adults eclosed after 4 days of exposure to 19–22 °C. Mature L4 consume many host buds, then wander, pupating away from the host plant under cover. Larvae will feed on leaves of *C. sericea* and *C. velutinus* if buds and fruits are not available. Scott (1986a) and Pyle (2002) report that larvae may be tended by ants. No larval nests are constructed, and protection is based primarily on camouflage. Frass is forcibly expelled by larvae, preventing contamination of feeding sites. Minute Pirate Bugs killed eggs and larvae in some of our rearings. There are 4 instars, and the pupa overwinters.

Description of Immature Stages

The egg is white, sometimes with a pale green tinge, contrasting with host buds. L1 is dull yellow-green, bearing many white setae. L2 is similar but the setae are proportionately shorter, imparting a shaggy appearance. In L3 the setae are shortened to stubble, and the yellow-green larva is marked with obscure chevrons of alternating green and white. L4 is highly variable but usually with reddish dorsal plates on each segment, resembling a marine chiton (Mollusca). Some larvae have a bold white ventrolateral stripe, others are pale green laterally, whitish below and punctuated with a large green spot on each segment. Contrasting dorsal plates are present in all populations, and the 3 anterior segments are darker than the other segments. Larvae of *Celastrina* with noncontrasting dorsal plates are usually *C. e. nigrescens* or *C. lucia*. Prepupal larvae become reddish purple. The pupa is light brown, becoming dark brown with blotchy black spots behind the thorax and toward the rear of the abdomen. Allen et al. (2005) presented images of 5 mature larvae of *Celastrina ladon*; of these the "Spring Azure Northern

Adult female

Eggs @ 0.5 mm

Pupa @ 10 mm

L1 @ 2 mm (late)

L2 @ 2.5 mm

L3 @ 5 mm

L4 @ 12 mm

L4 @ 14 mm

L4 @ 14 mm

Echo Blue *Celastrina echo echo* (W. H. Edwards)

population" appears to be the most similar to our *C. echo echo*. Miller and Hammond (2003) showed a very pale larva as *Celastrina argiolus*.

Discussion

We sometimes found it difficult to obtain eggs from females in captivity, possibly because they were unmated or egg-depleted. Females tend to be short-lived in captivity. However, females caged in small (12.5 × 12.5 cm) plastic cylinders with muslin lids, held at 17–25 °C in dappled sunlight and provided with *C. sericea* buds and lilac flowers, sometimes provide enough eggs to rear a cohort. Eggs may be found in spring by examining the terminal buds of host plants such as *C. sericea*, concentrating on buds on outer branches 4–7 feet above ground level. Once eggs are obtained, *C. echo echo* is relatively easy to rear, requiring only fresh host plant and regular cage hygiene. Host plants in spring often harbor natural enemies like predatory bugs that can rapidly kill eggs and larvae. Late instars may also be collected from the flowers of *C. velutinus* in prime fresh bloom by shaking branches with flowers into a net, then examining the contents. Overwintering the pupae successfully is difficult, with many desiccating and dying during hibernation. Color variation in larvae may be influenced by host plant. Little is known of the biotic and abiotic factors affecting the populations of this species.

Dusky Echo Blue   *Celastrina echo nigrescens* (Fletcher)

Adult Biology

Celastrina echo nigrescens, a subspecies of *C. echo,* flies in NE OR, E of the Cascades in WA, and in the Kootenay Lks region of BC; however, its range has not been clearly differentiated from that of *C. e. echo.* In areas where the 2 subspecies fly sympatrically, such as Bear Canyon, Yakima Co., WA, worn adults of *C. e. echo* may be seen with freshly eclosed *C. e. nigrescens,* indicating partially offset flight periods; see Warren (2005) for more examples of these staggered flights. The flight period for *C. e. nigrescens* is imperfectly known but extends at least from April 10 to May 29 in WA, and there may be a partial second brood. Known habitats include lower mountain canyons, open shrub habitat, and along mountain trails. *Celastrina echo nigrescens* utilizes some of the same host plants used by *C. echo echo,* although our database is small. In early April we found *C. e. nigrescens* eggs on *Cornus sericea* (Red Osier Dogwood) in Black Canyon, Okanogan Co., WA, and in early June we recovered probable *C. e. nigrescens* larvae from *Ceanothus velutinus* (Mountain Balm) at higher elevations in the same canyon. Warren (2005) documented fairly widespread use of *C. velutinus* and *Holodiscus discolor* (Ocean Spray) in OR. Where adults congregate, mud, and mate in the immediate vicinity of host plants, males can be numerous whereas females are reclusive and more difficult to find. Females inspect several terminal bud clusters of a host plant before placing a single egg, tucked between the buds as deeply as possible. Plants infested by aphids appear to be avoided by females.

Immature Stage Biology

We reared *C. e. nigrescens* twice from wild eggs collected from *C. sericea* in lower Black Canyon, April 10 and May 1 (different years), also from a gravid female obtained at Bear Canyon on May 29. Larvae from the Black Canyon eggs (3, 4) were reared on *C.*

sericea flower buds at 18–20 °C, 3 reaching adulthood. The Bear Canyon female laid ~20 eggs on flower buds of *Ceanothus sanguineus* (Redstem Ceanothus). Larvae were reared on buds and stems of *C. velutinus,* but all died by late L3. In both groups eggs hatched in 3–4 days and larvae grew rapidly, pupating 17–19 days after hatch; the adults eclosed 9 days later. Early L1 chew round holes through the sides of flower buds and hollow out the insides. Late instars feed heavily, requiring frequent food replacement, eating only buds and stems, refusing leaves, and avoiding aphids. *Cornus* bud clusters with several damaged buds are a good field sign for locating larvae. In one of our rearings, eggs of the Sunflower Aphid (*Aphis helianthi*) on dogwood stems hatched concurrently with *C. e. nigrescens,* and an attack by aphids on an L1 was observed. One inserted its mouthparts into the L1 while others attempted to. This appears to be the first report of an aphid attacking a butterfly larva (Keith Pike, WSU, pers. comm.). An egg that disappeared may also have been consumed by these aphids. The larvae did not construct nests and did not leave the host plant until pupation, unless buds were depleted. Our experience of development to adulthood in less than a month in captivity suggests that *C. e. nigrescens* may be double brooded in some locations, but field observations are needed for confirmation. Larvae are solitary but will withstand crowding in captivity; there are 4 instars, and overwintering occurs as a pupa. Protection is based on camouflage and concealment. Ant attendance has not been observed but likely occurs as it does in other *Celastrina* spp. (Ballmer and Pratt, 1991).

Description of Immature Stages

The eggs are compressed and spherical, greenish white with a well-defined micropyle. L1 is creamy brown with a shiny black head. Each segment has 10–12 brown tubercles, each with a long pale seta, most swept backward. L2 is yellowish green with indistinct white markings arranged in broken stripes. In some individuals a distinct white ventrolateral stripe is present. Setae are proportionally shorter but numerous, swept strongly backward dorsally, imparting a shaggy appearance. L3 is bright forest green with numerous tiny yellow backward-curled setae. There is a strong whitish ventrolateral stripe. L4 is bright green with reddish brown markings. The ventrolateral white stripe is bold with a reddish brown spot on each segment. Dorsally each segment bears bold white markings in a modified chevron pattern, each infused with reddish brown to a greater or lesser extent. The pupa is light brown, darkening

Adult male

Egg @ 0.5 mm

L1 @ 1.0 mm

Pupa @ 8.5 mm

L2 @ 3 mm

L3 @ 7 mm

L4 @ 12 mm

L4 @ 12 mm

Dusky Echo Blue *Celastrina echo nigrescens* (Fletcher)

later, with numerous small bristly blond setae covering the surface, and there is a dark dorsal stripe. The immature stages of *C. e. nigrescens* are very similar to those of other *Celastrina*, and the considerable intraspecific variation may preclude definitive identification; however, the larvae of *C. lucia* tend to have less strongly marked dorsal white chevrons in L3–L4 and an absence of red-brown markings. *Celastrina echo echo* differs in generally having bolder markings; the dorsal chevrons and ventrolateral white stripe usually have strongly contrasting reddish borders. Our images are the first published of immature stages of *C. e. nigrescens*.

Discussion

Eggs can be obtained by caging gravid females with *C. sericea* or *C. velutinus* twigs or branches with unopened flower buds. Eggs and larvae may also be found in known habitats on terminal clusters of buds on the same plants, ~3–6 feet above the ground. Larvae are relatively easy to rear as long as fresh food plants are provided and cages kept clean. Additional rearing and fieldwork is needed to determine relationships among *Celastrina* spp. in Cascadia. In particular, clarification of the differences described here in larval coloration between the species is needed. Studies are also needed on the possible influence of host plants on larval coloration. The unusual attack on an early instar by aphids warrants further study. The presence of a second brood in our area has been shown only in captivity and needs verification in the wild. The full extent of the range of this subspecies needs study. Little is known of the natural factors controlling populations.

Lucia's Blue   *Celastrina lucia* (W. Kirby)

Adult Biology

Celastrina lucia was recently described in Cascadia as a species level taxon (Warren, 2005). Formerly it was variously considered to be a subspecies of *C. argiolus*, *C. ladon*, or *C. echo*. *Celastrina lucia* has a flight period extending from early April to mid-July and is found only E of the Cascades, typically in isolated colonies in BC and WA. It normally flies in open forested areas, but a colony in Cowiche Canyon, Yakima Co., WA is in a unique riparian corridor in shrub-steppe habitat; at this locality, *C. lucia* is the only resident *Celastrina* species. In Cowiche Canyon, *C. lucia* typically appears April 1–15, flying for 4–6 weeks near large stands of *Cornus sericea* (Red Osier Dogwood). In some years the population is large, with up to several hundred counted during a few hours of observation along the trail. Several possible host plants have been investigated at Cowiche Canyon, but *C. sericea* appears to be the only host plant utilized; indeed, this is the only confirmed food plant in Cascadia. Before *C. sericea* flower buds open, females fly from cluster to cluster, testing each with their antennae, occasionally depositing a single egg directly on the buds, thrusting the egg as deeply as possible into the cluster head. Adults continue to fly and oviposit as long as there are unopened *C. sericea* flower buds.

Immature Stage Biology

We reared *C. lucia* 3 times. Eggs were obtained from gravid females twice and found on *C. sericea* buds on the third occasion. Eggs obtained April 4, May 3, and May 14 (different years) hatched after ~3–6 days at 19–22 °C. L1 exits without consuming the eggshell and immediately bores a round hole into a host bud (*C. sericea*), extending its head and long neck inside to feed. Larvae preferentially eat buds but will feed on young leaves of *C. sericea* in captivity if no buds are

Adult male (second brood)

Adult male (first brood)

available. Larval development is rapid, with pupation occurring 16–25 days after oviposition at 18–27 °C. In captivity, the prepupal larva wanders off the host plant before pupating. The majority of reared pupae formed in May oversummer and overwinter, but a small proportion (~30%) produce a second generation of adults after 7–13 days at 22–27 °C. Second-generation adults are distinctively marked, with connected black spots forming a distinct Y pattern rather than a large black patch on the ventral hindwing. No second generation has yet been observed in Cowiche Canyon or elsewhere in WA. Given that our rearing temperatures and daylengths were similar to natural conditions in E WA during April–May, the absence of a second generation in the wild is puzzling; however, Y-marked second-generation *C. lucia* may occur in BC (D. St. John, pers. comm.). Adults found in June and July at other locations in WA appear to be first-brood individuals delayed by higher-elevation temperatures. There are 4 instars, larvae do not make nests, and pupae overwinter.

Description of Immature Stages

The egg is greenish white, contrasting with green *Cornus* buds. L1 is green, with several long backward-pointing setae on each segment. The head is black, and each segment has a distinct vertical fold (pleat) down the middle. L2 is lighter green and has well-developed segments with vertical pleats; the setae are proportionately shorter but impart a shaggy appearance, and the head is shiny black. In L3 there is a yellow-white longitudinal line along the ventrolateral margin, and obscure light markings on the dorsal surface give the larva a mottled appearance. In L4 there is a bold white ventrolateral line, and in some individuals blurred white patches appear on the dorsal side of most segments. In some individuals, no

Egg @ 0.7 mm

L1 @ 2.5 mm (late)

Pupa @ 8 mm

L2 @ 3 mm

L2 @ 5 mm (late)

L3 @ 6 mm

L4 @ 12 mm

L4 @ 12 mm (prepupal)

white markings occur in L4, with the larva entirely green except for a much-reduced pale ventrolateral line. In white-marked forms, the thoracic segments bear small dark dorsolateral spots. Prepupal larvae lose markings and become dark greenish brown. The pupa is light brown with dark brown smudges, lacking the distinct paired abdominal spots of *C. e. echo*. Larvae are similar to those of *C. e. nigrescens* and *C. e. echo*; however, *C. lucia* are predominantly green and lack the contrasting multicolored dorsal segment plates that typify *C. e. echo*. There are no known published images of the immature stages of *C. lucia*. Allen (1997, pl. 10, row 5) shows an image of a univoltine adult "*Celastrina ladon ladon* form lucia male" from WV that is very similar to our captive second-brood *C. lucia*.

Discussion

Females oviposit sparingly in captivity, but larvae are reared fairly easily. Eggs and larvae can be found, although searching large patches of *C. sericea* may appear to be a daunting task. Searching the unopened terminal buds, particularly on protruding branch tips 3–6 ft high during a mid-May afternoon in Cowiche Canyon, we recovered 7 eggs, 2 L1, and 5 hatched eggs. Uneaten eggshells are a good clue that young larvae are probably nearby, and buds with holes in the sides are indicators of larval feeding. Further rearing studies are needed to confirm the differences in larvae of the *Celastrina* spp. we describe.

Lucia's Blue   *Celastrina lucia* (W. Kirby)

Dotted Blue   *Euphilotes enoptes* (Boisduval)

Adult Biology

Euphilotes enoptes occurs widely in W US, E to NV and AZ. In Cascadia it occurs patchily in W and central OR to S WA. When Warren (2005) elevated *Euphilotes columbiae* to full species, it was believed that *E. enoptes* did not occur in WA. All previous records of *E. enoptes* in WA were assigned to *E. columbiae*; however, in June 2005, T. L. and R. M. Pyle discovered a small population of *E. enoptes* in W Yakima Co., WA, associated with *Eriogonum nudum* (Barestem Desert Buckwheat), a plant of very limited distribution in WA. To date *E. enoptes* has been observed in this area from June 14 to July 3. Before *E. nudum* flower buds open, males perch and fly near the host plants, which typically grow on steep talus slopes. Females fly upslope, where the males intercept and court them in a rapid swirling flight. Females were observed in oviposition behavior on unopened flower buds of *E. nudum*. In captivity, the females preferred to oviposit on enclosing sepals of unopened flower buds, especially in tightly confined spaces between 2 adjacent buds, as well as on the petals of just-opened flowers. Eggs are laid singly. It is not known whether this species is single or double brooded in Cascadia.

Immature Stage Biology

We partially reared *E. enoptes* 3 times, once from field-collected eggs, once from gravid females, and once from several field-collected larvae. Two eggs were obtained on July 6 by searching *E. nudum* flowers; however, both were parasitized. On July 4 several gravid females were obtained and caged with the host plant, where they produced a dozen eggs. The eggs began hatching on July 9, 5 days post oviposition, and 1 larva reached L2 4 days later. The larvae did not survive past L2, but the site was revisited on July

Eriogonum nudum (host plant)

21, and 8 L3 and L4 were obtained by searching the host plants and by drying cut flower heads. During rearing, the larvae were not observed to leave the flower heads of the host plant until pupation, and no nests were constructed. When flowers of *E. nudum* senesce, larvae switch from eating them to feeding on the developing seeds. The larvae feed by eating round holes in the seeds and hollowing out the inside with their extendable necks. The variegated pink-and-white mature larvae are very well camouflaged against the pink-and-white blossoms of the host plant. Half the field-collected larvae survived to late L4, whereupon they wandered off the host plant to a covered area, rested for 2 days, and pupated on Aug 1. Development from egg-hatch to pupation is estimated at ~23 days based on partial rearings. The pupae did not eclose to second-brood adults and were overwintered; however, they died before the following spring. Scott (1986a) reported that *E. enoptes* is tended by ants and that the pupa is the overwintering stage. We saw some ants loosely associated with wild larvae but did not witness direct tending. There are 4 instars.

Description of Immature Stages

The eggs are greenish white and slightly compressed, with an odd faintly crinkled surface texture that lacks structural character. The micropyle is irregular in shape, somewhat obscure and green. L1 is pale and maggotlike, with a deep purple head and numerous small bullae on the body. There are 2 rows of long setae down the dorsum and a line of setae around the ventral margin. In L2, the dorsal setae disappear and a broken dorsal red line appears. There are sparse long setae around the ventral margin of each segment. L3 is similar except that numerous distinct black spots appear all over the body. A line of indistinct red dorsolateral spots supplements the red dorsal line, and pale setae occur around the basal periphery. In

Adult female

Egg @ 0.6 mm

L1 @ 1.1 mm

L2 @ 2.0 mm

L3 @ 4.5 mm

L4 @ 9 mm

Pupa @ 7 mm

L4, the red dorsal line is stronger and unbroken on the anterior third, and the dorsolateral spots develop into very strong red diagonal marks, 1 per segment. A ventrolateral row of red spots is also present, and the overall appearance is attractive. Throughout development, each larval segment has a vertical groove or fold in the side. The pupa is uniform light brown with obscure red dorsal marks and a line of brown spots along the abdomen; setae are entirely absent. At all stages this species is very similar to the closely related *Euphilotes columbiae*; these 2 species are best separated in the adult stage. *Euphilotes columbiae* has not yet been documented using *E. nudum* in WA, nor is *E. enoptes* known to use the host plants of *E. columbiae* in Cascadia. Other *Euphilotes* spp. are in the *battoides* group, with more sculptured eggs and late instars that lack segment side folds. Additionally *battoides*-group spp. are readily separable in the adult stage. Comstock and Henne (1965) reported on the life history of *Philotes enoptes dammersi* in CA, providing illustrations of a full-grown larva and pupa. The eggs of *Plebejus acmon* may also be found on *E. nudum* but are easily distinguished by their strongly ornamented surface.

Discussion

This species should be sought wherever *E. nudum* grows. Eggs and larvae may be found by carefully searching the flower heads of *E. nudum* where the species flies, or larvae can be obtained by drying flower heads in an enclosure and waiting for the larvae to wander out. Eggs can be obtained by caging gravid females, which fly in late June–early July, with fresh cuttings of the host plant. Egg parasitism occurs; research is needed to determine the identity of the parasitoids, along with other natural enemies affecting larvae and pupae. This species is little known in WA, and there are many research needs. The adult flight period needs better definition in our area, it is not known whether the species is single or double brooded, the overwintering stage is not known with certainty, and if the species is double brooded the host plant used by the second brood is not known.

Columbia Blue   *Euphilotes columbiae* (Mattoni)

Adult Biology

Euphilotes columbiae was long regarded as conspecific with *Euphilotes enoptes* (Dotted Blue), but a thorough analysis of the *Euphilotes* group by Warren (2005) revealed far more complexity than earlier believed. The known range of *E. columbiae* is centered on the Columbia Basin of E WA and NE OR; this is one of the two most common *Euphilotes* species in WA. *Euphilotes columbiae* is seldom found far from its larval hosts, deriving nectar and larval nourishment from the same plants. Mating pairs are often seen on *Eriogonum* hosts, where females search for oviposition sites. On June 5 at Elk Hts, Kittitas Co., WA, several females were observed ovipositing on *Eriogonum compositum* (Northern Buckwheat). The female pauses at a flower head, crawls about while feeling with her antennae, selects a fully open individual blossom, and bends her abdomen forward into the flower, placing a single egg deep inside. In addition to *E. compositum*, used early in the flight period, *E. elatum* (Tall Buckwheat) is used later. The single-brooded adults have a long flight period, from late April to mid-Aug, due at least in part to the ability to exploit 2 buckwheat species with offset flowering.

Immature Stage Biology

We partially reared this species on a number of occasions. Eggs were obtained from females on 4 occasions (different years) May 5–June 18; these developed to L1–L2 larvae. Wild L3–L4 larvae were collected May 28–Aug 26 on 5 occasions, with 20 reared to pupae and a few to adults. All were from Klickitat, Yakima, Kittitas, or Chelan Co., WA at elevations of 220–5,000 ft. Eggs hatch in 5 days, and L1 quickly move deep into the complex flowers, where they are well camouflaged and difficult to find. The

larvae feed only on the flowers and fruits of *Eriogonum* hosts and remain on the flowers during development. When immature fruits (seeds) are available, the larvae bore holes and hollow out the insides using their extendable necks. As the larvae grow, they become easier to find by carefully searching the flower heads or by watching for ants which sometimes attend them. Feeding damage (seeds with holes) often indicates that larvae are to be found nearby. The larvae continue to feed on the fruits after the flowers have senesced, and by pupation a significant percentage of the fruits are often eaten. Coloration of the larvae is closely linked to the color of the food plants; flowers with more pink coloration have brightly colored larvae, while pale flowers produce pale larvae. No nests are built at any stage. Based on partial rearings and multiple collecting trips, it appears that development from oviposition to pupa takes ~40 days. Overwintering occurs as a diapausing pupa.

Description of Immature Stages

The eggs are tiny, pale green, nearly smooth with undulating irregular surface ridges, lacking the well-organized rows of fine structural elements found on most other lycaenid eggs. L1 is greenish cream with a black head; each segment has a deep vertical groove laterally, a character that persists until pupation. Long setae occur in a band behind the head and around the basal perimeter of the caterpillar. The setae become proportionately shorter but persist in the same pattern throughout development; otherwise the larvae are notably hairless. L2 is pale and rather maggotlike in appearance. L3 is dark gray and strongly speckled in black. Distinctive red markings appear in some individuals in L4. The red markings include a bold diagonal bar laterally on each segment on a cream ground color, and an irregular bold red dorsal stripe. The cream ground color reddens prior to pupation. The pupa is light brown, lacks bristles, and bears a row of contrasting small dark brown spots along each side of the abdomen. The larvae of *E. columbiae* are like others of the "dotted blue" group; however, they may be separable from those of the closely related "square spotted" *E. battoides* group. The strong vertical fold laterally on each segment is much less prominent, becoming almost completely obsolete in final instars in the *battoides* group, but persists throughout development in *E. columbiae*. No other *Euphilotes* spp. appear to feed on the buckwheat species used by *E. columbiae*. Larvae of *Strymon melinus* may be confused with *Euphilotes* as they also occur on *Eriogonum* flowers; however, *S.*

Adult male

Egg @ 0.55 mm

L1 @ 1.0 mm

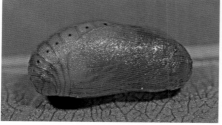

Pupa @ 7 mm

L2 @ 2.2 mm

L2 @ 2.8 mm (late)

L3 @ 5 mm

L4 @ 6 mm

L4 @ 11 mm (prepupal)

melinus grows to a considerably larger size, is only occasionally as brightly colored, and has a double stripe dorsally. *Plebejus acmon* larvae also occur on the same flowers but are densely hairy. Pyle (2002) shows an image of a late-instar larva attended by ants that appears similar to ours.

Discussion

Captive females are often reluctant to oviposit but sometimes produce a few eggs on host flowers. In captivity eggs may be laid on inert surfaces or leaves when flowers are not available, but newly hatched larvae will not feed on leaves. In captivity the eggs are quite difficult to find on the complex *Eriogonum* flowers and are easily lost. It is difficult to keep small cut portions of buckwheat flowers fresh, so many L1 and L2 fail to survive in captivity; once the larvae reach L3, they become much hardier. At this stage the larvae can also be found in the wild. One technique is to collect a sizable sample of buckwheat flowers in an area where the adults were common earlier. The flower heads can be searched manually by picking the tiny flowers apart; this is very time consuming and tedious but is often productive. Another technique is to place the cut flowers in a terrarium and allow them to dry for 2–3 days. Any larvae present usually wander onto the glass sides and are then easily recovered. Very little is known about the biology of this species, particularly the biotic and abiotic factors regulating populations.

Columbia Blue *Euphilotes columbiae* (Mattoni)

"Cascadia Blue" *Euphilotes* on *Eriogonum heracleoides* (undescribed)

Adult Biology

Euphilotes on *Eriogonum heracleoides* was previously considered conspecific with *E. battoides* (Square-spotted Blue); however, Warren (2005), following an exhaustive study of the *Euphilotes* spp. of OR, substantially revised this group. He concluded that true *E. battoides,* whose type locality is in CA, does not occur in OR or WA. Until this species is formally described, we propose the name "Cascadia Blue" for *Euphilotes* on *Eriogonum heracleoides* because its range appears to encompass much of Cascadia (N OR, WA, N ID, W MT, S BC). This species is very common in WA and OR, flying in foothill and mid-elevation areas E of the Cascades, absent from the driest parts of the Columbia Basin. The single brood flies late March–mid-July at 1,500–7,500 ft in OR (Warren, 2005). Favored habitats include any flowery areas where the host buckwheats grow, particularly in shrub-steppe canyons and adjacent hillsides; adult flights are timed to the budding and early blooming of their hosts. Mating pairs can often be seen on the flowers of the known host, *Eriogonum heracleoides* (Parsley Desert Buckwheat) and the suspected alternate host *Eriogonum douglasii* (Douglas' Buckwheat). At Mary Ann Creek, Okanogan Co. (3,600 ft), mating pairs were abundant on *E. heracleoides* flowers on June 28. *Euphilotes* on *E. heracleoides* appears to prolong its flight period by switching between *E. heracleoides* and *E. douglasii* as the season progresses. According to Warren (2005), adult "Cascadia Blues" in OR are distinguished from the closely related *E. glaucon* by tending to fly later and having a paler ventral ground color.

Immature Stage Biology

We reared or partially reared this species on several occasions, most frequently from partially grown larvae found on buckwheat flowers. On Aug 6, we collected 28 larvae (L2–L4), from the flowers of *E. heracleoides* at a high-elevation site in Chelan Co. When flowers of *E. heracleoides* became scarce, the larvae were moved to *Eriogonum umbellatum* (Sulphur-flower Buckwheat), which they readily accepted. Despite some cannibalism and other losses, 9 of the larvae pupated and 7 eclosed to adults (4 males, 3 females) the following spring. All males eclosed before the first female, and female eclosion averaged 7 days later than males. We estimate development period from egg-hatch to pupation to occupy ~25 days, based on a composite of partial rearings and field observations. During development, the larvae fed first on flowers, later on the developing seeds after the petals had senesced. We observed that a pale-colored L4 feeding on poor-quality (partly wilted) flowers developed bright red markings within 2 days after feeding on fresh, brightly colored *E. heracleoides* flowers. This species has 4 instars, and the pupa overwinters; no nests or shelters are made at any stage. The bright red coloration seen in many mature larvae is unlikely to be aposematic; instead, the larvae blend well with the pink-and-white flowers of the host plant. Occasionally we observed ants on plants hosting larvae, so ant attendance is likely but needs confirmation.

Description of Immature Stages

The egg is white with a distinct angular green recessed micropyle, and is ornamented with a latticework of horizontally stretched diamond shapes formed by low intersecting ridges. L1 is dull yellow with long setae mostly around the ventral fringes and on the head, plus a partial double row of setae along the posterior half of the dorsum. The head is mostly yellow, darker on the anterior. L2 is brightly colored in red and white, and has dorsolateral white markings shaped like the letter F with the short legs pointing toward the head. Short setae on the body impart a slightly shaggy appearance, while longer setae are limited to the ventral margins. L3 retains the red coloration and white F-shaped markings, but also has numerous tiny dark spots on the body. L4 is variable, ranging from greenish gray to much brighter red markings. Setae are present along the ventral periphery. The pupa is light brown, almost unmarked, and lacking setae. The most closely related species is *E. glaucon*, from which the larvae may not be readily distinguishable. In our rearings, the L2–L3 of *E. glaucon* are paler than those

Adult

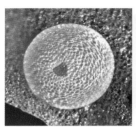

Egg @ 0.6 mm

L1 @ 1.1 mm

Pupa @ 5.5 mm

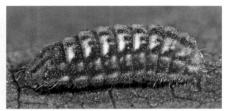

L2 @ 3.5 mm (late)

L3 @ 5 mm

L4 @ 9 mm

L4 @ 9 mm

"Cascadia Blue" *Euphilotes* on *Eriogonum heracleoides* (undescribed)

of *Euphilotes* on *E. heracleoides*, and the reverse is true in L4; however, these differences are likely due to diet. *Euphilotes* of the *enoptes* group differ by having less strongly ornamented eggs and a persistent vertical groove on each larval segment, and by lacking the F-shaped dorsolateral markings. There are no known published images of this species.

Discussion

The "Cascadia Blue," a common and widespread butterfly in the Pacific Northwest, is currently being studied by Warren and Kohler, who will formally describe it (A. Warren, pers. comm.). Research on biology and ecology, particularly host-plant preferences and utilization, is needed and will further clarify taxonomic issues. Of special interest is the apparent switch between 2 flowering buckwheat hosts, helping to extend the flight period, but confirmation of the alternate host plant is needed. Studies are also needed on biotic and abiotic factors influencing the distribution and abundance of this species. If ants attend this species, their importance and role in protecting larvae from natural enemies needs study. As with other *Euphilotes* spp., the "Cascadia Blue" is difficult to rear from the egg beyond L1 or L2; however, partly grown larvae can be found by searching host-plant flowers, and rearing these later instars is not difficult. Late-season larvae continue to develop while host plants are desiccating, and many likely perish at such times.

"Summit Blue"   *Euphilotes glaucon* (W. H. Edwards)

Adult Biology

Euphilotes glaucon was until recently considered a subspecies of *E. battoides*, but Warren (2005) found sufficient differences to elevate this taxon to the species level. Warren (2005) found that all *Euphilotes* spp. in OR that feed on *Eriogonum umbellatum* belong to *E. glaucon* and stated that where they occur together with *Euphilotes* on *E. heracleoides*, they differ by tending to fly earlier and by having a darker ventral ground color. Warren (2005) also reported a population of *E. glaucon* at Bird Creek Meadows near Mt Adams, Yakima Co., WA. In 2006 we discovered an additional colony of this species ~120 miles farther N, on Chumstick Mt, Chelan Co., WA, feeding on *E. umbellatum* at about 5,700 ft in subalpine habitat. Warren confirmed the occurrence of *E. glaucon* at this location after viewing photographs of adults (pers. comm., March 2007) and later collecting specimens on his own. Worn adults were found on July 24 and L1 on Aug 12, so the flight period appears to be about mid-July–mid-Aug at this locality. On July 18 a colony of *E. glaucon* was found at Bethel Ridge, Yakima Co., WA, roughly midway between the first 2 localities, forming a narrow N–S band along the E flanks of the Cascades at 5,800–6,000 ft. Beyond this zone, the distribution of *E. glaucon* is largely unknown in Cascadia and elsewhere (Warren, 2005). The only documented host plant in Cascadia is *Eriogonum umbellatum*. Males and females closely associate with the host, with females flying flower to flower, bending their abdomens to oviposit on the blossoms of the freshly opened flowers, but avoiding the more mature ones. Females lay their eggs singly, almost always on the inside parts of an open flower. We suggest a new English name for this species, the "Summit Blue," in

Adult female

recognition of its generally high-elevation occurrence in WA, although other subspecies of *E. glaucon* may occur at lower elevations in OR (Warren, 2005). *Euphilotes glaucon* is single brooded.

Immature Stage Biology

We reared this species from 21 larvae (L1–L4) collected Aug 6 from *E. umbellatum* var. *hausknechtii* at Chumstick Mt. This plant is a miniature variety of the Sulphur Flower (*E. umbellatum*), which grows larger at lower elevations. Many of the larvae were parasitized, but 2 developed to pupae, overwintered, and eclosed to adults the following spring. Additionally, 10 eggs were collected July 24 (following year), and others were obtained from captive females at the same time. Eggs hatched after 5 days, and L1 left eggshells uneaten. The eggs blend well with the pale yellow flowers of the diminutive host plant, providing effective camouflage. L1 were fragile and did not survive to L2 in this cohort. Field-collected larvae remained on the flowers, feeding on the petals and fruits until pupation. When feeding on the fruits (seeds), larvae cut small round holes and hollow out the insides with their extendable necks. Two larvae pupated Aug 24 and Sep 1; we estimate total development time from oviposition to pupation is ~30 days, based on a composite of partial rearings and observations. The larvae are solitary and do not construct nests. The primary survival strategy appears to be camouflage, although we also observed ants attending larvae in the field. There are 4 instars, and the pupa overwinters.

Description of Immature Stages

The egg is pure white and distinctly ornamented with a sculpture of horizontally elongated diamond shapes. The micropyle is recessed and indistinct. L1 is maggotlike and bears 2 dorsal rows of long pale setae, and other long setae are situated around the basal perimeter of the larva. The head is dark brown and bears no setae. L2 is different in appearance, the ground color pale gray and the entire body covered with numerous contrasting small dark spots. The setae almost entirely disappear along the dorsum but persist along the ventral periphery. L3 is pale cinnamon with a contrasting but broken red dorsal stripe. The numerous small spots persist but are lighter, proportionally smaller and, along much of the dorsum, become obsolete. L4, especially when prepupal, develops considerably more red coloration, concentrated in blurry diagonal dorsolateral marks, a dorsal stripe, and an indistinct red stripe along the ventral margin. The pupa is a handsome bright

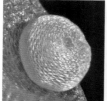

Egg @ 0.6 mm

Eriogonum umbellatum (alpine var.)

L1 @ 1.1 mm

L2 @ 3.5 mm

L3 @ 5.5 mm

L4 @ 11 mm

Pupa @ 7 mm

"Summit Blue" *Euphilotes glaucon* (W. H. Edwards)

cinnamon color and lacks setae; ornamentation is limited to an indistinct row of brown spots along the abdomen. Throughout development the larvae seem paler than related species, likely because of the pale color of the host buckwheat flowers on which they feed. Other *battoides* group species are similar; however, *E. glaucon* consistently has less red coloration. The eggs of the *enoptes* group species of *Euphilotes* are less ornamented, and their larvae have persistent folds laterally on the larval segments; these folds become obsolete in late instars of *E. glaucon*. The only other lycaenid that might be confused with this species is *Strymon melinus*; however, its larvae grow larger, have a double dorsal stripe, and bear setae all over the larva and the pupa. We know of no published images of immature stages of *E. glaucon* for comparison.

Discussion

This species, and *Euphilotes* spp. generally, are difficult to rear from eggs. Larger larvae can be found by carefully searching flower heads of the host plants. Collected larvae are hardier and more easily reared. Parasitism of larvae by small wasps appears to be common and deserves further investigation. This species is little known and is a strong candidate for further biological and ecological research. Some issues which need clarification include the extent of the adult flight period, whether other host plants are used (e.g., *E. ovalifolium*?), and whether other populations exist in WA. Additionally, larger collections of adults are needed for comparison with the sympatric *Euphilotes* on *E. heracleoides*. The suitability of the larger, low-elevation subspecies of *Eriogonum umbellatum* as a host plant also needs study.

Unnamed potential species

Adult Biology

In his detailed study of the butterflies of OR, Warren (2005) found that species of *Euphilotes* are closely associated with specific buckwheat host plants, with some species developing on only 1 species of *Eriogonum* while others may use 2. Consequently, if a population is discovered utilizing a different species of *Eriogonum*, it is possible that it represents a new species, a subspecies, or at least a host-plant race (Pratt and Emmel, 1998). This *Euphilotes* segregate feeds on *Eriogonum sphaerocephalum* (Round-headed Desert Buckwheat), a plant not previously reported as a host for any *battoides* group *Euphilotes* in Cascadia. We have determined by genetalic dissection that this segregate is of the *E. battoides* group, but it is unclear whether it represents a species or a host-plant race. The single brood of *Euphilotes* on *E. sphaerocephalum* flies from early May to mid-June, the flight period ending earlier than that of most other *Euphilotes* spp. in Cascadia. Flowers of *E. sphaerocephalum* senesce soon after the spring flight of this butterfly. Favored habitat is low- and mid-elevation shrub-steppe of E WA, at 1,000–2,700 ft, including E Kittitas and W Grant Cos. Mating pairs often rest on *E. sphaerocephalum*; females tend to remain on or near the host-plant flowers, whereas males fly between plants searching for receptive mates. Females lay their eggs singly on unopened buds or place them inside partially opened flowers, with no further oviposition occurring once flowers have fully opened.

Immature Stage Biology

We partially reared this segregate 4 times, twice from eggs and twice from larvae collected from *E. sphaerocephalum* flowers; however, the total number we have reared is low, ~15 eggs and 10 larvae. Eggs

Adult

Euphilotes on *Eriogonum sphaerocephalum*

were found throughout the first half of July, and the latest larva found was an L4 on Aug 2 at one of the higher-elevation sites. *Eriogonum sphaerocephalum* is also used by other lycaenids, including *Plebejus lupini*, *Callophrys affinis*, and probably *Lycaena heteronea* as well as *Apodemia mormo*, so care must be taken in identifying collected eggs. The eggs hatched after ~5 days, and 3 days later, 1 larva reached L2. L1–L2 fed initially on flowers, later switching to fruits (seeds) as the blossoms senesced. The larvae ate round holes in the seeds, hollowing out the insides with their extendable necks. On July 18 we collected L3 and L4 at a higher-elevation site (2,300 ft, nr Ellensburg, WA), and by July 30, 1 had pupated. Development time is uncertain; estimating from partial rearings and fragmentary evidence, it may be as many as ~90 days from egg-hatch to pupation, but more complete rearing will likely show it to be shorter. We did not see ant associations with this species, but in common with other *Euphilotes* spp., they are likely to exist. There are 4 instars, no nests are constructed, and the pupa overwinters.

Description of Immature Stages

The egg has a white exterior with low intersecting ridges forming a delicate, horizontally stretched diamond pattern on the sides. The micropyle is recessed and green, and through the latticework pattern, the green interior of the egg is visible. L1 is pale and maggotlike, bearing very long pale setae in a double row along the dorsum and a row along the ventral periphery. The head is dark chestnut and bears a few short setae. L2 is pale gray but unspotted, and the 2 dorsal rows of setae persist, as do those along the ventral periphery. An irregular vertical groove is present laterally on each segment. L3 is very pale for this genus, light yellow with no red coloration. By this stage the setae are very reduced, limited to the first segment and along the ventral margins. The head is black and shiny, bearing only a few short setae, and each segment still bears a deep vertical groove laterally. All *Euphilotes* spp. have eversible tubercles near the posterior end of the abdomen (Ballmer and Pratt, 1988), but they are only occasionally seen extended; our image of L3 shows these tubercles in the extended position. These tubercles are known to produce chemicals attractive to attending ants. Red coloration appears in L4 in a typical *Euphilotes* pattern, including a broken dorsal stripe and broad diagonal slashes on the dorsolateral part of each segment. The setae are further reduced to remnants around the head and the posterior end, and a pale bluish white

Egg @ 0.6 mm

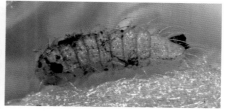

L1 @ 2.0 mm (late)

thoracic shield is present behind the head. The pupa is pale yellow with a glossy surface, lacks setae, and is almost unmarked. This segregate is most closely related to *Euphilotes* on *E. heracleoides* and *E. glaucon*. In *Euphilotes* on *E. heracleoides*, L1 apparently has shorter and more localized setae, and in L2 and L3 the red coloration is much more developed; however, this may be related to diet. Also, the pupa has darker segment lines and more distinct abdominal spiracles. In *E. glaucon*, L2 and L3 differ in being covered with numerous dark speckles, and the pupa is less constricted and has a more granular surface texture. As *Euphilotes* on *E. sphaerocephalum* is of uncertain affinities, there are no published accounts or images in the literature with which to compare it.

L2 @ 2.0 mm (early)

L3 @ 5 mm

Discussion

While this segregate appears to have some unique features and uses a unique host plant, it is uncertain whether it represents a separate species or is simply a host-plant or seasonal race of a known taxon. Further rearing is warranted to verify the apparent differences we have cited with other *Euphilotes* spp. Pratt and Emmel (1998) discuss the importance of host-plant and seasonal races in CA *Euphilotes* spp., and similar taxonomic and biological research on this genus is needed in Cascadia. Of special note, prior to submission of this manuscript we discovered that an even earlier *Euphilotes* population, flying in mid-April in low-elevation E Kittitas Co., WA, uses *Eriogonum thymoides* (Thymeleaf Buckwheat) as its larval host, a previously unreported host plant for *Euphilotes* spp. More than 20 larvae in L1–L4 were collected on *E. thymoides* on May 19 in Schnebley Coulee. This population was previously known from small numbers of adults, but the host plant had not been determined. Additionally, *Euphilotes* larvae have recently been recovered from *E. ovalifolium* (Aug 18, Chumstick Mt, Chelan Co.) in Washington. Further study is needed to clarify the affinities of the various *Euphilotes* populations.

L4 @ 8 mm

Pupa @ 6.5 mm

Unnamed potential species   *Euphilotes on Eriogonum sphaerocephalum*

Silvery Blue   *Glaucopsyche lygdamus* (E. Doubleday)

Adult Biology

The Silvery Blue is common in Cascadia, occurring throughout the region except for a narrow strip along the Pacific Ocean. It ranges over much of N North America from AK to NM and E to NF, S to the Great Lakes. The single brood has a very long flight period, late Feb–mid-Aug depending on elevation and seasonal conditions. Numerous legumes have been reported as larval hosts, including *Lupinus* (lupine), *Lathyrus* (wild pea), *Lotus* (deervetch), *Vicia* (vetch), and *Astragalus* (rattlepod). Carey (1994) showed that greatest densities of *G. lygdamus* occur where multiple host plants are utilized. *Glaucopsyche lygdamus* occurs in a wide variety of habitats, including meadows, roadsides, clearings, riparian areas, grasslands, and alpine areas. Mating pairs are often seen on host plants. Ovipositing females are also commonly seen laying eggs singly on unopened new-growth leaves or buds. A female was observed pressing her abdomen into the coiled, unopened new-growth buds of several *Vicia cracca* plants, eventually placing a single egg well out of sight. On other hosts such as *Lotus crassifolius*, the eggs are placed on the tips of terminal leaves, while on *Lupinus sericeus*, eggs are pressed between unopened flower buds. At low elevations oviposition usually occurs in May.

Immature Stage Biology

We reared this species several times from eggs and gravid females. Larvae were frequently found and reared. Feeding occurs on the leaves of hosts like *L. crassifolius* and *V. cracca*, but on lupines only unopened leaf and flower buds are used, with feeding continuing on flowers. Legume peas provide an additional favored food source. Larvae bore round holes into the pea pods, often boring multiple holes

into a single pod. Like many lycaenids, the Silvery Blue has a symbiotic relationship with ants. The ants feed on sugary secretions produced by larval honey glands, lenticles, and eversible tubercles. The ants provide protection from predators and parasitoids (Fraser, 2001; Spomer and Hoback, 1998). Ballmer and Pratt (1991) showed experimentally that larvae of *G. lygdamus* ranked second out of 59 lycaenid species for attendance by *Formica pilicornis* ants. Larvae may be found in late May–June at low elevations, and as late as Aug at higher localities. Scott (1986a) reported that *Lupinus* hosts used by *G. lygdamus* are high in alkaloid chemicals; as mature larvae are diurnal and often found in open view on lupines, they may utilize alkaloids for defense. Eggs hatch in 4–6 days at 22–27 °C, and the larvae molt to second instars 3–4 days later. Larval development is rapid, taking just 20–21 days from oviposition to pupation in 2 cohorts we reared; Guppy and Shepard (2001) reported a developmental period of 26 days. In captivity, larvae wander off the host plant to pupate in refugia under the host plant. Bower (1911) and Ballmer and Pratt (1988) reported that larvae may have 4 or 5 instars, although we found only 4 in our rearings. Larvae do not build nests and are solitary; larvae sometimes disappear during rearing, indicating cannibalism, as reported by Bower (1911). The pupal stage lasts up to 10 months.

Description of Immature Stages

The white egg has a green interior and is ornamented with a matrix of narrow raised white ridges intersecting numerous times in delicate triangular and diamond-shaped patterns. Where the ridges intersect on the sides of the egg, there are raised rounded nodes. L1 is brown becoming lavender brown, with long pale brownish setae in 2 dorsal rows, also along the ventral margins. The head is black and gray, becoming shiny black prior to molting. L2 is gray to purplish brown with white markings. There is a dark dorsal stripe, broad at the front, tapering narrowly and fading to the rear. White stripes border the central dark stripe, with alternating variable white and gray diagonal stripes laterally. L3 is peppered with small black dots, the stripes and diagonals are more pronounced, and a distinct white longitudinal stripe is present ventrolaterally. Coloration is extremely variable in L3 and L4, ranging from brown to gray, green, purple, pink, or red. Layberry et al. (1998) wrote, "The larval colour depends on their food, varying from whitish or purple, if feeding on flowers, to green when eating leaves." In L4 the black dots are indistinct, and the setae are reduced in length but numerous, imparting a slightly shaggy appearance. The tapered dark

Adult

Pupa @ 9.5 mm

Egg @ 0.7 mm

dorsal line may be very distinct to obscure. Pupal markings are distinct and characteristic, making this arguably the most easily identified immature stage. The ground color is light brown with black blotches, speckles, and a broken dorsal stripe. Larvae are generally larger than most other blues, but extreme variability of color can cause confusion. The most similar species is *Glaucopsyche piasus*, but its larvae are gray and usually paler, typically have a less distinct dorsal stripe, and have a much paler pupa. Images of *G. lygdamus* larvae are common in the literature, including Allen (1997), Allen et al. (2005), and Wagner (2005, with ants); additionally Guppy and Shepard (2001) show both the L4 and pupa. All these images agree closely with ours.

Discussion

Late instars are readily found by searching lupines for ant activity or for damage to the peas, and gravid females oviposit well when confined with *Lupinus* flower heads. Rearing early instars is difficult because of difficulties in keeping hosts fresh. L1 and L2 usually remain well hidden among or in flowers and pods and are difficult to transfer when hosts are replaced; however, simply placing old host parts on new hosts will often result in larvae transferring themselves. Natural enemies, including parasitoids, are poorly known, although *Apanteles* wasps (Braconidae) were reported by Fiedler et al. (1992). Study is needed to determine whether some food plants, especially lupines, provide chemical defense for the larvae, and also to determine if larvae in Cascadia may develop through 5 instars under some conditions. Local associations with ants should be studied to determine the ant species involved and benefits to *G. lygdamus*. In areas where *G. piasus* and *G. lygdamus* co-occur, research on comparative biology and ecology may provide insight on competition and outcomes.

L1 @ 2.2 mm (late)

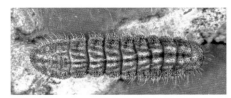

L2 @ 5 mm

L3 @ 8 mm

L4 @ 13 mm

L4 @ 16 mm

Silvery Blue *Glaucopsyche lygdamus* (E. Doubleday)

Arrowhead Blue *Glaucopsyche piasus* (Boisduval)

Adult Biology

Glaucopsyche piasus is widely distributed E of the Cascades but is relatively scarce or absent in many areas. Elsewhere it ranges from S BC and S AB south to Baja and NM, E to MT, WY, and CO. The single generation flies from mid-April to early Aug depending on elevation and seasonal conditions. Preferred habitats are lupine patches in desert, shrub-steppe, coniferous forests, canyons, and mountain meadows from ~200 to 8,000 ft (Warren, 2005). Both sexes visit flowers, and males visit mud and damp ashes. Adults are seldom found far from the lupine larval hosts. *Lupinus sericeus* (Silky Lupine), *L. laxiflorus* (Spurred Lupine), and *L. wyethii* (Wyeth's Lupine) are confirmed hosts in Cascadia. Other lupines such as *L. albifrons* (Silver Bush Lupine), *L. excubitus* (Grape Soda Lupine), and *L. argenteus* (Silvery Lupine) have been recorded elsewhere (Emmel and Emmel, 1973; Scott, 1986a), and it is likely that additional species are used in Cascadia. After mating, females oviposit on the tight, unopened flower heads of *Lupinus*, tucking eggs singly down between the buds or on the sides of the unopened racemes. Scott (1992) reported similar oviposition on *L. argenteus* in CO.

Immature Stage Biology

We reared or partially reared this species several times. Females obtained from Columbia, Franklin, and Chelan Cos., WA, on May 15 and 25, June 12, and July 3 readily laid eggs on unopened flower buds of *L. sericeus* and *L. lepidus* (Pacific Lupine). Eggs hatched after 2–3 days at 18–30 °C, and L1 immediately burrowed into the flowers or buds, but on all occasions L1 died because of deteriorating hosts. Eggs (*n* = 25) were collected June 24 and July 3 (different years) at No. 2 Canyon, Wenatchee, WA; hatch success was high, but the larvae failed to survive past L1. On July 10 and 22 (different years), also No. 2 Canyon, 17 larvae (L2, L3, and L4) were collected on *L. sericeus* and reared to 8 pupae on the peas of this host. The pupae were kept at ambient outdoor conditions until late Aug, when they were moved to 4 °C for overwintering. In mid-March they were exposed to 20–22 °C but failed to hatch. One pupa was parasitized, yielding an ichneumonid wasp. Remaining pupae (6) were overwintered again. In late April they were exposed to 20–22 °C, and 3 adults eclosed after 2 weeks. Oviposition is timed to enable larvae to exploit new blossoms, as indicated by hatched eggs on blooming flowers, but from L2, larvae are strictly associated with fruits (peas). Development from L2 to pupa took 19 days. The simultaneous presence of L2–L4 larvae suggests that oviposition occurs over 2–4 weeks. All larvae found were attended by ants, some tentatively identified as *Camponotus* sp. (carpenter ant). Larvae were readily found by searching for attending ants (2–17 per larva) on *L. sericeus* peas. In all cases the larvae were on or immediately adjacent to peas, which had neat round holes bored into them; some larvae were found with their heads and necks extended into the holes. Larvae often bore into a pod from multiple sites. Scott (1986a) reported that leaf feeding "sometimes" occurs in this species. Among larvae in captivity, we witnessed aggressive encounters suggestive of cannibalism, and the larvae generally avoided contact with each other. There are 4 instars, no nests are constructed, and the pupa overwinters, sometimes for >1 year.

Description of Immature Stages

The egg is pale greenish with a delicate latticework of white ridges. L1 is light brown grading to dull purple at each end and has a black head. L2 is creamy with a purplish cast, and there are many distinct, small brown spots. L3 is gray with a purplish cast; many small dark spots, concentrated anteriorly, cover the body. The segments are strongly divided, most bearing pairs of indistinct diagonal pale lines. A dark dorsal line tapers, widest at the anterior end, but is splotchy and indistinct. L4 is pastel gray with purplish and bluish shades, and there are paired diagonal side stripes. Short inconspicuous setae are numerous, and the tapered dorsal stripe is indistinct and grainy. The pupa is medium brown with pale greenish wing cases and dull blotchy gray spots. The most similar species is *Glaucopsyche lygdamus*, which differs in having a darker, more distinct tapered dorsal stripe. *Glaucopsyche piasus* larvae are usually paler than those of *G. lygdamus*; however, Allen et al. (2005)

Adult

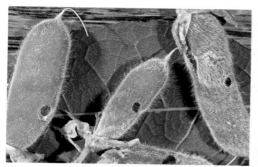

Lupinus sericeus **with feeding marks**

Eggs @ 0.5 mm

L1 @ 1.5 mm

suggested the 2 species may be indistinguishable as larvae. Ballmer and Pratt (1988) found the 2 species similar but noted several differences in the setae. The pupa of *G. piasus* is paler, with more subtle, smoother markings than that of *G. lygdamus*. Guppy and Shepard (2001) illustrate a mature larva and pupa that are similar to ours, and Allen et al. (2005) illustrate a mature larva that is similar except for a more distinct dorsal stripe.

Discussion

Gravid females readily oviposit on potted or cut lupines, but cut plants are unsuitable for L1 because of rapid deterioration. L1 feed poorly in captivity and are easily lost. Captured females are often infertile, suggesting that mating may not occur immediately after the females eclose. Eggs can be found by searching unopened flower buds of the host plant in areas where *G. piasus* flies; however, other lycaenids, especially *Plebejus icarioides,* may oviposit on the same plants. Later instars are easily found where *G. piasus* is common, betrayed by the presence of tending ants. Ballmer and Pratt (1991) considered larvae of *G. piasus* in CA to be facultatively myrmecophilous, tended by *Formica francoeuri* and a species of *Conomyrma.* Older larvae are easily reared but should be isolated to prevent cannibalism. Research is needed on the full range of lupines used by this species in Cascadia. The possibility that *Astragalus* spp. (milkvetch) may be utilized as a host, as suggested by Emmel and Emmel (1973), deserves study. The incidence of pupae diapausing for more than 1 winter, reported for the first time here, needs study. Larvae may be parasitized, especially by ichneumonid wasps, although no information is available on the identity or impact of parasitoids.

L2 @ 4 mm (late)

L3 @ 10 mm (late)

L4 @ 18 mm

Pupa @ 10 mm

Arrowhead Blue *Glaucopsyche piasus* (Boisduval)

Northern Blue   *Plebejus idas* (L.)

Adult Biology

Plebejus idas is a circumboreal species found in Eurasia and across North America (AK, Canada, Great Lakes, Rockies). In Cascadia it occurs in SE BC, NE WA, N ID, and NE OR. Favored habitats include moist montane meadows, alpine slopes, and trailsides above 3,000 ft. In BC and WA a single brood of *P. idas* flies relatively late in the season, from mid-June to mid-Oct. Guppy and Shepherd (2001) reported *Ericaceae* (heath family) as the host plants in BC, likely including *Vaccinium caespitosum* (Dwarf Bilberry), but we have not confirmed this usage in WA or OR. Pyle (2002) noted that *Empetrum* (crowberry), *Ledum* (Labrador tea), and *Kalmia* (laurel) are also reported as host plants elsewhere. In WA we found *P. idas* closely associated with *Lupinus sericeus* (Silky Lupine) in E Okanogan Co. and successfully reared it on this plant, as well as *L. lepidus* (Pacific Lupine). *Astragalus canadensis* (Canada Milkvetch) is another host plant recorded for WA (J. P. Pelham, in Pyle, 2002), and we also reared the species on *Astragalus purshii* (Woolly Milkvetch) but with limited success. Males eclose earlier than females in a partially offset flight period in which the sexes appear to fly together for as few as 7–10 days.

Immature Stage Biology

We reared this species 3 times from gravid females. Oviposition typically occurs in late July at mid-elevations such as Moses Meadows (~3,000 ft) in Okanogan Co. Seven females from that locality oviposited ~150 eggs on *L. sericeus*, placing ~60% on dried foliage, 20% on fresh leaves, and 20% on inert surfaces. The eggs diapause and overwinter. Embryonic development occurs prior to overwintering, and hatching occurs within a few days of introduction to warm conditions. L1 do not feed on eggshells. Larvae reared on *A. purshii* molted to L2 after 3 days at 20–22 °C. L2 ate small round holes halfway through leaves, hollowing them out and leaving distinctive yellow spots. Later instars showed a strong preference for the flower buds of *Astragalus* and eventually would not feed on other parts of the plant. Poor survival occurred on this host, with only 2 of 5 reaching L4 after 25 days. One larva pupated but died. A second batch of overwintered eggs produced several larvae, reared on *L. sericeus*. This batch matured more quickly and fed only on leaves, leaving the same yellow feeding spots seen on *Astragalus*. Rearing was trouble-free, 4 larvae producing pupae and adults. A third cohort of overwintered eggs hatched within 24 hrs at 22–28 °C and fed well on *L. lepidus* leaves during development, taking 30 days from egg-hatch to pupation (22–28 °C). Pupation occurred on upper surfaces of host leaves, and adults eclosed in ~12 days. Larvae of *P. idas* appear to prefer legumes with hairy leaves (*Lupinus* and *Astragalus*), build no nests, and have 4 instars. Larvae may be found by searching *L. sericeus* where *P. idas* is common, looking for yellow spots on the leaves about 2–3 weeks prior to the flight period. Scott (1986a) reported that *P. idas* is attended by ants, and in Europe often pupates in ant nests.

Description of Immature Stages

The egg is white with a greenish cast, the surface covered with a latticework of numerous intersecting ridges. L1 is pale green apart from the head and tip of the abdomen, which are medium gray. Tiny but distinct dark bullae are fairly numerous on the larva, each giving rise to a long blond seta (see fig. 48 in Scott, 1986a). L2 is green but develops a distinct dark dorsal line. There are also 2 broken dorsolateral lines and between them the larva is darker blue-green. L3 is bright grass green with a fairly distinct white ventrolateral line near the spiracles. There are faint diagonal white lines laterally on most segments, and small dark bullae carry long unbranched blond setae. L4 is green with a very distinct bold white lateral line. There is a distinct dark green line dorsally bordered on each side by broken white dashes. The black head bears numerous short setae. The pupa is bright green, with sparse short setae and a row of small pale brown spots along each side of the abdomen. The immature stages of *P. idas* and *P. anna* are very similar, with only slight differences in coloration. The immature stages of *P. melissa* are also similar, although L3 is dull brown, while this stage is bright green in *P. idas*. The pupae of *P. melissa* are usually darker than those

Adult

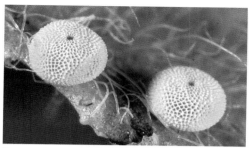

Eggs @ 0.6 mm

Pupa @ 9 mm

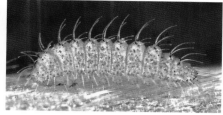

L1 @ 2 mm

L2 @ 3 mm

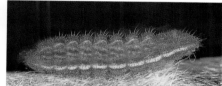

L3 @ 6 mm

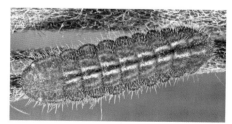

L4 @ 7 mm (early)

L4 @ 12 mm

Northern Blue *Plebejus idas* (L.)

of *P. idas*; however, there is variation within each species, and it is doubtful that the larvae of these species can be separated with certainty. Guppy and Shepard (2001) provided an image of a mature larva of *P. idas* that agrees closely with ours.

Discussion

Females should be obtained for eggs during a narrow window, perhaps little more than 7–10 days, when they are gravid and will oviposit. In our experience this is about the fourth week of July at mid-high elevations such as Moses Meadows, Okanogan Co. and S BC (Apex Mt). They should be confined with cut stems and foliage of their host plant, especially *L. sericeus*, on which they will oviposit. Many eggs are laid on the undersides of the foliage. The eggs should be kept at ambient temperature for some time (~2 weeks) for the embryos to develop before they are placed in cold storage for overwintering. As the larvae feed only on *Lupinus* leaves, it is not necessary to have the flowers available, but this may be different if using *Astragalus*. Further research on host-plant use by *P. idas* and *P. anna* in Cascadia is needed. Little is known about the biology and ecology of this species, including possible ant associations. The possibility of pupation in ant nests as reported from Europe appears unlikely in Cascadia but should be investigated. Biotic and abiotic factors influencing abundance and distribution also deserve study.

Anna's Blue   *Plebejus anna* (W. H. Edwards)

Adult Biology

Previously regarded as a subspecies of *Plebejus idas* (Northern Blue), *Plebejus anna* was elevated to species level by Guppy and Shepard (2001). The range of *P. anna* extends from S BC along the Cascades through WA and OR to N CA and along the Sierra Nevadas to S CA; it also occurs on Vancouver Is, in the Olympics of WA, and in the coast ranges of OR and CA. *Plebejus anna* flies in Cascadia in a single brood mid-June–mid-Oct and is found in cool, moist mountain meadows, along trails, on slopes, and in coniferous forest openings, usually above 3,000 ft, but ranging from 1,300 to 8,200 ft in OR (Warren, 2005). Larval host-plant records are confused with those of *P. idas*, but include *Lupinus arcticus* (Subalpine Lupine), *Lathyrus torreyi* (Torrey's Pea), *Astragalus* spp. (milkvetch), *Lotus* spp. (deervetch), and in OR, *Vicia ludoviciana* (Deerpea Vetch) (Warren, 2005). Additionally we observed close associations with *Trifolium* spp. (clovers) in Chelan, Okanogan, and Skamania Cos., WA, including oviposition, and a captive female from Pierce Co., WA laid eggs on *Trifolium medium* (Zigzag Clover). Males patrol all day near hosts and are often seen visiting mud. Both sexes visit flowers, including yarrow, aster, clover, and Pearly Everlasting. Females lay eggs singly on the stems and leaves or debris (Scott, 1986a).

Immature Stage Biology

We reared this species twice. Females collected July 30 at Reecer Canyon, Kittitas Co., WA produced 11 eggs which were held under ambient conditions until transferred to ~5 °C for overwintering on Oct 4. Five females collected from Sourdough Gap, Pierce Co., WA on Aug 27 laid ~100 eggs in 3 days on dried lupine foliage. Eggs were held at 22–27 °C until Oct 1, when they were transferred to an unheated garage for overwintering. Kittitas Co. eggs were transferred to 18–20 °C on May 10, and larvae were reared on a commercial lupine cultivar. Development took 28 days from L1 to pupation, with each instar averaging ~6 days. Pierce Co. eggs were transferred to 25 °C and continuous lighting on March 29. A few eggs hatched in 5 days, but most (>95%) remained dormant. Examination showed the eggs contained viable larvae for at least 2 months after removal from overwintering, suggesting that some eggs may overwinter more than once. L1 were fed on a commercial lupine cultivar before transfer to *Lupinus lepidus* (Pacific Lupine). Development from L1 to pupation took 36 days, with 7–9 days each in L1–L3 and 13 days in L4. Adults eclosed after 13 days. L1–L2 excavate round holes into leaves; later instars graze on leaf surfaces, creating characteristic pale areas. Pupation usually occurs on the host, often on the upper side of a leaf. Larvae are solitary, make no nests, and rely on camouflage and concealment for protection. Larvae are facultatively myrmecophilous and were attended by the ant *Formica pilicornis* for 68% of the time in a 5-min period in experiments reported by Ballmer and Pratt (1991). We have not observed ant attendance in Cascadia populations. There are 4 instars, and eggs overwinter.

Description of Immature Stages

The egg is a bluish white compressed sphere with a finely reticulate surface patterned in irregular polygons. L1 is creamy tan with a distinct rufous lateral stripe. Each segment has 25–30 brown or black spots bearing long, pale, backswept setae. The head is pale–medium brown and the true legs are black; there is a dark sclerotized plate dorsally on the last segment. L2 is uniformly green with numerous small brown-black spots supporting medium-length pale setae. There is an indistinct darker green dorsal stripe and darker green infusion laterally. L3 is brighter green with reduced spotting and proportionately shorter setae. The dark green middorsal stripe is prominent, and there is an indistinct whitish ventrolateral stripe that becomes more distinct with maturity. L4 is yellowish green to bright green with a distinct white ventrolateral stripe and a darker green dorsal stripe. Numerous tiny white spots and short white setae pepper the body. Indistinct darker green diagonal markings are sometimes present laterally, and the head is black. The pupa is variegated in pastel colors, greenish on the abdomen, pink, yellow, and green on the wing cases, and pale pinkish green around the head. A number of Cascadian lycaenids have stages

Adult

Egg @ 0.6 mm

L1 @ 1.5 mm

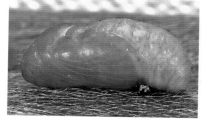

Pupa @ 10 mm

L2 @ 3 mm

L3@ 6 mm

L4 @ 9 mm

L4@ 9 mm (on lupine)

L4 @ 11 mm

Anna's Blue *Plebejus anna* (W. H. Edwards)

resembling those of *P. anna*. Mature larvae of *Callophrys perplexa* are similar, but the pupae are brown. The immature stages of *P. melissa* and *P. idas* are similar and may not be reliably separable from *P. anna*. *Plebejus idas* appears to have more distinct white dorsal dashes in L4, also a more distinct white lateral stripe, but these characters are probably variable. Ours appear to be the first published images of immature stages of *P. anna*. Guppy and Shepard (2001) and Allen et al. (2005) illustrate mature larvae of *P. idas*, and both are similar to our images of *P. anna*.

Discussion

The association we observed between *P. anna* and *Trifolium* (clover) needs investigation. It is possible that clover may be an unrecognized larval host plant in Cascadia; if so, the relationship between *P. anna* and *P. saepiolus*, which also uses clover and flies in similar habitats, deserves investigation. The differences between *P. anna* and *P. idas* also need further study, particularly as they relate to the biology and ecology of immature stages. The slight differences we observed in coloration and markings of larvae of the 2 species need confirmation by further rearing. The role of ants in the ecology of *P. anna* in Cascadia needs clarifying. The possibility that eggs in some populations of *P. anna* may remain dormant for extended periods needs study.

Melissa Blue   *Plebejus melissa* (W. H. Edwards)

Adult Biology

Plebejus melissa is a common Cascadia butterfly, found mostly in hot, dry, lower-elevation shrub-steppe habitats but also at higher elevations up to 4,000 ft (Warren, 2005). This butterfly has a long flight period of early May–late Sep, and is at least double and possibly triple brooded. Higher-elevation populations may be univoltine (Warren, 2005). *Plebejus melissa* is found E of the Cascades; elsewhere its range extends from the Cascade and Sierra Mts E to the Great Plains, and from middle Canada to Mexico. Both sexes feed on a wide variety of flowers, and males visit mud. Males patrol for females, which are found mostly near host plants. A wide variety of legume hosts are used, including *Astragalus* (milkvetch), *Lupinus* (lupine), *Oxytropis* (locoweed), *Medicago sativa* (Alfalfa), and *Lotus* (deervetch). Favored hosts in WA include *Astragalus purshii* (Woollypod Milkvetch), *A. lentinginosus* (Freckled Milkvetch), *A. canadensis* (Canadian Milkvetch), *Lotus nevadensis* ("Cowpie" Deervetch), and *Lotus sericeus* (Silky Lupine) (Warren, 2005). *Astragalus* is preferred early in the season, and in late summer the occasional still-fresh *L. sericeus* is used by the second brood, with eggs placed on the undersides of leaves. Scott (1992) observed wild eggs on a number of CO host plants, placed on leaves, stems, and basal debris.

Immature Stage Biology

We reared or partially reared this species on several occasions. On June 16 a female from Waterworks Canyon, Yakima Co., WA produced ~50 eggs that developed to 10 pupae 24 days after oviposition. Reared at 22–27 °C on *Lupinus lepidus* (Pacific Lupine), eggs hatched after 8 days and development from L1 to pupation occupied 16 days, with ~4 days in each instar. Adults eclosed after 8 days (July 22–24). On Sep 4, 3 females were collected from Wenas Lk, Yakima Co. They produced 4 eggs that diapaused and were overwintered with 13 eggs collected the same day on *L. sericeus*. The eggs hatched 3 days after removal in May, were reared on *A. purshii*, and molted to L2 and L3, 11 and 18 days, respectively after egg-hatch; this cohort was not documented further. On May 16, 3 L4 and 4 pupae were collected at Wenas Lk under *A. purshii* and reared to adults. On May 31 several pupae and 2 L4 were collected from the same locality and situation, and reared to adults. Eggs were found on several occasions, always on the undersides of host leaves, especially those of *L. sericeus*. Eggs hatch in early spring as soon as host plants are available, the larvae maturing quickly. Developing larvae eat round holes halfway through leaves, producing characteristic yellow feeding marks, recognizable in the field. The larvae feed nocturnally, retreating under the base of the host plant for protection during the day. Larvae found on host-plant leaves were often attended by ants, as were those found resting under host plants. Larvae of *P. melissa* were ranked highly for ant attendance by Ballmer and Pratt (1991), and Neill (2007) provided a photo of ants attending a mature *P. melissa* larva. Pupation occurs under the host plants; we found pupae under basal leaves and in soil voids utilized by ants. Larvae do not construct nests although they may occupy (or be carried to?) ant holes. This species is solitary and may be partially cannibalistic in crowded conditions. There are 4 instars, and the egg overwinters. The survival strategy is based on camouflage, concealment, ant-attendance, and nocturnal behavior.

Description of Immature Stages

The egg is a white compressed sphere with a flat top, the surface covered with a delicate pattern of numerous interconnected rounded polygons. L1 is pale green and bears 2 long and 2 short pale dorsal backward-curving setae on segments 1–8. A single row of setae occurs around the basal periphery, and there are indistinct rows of small dark spots. L2 is chartreuse green with an indistinct pattern of alternating green and whitish broken lines laterally and 2 indistinct pale stripes dorsally separated by a dark line. Pale setae of varied lengths impart a shaggy appearance, and there is a white line low on each side. L3 is similar except that dorsal markings are prominent, with a central dark line bordered on each side with white stripes. Setae are numerous,

Adult

Egg @ 0.6 mm

L1 @ 2 mm

L2 @ 5 mm

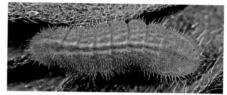

L3 @ 8 mm

L4 @ 12 mm

L4 @ 16 mm

Pupa @ 9 mm (pharate)

Melissa Blue *Plebejus melissa* (W. H. Edwards)

white, and medium length, and the ventrolateral line is yellowish. In L4 the dorsal and ventrolateral lines are distinctly yellowish, and numerous tiny setae adorn the body, each bearing a whitish yellow "club," giving the larva a fuzzy appearance. The pupa is smooth-ovoid, green, becoming light brown and green with maturity, with an indistinct dark broken dorsal line. Setae are numerous but tiny and slightly fuzzy in appearance. The spiracles are pink. The most similar immature stages are those of *Plebejus anna* and *P. idas* and, while both tend to have whiter lateral stripes in mature larvae, the 3 species are probably not reliably separable. *Plebejus anna* occurs in more mountainous areas, and *P. idas* shuns xeric shrub-steppe habitats favored by *P. melissa*. Published images include mature larvae in Ballmer and Pratt (1988), Guppy and Shepard (2001), Miller and Hammond (2003), and Neill (2007), all with distinct lateral white stripes, and Allen et al. (2005), in which the larva has a more whitish, hoary appearance imparted by setae.

Discussion

Plebejus melissa is easily reared from eggs, which are readily obtained from gravid females or by searching host plants. Larvae are also readily found using yellow feeding spots as a field mark. Ants often signal the presence of the larvae of this and other lycaenid species, but the ants attending *P. melissa* have not been identified in Cascadia; Ballmer and Pratt (1991) identified *Formica neogagates* attending *P. melissa* in CA. Braconid wasps often parasitize *P. melissa,* emerging from many field-collected pupae, but little is known of other parasitoids, diseases, or predators. Utilization of Alfalfa as a host has likely enabled *P. melissa* to expand its range; however, research has shown that *P. melissa* reared on Alfalfa produces smaller adults than those reared on *Astragalus*. *Plebejus melissa* seems particularly vulnerable in unusually cold spring weather during which populations may suffer.

Greenish Blue   *Plebejus saepiolus* (Boisduval)

Adult Biology

Plebejus saepiolus is common in Cascadia, sometimes abundant in prime habitat, flying in moist areas mostly E of the Cascades in WA, OR, ID, and BC, but scattered in coastal areas as well. Beyond Cascadia it ranges from AK to CA, throughout the W, and across much of Canada to the maritime provinces. It is a species of cool mountain meadows, found wherever moist seeps and clovers occur together, flying in a single brood mid-May–mid-Aug. Both sexes visit flowers, including clovers, yarrow, and asters, and males visit mud. Males patrol near hosts, flying close to the ground seeking females. Larval hosts are native and nonnative clovers, including *Trifolium longipes* (Long-stalked Clover), *T. wormskjoldii* (Springbank Clover), and *T. pratense* (Red Clover), as well as *Lotus corniculatus* (Bird's-foot Trefoil) and *Hedysarum boreale* (Northern Hedysarum). Scott (1992) reported numerous eggs on 2 species of *Astragalus* (milkvetch) in CO. Females place their eggs singly, deep in blooming clover heads, attached to the base of a seed or the side of a petal. Clover flowers at the time of oviposition may be brown externally, but the seeds inside are usually still fresh and green; females will not oviposit after the clover seeds have browned or senesced, nor will they oviposit on leaves.

Immature Stage Biology

We partially reared this species several times. Eggs were obtained 3 times from gravid females collected during June–July in Yakima, Pend Oreille, and Columbia Cos., WA. In one instance, all L1 were killed by Minute Pirate Bugs (*Orius tristicolor*), which commonly reside in clover flower heads. On the other occasions larvae were reared on clover flowers or leaves until L2. These entered dormancy, overwintered (5 °C) but died post overwintering. On July 16 near Quartz Mt, Kittitas Co., WA, 6 eggs were collected from clover flowers. Larvae were reared on clover flowers to L2, when they entered dormancy and were overwintered (Aug 15). In early May the larvae were transferred to 18–20 °C. Larvae fed on clover flowers and leaves but died in L3. On May 7, 2 overwintered L3 were collected from Taneum Creek, Kittitas Co., WA. One was in debris at the base of clover plants and the other on a stem. The larvae grew slowly, feeding on leaves and flowers, and reached L4, but died shortly thereafter. On May 29 a pink prepupal larva was collected from under clover at Taneum Creek; the clover was not blooming, indicating that postdiapause larvae develop on leaves. The larva pupated June 1 and eclosed June 13. On Aug 5 at Haney Meadows, Kittitas Co., when worn adults were still flying, an L3 was found on a clover flower, indicating that egg laying may be prolonged and that larvae may also overwinter as L3. Development from egg to L2 takes 2–3 weeks then L2/L3 enter diapause. Postdiapause development to pupation occupies 5–6 weeks at 20–25 °C, with adults eclosing after ~2 weeks. Larvae overwinter in vegetation/detritus at the base of host plants. Postdiapause larvae feed nocturnally on leaves, creating holes, and retreat daily to the base of the plant. We have not observed ant associations in this species. Overwintering occurs as L2 (Ballmer and Pratt, 1988) or L3, and there are 4 instars. Larvae do not build nests, relying on camouflage, nocturnal activity, and concealment for survival.

Description of Immature Stages

The pale greenish or bluish white egg is a compressed sphere, the surface coarsely patterned with ridges intersecting in irregular polygons. L1 is creamy tan, developing orange or red longitudinal stripes with maturity, and has a shiny black head and sparse medium-length pale setae arising from small brown spots. L2 is cinnamon brown with 3 darker brown dorsal stripes and numerous pale brown setae. Small dark brown spots are present around the periphery, and paler spots occur dorsally. L3 is initially patterned in variegated colors, yellow-brown laterally, charcoal blackish dorsally, with a bright, broad magenta-brown stripe ventrolaterally. There is a broken but bold black dorsal stripe, and numerous black setae arising from black spots. With maturity L3 is pastel green with subdued dark spots and paler setae, and the ventrolateral magenta stripe is bordered below with white. L4 is forest green infused with black anteriorly,

Adult male

Egg @ 0.5 mm (side and top views)

L1 @ 2 mm

Pupa @ 11 mm

L2 @ 2.5 mm

L3 @ 4 mm (early)

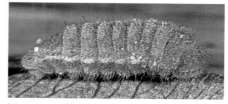

L3 @ 7 mm (late)

L4@ 10.5 mm

L4 @ 13 mm (prepupal)

with a contrasting white stripe ventrolaterally. Pale setae are abundant but fairly short, and there are numerous tiny black and white spots. The prepupal L4 is fuchsia pink. The pupa is rather elongate and bright muddy pink, the abdomen darker than the thorax and wing cases. A number of Cascadian lycaenids are similar to *P. saepiolus* in some stages; however, the multi-toned pink and green L3 and the elongate pinkish pupa are distinctive. Published images of mature *P. saepiolus* larvae are found in Guppy and Shepard (2001) and Allen et al. (2005), both very similar to our mature green larva image. Scott (1986a) shows a differing mature larva, very pale with a partial dorsal red stripe and indistinct diagonal side stripes.

Discussion

Eggs of this species are easily found by searching clover flowers in July and Aug in suitable habitat. Prediapause larvae can be reared on clover flowers or leaves. Clover in inland areas often harbors Minute Pirate Bugs, which can destroy larvae rapidly in captivity and likely exert significant mortality on wild populations. Larvae overwinter well, but rearing postdiapause larvae is difficult, with most feeding poorly. The use of nonnative clovers has favored this species; it would be interesting to compare the ecology on native and nonnative hosts. Why this species is apparently not ant-attended is unknown (due to preferred wet habitats?), but this would make an interesting study. Apart from Minute Pirate Bugs, little is known of the natural enemies regulating populations of this species.

Greenish Blue *Plebejus saepiolus* (Boisduval)

Boisduval's Blue   *Plebejus icarioides* (Boisduval)

Adult Biology

This large blue is common and widespread in Cascadia, with males often seen in congregations imbibing at mud. Its range extends from the Great Plains to the W Coast and from S BC to Baja. There are a number of recognized subspecies, including the federally listed *P. i. fenderi* in W OR. Information on the biology of this subspecies is provided by Schultz et al. (2003). The flight period is mid-April–mid-Sep, although Guppy and Shepard (2001) reported that in any one locality the flight is only ~30 days. Typical habitat includes forest openings, subalpine meadows, seeps, and shrub-steppe. Both sexes nectar on flowers, including lupines, thistles, and asters, and females oviposit on lupines. Scott (1986a) reported that eggs are laid singly on leaves, stems, flowers, and pods, favoring new growth. Larval hosts in Cascadia include *Lupinus sericeus* (Silky Lupine), *L. lepidus* (Pacific Lupine), *L. laxiflorus* (Spurred Lupine), *L. latifolius* (Broadleaf Lupine), and *L. sulphureus* (Sulphur Lupine). Nearly 40 species of lupine are recorded elsewhere, with hairier species preferred (Scott, 1986a). *Plebejus icarioides* is univoltine, although Scott (1986a) reported an apparent second brood in CO.

Immature Stage Biology

We reared or partially reared this species several times. On April 17, an L4 was collected on Whiskey Dick Ridge, Kittitas Co., WA, and on April 30 an L3 was collected from the same site. Both pupated; 1 was parasitized and the other eclosed. On May 20, a female from Reecer Canyon, Kittitas Co., WA, produced several eggs. Three larvae reached L2 ~20 days post egg-hatch and entered diapause. They were placed in 3–5 °C in mid-Aug and removed May 13;

they grew to L4 within a week but then died. Females from Benton and Yakima Cos., WA, during May of 2 different years laid ~150–200 eggs on each occasion. Eggs hatched after 5 days at 18–25 °C (up to 9 days at 20–22 °C) and reached L2 after 14 days (June 1–15), feeding on *L. lepidus*. Diapause, as evidenced by the larvae becoming purplish brown, seeking refuge, and ceasing activity, occurred in L2. Larvae were held at ambient summer temperatures until transferred to 5 °C on Sep 5. After 110–205 days, L2 were transferred to 18–22 °C and supplied with *L. lepidus*. Development to pupation took 40 days, with 15, 20, and 15 days spent in L2, L3, and L4, respectively. Adults eclosed after 10–14 days. Larvae fed on leaves, eating holes halfway through and leaving round yellow scars. Scott (1986a) reported that larvae eat leaves, then transfer to flowers and fruits, in spring eating young shoots. Larvae are tended by ants, and sometimes rest by day in holes dug by ants below the plant. Attendance by the ant *Formica pilicornis* was among the highest recorded in a group of 59 lycaenid species tested by Ballmer and Pratt (1991). We confirmed ant attendance in Cascadia. There are 4 instars, the larvae do not construct nests, and overwintering occurs as a diapausing L2 at the base of the host plant. The larvae are cannibalistic in crowded conditions. Camouflage, diurnal concealment, and ant attendance are likely important features of defense.

Description of Immature Stages

The egg is pale greenish white, a compressed sphere covered with intersecting ridges forming numerous rounded polygons. L1 is pale green with a black head. Small dark green spots support long pale setae, ~20 per segment. L2 is pale yellowish green; black speckling is prominent, and the setae are yellowish and numerous. Dorsally there is an indistinct green stripe with pale yellow edging. This instar darkens and shrinks prior to diapause. Postdiapause L3 remain purplish brown or are forest green with a pale purplish dorsal stripe, bordered indistinctly with white. A pale lavender band encircles the base, bordered with white below. Black speckling remains and setae are numerous and short. Laterally each segment has 2 indistinct pale diagonal stripes. This pattern extends to L4; a middorsal purple stripe becomes more distinct, and some larvae become more solidly green. A contrasting pale yellow or white ventrolateral stripe is present. Some larvae remain purplish brown until midway through L4, when they become greener. The ovoid pupa is bright cinnamon brown on the abdomen and dull green on the head, thorax, and wing case. The larvae of *Plebejus idas*

Adult

Egg @ 0.7 mm

L1 @ 2.0 mm

L2 @ 3.5 mm

Pupa @ 9 mm

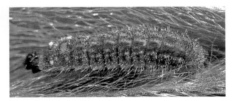

L2 @ 4 mm (diapause colors)

L3 @ 5 mm

L4 @ 9 mm

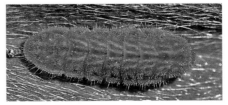

L4 @ 10 mm

are similar. The larvae of *P. saepiolus, P. melissa, P. lupini, and P. anna* may also be confusing, as well as some *Satyrium* hairstreaks. Attention to host plant, timing, and larval patterns will aid identification. Photographs of larvae appear in Guppy and Shepard (2001), Miller and Hammond (2003), and Allen et al. (2005) but do not closely resemble ours. The diapausing and mature larvae pictured by Neill (2007) and the mature larva and pupa shown by Pyle (1981) are more similar.

Discussion

Adults are easily found, and females oviposit well in captivity. Larvae should be held until they are in diapause (brown, immobile) before overwintering, but winter survival is often poor. Larvae can be found in April with thorough searching, concentrating on feeding marks on lupine and the presence of ants, then searching under plants for the nocturnally active larvae. Parasitism is common; Guppy and Shepard (2001) mentioned egg parasitism by *Trichogramma* (Hymenoptera), and larval parasitism by *Apanteles theclare* (braconid wasp) and tachinid flies. The factors determining onset and termination of diapause have not been established. The impact of environmental conditions during development, particularly temperature and moisture, on adult phenotype needs study.

Boisduval's Blue *Plebejus icarioides* (Boisduval)

Acmon Blue   *Plebejus acmon* (Westwood)

Adult Biology

Plebejus acmon was long believed to be widespread and common in WA; however, Warren (2005) determined that the predominant acmonoid species in Cascadia is in fact *Plebejus lupini*. Warren discovered a small lowland population of *P. acmon* in S WA along the Columbia R, and we have found other lowland populations in the Columbia Basin. In OR, *P. acmon* is widespread and sometimes common (Warren, 2005) where first-generation adults fly late April–early May at lowland sites; second-generation adults occur at Columbia Basin sites during Aug–Sep. Warren (2005) reported that *Eriogonum* as well as legumes are used by this species in OR. The Columbia R population in WA is associated with *Eriogonum compositum* (Northern Buckwheat), the probable but unconfirmed larval host there. Warren (2005) reported that montane populations of *P. acmon* also occur in OR at elevations of 4,000–6,500 ft, where the host plant is *Eriogonum nudum* (Barestem Desert Buckwheat); these populations are single brooded, flying June–July. In WA we discovered a similar montane population in W Yakima Co., WA, near Rimrock Lk at 3,700 ft. As with the OR mountain populations, the larval host is *E. nudum*, a plant with very limited distribution in WA; both sexes fly together in the proximity of *E. nudum*, which serves as both a nectar source and larval host. Females oviposit on the flower heads at the prime of their bloom, placing the eggs on sepals below the individual flowers.

Immature Stage Biology

We partially reared this species 3 times. On May 5, 5 females were collected at Dallesport, WA, but eggs

Adult

obtained were infertile. On July 2, 2 L2 and several gravid females were collected from near Rimrock Lk, laying ~30 eggs on *E. nudum* flowers. Eggs hatched after 6 days, and larvae were reared at 20–22 °C; the field-collected L2 reached L3 in 9 days, but none survived further. On a return visit to Rimrock Lk on July 21 (same year), 5 L3 and L4 were collected from *E. nudum* flowers and reared on this host at 20–25 °C; 4 reached pupation beginning July 27, and adults eclosed by Aug 13. At all stages the larvae fed only on flowers and fruits (seeds) of *E. nudum*, which retains fresh flowers well after the leaves have senesced. Leaves were senescing in early July and were senesced before the larvae were half-grown in mid-July; thus *P. acmon* larvae on *E. nudum* appear to be obligate flower feeders. The overwintering stage was not determined as all individuals eclosed to second-brood adults, possibly an artifact of captive rearing as montane populations do not appear to produce a second generation. Development from oviposition to pupation appears to be ~4 weeks, and adults eclosed ~7 days after pupation. The well-camouflaged larvae did not build nests and typically remained hidden deep in the host flower heads feeding on the seeds. In captivity the larvae wandered off the host plant before pupating. There are 4 instars and the survival strategy is based on camouflage and concealment.

Description of Immature Stages

The egg is pale green and finely textured with dots arranged in irregular rows. L1 is pale dusky yellow with long white setae, and the head capsule is purplish black. In L2 the head color matches the pale greenish cream body color, and long setae persist, imparting a shaggy appearance. In L3 the setae are proportionately shorter, and a pattern of indistinct alternating light and dark diagonal lines appears on each segment. Pink coloration develops in this instar, becoming brighter late in the stage and developing pale and bright pink alternating patterns in some individuals. The intensity and pattern of the coloration is variable, perhaps depending on host-plant pigment. L4 has short bristly setae, alternating pale and darker diagonal lines on the sides, and a single dark pink dorsal line. The pupa is smooth and tan, bearing numerous very fine setae but lacking distinct markings. Larvae of this species may occur with those of *Euphilotes enoptes*, but *P. acmon* larvae are shaggy with numerous setae while those of *E. enoptes* bear setae only along the basal periphery in later instars. The eggs of *Euphilotes* spp. lack the well-defined rows of textured dots of *P. acmon*. *Plebejus lupini*, a sibling species often confused with

Egg @ 0.5 mm

L2 @ 2.5 mm

L1 @ 1.1 mm

L3 @ 5 mm

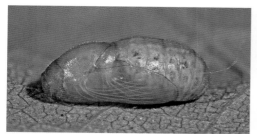

Pupa @ 8 mm

L3 @ 5.5 mm

L4 @ 9 mm

L4 @ 9 mm

Acmon Blue *Plebejus acmon* (Westwood)

P. acmon, is not known to use *E. nudum* but feeds on other *Eriogonum* species. *Plebejus lupini* larvae are green with no pink coloration in populations we have observed. Allen et al. (2005) show a mature larva from CA which is green and lacks a distinct dorsal line; Guppy and Shepard (2001) illustrate a pupa similar to ours, also a green larva with pink highlights.

Discussion

Gravid females produced relatively few eggs on host-plant cuttings; mortality of early instars was high, and it was difficult to rear larvae past L1. Later instars can be collected from host-plant flowers by careful examination in the field, or by collecting flower heads and waiting for larvae to wander off. Much research is needed to better define differences in the morphology, biology, and ecology of immature stages of *P. acmon* and *P. lupini*. A better understanding of the distribution of these 2 species in Cascadia is also needed, and host plants used by low elevation populations need confirmation. It is very likely that additional low-elevation populations of *P. acmon* will be found along the Columbia and Snake rivers.

Lupine Blue   *Plebejus lupini* (Boisduval)

Adult Biology

Adults of *Plebejus lupini* and *P. acmon* are very similar and have long been confused in Cascadia. In the currently accepted distribution, *P. lupini* is widespread and common in S BC and WA E of the Cascades and is locally present in the Olympics, while *P. acmon* is uncommon, found in small colonies along the Columbia R, in W Yakima Co., WA and along the lower Snake R. In OR both species are widespread, extending S to Baja CA and E to the Midwest. A range of habitats are occupied from low to high elevations, including shrub-steppe, forest openings, canyons, and alpine slopes. Larval hosts are buckwheats, including *Eriogonum compositum* (Northern Buckwheat), *E. niveum* (Snow Buckwheat), and *E. sphaerocephalum* (Round-headed Buckwheat). Pyle (2002) reported *E. ovalifolium* (Oval-leaf Buckwheat) and in OR, Warren (2005) added *E. umbellatum* (Sulphur Flower) and *E. heracleoides* (Parsley Desert Buckwheat). Pratt and Ballmer (1992) reared *P. lupini* on *Lotus scoparius* (Deervetch) in the laboratory, but legumes are not reported as natural hosts. Females lay eggs singly on host flowers, leaves, or bracts. The flight period is mid-April–late Sep, and Warren (2005) reported that *P. lupini* is univoltine in OR, but it appears to be facultatively bivoltine in WA.

Immature Stage Biology

We partially reared this species several times. Four postdiapause L4 and 1 L3 from Cowiche Canyon, Yakima Co., and Reecer Canyon, Kittitas Co., WA during April–May (3 years) were reared to adults that eclosed in May. Postdiapause larvae are typically found on leaf undersides, where holes are eaten halfway through the leaf, leaving yellow blisters. Spring (May) females from near the Snake R, Franklin Co., WA laid 20–50 eggs on *E. niveum* on 3 occasions. Eggs hatched in ~7 days at 22–26 °C. L1 were fed *E. niveum* leaves, reaching L2 and L3 after 8 and 15 days, respectively. Under natural daylengths and at 22–28 °C, larvae stopped feeding in early–mid-June and entered diapause. On June 12 near Ellensburg, Kittitas Co., 6 eggs and 4 L1 were collected from flowers of *E. sphaerocephalum*. Larvae developed to L3 by July 23 and pupated by July 29, eclosing to adults in Aug. On July 16, 1 egg and 4 L2 collected from *E. sphaerocephalum* at Ellensburg diapaused as L2 and L3. One female collected near the Tucannon R, Columbia Co., WA on July 10 laid ~50 eggs on *E. heracleoides* and *E. umbellatum*. L2 entered diapause in late July and were overwintered at 5 °C from Sep 14. Larvae were exposed to 19–21 °C and natural daylengths on March 26 and fed on *E. elatum* and *E. heracleoides*. Most died, but 1 molted to L3 and L4 after 15 and 27 days. On entering L4 it was exposed to continuous lighting. Feeding ceased; the larva turned pinkish brown and remained dormant through the summer. On Oct 15, the larva was overwintered a second time (5 °C) for 101 days then transferred to 20–26 °C, 16 hr light. Feeding resumed on *E. elatum*, but the larva died after 24 days without pupating. Two females obtained from Waterworks Canyon, Yakima Co. on Sep 13 laid 10 eggs on *E. elatum* which developed to diapausing L3 by Oct 8. In captivity pupation occurred under cover on the cage floor, and in the wild probably occurs under debris at the base of host plants. L1–L2 feed on leaves, flowers, and seeds and develop slowly. The larvae are solitary and may be cannibalistic in close confinement. There are 4 instars, and larvae rely on concealment and camouflage for survival. We have not observed ant-larvae associations in spring but have seen associations in summer. Peterson (1993) reported that *P. lupini* is facultatively myrmecophilous in Yakima Co., tended by *Tapinoma sessile* and *Formica neogagates*.

Description of Immature Stages

The pale greenish egg is covered with a fine lattice of intersecting ridges. L1 is yellow-tan with small brown spots, each sporting a long white seta. L2 is uniformly pale yellowish with numerous pale setae of varying lengths, appearing shaggy. L3 is pale greenish, maturing to green, with dense white setae continuing to impart a shaggy appearance; numerous white spots produce a silvery sheen. In some individuals there is a pale ventrolateral stripe. L3 may be reddish if feeding on flowers. L4 is uniformly medium-dark green, densely covered with contrasting white setae

Adult

Eggs @ 0.5 mm

L1 @ 2.0 mm

Pupa @ 8 mm

L2 @ 2.5 mm

L3 @ 6 mm

L4 @ 13 mm

L4 @ 13 mm

Lupine Blue *Plebejus lupini* (Boisduval)

(star-shaped on the tips), a bold ventrolateral white stripe, and indistinct pale diagonal marks laterally. Larval coloration varies according to host-plant species and whether leaves or flowers are eaten. The pupa is brownish near the head, the wing cases are green, and the abdomen is mottled yellow and green. A number of lycaenid larvae are similar to *P. lupini*; however, the shaggy appearance is unique in our area. *Plebejus acmon* is most similar but was pinkish in our rearings, likely because of flower feeding. L4 of *P. acmon* are more strongly marked, but the early instars are similar. Our images appear to be the first published of *P. lupini* immature stages.

Discussion

Spring larvae are not difficult to find by searching *E. compositum* leaves with yellow feeding scars in April. Eggs can be found in June by carefully searching buckwheat flowers, and gravid females oviposit readily in captivity. Voltinism is uncertain in Cascadia. Our observations suggest that some low–mid-elevation populations in WA (Kittitas and Yakima Cos.) have 2 broods (April–Jun, Aug–Sep) while others have a single spring-summer brood (Columbia and Franklin Cos.). In the former case the species may be truly double brooded, or some larvae that aestivate for a period in summer before resuming development may potentially produce an extended single brood. Our rearings also indicate that larvae have the potential to overwinter twice (L2/L3 and L4). More research is needed to clarify voltinism and the determinant environmental cues. The impact of flower-feeding on larval coloration needs attention, and further comparisons with *P. acmon* are needed. Rearing on *Lotus* might produce interesting results.

Arctic Blue   *Plebejus glandon* (de Prunner)

Adult Biology

Plebejus glandon is a Holarctic species, found in N latitudes in Eurasia. It is found throughout Canada, extending S into the US along the Rocky Mt corridor and through the Sierras to S CA and NM. In Cascadia *P. glandon* is found in the Olympic and Cascade Mts of S BC and WA. *Plebejus glandon* flies in alpine areas in a single brood from late June to late Aug. The preferred habitat is high, windswept rocky ridges and scree slopes, although it has been found at mid-elevation in Okanogan Co., WA. This little blue flies close to the ground, males perching on rocks waiting for females; the females are mostly seen nectaring. The only known larval host in WA is *Saxifraga bronchialis* (Spotted Saxifrage), although *S. tolmei* (Alpine Saxifrage) is probably used in the Olympic Mts, the Goat Rocks wilderness, and N of Mt Baker, and is confirmed in BC. Scott (2006) observed oviposition on *Androsace septentrionalis* (a primrose) in CO. Females lay eggs singly on the spiny leaves of the host, tucking them into crevices.

Immature Stage Biology

We reared or partially reared this species 6 times; in all cases stock was from Slate Peak, Okanogan Co. On July 27, 3 females produced many eggs; larvae were reared on *S. bronchialis*. One reached L2 and overwintered together with 4 late L1. Three survivors refused to feed on a *Saxifraga* cultivar in the spring and died. On Aug 6, 3 eggs were collected; 1 hatched, developed to L2, and diapaused, but failed to overwinter. On July 30, 5 females produced eggs; larvae were placed on potted *S. bronchialis* but none overwintered. On July 15, an L4 collected on *S.*

Saxifraga bronchialis (host plant)

bronchialis pupated within 3 days. On Aug 4, 20 eggs were collected from *S. bronchialis*. Most were dead or parasitized, but 5 hatched ~6 days post oviposition; larvae were reared on *S. bronchialis* at 18–20 °C. Four died in L2 and L3 from mold, but 1 pupated on Sep 18 and eclosed on Oct 2. Development from L1 to pupation was 39 days, with ~9–10 days in each instar; eclosion occurred 14 days later. Five females obtained on Aug 6 produced ~40 eggs which hatched in 8 days at 21–27 °C. Development from L1 to L3 took ~18 days. Most L2 and all L3 entered dormancy in early Sep and were transferred on Sep 17 to 5 °C for overwintering. Thirty percent of larvae (all L2) survived and were exposed to 20–22 °C on April 19. Feeding on *S. bronchialis* commenced immediately, with L3 reached after 8 days. Thereafter, development slowed and larvae appeared to become dormant, with none surviving to adulthood. Oviposition in captivity always occurred on green parts of *S. bronchialis*. The larvae fed nocturnally on *Saxifraga* leaves, especially near the tips, hiding by day in the complex, spiny leaves. Survival is based on camouflage (red larva on red flower stalks) and nocturnal activity. The larvae do not build nests, have 4 instars, and overwinter opportunistically as L1–L3, and possibly in other stages.

Description of Immature Stages

The egg is a compressed white sphere with an ornamented surface of small rounded polygons. L1 is pale tan with a shiny black head and dark sclerotized dorsal plates on the first and last segments. Each segment has ~16 dark spots, most bearing a short–medium-length coarse seta. L2 is dark reddish brown with a broken dark dorsal stripe. L3 is purplish red with a broad black dorsal stripe and black diagonal marks laterally. L4 is dark purplish red with a contrasting white ventrolateral stripe, a black dorsal stripe

Adult

Egg @ 0.7 mm

L1 @ 1.1 mm

L2 @ 2.2 mm

L3 @ 6 mm

L4 @ 10 mm

L4 @ 10 mm

Pupa @ 9 mm

Arctic Blue *Plebejus glandon* (de Prunner)

bordered white on each side, and black diagonal marks dorsolaterally. Setae are numerous, short, and dark. The pupa is ovate; the abdomen is bright reddish with a dark dorsal stripe bordered white on each side. Other Cascadian lycaenids may have reddish or purplish instars (e.g., *Callophrys mossii, Strymon melinus, Satyrium californica, Apodemia mormo*); however, none have the persistent deep magenta-purple colors of *P. glandon* and all should be easily separated, particularly when habitat and host plants are considered. Allen et al. (2005) illustrate a mature larva from CA which is dark green with only a trace of purple around the base. Similarly, Ballmer and Pratt (1988) illustrate a green L4 of *Agriades franklinii* from CA, a subspecies of *P. glandon* (see Pelham, 2008). Neill (2007) illustrates an L4 from Cascadia identical to ours.

Discussion

Gravid females oviposit reasonably well in captivity. Eggs are fairly easily found by examining host plants late in the flight period, searching for white eggs contrasting with dark green leaves; however, many eggs are parasitized and others predated (collapsed in situ). Rearing larvae requires frequent attention, and mold can be a problem. Obtaining a reliable source of the host plant at lower elevations is difficult, and overwintering survival is poor. Larvae produce considerable frass, which may provide a search pattern in the field. Little has been published on the life history of this species and host plants are imperfectly known in Cascadia. Development to adult in 1 season, as occurred in one of our rearings, is unlikely to occur in the wild because of the short growing season at higher elevations where this species occurs. The reluctance of larvae to develop during late spring in one of our rearings supports this hypothesis. The effect of temperature and daylength on development of pre- and postdiapause larvae and on overwintering stage and survival needs study. Confined to cool, high-elevation areas, *P. glandon* may be adversely affected by climate warming.

Mormon Metalmark   *Apodemia mormo* (C. & R. Felder)

Adult Biology

Apodemia mormo is the only metalmark found in Cascadia, occurring along the E flanks of the Cascades, also along the Snake R, and near the Ochoco Mts in OR. A population also occurs in the Similkameen R Valley in S BC. Elsewhere it ranges through CA to Mexico and E to the desert SW, S Rockies, and Black Hills of SD. Favored habitats include canyons, arid flats, and roadsides in dry, hot sagebrush country. The single brood flies from early Aug to late Sep. Several species of buckwheats are used as larval hosts in Cascadia, including *Eriogonum elatum* (Tall), *E. compositum* (Northern), *E. sphaerocephalum* (Round-headed) and *E. niveum* (Snow Buckwheat). Others have been reported elsewhere (Scott, 1986a). A female was observed ovipositing Aug 28 near Ellensburg, WA on a nearly leafless *E. sphaerocephalum* plant. She laid a single egg on a woody branch tip near a terminal cluster of leaves. Captive females with buckwheat cuttings oviposited about equal numbers of eggs on leaves, flowers, stems, and inert surfaces.

Immature Stage Biology

We partially reared this species several times. The Aug 28 egg (above) overwintered and was reared to L2 the following spring. On Aug 31, 6 females from Kittitas Co., WA produced 47 eggs. On 4 occasions, 5–20 females from Yakima Co., WA produced very few eggs, 7–29 Sep. Dissection of these females revealed few eggs, suggesting most eggs are laid earlier in the flight period. Eggs held in warm conditions for ~1 month develop embryos then diapause and overwinter. Overwintered larvae hatched in 4–10 days under warm (20–26 °C) conditions and were reared on *E. sphaerocephalum, E. elatum,* or *E. niveum.* Larvae developed slowly, feeding intermittently and spending much time sheltering in apparent dormancy.

Adult

Larvae chewed tiny holes halfway through the middle part of *Eriogonum* leaves, usually from the upper side. Most died in L1, some molting to L2 after 7 days at 25 °C. Larvae fed unpredictably during daylight hours, although nocturnal feeding is reported (Scott, 1986a). Following the molt to L2, slow development continued and most died, only 1 individual reaching L3. On May 14 an L3 was found on *E. compositum* W of Spokane, WA and reared to L5. At 16 mm it molted to L6 (June 20), shrinking to 12 mm, thereafter regrowing slowly to 17 mm. The larva died in early July without pupating. On July 29 near Wenatchee, WA we found a prepupal L5 (?) in a very large messy silk nest incorporating several large basal leaves of *E. elatum*; it pupated and an adult eclosed. Scott (1986a) reported that "older larvae…live in a nest of leaves silked together." As overwintered eggs hatch soon after exposure to warm temperatures and adults do not fly until Aug, larvae clearly spend much of the spring–summer in a semidormant state. Discovery of a prepupal larva in late July suggests that *A. mormo* does not pupate until shortly before the flight period. We suggest that eggs hatch in March–April, the larvae develop very slowly during April–July, pupating in late July and eclosing to adults in Aug. Pyle (2002) and others report that *A. mormo* overwinters as a young larva. Shapiro and Manolis (2007) state that eggs of *A. mormo* in CA hatch during winter, larvae feeding in late winter or early spring; however, we found eggs to contain fully developed embryos and to be the overwintering stage in WA. We have not observed larval-ant associations. Ballmer and Pratt (1988) reported 5 instars for CA *A. mormo*, but also stated this number was variable. Our observations suggest that 4 or 5 instars occur in Cascadia. The survival strategy appears to be based on concealment in refugia, but the bright purple and gold larval colors may be aposematic.

Description of Immature Stages

The egg is white with a pink cast, with numerous rounded indentations and a distinct micropyle. All instars are purple in color, a character unique among our lycaenids. L1 has bold, paired black dorsal bullae while black lateral bullae sport long white setae. In L2 the setae are proportionally shorter, pale yellow-orange spots appear between the paired dorsal bullae, and yellow-orange spots surround the lateral bullae. In L3 this pattern is accentuated and bold, the yellow spots bright gold and contrasting. Laterally, a secondary line of smaller gold spots appears above the larger row. The basic pattern is the same in L4 and

Egg @ 1.0 mm

L1 @ 1.5 mm

Pupa @ 12 mm

L2 @ 3 mm

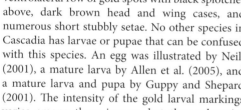

L3 @ 7.5 mm

Mormon Metalmark *Apodemia mormo* (C. & R. Felder)

L5, with increasing color contrast. The lateral white setae become long, while the dorsal black setae are shorter. The unique pupa is bright purple with a dark dorsal line, a dorsal row of butterfly-shaped gold spots bordered distally with dark splotches, a ventrolateral row of gold spots with black splotches above, dark brown head and wing cases, and numerous short stubbly setae. No other species in Cascadia has larvae or pupae that can be confused with this species. An egg was illustrated by Neill (2001), a mature larva by Allen et al. (2005), and a mature larva and pupa by Guppy and Shepard (2001). The intensity of the gold larval markings varies among these images, but they otherwise agree well with ours. The illustrated pupa appears to lack the gold spots that are prominent in ours.

L4 @ 10 mm

Discussion

Much remains uncertain in the life history of *A. mormo*, and more research is needed on many aspects of the biology and ecology of this species. Overwintering eggs are susceptible to mold and larvae are difficult to rear as they take a long time to mature and must be supplied continuously with fresh host plant. Finding eggs or larvae is very challenging. The occurrence of an L5 in one of our rearings is intriguing, and studies are needed to determine the number of instars and instar variability in Pacific Northwest populations. Why eggs overwinter in Cascadia but young larvae do so in other areas warrants research. Little is known concerning natural enemies, or of the strategies employed for defense.

L5 @ 17 mm

L5 @ 15 mm (prepupal)

Family Nymphalidae
Subfamily Heliconiinae
FRITILLARIES

Sixteen species of fritillaries occur in Cascadia, including 9 greater fritillaries (*Argynnis*), 6 lesser fritillaries (*Boloria*), and *Euptoieta claudia*. All fritillaries are orange dorsally with complex markings of black lines and spots. Ventrally, greater fritillaries bear large oval spots (silvered or not) on the hindwings, and lesser fritillaries have violet- or rust-tinted hindwings with white or yellow non-oval spots or markings, and *Euptoieta* has drab banded markings.

Greater fritillaries are found in all montane habitats and many moist lowland habitats. They are single brooded with long-lived adults; females of most species enter reproductive diapause and aestivate in midsummer. Males eclose before females, and eggs are laid singly in late summer and early fall near or on dried remnants of violets. Lesser fritillaries (except *B. selene*) also have a single generation, most are restricted to mid–high elevations, and adults do not aestivate.

Fritillary eggs are conical to barrel-shaped with strong vertical ribs, white-yellow maturing to orange or red. The greater fritillaries overwinter as unfed first-instar larvae and develop during spring, passing through 6 instars. The larvae are generally gray-black, adorned with branched spines, some with a lighter dorsal stripe and many with orange spine bases. Pupae are brown and hang vertically from a cremaster.

Lesser and greater fritillaries have similar eggs; the larvae are also similar, although those of lesser fritillaries are smaller, sometimes with subtle spotting and black or orange spine bases. All *Boloria* have 5 instars, 4 species overwintering as partly grown larvae and 1 as an unfed first instar. *Boloria astarte* is biennial, overwintering twice. Larval hosts include violets, bilberry, Kinnikinnick, saxifrage, bistort, and willow. The pupae are similar to those of *Argynnis* but vary in darkness and girth, and 3 have spiny abdominal processes and pearly nodes.

Most fritillary larvae are active nocturnally. All possess an eversible ventral gland on the first segment that emits chemicals likely used in defense against predators.

Variegated Fritillary   *Euptoieta claudia* (Cramer)

Adult Biology

Euptoieta claudia is a primarily tropical species intermediate between *Argynnis* and *Heliconius*. In the US it is a permanent resident from S AZ E to NC, flying all year with multiple generations (Scott, 1986a). Highly migratory and opportunistic, *E. claudia* ranges far to the N in summer, reaching SE BC and across Canada to the Atlantic. Unable to withstand freezing, N populations die out each fall, although a population has survived in Quebec for at least 10 years (Layberry et al., 1998). This species has not been found in WA or OR but has been reported in BC close to the NE WA border, and there are records from the ID panhandle. During invasive years *E. claudia* reaches its N limits in late summer; records in BC occur from July 17–Sep 4 (Guppy and Shepard, 2001). In its breeding range the preferred habitat includes open sunny fields, prairies, and roadsides (Tveten and Tveten, 1996), but it can be found everywhere but deep forests (Pyle, 1981). Larvae feed on a wide variety of host plants, including *Viola* (violet), *Linum* (flax), *Plantago* (plantain), *Desmodium* spp. (beggarticks), *Portulaca* spp. (purslane), and *Passiflora* (passion vine). Males patrol all day just above the ground for mates; a receptive female holds her wings mostly closed and abdomen raised slightly. The male lands behind and flutters his wings, then mating occurs (Scott, 1986a). Eggs are laid singly on stems and leaves of the host.

Immature Stage Biology

We reared this species once, from eggs from a female collected W of Golden, Jefferson Co., CO. After 12 days caged with *Viola glabella* (Stream Violet), *Plantago major* (Common Plantain), and *Passiflora incarnate* (Purple Passionflower), she laid 58 eggs on the paper cage liner. The eggs hatched in 3 days, and L1 were placed on the above hosts plus *Sedum lanceolatum* (Lanceleaf Stonecrop). The larvae fed only on *V. glabella*, which was replaced with *V. adunca* for the remainder of the rearing. Fewer than half

the eggs hatched, and 9 survived to L2, but survival thereafter was 100%. L1 consumed eggshells and were reared at 20–22 °C, initially in Petri dishes with cut leaves. Early instars avoided leaf edges, grazing the midleaf surface halfway through in irregular patches, breaking through in places, leaving translucent windows elsewhere. From L3, larvae consumed whole leaves, producing large jagged holes from the middle or edges. The rate of development was staggered; some reached L3 while others were still L1. In L3 and L4, larvae mostly rested on the underside of leaves during the day, feeding nocturnally. By late L4 and L5, larvae often wandered off the host, seeking sunny spots for basking, returning to feed mostly nocturnally. At this stage entire leaves were consumed, leaving only the stems. When L5 reached ~40 mm, feeding ceased, wandering increased, and pupation occurred at the top of the cage. The time from egg-hatch to pupation was 19–27 days and adults eclosed 9 days later. There are 5 instars and no nests are constructed. Mature larvae make little attempt at concealment and may be aposematic, although they are eaten by birds (Allen et al., 2005). Early instars derive protection from setal droplets (L1), concealment, and nocturnal feeding, and later instars possess spines, intimidatory anterior horns, and a large black ventral gland that may emit defensive chemicals. Tveten and Tveten (1996) reported that adults overwinter in TX, perhaps in reproductive diapause (Cech and Tudor, 2005).

Description of Immature Stages

The egg is a white, strongly tapered cone with ~14 strong vertical ribs that merge to ~9 at the apex around the micropyle. L1 has a shiny black abdomen tip and bifurcated head, and the body is white infused with considerable yellow laterally and dorsally. Each segment has ~8 black bullae, each sporting a long black droplet-tipped seta. L2 is brown, and there are white dorsolateral and ventrolateral stripes. The bullae are tall black conical spines, each with multiple black setae; the anterior-most pair constitute forward-projecting black conical horns. The head is shiny black, remaining so until pupation. L3 is similar, but the ground color is orange and the horns are proportionally longer. The 2 white stripes are strongly developed and divided vertically into 3–5 pieces per segment. The spines and setae are shiny black, remaining so to pupation. L4 is brighter orange; the white longitudinal stripes are bold and edged in black. The anterior horns are twice as long as other spines and there is a squarish black spot between them. L5 is similar, the anterior horns ~3 times longer than

Pupa @ 17 mm (lateral, *left*, dorsal, *right*)

Egg @ 0.8 mm

L1 @ 1.6 mm

L2 @ 4.0 mm

L3 @ 11 mm

L4 @ 16 mm

L5 @ 36 mm

other spines. True legs and prolegs are shiny black. The pupa is ornate, initially bright orange dorsally, maturing to white with black spots and metallic gold conical spines in lateral and dorsolateral rows. The wing cases are white with strongly contrasting black spots and oblong markings. The larva of *E. claudia* is unique in Cascadia and should not be confused with any other species, although the pupa resembles those of *Euphydryas* spp. Published images include mature larvae in Tveten and Tveten (1996), Allen et al. (2005), and Minno et al. (2005), and mature larvae and pupae in Pyle (1981), Scott (1986a), Allen (1997), and Wagner (2005). All images agree with ours, although FL larvae have orange heads with black markings (Minno et al., 2005).

Discussion

This species is easily reared, provided eggs are obtained in sufficient numbers to accommodate high egg and L1 mortality. Adults should be sought in far NE WA, where they will likely eventually be found. Why and how 1 population persists in Quebec City, 1,000 miles north of its normal overwintering range, would make a fascinating research project. The benefits derived from great summer range expansion each year followed by invariable winter contraction are unclear and would make an interesting study. The possible aposematism of mature larvae and possible defensive function of the L1 setal droplets and ventral gland would provide excellent chemical ecology research subjects.

Great Spangled Fritillary   *Argynnis cybele* (Fabricius)

Adult Biology

Argynnis cybele, the largest Cascadian fritillary, is found in most low–mid-elevation mountainous and hilly areas, absent W of the Cascades in BC and from the driest regions. Beyond Cascadia the range includes most of S Canada and N US. Preferring moist woods, meadows, and riparian areas, *A. cybele* is widely distributed, occurring from sea level to 6,000 ft (Warren, 2005). The single brood flies early June–early Oct. Larval hosts include *Viola glabella* (Stream Violet), *V. adunca* (Blue Violet), *V. nuttallii* (Nuttall's Violet), and *V. sempervirens* (Evergreen Violet). Males eclose 4–5 days before females (James, 2008a), although Allen (1997) reported their appearance 2–3 weeks before females in E US. Females mate, enter reproductive diapause, and aestivate for 3–5 weeks after eclosion (James, 2008a), and are rarely seen before Aug. Flight is strong, and males patrol for females often in a territorial circuit (Douglas and Douglas, 2005). Both sexes nectar on thistles, milkweed, and dogbane. Populations may be sedentary (Miller and Hammond, 2007). In Aug and Sep females lay eggs singly and haphazardly near the bases of dead and dying *Viola* spp. (Scott, 1986a; Douglas and Douglas, 2005).

Immature Stage Biology

We reared this species 3 times from gravid females collected mid–late Aug in Kittitas and Asotin Cos., WA. Eggs hatched in 10–12 days at 20–22 °C, then the larvae congregated under cover and diapaused as unfed L1. Overwintering for ~80 days at 5 °C was sufficient for L1 to break diapause and begin feeding on *V. adunca* leaves within ~4 days at 25 °C (James, 2008a). Development from postdiapause L1 to pupation took an average of 44 days, females ~1 week longer than males, and adults eclosed 13–19 days after pupation (James, 2008a). Larvae feed

Adult female

nocturnally, seeking cover during the day under leaves, rocks, or debris. Violet leaves are eaten from the edges, producing large scooped-out holes. Larvae do not make nests. Postdiapause larvae are solitary but tolerate each other well in captivity. L2 and older larvae possess a ventral gland on segment 1 that everts if the larva is threatened, emitting a musky odor, likely for defense (James, 2008a). Larvae nearing pupation silk a number of leaves together to form a "pupation tent," often very close to and sometimes touching the ground. Pupae are noticeably more "wriggly" than other *Argynnis* spp. Protection is based on nocturnal feeding, spines, possible chemical protection, and camouflage. There are 6 instars.

Description of Immature Stages

The pale yellow egg matures to pinkish red with darker spots, and is truncated conical. It bears ~20 strong vertical ribs that merge to 9–10 at the top. L1 is reddish brown or grayish, with 8 distinct black bullae in a transverse row across each segment; each bulla has a long brown seta with a droplet at the tip. There is a gray dorsal stripe that persists through L3. L2 is gray with black longitudinal lines along the spine bases. The bullae are tall conical black spines bearing 10–12 black setae. L3 is similar with black lines along the spine rows, except for the ventral spines, which have bright orange bases. Dull cream and orange patches appear at the bases of other spines. L4 is similar but is darker and has more distinct orange bases to lateral spines. The head has paired dorsal orange-brown patches. L5 is dark gray-black with no dorsal stripe but indistinct dark lines laterally. The black spines have large bright orange bases and the orange-brown head patches merge dorsally. In some populations the dorsal spines lack orange bases. L6 is jet black with an orange head dorsally. Paired spines on segment 1 point forward over the head, which is black on the "face." Spines are pale to bright orange with black tips, or all black in the 2 dorsal rows. True legs are black, prolegs brown. The pupa is shiny dark brown, and the wing cases are slightly lighter with black vein lines. The shape is robust, swollen in the middle with a large girth, and the abdomen has parallel rows of orange bullae. Immature stages of this species are similar to those of other *Argynnis* spp., especially *A. egleis*, *A. hydaspe*, and *A. hesperis*, all of which have dark larvae. *Argynnis cybele* differs by having oranger spines and lacking a dorsal stripe. The heads of mature *A. cybele* larvae have more extensive orange markings than those of other *Argynnis* spp. The pupa of *A. cybele* is more robust and more strongly textured than those

Egg @ 1.0 mm

L2 @ 4 mm

L3 @ 8 mm (premolt)

Pupa @ 30 mm   L1 @ 1.8 mm

L4 @ 10 mm

L5 @ 21 mm

L6 @ 45 mm

L6 @ 45 mm

Great Spangled Fritillary *Argynnis cybele* (Fabricius)

of other *Argynnis* spp. (James, 2008a). Larvae and/or pupae were illustrated by Allen (1997), Layberry et al. (1998), Guppy and Shepard (2001), Wagner (2005), Allen et al. (2005), and Miller and Hammond (2003, 2007). James (2008a) illustrated the entire life cycle. All images are similar to ours.

Discussion

Additional information on immature stages is available in James (2008a). Larvae are difficult to find, but eggs are readily obtained from gravid females. One wandering larva we found was parasitized. Douglas and Douglas (2005) captured larvae in pitfall traps placed at the bases of host plants showing leaf damage. Larvae are easily reared on violet leaves, with native species preferred; late-instar *A. cybele* fed poorly on pansy (James, 2008a). Research is being conducted on the chemistry and defensive function of the ventral osmeterium-like gland (James, 2008a), and similar studies are needed for the droplets on L1 setae. This species is sensitive to forest management, with populations sometimes negatively affected by insecticide applications (Miller and Hammond, 2007).

Coronis Fritillary   *Argynnis coronis* (Behr)

Adult Biology

Argynnis coronis ranges throughout much of W US from WA and SD to CO, AZ, and S CA. In Cascadia it occurs primarily along the Cascades and foothills but also E of the Cascades in OR. Favored habitats include canyons, hillsides, shrub-steppe, forest margins, and mountain meadows from 1,000–8,000 ft. A single generation flies May–Oct, with males emerging before females at relatively low elevations (< 2,000 ft). Mating occurs, but females enter reproductive diapause, migrating upslope. Males remain largely at low elevations. Populations seen migrating upslope near Rimrock Lk, Yakima Co., WA (3,000 ft) on June 4 were entirely female; they spend June–Aug at high elevations (5,000–8,000 ft) seeking nectar. Reproductive diapause in females terminates by early Aug in WA (James, 2008a), and they move downslope seeking host violets at lower elevations. This altitudinal migration is most marked in the E Cascades and Yakima Valley of WA. Larval hosts are *Viola trinervata* (Sagebrush Violet), *V. beckwithii* (Great Basin Violet), *V. nuttallii* (Nuttall's Violet), *V. praemorsa* (Prairie Violet), *V. purpurea* (Goosefoot Violet), and *V. douglasii* (Douglas' Violet) (Scott, 1986a; Pyle, 2002). Eggs are laid singly on or near senesced violets in late Aug–Oct.

Immature Stage Biology

We reared or partially reared this species several times. Gravid females obtained from Yakima Co., WA from July 30 to Sep 29 over 4 years produced 100–1,000 eggs per cohort, 2–5% of which were reared to adults. Oviposition occurred in partial sunlight at 21–35 °C on dried leaves and stems of *Viola labradorica* (Labrador Violet). No oviposition occurred on live violets. Females obtained on July 30 took 7 days to oviposit; later-caught females oviposited within 2 days of caging. Eggs hatched in 14–20 days, and L1 sought refuge in curled leaves and dried seedpods without feeding. L1 were overwintered at 4–5 °C for 10–20 weeks before exposure to 25–27 °C and constant light. Short daylengths (12 hr) failed to break dormancy despite warm temperatures. Postdiapause L1 took 3–5 days to commence feeding on *Viola adunca* (Blue Violet) and *Viola glabella* (Stream Violet). L1 molted to L2 after 8–9 days, and development from L2 to pupation took 31 days, with a further 12 days before adult eclosion. Postdiapause larvae preferentially feed on flowers and young leaves at first, then leaves only, eating holes from the edges. Larvae are solitary with no nests, but tolerate high densities well in captivity. Most feeding is nocturnal, with larvae often resting under leaves during the day. L2–L6 possess an eversible, ventral gland that produces a musky odor when the larva is disturbed (James, 2008a) and is likely defensive. Mature larvae often silk together leaves as "pupation tents," and pupation always occurs close to the ground. Larval defense is based on concealment (nocturnal feeding, pupa), chemical protection (ventral gland, setal droplets in L1), and spines.

Description of Immature Stages

The egg is conical, white, developing red spots and dashes with maturity, becoming dark before hatching. There are 20–22 prominent vertical ribs terminating around a well-defined micropyle. L1 is cream or yellowish orange with a shiny black setaceous head and 8 dark bullae on each segment; each bulla sports a long seta with a droplet at the tip. Postdiapause L1 develop a reddish tinge. L2 is blackish gray with an indistinct middorsal dark stripe and tall conical black spines sporting ~10 setae. Bullae in the midlateral row on segments 4–11 have orange bases. L3 is mostly black with 2 intermittent middorsal white stripes. L4 is similar but grayer with distinct orange-based spines laterally and a prominent middorsal pair of white stripes. The head is shiny black. L5 is darker, mottled with white and gray especially laterally, and the midlateral spines become orange. L6 is similar; white markings on the body, including the dorsal stripes, are larger and more distinct, giving the larva an overall grayish appearance. The head is black with no orange markings. The pupa is medium brown with dark brown-black markings, including segment joints. Most *Argynnis* spp. are similar, but *A. h. brico*, *A. h. dodgei*, *A. egleis*, *A. cybele*, and *A. hydaspe* larvae are darker. *Argynnis callippe* has a broad pale dorsal stripe,

Adult

Egg @ 1.0 mm

Pupa @ 21 mm

L3 @ 8 mm

L4 @ 15 mm

L1 @ 1.5 mm

L5 @ 20 mm

L2 @ 3 mm

L6 @ 35 mm

A. mormonia is smaller, and *A. atlantis* has more complex side markings. *Argynnis zerene* is similar but tends to be lighter. Neill (2007) illustrated a larva similar to ours, and James (2008a) illustrated all immature stages.

Discussion

Additional observations and data are reported in James (2008a). L1 overwinter well in captivity if sufficient humidity is provided (>60%). Although eggs were laid on dried *V. labradorica* and L3–L6 fed on this NE US native, L1 and L2 did not readily accept this host. Similarly, *Viola tricolor* (Pansy) was accepted only by L5–L6. The W native violets *V. adunca* and *V. glabella* were very acceptable hosts. The reproductive ecology of this species with well-defined adult and larval diapauses deserves further study. Of particular interest are the altitudinal migration, environmental cues determining movement, distances traveled, and whether males migrate.

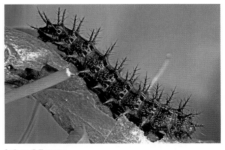

L6 @ 35 mm

Coronis Fritillary *Argynnis coronis* (Behr)

Zerene Fritillary   *Argynnis zerene* (Boisduval)

Adult Biology

Argynnis zerene is a variable species with 15 subspecies (Pelham, 2008). It occurs in most mountain and foothill areas of Cascadia and in some prairie and shrub-steppe, but is absent from most of the Cascades in N WA. Beyond Cascadia it ranges from N BC to central CA, and the Pacific coast to MT and CO. The single brood flies from mid-May to mid-Sep, with males eclosing 4–5 days before females (James, 2008). Habitats include prairies and foreshores W of the Cascades, mountain clearings, logging roadsides, and shrub-steppe from sea level to 9,000 ft (Warren, 2005). Coastal populations are threatened; *A. z. hippolyta* (Oregon Silverspot) is federally and state listed, with captive breeding and reintroduction programs in progress (Pyle, 2002; Miller and Hammond, 2007). Larval hosts are *Viola adunca* (Blue Violet), *V. glabella* (Stream Violet), *V. palustris* (Marsh Violet), and *V. nuttallii* (Nuttall's Violet) in Cascadia. Mating occurs immediately after female eclosion (Guppy and Shepard, 2001), and females do not mature their ovaries for 3–5 weeks (Sims, 1984; Scott, 1986a), spending the summer at higher elevations, returning to areas with violets to oviposit in Aug and Sep. Reproductive diapause in females terminates by early Aug in WA (James, 2008a). Oviposition occurs on senesced violets; Pyle (2002) suggested that females hunt over turf for violet scent.

Immature Stage Biology

We reared this species 5 times. Gravid females from Kittitas, Columbia, and Pend Oreille Cos., WA during Aug over 5 years produced 100–1,000 eggs per cohort, 2–5% of which were reared to adults. Oviposition occurred in a variety of enclosures with fresh or dried violets, in partial sunlight at temperatures of 20–35 °C. Eggs hatched in 13–19 days according to temperature,

Adult

and L1 overwintered (at 4–5 °C) in clusters among leaf debris without feeding. Postdiapause L1 took 5 days at 25 °C to commence feeding. Development from L2 to pupation took 45 days at 20–22 °C on *V. nuttallii* and 34 days at 25 °C on *V. adunca*. L5 and L6 also fed well on *V. labradorica* (Labrador Violet) and *V. tricolor* (Pansy). Adults eclosed 13–21 days after pupation. Larvae initially fed on flowers and leaves, then leaves only, eating holes from the edges. Larvae are solitary with no nests, but tolerate siblings in captivity. Most feeding is nocturnal, with larvae often resting under leaves during the day. L2–L6 possess an eversible, ventral gland on segment 1 that produces a musky odor when the larva is disturbed (Scott, 1986a; McCorkle and Hammond, 1988; James, 2008a), and is likely defensive. Scott (1986a) and James (2008a) reported that silked-together leaves are often used as "pupation tents." Larval defense is based on concealment (nocturnal feeding), chemical protection (ventral gland, setal droplets?), and physical protection (spines).

Description of Immature Stages

The egg is white, narrowed, and truncated at the top. There are ~20 strong vertical ribs, merging to ~10 at the top. L1 is cream-reddish, with 8 dark bullae on each segment; each bulla sports a long dark seta with a droplet at the tip. L2 is gray-brown; bullae are conical spines with ~10 setae. In L3 the spines are larger with more setae. The body is mostly gray, with black stripes and orange spots at the bases of the spines. L4 is gray mottled with black stripes. There is a broad gray dorsal stripe with a dark central line. The black head has paired brown dorsal spots, and the rows of orange spots persist along the body until pupation. L5 is similar; the black spots at the bases of dorsal spines are bold. In L6 the ventrolateral spines are orange, the others remaining black. Overall the color is light, gray-brown with black splotches. The pupa is medium brown with black bands at segment joints and the wing cases are gray-brown with black veins. All *Argynnis* spp. larvae are similar, but *A. h. dodgei*, *A. h. brico*, *A. egleis*, *A. cybele*, and *A. hydaspe* are darker. *Argynnis callippe* has a broad pale dorsal stripe, *A. mormonia* is smaller, and *A. atlantis* has more complex side markings. *Argynnis coronis* is most similar; however, it tends to be darker with dorsolateral orange spots split into 2 lines. Guppy and Shepard (2001), Allen et al. (2005), and Neill (2007) published images of larvae from AK, CA, and Cascadia that are darker than ours. Miller and Hammond (2003, 2007) and James (2008a) show images similar to ours.

Egg @ 0.9 mm

Pupa @ 23 mm

L3 @ 7 mm

L4 @ 13 mm

L5 @ 22 mm

L1 @ 1.5 mm

L2 @ 4 mm

L6 @ 38 mm

L6 @ 38 mm

Discussion

Additional observations and data are reported in James (2008a). Eggs are easily obtained from closely confined females as long as violet detritus is provided. L1 overwinter well if sufficient humidity is provided (>60%). Substantial mortality occurs under dry conditions, and larvae are subject to mold if RH is higher than 90%. Overwintered larvae feed well on succulent fresh violet blossoms and leaves; tough species like *V. sempervirens* and *V. tricolor* may be rejected by early instars. Larvae are very difficult to find. Research is being conducted on the chemistry and defensive function of the ventral osmeterium-like gland, and similar studies are needed for the droplets on L1 setae.

Callippe Fritillary *Argynnis callippe* (Boisduval)

Adult Biology

Argynnis callippe flies in a single brood mid-May–mid-Aug E of the Cascades in much of Cascadia, also W of the divide in S OR, but is absent from the driest parts of the Columbia Basin. Beyond Cascadia it ranges over much of W North America from S BC to the prairie provinces, S to CO and CA. Preferred habitats include montane canyons, open moist wooded areas, and shrub-steppe prairie at elevations of 300-9,700 ft. The flight period is protracted in S BC and WA, females entering reproductive diapause soon after eclosion, moving to high elevations during June–July before returning to lay eggs in mid–late Aug (Guppy and Shepard, 2001). Scott (1992) suggested that *A. callippe* undergoes little or no reproductive diapause in CO, dying before late summer, which also appears to be the case for populations in N CA (Shapiro and Manolis, 2007). Larval hosts confirmed in Cascadia are *Viola nuttallii* (Nuttall's Violet) and *Viola sempervirens* (Evergreen Violet), although other *Viola* spp. are likely used as well. Males frequent hilltops as well as patrolling hillsides near the ground to seek females (Scott, 1986a). Both sexes visit thistles and mints, and males are avid puddlers. Females crawl into shady hollows where they lay eggs singly, often in litter where no violets are apparent but are present in spring (Scott, 1986a, 1992).

Immature Stage Biology

We reared this species several times from gravid females. A female collected June 29 in Yakima Co., WA, lived for 6 weeks but did not lay eggs until early Aug. Larvae died during overwintering. Females collected July 24 (Yakima Co.), July 24, July 26 (different years) (Kittitas Co., WA), produced ~150, 100, and 100 eggs, respectively. Eggs hatched after ~11 days. L1 sought refuge and were overwintered at 4–5 °C from late Aug. Winter survival was poor if

relative humidity was <70%. After ~100 days, 30 L1 were exposed to 26 °C and continuous light. Feeding on *Viola labradorica* (Labrador Violet) commenced after 3 days with L2 reached 7 days later. Development from L2 to pupation took 47 days at 22–26 °C. Adults eclosed after ~35 days at 18–21 °C. On March 17, 7 L1 were exposed to 20–22 °C at natural daylengths and supplied with *Viola tricolor* (Pansy) flowers, then leaves of *Viola odorata*, a nonnative violet. One individual reached L6 before dying in April. On April 18, 2 L1 were exposed to the same conditions and reared on *V. nuttallii*. These grew faster, pupating 34 days after first feeding, adults eclosing after a further 13 days. L1 fed on violet petals or leaves. Later instars ate only leaves, feeding nocturnally and eating holes from the leaf edges. At maturity the larvae wandered off the host plant to pupate close to the ground under lightly silked leaf tents. There are 6 instars and no nests are constructed. Protection is based on nocturnal feeding, spines, and likely chemical protection from a ventral gland on segment 1 (L2–L6) and setal droplets in L1.

Description of Immature Stages

The egg is pale yellow, ovate at the base, narrowed and truncated at the top. There are 20–26 strong vertical ribs. L1 is brown with 8 black bullae in a transverse row on each segment; each bulla sports a long dark seta with a droplet at the tip. L2 is gray-black with tall black conical spines, each with ~10 setae. A prominent broad gray dorsal stripe with a thin black median line is present. L3 is similar, with complex mottled orange markings laterally, and black stripes follow the lines of conical spines. In L4 and L5 the patterns are more complex, the dorsal stripe remaining prominent. The lateral spines have orange bases, and there are smaller dorsolateral orange markings. L6 is gray-brown with complex black markings in broken stripes following the rows of spines. The dorsal stripe is prominent and broad, a dull creamy-bronze, and the central dark line is almost obsolete in some individuals. Dorsally the head bears a transverse brown mark with small black speckles, and the spines are cream-white with orange tips; those on the sides also have orange bases. The pupa is rich reddish brown, the abdomen with broad coppery bands alternating with black markings. Larvae are paler than those of *A. h. dodgei*, *A. h. brico*, *A. hydaspe*, *A. mormonia*, and *A. cybele*. The larvae of *A. atlantis* have complex white lateral lines. *Argynnis coronis* and *A. zerene* are similar, but the dorsal stripe in *A. callippe* is broader with a fainter median line. Guppy and Shepard (2001) show a mature larva and

Adult

Egg @ 1.0 mm

Pupa @ 22 mm

L3 @ 10 mm

L4 @ 11 mm

L1 @ 2.5 mm

L5 @ 15 mm

L2 @ 3 mm

L6 @ 30 mm

L6 @ 35 mm

Callippe Fritillary *Argynnis callippe* (Boisduval)

pupa from WA similar to ours. Allen et al. (2005) picture a mature larva from TX that is dark with a narrow dorsal stripe and a strong black central line.

Discussion

Larvae of *A. callippe* are difficult to find, but eggs are readily obtained from females confined with violet detritus. Overwintering L1 is difficult, with dry or excessively moist conditions causing high mortality. Post-winter larvae feed well on a number of *Viola* spp., although *V. sempervirens* appeared unfavorable. Research is needed on the chemistry and defensive function of the eversible ventral gland and the setal droplets in L1. Natural mortality factors and their impact on population dynamics deserve study, as do the dynamics of reproductive diapause and associated environmental cues.

Great Basin Fritillary   *Argynnis egleis* (Behr)

Adult Biology

Argynnis egleis is a W species ranging from WA, ID, and MT S to WY, OR, NV, CA, and UT, occurring mostly at mid–high elevations (2,000–10,000 ft) in forest clearings and meadows, and on dry rocky ridges and hilltops. It is sparsely distributed in WA but more readily found in the Wallowa-Blues, Ochocos, Warners and Siskiyou ranges of OR. Adults fly in 1 generation mid–June–early Sep and can be locally common. Males emerge a week or two before females, patrol and sometimes hill-top. No evidence has been presented that females enter reproductive diapause; this strategy may be unnecessary in these higher-elevation habitats. Later in the flight period females are found mostly in wooded areas seeking host plants. Both sexes are avid flower visitors on Mountain Balm, thistles, asters, yarrow, and rabbitbrush; males visit mud. Recorded host plants are *Viola nuttalli* (Nuttall's Violet), *V. praemorsa* (Prairie Violet), *V. adunca* (Blue Violet), *V. purpurea* (Goosefoot Violet), and *V. walteri* (Prostrate Blue Violet) (Scott, 1986a; Pyle, 2002). Eggs are laid singly on or near senescing violets.

Immature Stage Biology

We reared this species twice, obtaining eggs from 4–7 gravid females collected near Mt Misery, Garfield Co., WA on Aug 3 and 14 (different years). Females were caged with potted *Viola labradorica* (Labrador Violet) and 50–100 eggs were laid over 2 weeks. Initially (48 hrs) eggs were laid on the living plant, thereafter on inert surfaces and senesced leaves and stems. Eggs hatched after 15 days at 21–23 °C, and unfed L1 sought refuge in curled leaves and dried seedpods. L1 were held at 20–25 °C until early Oct then transferred to an unheated garage. One month later they were transferred to 5 °C for overwintering. After 61, 69, 77, and 133 days, 9, 3, 12, and 22 L1 were transferred to summer-like rearing conditions, 25 °C, constant illumination. L1 in the first 2 cohorts took 10–12 days

to commence feeding on *V. adunca*; the later cohorts took 48 hrs. L2 was reached 8 days after transfer, and development from L2 to pupation took 34 days with a further 12 days to eclosion. All instars fed on *V. adunca*, *V. labradorica*, and *V. glabella* (Stream Violet). Postdiapause larvae fed on young leaves at first, then older leaves, eating holes from the edges. L6 accepted *Viola tricolor* (Pansy) as a host, but L1 were unable to feed on this plant. *Viola trinervata* (Sagebrush Violet) was refused by L3. Larvae are solitary with no nests but tolerate high densities well in captivity. Most feeding is nocturnal, with larvae resting under leaves during the day. L2–L6 possess an eversible, ventral gland on segment 1 that produces a musky odor when the larva is disturbed (James, 2008a) and is likely defensive. Mature larvae silk together leaves as "pupation tents" close to the ground in limited space. Many pupae fall to the ground. Larval defense is based on concealment (nocturnal feeding, pupa), chemical protection (ventral gland, setal droplets in L1), and physical protection (spines).

Description of Immature Stages

The egg is white and conical with 24–26 prominent vertical ribs merging to ~15 at the top. L1 is reddish brown, paler laterally, with a shiny black setaceous head and 8–10 prominent black bullae across each segment. Each bulla sports a long dark seta with a droplet at the tip. L2 is dark brown with a wide middorsal pale stripe and a thin central dark stripe. There are 6 prominent conical black spines across each segment, each bearing ~10 setae. Lateral spines on segments 4–12 have vaguely orange bases. L3 is blackish with larger spines bearing more setae and distinct orange bases to lateral spines on segments 4–12. L4 is black mottled with white, creating a grayish appearance. The orange-based lateral spines are distinct, and 2 parallel dorsal white stripes are separated by a thin black stripe. L5 is similar, darker with brown head markings dorsally and laterally, distinct orange-based lateral spines, and indistinct orange bases to the other spines. Dorsally, white markings on each segment are concentrated posteriorly. L6 is black mottled with brown and has 2 indistinct thin white lines dorsally. Spines in the lateral row are whitish yellow, as are bases of the dorsolateral black spines. The black head is like that of L5 and the spiracles are black. The pupa is brown with black markings limited to intersegmental banding, the eyes, and wing venation. There are 2–3 lateral rows of indistinct orange spots on the abdomen. Most *Argynnis* spp. larvae are lighter. *Argynnis hydaspe*, *A.*

Adult

Egg @ 1.0 mm

Pupa @ 22 mm

L3 @ 5 mm

L4 @ 12 mm (late)

L1 @ 1.5 mm

L5 @ 15 mm (early)

L2 @3.0 mm

L6 @ 35 mm

h. brico, *A. h. dodgei*, and *A. cybele* have dark larvae, but *A. h. dodgei* has a wide pale dorsal band, and *A. h. brico* and *A. cybele* have vivid orange spines. The mature larva of *A. hydaspe* is most similar, but the head usually has no brown markings. James (2008a) illustrated the life cycle of *A. egleis* and provided further information on immature stages.

L6 @ 35 mm

Discussion

Similar to other *Argynnis* spp., *A. egleis* oviposits readily in captivity and is easily reared. Overwintering L1 require at least 60–70% humidity and must experience ~2.5 months of cold temperatures before being able to commence feeding when exposed to warm temperatures. Three previously unreported violet hosts were accepted by all instars, and pansies were acceptable to final instars. Studies are needed on the reproductive ecology of this species, which may vary between subspecies at different elevations. Little is known of natural enemies and other population-regulating factors.

Northwestern Fritillary   *Argynnis hesperis dodgei* (Gunder) (unsilvered form)

Adult Biology

Argynnis hesperis is considered by some authors a subspecies of *A. atlantis* (e.g., Pyle, 2002; Miller and Hammond, 2007). We follow Guppy and Shepard (2001), Warren (2005), and Pelham (2008) in treating *A. hesperis* as a distinct species. The Northwestern Fritillary ranges from AK to MB, S to NM and CA. In Cascadia it occurs in BC, WA, OR, and ID. *Argynnis hesperis dodgei* is unsilvered and occurs in the S Cascades of WA and most mid–high-elevation areas in OR. In N WA and BC it is replaced by the silvered *A. h. brico*. In SE WA and OR *A. h. dodgei* is found in cool, forested and open, dry subalpine habitats (1,500–8,000 ft) and can be locally common (Dornfield, 1980; Warren, 2005; Miller and Hammond, 2007). The single brood flies mid-June–Sep, males emerging a week or so before females. Males patrol meadows, hillsides, and ridgelines and visit mud and animal scat. Both sexes feed at flowers, particularly Mountain Balm, thistles, asters, and yarrow. Larval hosts are violets, with *Viola pupurea* (Goosefoot Violet), *V. adunca* (Blue Violet), *V. nuttallii* (Nuttall's Violet), and *V. praemorsa* (Prairie Violet) confirmed (Scott et al., 1998; Warren, 2005). Eggs are laid singly on or near senescing violets.

Immature Stage Biology

We reared this species 3 times from gravid females obtained on Aug 14 near Mt Misery, Garfield Co., WA and Aug 20 and Aug 21 near Mt Howard, Wallowa Co., OR. Confined with dried foliage of *Viola labradorica* (Labrador Violet) in a plastic container with a muslin lid (32 × 20 × 9 cm) and held at 23–26 °C in dappled sunlight, females laid 15, 50, and 65 eggs, respectively, during 3–7 days. Eggs hatched after 12–13 days, and L1 sought refuge in curled leaves and dried seedpods without feeding. L1 were held at 20–25 °C until

Adult female

mid-Sep, then transferred to 5 °C for overwintering. After 92–107 days, L1 were placed on *V. adunca, V. labradorica* (Labrador Violet), or *V. odorata* (Sweet Violet) leaflets at 27 °C with continuous lighting. Feeding commenced after 4 days, and L2 was reached 2 days later. Development from L2 to pupation took 30–40 days, with adults eclosing 19–28 days later (pupae held at 19–21 °C). L1–L5 fed well on the 3 *Viola* spp. provided. Postdiapause larvae fed on young leaves at first, then older leaves, eating holes from the edges. Larvae are solitary with no nests and most feeding is nocturnal, with larvae resting under leaves during the day. L2–L6 possess an eversible ventral gland that produces a musky odor when the larva is disturbed (James, 2008a) and likely defensive. Mature larvae silk together leaves as "pupation tents," and pupation occurs close to the ground in limited space. Larval defense is based on concealment (nocturnal feeding, pupa), chemical protection (ventral gland, setal droplets in L1), and physical protection (spines).

Description of Immature Stages

The egg is creamy white, developing distinctive reddish brown lines and spots. There are 22–23 prominent vertical ribs merging to ~15 at the top. L1 is medium-dark brown, paler laterally with a shiny black setaceous head and 8–10 black bullae across each segment. Each bulla sports a long dark seta with a droplet at the tip. L2 is dark brown-black with an indistinct middorsal pale stripe. Six prominent conical black spines traverse each segment, each bearing ~10 setae. Bases of spines in the lower lateral row are pale. L3 is black with paler areas, especially laterally. The black spines are larger, and their bases are orange in the sublateral row on segments 4–11. L4 is similar but darker, with orange-based sublateral black bullae and a shiny black head. An indistinct pair of intermittent white stripes is present dorsally. L5 is black with a pair of indistinct gray stripes dorsally, pale markings laterally, and a shiny black head. Bases of the spines in the sublateral row are dull red and indistinct. L6 is black with a pale middorsal stripe but no lateral pale markings. Initially spines in the sublateral row are orange with a black tip but fade with maturity to whitish tan. Setae on all spines are orange-brown. The black head has few or no brown markings. The pupa is mostly black dorsally with mahogany brown wing cases marked in black primarily following venation. Most *Argynnis* spp. larvae are similar but lighter in color. *Argynnis hydaspe, A. egleis, A. cybele,* and *A. h. brico* have dark larvae, but *A. egleis* and *A. h. brico* have a pair of dorsal white lines, and *A. cybele* has vivid orange spines. Larvae of *A. hydaspe* may be indistinguishable. The

Egg @ 0.9 mm

Pupa @ 17 mm

L3 @ 7 mm

L4 @ 12 mm

L5 @ 16 mm

L1 @ 1.5 mm

L2 @ 4 mm

L6 @ 34 mm

L6 @ 34 mm

Northwestern Fritillary *Argynnis hesperis dodgei* (Gunder) (unsilvered form)

larva shown in Miller and Hammond (2007) (*S. atlantis dodgei*) is similar to ours.

Discussion

The substantial difference between *A. hesperis* (both subspecies) and *A. atlantis* larvae, first reported by Scott (1988) and Scott et al. (1998) and illustrated here, supports the specific status of these 2 entities. Additional studies on the biology of immature stages of these species should be conducted and may reveal further differences. Gravid females appear to oviposit well in captivity, as do other *Argynnis* spp. The full host range of this species remains to be determined; *Viola glabella* (Stream Violet) occurs commonly in the SE WA and NE OR habitats occupied by *A. h. dodgei* and is a likely host. The identity and impact of natural enemies on population dynamics of this localized but sometimes common species are unknown and deserve study. Also worthy of study is the possible existence of reproductive diapause in newly eclosed females.

Northwestern Fritillary (silvered form)
Argynnis hesperis brico
Kondla, Scott & Spomer

Adult Biology

Argynnis hesperis brico is a subspecies of the Northwestern Fritillary, occurring in S BC and N WA from Okanogan to Pend Oreille Cos. *Argynnis hesperis brico* ranges across W Canada into MT and CO. We separate the subspecies of *A. hesperis* because the adults are readily distinguished and there are no known blend areas in Cascadia where the two meet or merge. *Argynnis hesperis brico* has affinities to the N and E and the Rocky Mts, while *A. h. dodgei* resembles relatives to the S. A single brood flies from mid-June to early Sep, occupying partially wooded, moist meadow habitats, often near watercourses. Larval host plants include *Viola adunca* (Blue Violet), *V. canadensis* (Canadian Violet), *V. nuttallii* (Nuttall's Violet), and perhaps *V. sempervirens* (Evergreen Violet). Males patrol meadows, ridges, and hilltops and visit mud. Courtship is similar to that of other *Argynnis* spp., females ovipositing throughout Aug near senesced violets after most of the earlier-flying males have died. It is likely, although unconfirmed, that females are in reproductive diapause from eclosion to late July, as in other *Argynnis* spp. (James, 2008a).

Immature Stage Biology

We reared this subspecies once. Females obtained on Aug 22 near Tiger Meadows, Pend Oreille Co., WA produced 86 eggs. Eggs hatched after 5–18 days, and ~70 L1 sought refuge under dead leaves and entered diapause without feeding. Larvae were overwintered at 4 °C from Sep 23. On May 17, larvae were exposed to 20–22 °C under natural daylengths and provided with *V. nuttalli* and *V. glabella* (Stream Violet). The larvae rejected *V. nuttallii* but readily accepted *V. glabella* and commenced feeding within a day or

two. Development from commencement of feeding to pupation was rapid, taking 22 days (June 8), and adults began eclosing ~17 days later. Many larvae died in early instars, but 5 survived to L6. Further mortality occurred before pupation from disease; only 3 pupated successfully and eclosed to adults. Larvae fed nocturnally on *Viola* leaves, eating from edges; they were solitary and did not construct nests. Camouflage coloration, nocturnal feeding activity, spines, and likely chemical protection (from a ventral gland in L2–L6 and setal droplets in L1) contribute to the defense of these larvae from natural enemies.

Description of Immature Stages

The eggs are yellow maturing to red, with ~15 strong vertical ribs crossed by ~20 weaker lateral ribs. L1 is dark magenta with a shiny black head and ~10 dark bullae per segment, each with a long dark seta. L2 is purplish brown; tall conical spines replace bullae, each bearing several dark setae. L3 is darker, the spines black with bristly setae, and the black head has distinct brown dorsal patches. L4 is shiny black except for small orangish patches at the lateral spine bases. L5 is similar to L4, the brown head patches prominent, with 2 dorsolateral stripes darker than the blackish body, and a pair of thin pale dorsal stripes. In L6 the pale dorsal stripes remain, and all spines except dorsal rows are pinkish with black tips. The shiny pupa is rich chestnut brown with black splotches. The eggs of *A. h. brico* and *A. h. dodgei* are generally similar, but in L1 *A. h. brico* is magenta while *A. h. dodgei* is more olive. In L3 *A. h. brico* develops a pair of brown patches on the top of the black head, while in *A. h. dodgei* the head is often solid black. During L4–L6 the brown markings on the head capsule of *A. h. brico* enlarge, while in some individuals of *A. h. dodgei* the head remains black throughout development. The pupae of the subspecies are similar, with *A. h. brico* more reddish, especially on the abdomen. Larvae of *A. hydaspe* are similar but usually lack brown markings on the head, or if they occur they are very minor. Mature larvae of *A. h. brico* tend to be shiny black with brighter orange markings than those of either *A. h. dodgei* or *A. hydaspe*. Other *Argynnis* spp., except *A. cybele* and *A. egleis*, have paler larvae, the former lacking dorsal stripes and with more extensive orange spines, the latter with paler sides. We know of no published images of immature stages of *A. h brico*.

Discussion

Argynnis h. brico is found close to and synchronously with populations of *A. atlantis* in Pend Oreille Co. in NE WA. Adults are often very similar, particularly

Adult

Egg @ 0.8 mm (fresh, *left*, mature, *right*)

L2 @ 3.7 mm

L1 @ 1.2 mm

L3 @ 8.5 mm

Pupa @ 20 mm

L4 @ 11 mm

L5 @ 18 mm

Northwestern Fritillary (silvered form)   *Argynnis hesperis brico* Kondla, Scott & Spomer

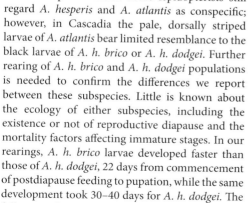

L6 @ 30 mm

L6 @ 30 mm

when worn; however *A. h. brico* occupies drier habitats while *A. atlantis* is found mostly near bogs and wet riparian areas. Some lepidopterists still regard *A. hesperis* and *A. atlantis* as conspecific; however, in Cascadia the pale, dorsally striped larvae of *A. atlantis* bear limited resemblance to the black larvae of *A. h. brico* or *A. h. dodgei*. Further rearing of *A. h. brico* and *A. h. dodgei* populations is needed to confirm the differences we report between these subspecies. Little is known about the ecology of either subspecies, including the existence or not of reproductive diapause and the mortality factors affecting immature stages. In our rearings, *A. h. brico* larvae developed faster than those of *A. h. dodgei*, 22 days from commencement of postdiapause feeding to pupation, while the same development took 30–40 days for *A. h. dodgei*. The full range of violet host plants used by *A. h. brico* in Cascadia is unknown and needs study.

Atlantis Fritillary   *Argynnis atlantis* (W. H. Edwards)

Adult Biology

Argynnis atlantis occurs in the NE US and E Canada, W to E BC and S to MT, SD, CO, and NM. In Cascadia it is found in SE BC, N ID, NE WA, and NE OR. This species was formerly combined with *A. hesperis*. Miller and Hammond (2007) treat the Cascadia subspecies of *A. atlantis* as a distinct species, *A. hollandi*. In Cascadia, *A. atlantis* is found in cool, moist to boggy boreal habitats. Never common, it sometimes flies with *A. hesperis brico* in NE WA and *A. hesperis dodgei* in NE OR. The single brood flies late June–late Aug, and males emerge before females. Egg-laying does not occur until mid-Aug in WV, suggesting reproductive diapause (Allen, 1997). Males patrol trails and roads for females, and both sexes nectar on asters, Pearly Everlasting, mints, vetches, yarrow, and dogbane. Males visit mud and animal scat. *Viola canadensis* (Canada Violet) is the only larval host confirmed for Cascadia (Miller and Hammond, 2007), but other violets are recorded elsewhere. Eggs are laid singly near senescing violets.

Immature Stage Biology

We reared this species from females obtained on Aug 14 and 28 at Bunchgrass and Tiger Meadows, Pend Oreille Co., WA. Confined with dried *Viola labradorica* (Labrador Violet) or *V. glabella* (Stream Violet), females laid ~165 eggs. Eggs hatched after 13–14 days at 21–25 °C, and L1 sought refuge in curled leaves without feeding. One group of L1 held at these temperatures and 30–40% RH suffered major mortality, with ~70% desiccating within a few days. The survivors imbibed moisture and expanded in size. Larvae were transferred to 5 °C, 70–80% RH on Sep 24 for overwintering, 1 group in dead leaf refugia, the other on filter paper. After 101 days, larvae were inspected; 100% and 48% of the dead leaf

and filter paper L1 survived, respectively; all were exposed to 15–24 °C, 12 hr light, and provided with *V. labradorica* and *V. adunca* (Blue Violet). Feeding commenced after 3–5 days (only on *V. labradorica*), and L2 was reached 7–9 days later. Another cohort (*n* = 25) was held at 5 °C for ~250 days, then exposed to 20–22 °C with *V. glabella*; they fed sparingly, but all died by L2. In the successful cohort, development from L2 to pupation took 35–39 days, with adults eclosing after a further 12 days (21–26 °C). Larvae fed readily on *V. labradorica* and *V. odorata* (Sweet Violet), rejecting *V. adunca*. Postdiapause larvae fed on young leaves at first, then older leaves, eating from the edges. Larvae are solitary with no nests and most feeding is nocturnal, with larvae resting under leaves during the day. L2–L6 possess an eversible, ventral gland on segment 1 that produces a musky odor when the larva is disturbed (James, 2008a) and is likely defensive. Mature larvae silk together leaves as "pupation tents" close to the ground. Larval defense is based on concealment (nocturnal feeding, pupa), chemical protection (ventral gland, setal droplets), and spines.

Description of Immature Stages

The egg is ovoid, yellowish white becoming red. There are 20–22 prominent vertical ribs merging to ~14 at the top. L1 is dark brown, with 8–10 prominent black bullae across each segment. Each bulla sports a long seta with a droplet at the tip. L2 is dark brown to black with a middorsal pale stripe. Six conical black spines traverse each segment, each bearing ~10 setae. Late in L2, patches of white appear on the body, and spine bases become orange. L3 is black with a distinct pair of dorsal white stripes and grayish white patches. Orange-based black spines bear ~15 setae. L4 is similar to late L3, but the shiny black head has brown markings laterally. Late L4 have increased white lateral markings and prominent dorsal white stripes. L5 is darker with distinct dorsal white stripes and a pattern of white lines, reminiscent of crocodile skin. L6 is similar, with intensified white markings. The sublateral row of spines is bright orange, becoming whitish. The lateral white markings coalesce to form "snakeskin" patches separated by black areas. The black head is distinctly marked with dorsolateral brown markings. The pupa is medium–dark brown with black markings on the head, thorax, and abdomen, black-banded intersegmentally, with a pair of dorsal black spots on each segment. Most *Argynnis* spp. larvae are similar but mature larvae of *A. h. brico*, *A. h. dodgei*, *A. hydaspe*, *A. egleis*, and *A. cybele* are darker, with little or no white markings. The lighter

Egg @ 0.8 mm (fresh)   **Egg @ 0.9 mm (mature)**

L3 @ 9 mm

L1 @ 1.6 mm

L4 @ 13 mm

L2 @ 5 mm

L5 @ 23 mm

L6 @ 36 mm

Atlantis Fritillary   *Argynnis atlantis* (W. H. Edwards)

larvae of *A. zerene, A. coronis, A. mormonia,* and *A. callippe* are similar, but none has the vivid dorsal white stripes and complex lateral white markings seen in *A. atlantis.* The black spots on the pupa abdomen also appear to be unique. Images of mature larvae appear in Allen (1997), Wagner (2005), and Allen et al. (2005) and agree with ours.

L6 @ 36 mm

Discussion

The need for humid conditions (>70%) to ensure good survival of diapausing L1 likely reflects adaptation to the moist habitats occupied by *A. atlantis* in Cascadia. The violet species used in Cascadia need documentation. The rejection of *V. adunca* in our rearing is surprising, as this species is present in habitats used by *A. atlantis.* The existence or not of reproductive diapause in females deserves study, as does the ecology of immature stages. Miller and Hammond (2007) considered this species to be local and sedentary, thus sensitive to forest management.

Pupa @ 19 mm

Hydaspe Fritillary   *Argynnis hydaspe* (Boisduval)

Adult Biology

Argynnis hydaspe is widely distributed and common, found in a variety of forested habitats, particularly moist woodlands, meadows, and flowery scree slopes. The range includes nearly all of Cascadia except the drier parts of the Columbia Basin and SE OR. Beyond Cascadia it is found from central BC to central CA, E to CO and UT. Found from near sea level to 8,300 ft (Warren, 2005), the single brood flies mid-May–late Sep. Larval host plants include *Viola glabella* (StreamViolet), *V. adunca* (Blue Violet), *V. nuttallii* (Nuttall's Violet), and *V. sempervirens* (Evergreen Violet). Other violets (e.g., *V. orbiculata*, *V. purpurea*) are used elsewhere (Scott, 1986a). Males hill-top and are often seen at mud; both sexes visit thistles, asters, penstemons, and dogbane. Males begin flight earlier than females, and after mating the females enter summer aestivation for 3–5 weeks before becoming gravid in early Aug (James, 2008a). Oviposition occurs in Aug–Sep in the vicinity of senesced violets. Pyle (2002) observed females ovipositing on rocks at the base of *V. adunca* in the Olympic Mts of WA.

Immature Stage Biology

We reared this species 4 times. On Aug 1, 3, 10, and 30 (different years) gravid females were obtained from Yakima, Kittitas, and Garfield Cos., WA, and Grant Co., OR, respectively. On each occasion large numbers of eggs (~100–500) were laid on dried *V. glabella* and *V. labradorica* (Labrador Violet) leaves and stems and inert surfaces (paper, cage sides, etc.). The Yakima Co. cohort oviposited only on potted *V. labradorica* during the first 2 days, ignoring dried foliage. Eggs hatched in ~9 days at 25 °C; L1 partially

consumed eggshells, sought refuge under leaves and debris, and became dormant. L1 were overwintered at 4–5 °C from late Sep/early Oct. Survival of L1 during overwintering varied with relative humidity; high mortality occurred at 60–70% RH, low mortality at 80–90% RH. Larvae were exposed to warm (20–26 °C) temperatures and continuous light or natural daylengths after 90–180 days and reared on *V. nuttalli, V. adunca, V. glabella*, or *V. labradorica*. Development from start of feeding to pupation took 30–35 days, with adults eclosing after a further 12 days. This is significantly faster than the rates recorded for *A. coronis, A. zerene, A. egleis*, and *A. cybele* (James, 2008). Feeding by L1 was initially on flowers or small leaves, the earliest feeding causing "windowpanes" in leaves. L2–L6 fed on larger leaves, eating from the edges. Larvae fed well on all violets except *V. labradorica* when provided with alternative hosts; however, L4–L6 in a no-choice test with *V. labradorica* developed to adulthood. No nests are made and larvae are nocturnal, retreating into refugia during the day. There are 6 instars and protection is based on diurnal concealment, chemical protection (ventral gland on segment 1, setal droplets in L1), and spines.

Description of Immature Stages

The egg is yellowish white turning purplish red, with ~20 strong vertical ribs merging to ~12 around the micropyle. L1 is reddish brown with a transverse row of 8 black bullae on each segment, each with a single long black seta with a droplet at the tip. L2 is purplish brown with black conical spines, each with ~10 dark setae. There are vague dark longitudinal lines and the subspiracular bullae have yellowish bases. L3 is darker; subspiracular spines sometimes have bright orange-red bases. L4 is black with 2 pale dorsal stripes and sometimes similar orange markings. L5 is similar, but the posterior half of each segment is lightly corrugated with several small folds, and the black head usually has some orange-brown markings. L6 is uniformly charcoal black; the subspiracular spines are orange, and the bases of dorsolateral spines may be slightly orange. Dorsolateral spines have a squarish black patch around the bases, forming 2 longitudinal bands; between them are 2 narrow indistinct gray lines. Orange-brown markings on the head, if present, are limited. The pupa is dark reddish brown with black dorsal markings and rufous intersegmental bands. The wing cases are brown with black smudged markings. All *Argynnis* spp. have similar larvae (James, 2008a), but several are considerably lighter than *A. hydaspe*, including those of *A. coronis, A. zerene, A. mormonia*,

Adult

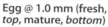

Egg @ 1.0 mm (fresh, *top*, mature, *bottom*)　Pupa @ 22 mm

L2 @ 5 mm

L3 @ 8 mm

L4 @ 15 mm

L1 @ 1.5 mm

L5 @ 19 mm

A. *atlantis*, and *A. callippe*. The larvae of *A. egleis* are dark with pale dorsal lines. In *A. cybele* the orange spines are numerous and prominent, and the head is marked with brown dorsally. *Argynnis hesperis brico* and *A. h. dodgei* are very similar, although *A. h. brico* may have more extensive brown head markings. Photos of larvae and pupae in Woodward (2005) and Neill (2007) and a larva in Guppy and Shepard (2001) are similar to ours. James (2008a) illustrated the complete life history.

Discussion

Gravid females lay eggs readily even in small containers as long as violet debris is present; however, overwintering of L1 can be difficult, with substantial mortality if conditions are too dry or too moist. Details of female aestivation (e.g., timing, behavior) need elucidation, as do the factors that shape the abundance and distribution of this species. Little is known of the natural enemies affecting *A. hydaspe*.

L6 @ 32 mm

L6 @ 37 mm

Hydaspe Fritillary *Argynnis hydaspe* (Boisduval)

Mormon Fritillary   *Argynnis mormonia erinna* (W. H. Edwards)

Adult Biology

Argynnis mormonia ranges from AK S to AZ and E to MB, SD, UT, and CO, occupying mid–high-elevation habitats. Two subspecies are found in Cascadia; *A. m. erinna* is found in extreme S BC, in NE and SE WA, and in the Cascades, Ochocos, Wallowas, and Blue Mts of OR. *Argynnis mormonia erinna* flies in moist alpine and subalpine meadows, also drier hillsides, from ~2,400 to 7,000 ft (Warren, 2005). The single brood flies from late June to mid Sep but is most common during July–Aug. Boggs (1987), studying a CO population of *A. mormonia*, found males emerge 2–3 weeks before females and mate more than once. Females mate only once and do not appear to undergo reproductive diapause. Dispersal is limited, and males patrol for females. Mated females spend much time crawling through ground vegetation looking for oviposition sites. Fecundity is directly related to adult nutrition (Boggs and Ross, 1993). Both sexes nectar on flowers, including asters, yarrow, and Pearly Everlasting; males and older females also visit mud, carrion, and animal scat. Larval hosts in Cascadia include *Viola palustris* (Marsh Violet), *V. adunca* (Blue Violet), *V. nephrophylla* (Northern Bog Violet), *V. nuttalli* (Nuttall's Violet), and *V. sempervirens* (Evergreen Violet) (Pyle, 2002; Warren, 2005; Miller and Hammond, 2007). Eggs are laid singly on or near senescing violets.

Immature Stage Biology

We reared or partially reared this species from females collected in Pend Oreille Co., WA and Wallowa Co., OR during Aug. Oviposition occurred immediately when females were provided with dried foliage of *V. labradorica* (Labrador Violet) in plastic boxes with muslin lids (32 × 20 × 9 cm). Eggs were hidden underneath foliage and also laid on living foliage when provided. Eggs hatched after 10 days at 21–25 °C, and L1 sought refuge in curled leaves without feeding on eggshells or leaves. L1 were overwintered at 5 °C from Sep 24 for 109 days with ~90% survival. Feeding commenced on *V. adunca*, *V. labradorica*, and *V. odorata* (Sweet Violet) after 5 days at 20–25 °C, 12–15 hr light, and development from L2 to pupation took 68 days. Pupae held at 17–21°C eclosed in 22 days. Larvae fed on young leaves at first, then older leaves, eating from the edges mostly at night. L1–L3 tend to rest and feed communally in captivity. L2–L6 possess an eversible ventral gland on segment 1 that produces a musky odor when the larva is disturbed (James, 2008a) and is likely defensive. Mature larvae silk together leaves as "pupation tents," and pupation occurs close to the ground. If hanging leaves are not available, larvae gather leaves and silk them onto a substrate to form a "tent." Larval defense is based on concealment (nocturnal feeding, pupa), chemical protection (ventral gland, setal droplets in L1), and physical protection (spines).

Description of Immature Stages

The broad-based conical egg is creamy white turning orange-brown with indistinct red spots. There are 24–26 prominent vertical ribs merging to ~12 at the top. L1 is medium–dark brown, with 8–10 prominent black bullae across each segment. Each bulla sports a long dark seta with a droplet at the tip. L2 is black with a wide white stripe dorsally, bisected by a thin brown line. Six conical black spines with pale orange and white bases traverse each segment, each bearing ~10 setae. L3 is similar, with well-developed lateral white patches. L4 has increased white areas, and the larva appears white mottled with black and orange markings. The head is black with indistinct brown dorsolateral markings. L5 is similar, black with distinct white stripes dorsally separated by a black line, and extensive white vermiform markings laterally. Spines are black except for the ventrolateral row, which is yellowish orange; all spine bases are orange. L6 is black with orangish white spines. The dorsal orangish white stripe bisected by a black line is prominent. Lateral white vermiform markings are distinct, and some coalesce to form oblique forward-facing dashes from spine bases in the upper 2 lateral rows. The black head is covered dorsally with brown markings dotted with black. The pupa is light brown with black markings on the head, thorax, and abdomen. All *Argynnis* spp. larvae are similar,

Adult

Egg @ 0.8 mm

L1 @ 1.5 mm

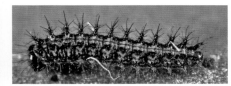

L2 @ 4 mm

L3 @ 7 mm

L4 @ 11 mm

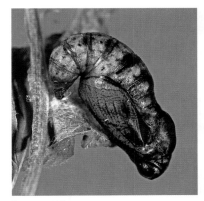

Pupa @ 16 mm

L5 @ 18 mm

L6 @ 32 mm

L6 @ 32 mm

Mormon Fritillary *Argynnis mormonia erinna* (W. H. Edwards)

but mature larvae of *A. h. brico*, *A. h. dodgei*, *A. hydaspe*, *A. egleis*, and *A. cybele* are darker with limited or no white markings. The lighter mature larvae of *A. zerene*, *A. coronis*, *A. atlantis*, and *A. callippe* are most similar, but only *A. atlantis* and *A. coronis* have bisected dorsal stripes as prominent; however, both subspecies of *A. mormonia* are the only entities with all spines whitish in L6. Miller and Hammond (2007) show an image of a mature larva that appears similar to ours.

Discussion

Both subspecies of *A. mormonia* are easily reared from gravid females. In our rearings, *A. m. erinna* L3–L6 were lighter and had more extensive white markings laterally than *A. m. washingtonia*. The dorsal white stripe was also more prominent and the overall appearance of late instars resembled that of *A. atlantis*. Dornfield (1980) described the larva of *A. m. erinna* in OR as paler than other *Argynnis* spp., but rearing of this subspecies from other locations (e.g., high-elevation habitats in OR) is needed to determine whether lighter coloration is diagnostic in separating *erinna* from *washingtonia*.

Mormon Fritillary   *Argynnis mormonia washingtonia* (Barnes & McDunnough)

Adult Biology

Argynnis mormonia ranges from AK S to AZ and E to MB, SD, UT, and CO, occupying mid–high-elevation habitats. Two subspecies that differ substantially as adults occur in Cascadia. *Argynnis mormonia washingtonia* is found in S BC and in the Cascades of WA (Pyle, 2002) and flies in moist alpine and subalpine meadows, also drier hillsides, from ~2,500 to 7,000 ft. The single brood flies from early June to mid-Oct but is most common during July–Aug. Females mate only once and do not appear to undergo reproductive diapause. Dispersal is limited, and males patrol for females. Mated females spend much time crawling through ground vegetation looking for oviposition sites. Both sexes nectar on flowers, including asters, yarrow, and Pearly Everlasting; males and older females also visit mud, carrion, and animal scat. Larval hosts include *Viola palustris* (Marsh Violet), *V. adunca* (Blue Violet), *V. nephrophylla* (Northern Bog Violet), *V. nuttalli* (Nuttall's Violet), and *V. sempervirens* (Evergreen Violet) (Pyle, 2002; Warren, 2005; Miller and Hammond, 2007). Eggs are laid singly on or near senescing violets.

Immature Stage Biology

We reared this species several times. Females obtained from Yakima Co., WA July 29–30 (2 years) oviposited immediately when provided with dried foliage of *V. labradorica* (Labrador Violet) in plastic boxes with muslin lids (32 × 20 × 9 cm). Eggs were hidden underneath foliage and also laid on living foliage when provided. Eggs hatched after 10 days at 21–25 °C, and L1 sought refuge in curled leaves without feeding on eggshells or leaves. L1 held at 21–25 °C, 30–40% RH for 5–10 weeks before overwintering suffered 95–100% mortality, but limited exposure (7 days) resulted

in <5% mortality. L1 were overwintered at 5 °C for 88 days with ~90% survival. Feeding commenced on *V. adunca*, *V. labradorica*, and *V. odorata* (Sweet Violet) after 2 days at 27 °C, 12–15 hr light. Development from L2 to pupation took 44 days. Pupae held at 17–21 °C eclosed in 18 days. Larvae fed on young leaves at first, then older leaves, eating from the edges and mostly at night. L1–L3 tend to rest and feed communally in captivity. L2–L6 possess an eversible ventral gland on segment 1 that produces a musky odor when the larva is disturbed (James, 2008a) and is likely defensive. Mature larvae silk together leaves as "pupation tents," and pupation occurs close to the ground. If hanging leaves are not available, larvae gather leaves and silk them to a substrate to form a "tent." Prepupae that fall to the ground are able to pupate successfully. Larval defense is based on concealment (nocturnal feeding, pupa), chemical protection (ventral gland, setal droplets in L1), and physical protection (spines).

Description of Immature Stages

The conical egg is creamy white becoming brown with indistinct red spots prior to hatching. There are 24–26 prominent vertical ribs merging to ~12 at the top. L1 is medium–dark brown with 8–10 prominent black bullae across each segment. Each bulla sports a long dark seta with a droplet at the tip. L2 is dark brown-black with a wide pale stripe dorsally and a shiny black head. Six prominent conical black spines traverse each segment, each bearing ~10 setae. L3 is black with a pale dorsal stripe bisected by a dark line. Orange-based black spines are larger and bear ~12 setae. L4 is black mottled with gray vermiform markings and a dorsal grayish white stripe bisected by a dark line. Orange bases of spines are larger, and the head is black with distinct brown markings dorsolaterally, peppered with black dots. L5 is similarly marked but darker with a distinct white stripe bisected by a dark line. Ventrolateral spines are yellowish orange, and the brown markings on the head are more extensive. L6 is dark brown-black with orangish white spines and a dorsal orangish white stripe bisected by a dark line. Lateral pale markings are limited. The black head is covered dorsally with brown markings dotted with black. The pupa is medium–dark brown with darker markings on the head, thorax, and abdomen. All *Argynnis* spp. larvae are similar, but mature larvae of *A. h. brico*, *A. h. dodgei*, *A. hydaspe*, *A. egleis*, and *A. cybele* are dark with limited or no white markings. The lighter and dorsally striped mature larvae of *A. zerene*, *A. coronis*, *A. atlantis*, and *A. callippe* are most similar,

Adult

Mormon Fritillary   *Argynnis mormonia washingtonia* (Barnes & McDunnough)

Eggs @ 0.9 mm

L2 @ 3 mm

L1 @ 1.5 mm

L3 @ 6 mm

L4 @ 12 mm

Pupa @ 17 mm

L5 @ 16 mm

L6 @ 26 mm

but only *A. atlantis* and *A. coronis* have prominent bisected dorsal stripes; however, both subspecies of *A. mormonia* are the only entities with all spines whitish in L6. Guppy and Shepard (2001) show an image of a mature larva of ssp. *opis* that appears similar to ours.

Discussion

Both subspecies of *A. mormonia* are easily reared. We found the larvae of *A. m. washingtonia* were darker and did not have the extensive white lateral markings of *A. m. erinna*; however, coloration and markings may vary with habitat and geography, and more rearing is needed to confirm this difference. The need for sufficiently humid conditions (>70%) to ensure good survival of pre-overwintering, diapausing L1 likely reflects adaptation to the moist habitats occupied by *A. m. washingtonia*. High humidity is also needed for good survival of overwintering L1.

L6 @ 26 mm

Silver-bordered Fritillary   *Boloria selene* ([Schiffermuller])

Adult Biology

Boloria selene is primarily a N species found E of the Cascades in BC and in N WA, also in small remnant populations scattered in central and SE WA, NE OR, and the ID panhandle. It is circumboreal, occurring in Europe and Asia. In North America it occurs in AK and much of Canada, S into N US, and along the Rocky Mts to NM. The preferred habitats are bogs, fens, and moist riparian areas between 1,300 and 3,600 ft and it is locally common in some areas. The flight period in Cascadia is mid-April–early Sep with 2 broods. Guppy and Shepard (2001) reported an apparent partial third brood in early Sep in S BC. The larval food plants include *Viola palustris* (Marsh Violet), *V. nephrophylla* (Northern Bog Violet), and *V. glabella* (Stream Violet). Males patrol conspicuously over grassy, boggy areas near violets, searching for females. After mating, females remain concealed in the vegetation. In captivity, oviposition occurs on host leaves or inert surfaces. On Aug 15 in Pend Oreille Co., WA, we saw a second-brood female ovipositing on a hard edge of an unsurfaced road; *Viola* was present a few hundred feet away. A population of *B. selene* at Moxee Bog, Yakima Co., WA (Pyle, 2002) may have been extirpated, possibly because removal of livestock allowed unbrowsed vegetation to choke out the *Viola*. Scott (1992) reported the extinction of a small population in CO, when horse grazing was stopped. The survival of *B. selene* appears dependent on herbivore grazing in at least some situations.

Immature Stage Biology

We reared this species 3 times from gravid females from the first and second broods. On June 24, 4 females from Mary Ann Creek, Okanogan Co., WA, produced ~50 eggs. Larvae provided with *Viola glabella* and *V.*

Adult

sempervirens preferred *V. glabella*. Thirteen larvae reached L2, 10 pupating and producing adults. Females from Tiger Meadows, Pend Oreille Co., WA, on July 1 and Aug 28 produced ~100 eggs during 5 days. Eggs hatched in 5–6 days at 21–25 °C, and larvae were provided with *V. glabella*, *V. labradorica* (Labrador Violet), and *V. tricolor* (Pansy), which they accepted equally. Development from L1 to pupation took 25–30 (first brood) or 30–34 days, with eclosion after 10–20 days. Guppy and Shepard (2001) reported that some larvae in the first brood in BC develop very slowly. Some (30%) of our July 1 first brood became dormant in L3 despite long daylengths. Autumn-reared larvae were exposed to naturally declining daylengths, and development ceased in L2. A subgroup was transferred to 26–30 °C and continuous lighting and immediately recommenced feeding. Overwintering is reported to occur as L2–L4 (Scott, 1986a). Larvae feed nocturnally on violet leaves, with early instars feeding on the undersides (Scott, 1986a). L3–L5 eat large holes from the edges inward. There are 5 instars, and no nests are made. L2–L5 possess a ventral gland on segment 1 that secretes chemicals likely used to repel predators. The survival strategy employs diurnal concealment, protective spines, and chemical defense.

Description of Immature Stages

The egg is whitish turning straw yellow with ~18 vertical ribs crossed by 22–24 lesser horizontal ribs, creating a cancellate pattern. L1 is marbled rusty, cream, and greenish. There are paired dorsal brown patches on the 4th, 6th, 8th, and 10th segments. All segments bear 8–12 shiny black bullae, giving rise dorsally to long black simple setae and laterally to pale yellowish setae. The bullae in L2 are raised into black cones in uniform rows, each bulla bearing a cluster of dark setae. The body is streaked longitudinally in a pattern of alternating dark and light stripes. In L3 the dark stripes are reduced, the black bullae contrast with lighter colors at their bases, and considerable orange suffusion occurs laterally and dorsolaterally. Ventrolaterally there is a broad blackish stripe and the head is shiny black with dark setae. The 2 dorsal bullae on segment 1 are longer and point forward like spiny horns. L4 is a mottled jumble of gray, orange, and black, with gray dominating dorsally. The horns on segment 1 are longer, and all bullae are black, bearing numerous setae. L5 is purplish gray, mottled with numerous black splotches and yellow spinelike bullae bearing setae. The anterior 3 segments are

Egg @ 0.9 mm

Pupa @ 15 mm

L1 @ 2.0 mm

L2 @ 5 mm

L3 @ 10 mm

L4 @ 15 mm

L5 @ 23 mm

L5 @ 23 mm

Silver-bordered Fritillary *Boloria selene* ([Schiffermuller])

black, and the horns on the first segment are long and black with yellow bases. The pupa is mottled brown, black, and reddish with spiny projections; there are 5 pairs of shiny cream-colored cones on the head and thorax. The larvae of other *Boloria* spp. are similar; however, the contrasting yellow spines, long anterior horns, and black anterior segments of *B. selene* in L5 are diagnostic. There are numerous published images of larvae, the most recent including Allen (1997) and Guppy and Shepard (2001) (both with pupa images), and Wagner (2005), Allen et al. (2005), and Neill (2007). All images agree with ours.

Discussion

Viola host plants grow very low, often under other vegetation where the cryptically colored, nocturnal larvae are very difficult to find, but eggs can be obtained from gravid females. Larvae may be fragile in L1, but later instars are hardier and will accept a number of *Viola* spp., including commercial Pansy. Most eggs from first-brood females develop directly to adults, but some midstage larvae diapause. All second-brood larvae overwinter as diapausing L2–L4. The environmental cues for inducing diapause in the second generation are unknown, but we suspect declining daylength to be important; we were able to avoid diapause by exposing L2 to continuous light. A proportion of the population may be obliged to diapause as larvae regardless of environmental conditions. Little is known of natural enemies; however, Pyle (2002) found that ambush bugs prey on adults visiting flowers. Research is needed to clarify the relationship between the ecology and persistence of *B. selene* and grazing by herbivores.

Meadow Fritillary   *Boloria bellona* (Fabricius)

Adult Biology

Boloria bellona ranges across Canada and NE US with disjunct populations in Cascadia and the Rocky Mts. In Cascadia small populations occur in N WA (Okanogan and Ferry Cos.) and NE ID. Populations previously existed in the WA Blue Mts and in Umatilla Co., OR but have not been seen at these locations for more than half a century. E populations are often found in disturbed environments, but in the W the habitat is restricted to mid–high-elevation meadows. Adults fly from late May to early July where violet host plants grow, particularly the favored *Viola canadensis* (Canada Violet). Males patrol all day, conspicuously searching for females where open forest areas adjoin grassy meadows and boggy streams. Females generally remain low in the vegetation. During mating the pair alternately perches and flies in tandem, avoiding other patrolling males. In mid to late June, females oviposit on or near host plants, in damp boggy areas or adjacent woodland meadows. Scott (1992) observed a female landing on a non-host plant, crawling down to the base, and ovipositing on a dead grass blade. On June 23 we collected a mating female, and 3 days later she laid 85 eggs. The eggs were laid singly, 12 on live violets, the remainder on inert surfaces. The Meadow Fritillary is reported to be double- and even triple-brooded in more E and S areas (Scott 1986a). Larvae often produce a second brood of adults in captivity, and Pyle (2002) reported 2 generations in Cascadia; however, records of second-brood adults in WA appear to be lacking. Scott (1992) reared a single brood in CO in which 19 adults eclosed but 23 larvae diapaused, producing adults after a period of dormancy.

Immature Stage Biology

We reared this species once. Eggs from gravid females obtained in Okanogan Co. hatched after 5–7 days, with L1 consuming eggshells before feeding on *Viola glabella* (Stream Violet) leaves. Larvae matured quickly, eating holes in leaves from the edge inward. Feeding throughout development was mostly nocturnal, and larvae often wandered off host plants to rest. The larvae did not build nests and typically rested on inert surfaces or under cover. Larvae of *B. bellona* are not gregarious but are tolerant of each other. A cohort of 12 L4 and L5 were transferred to *Viola labradorica* (Labrador Violet) and readily accepted this commercially available violet species. Development from egg-hatch to pupation took ~17–20 days, with adults eclosing 24–34 days post egg-hatch. All captive larvae produced second-brood adults. According to Scott (1992), *B. bellona* overwinters as freshly molted (unfed) L4, and it is likely that environmental and host-plant conditions experienced by L2–L3 determine whether direct development or diapause occurs. Pyle (2002) reported that overwintering occurs in L3 or L4.

Description of Immature Stages

The pale greenish yellow egg bears 18–20 vertical ribs, and many small lateral cross-grooves. L1 has a shiny black head and many simple, long setae on the body and head. The anterior 3 segments are creamy gray with posterior segments alternately brown and gray. In L2 there is a staggered row of about 8 small dark tubercles across each segment, each of the tubercles giving rise to 2–3 simple setae; the body is cream, brown, and green. In L3 the tubercles are cone shaped and black, each giving rise to 10–15 unbranched black spines. The body is variegated in brown and cream, with a distinct dark brown lateral stripe and a broken cream stripe above it. The head in this and all instars is shiny black, bearing numerous setae. L4 is grayish, with 2–3 faint rings circling each segment between tubercles and the black side stripe. L5 is similar except the ground color is magenta-gray, and the tubercles are black at the tip and pink at the base. An additional white longitudinal stripe is present along the basal periphery. The pupa is brownish gray and angular. A vertical column of 3 strong V-shaped chevrons dominates the posterior end of the abdomen, and there are several reflective metallic spots. Larvae of other *Boloria* spp. are similar to *B. bellona* and are best separated by comparison of images. Larvae of *Boloria epithore* lack the well-defined side stripes and differ in details of the complex body mottling; also spine-bearing bullae are orange. *Boloria selene* often co-occurs with *B. bellona* and feeds on the same host

Adult

Eggs @ 1.0 mm

L2 @ 3.1 mm

L1 @ 1.7 mm

L3 @ 6 mm

L4 @ 13 mm

L5 @ 24 mm

Meadow Fritillary *Boloria bellona* (Fabricius)

plants; however, mature larvae of this species are solid black anteriorly, mottled gray posteriorly, with contrasting orange tubercles. Scott (1992) provided a description of the immature stages of *B. bellona* that agrees well with our images. Images of mature larvae shown by Allen et. al. (2005) and Wagner (2005) are similar to ours; however, the image of a mature larva from WV in Allen (1997) differs in being brown with white markings and has a red-brown head.

Discussion

The scarcity of *B. bellona* in Cascadia presents a challenge for obtaining gravid females. Timing appears to be important, with newly mated females likely to provide the best chances of obtaining fertile eggs. Females hide in grass and shrubs, and walking through prime habitat to flush them into flight may assist in finding them. Once eggs are obtained, *B. bellona* is relatively easy to rear as long as violet hosts are available. Commercially available *V. labradorica* and native *V. glabella* are excellent substitutes for *V. canadensis*. The full host range in Cascadia is unknown and deserves study. It is difficult to find larvae as host plants are numerous, small, and typically well hidden under taller grasses and shrubs; also the larvae are nocturnal feeders and likely spend daylight hours in protected refugia. Very little is known of the biology and ecology of this species, including the biotic and abiotic factors affecting abundance and distribution in Cascadia.

Pupa @ 15 mm

Western Meadow Fritillary · *Boloria epithore* (W. H. Edwards)

Adult Biology

Boloria epithore flies in a single brood from early May to late Sep in Cascadia (depending on elevation), favoring meadows, riparian areas, logging roads, and moist openings in forests and mountains from sea level to ~8,000 ft (Guppy and Shepard, 2001; Warren, 2005). Guppy and Shepard (2001) found that populations fly for ~3 weeks. This species is locally abundant and is found W of the Cascades throughout Cascadia and along the E slopes as well. It also occurs in the mountains of NE WA, and in the Blue and Wallowa Mts of SE WA and NE OR. Beyond Cascadia it ranges from SW YT and SW AB, S through BC, MT, WA, ID, WY, and OR to central CA. Larval host plants are violets including *Viola adunca* (Blue Violet), *V. glabella* (Stream Violet), and *V. sempervirens* (Evergreen Violet); Warren (2005) and Shapiro and Manolis (2007) reported *V. ocellata* is used in CA, and Scott (1986a) suggested that *V. nephrophylla* is also a likely host. Both sexes visit flowers, including Pearly Everlasting, asters, and thistles, and males patrol relentlessly in search of females. In captivity, females lay eggs singly, mostly on inert surfaces but also on cut and living *Viola* plants if provided.

Immature Stage Biology

We reared or partially reared this species 5 times, in all cases from eggs from captive females. On July 27, 4 females from Tuquala Lk, Kittitas Co., WA, produced 45 eggs resulting in 31 overwintering L4, and on July 16 (different year, same locality), 3 females produced 20 eggs resulting in several overwintering L4, but none survived winter in either case. Two females obtained July 31 near Swauk Pass, Kittitas Co., produced many eggs; 21 L4 overwintered, with 3 achieving adulthood. Females obtained Aug 1 from Bear Creek Mt (Yakima Co., WA) laid numerous eggs on potted *V. glabella* resulting in 8 overwintering

Adult

L4. Three survived and molted to L5 but died before pupation. Three females from Sawtooth Ridge (Columbia Co., WA) on July 8 produced ~100 eggs on potted *V. glabella* and *V. labradorica* (Labrador Violet) resulting in 2 L4 overwintering and producing adults. Eggs hatched in 4–7 days and development was rapid on *V. glabella*, 17–18 days from L1 to L4. L4 continued feeding for 3–4 days before wandering off the plant, finding shelter in dead leaves, and entering diapause, often in groups. Diapause was terminated after 3–8 months by exposure to warm (15–26 °C) conditions. Feeding commenced in 2–6 days, and larvae molted to L5 after 7–24 days, with pupation occurring 10 days later. Delayed termination of overwintering resulted in faster postdiapause development, with pupation occurring within 9–12 days. Newly hatched larvae refused to feed on the commercial *V. odorata* (Sweet Violet), fed poorly on *V. labradorica*, but fed readily on *V. glabella* and *V. adunca* leaves, eating scalloped holes from the edges. Leaves were strongly preferred over flowers, and when not feeding the larvae rested on the undersides of leaves. Survival was good until diapause, but overwintering mortality was substantial. Protection is based on concealment (nocturnally active), a ventral gland, and spines to dissuade predators. There are 5 instars, and nests are not constructed.

Description of Immature Stages

The egg is white with 20–22 prominent vertical ribs at the base, merging to ~8–12 at the apex. L1 is greenish gray with a shiny black setaceous head that persists throughout development. The 4th, 6th, 8th, and 10th segments are rusty colored dorsally; tiny dark tubercles on the body give rise to simple long pale setae that bear small droplets at the tips. The body darkens in L2 to purplish with indistinct gray markings. The black bullae are conical, organized into 3 longitudinal rows laterally, each bulla bearing a cluster of black spines. Dorsally there is a broken narrow purplish line bordered on each side with gray markings. L3 patterns are similar, but the conical bullae are taller, and the setae borne by them are more numerous, imparting a spinier appearance. Ventrally there is more orangish coloration and the whitish-brown stripe is straight. L4 is intermittently striped longitudinally in black, gray, or white with a ventrolateral orange stripe. The conical spines on segments 1–3 are black, but others are half orange. In L5 the conical spines are entirely orange except on the terminal posterior segments and on the first 2 or 3 segments, where they are black. The ventrolateral orange line is prominent, and there

Egg @ 1.0 mm

Pupa @ 15 mm

L1 @ 2 mm

L2 @ 6 mm (late)

L3 @ 10 mm

L4 @ 13 mm

L5 @ 21 mm

are numerous white spots near the white circled spiracles. Dorsally L5 varies from gray to black and there is a narrow darker line down the center, surrounded by pale spots. The pupa is dark brown, sometimes with a purplish tint, bearing 5 pearly conical processes in an irregular row on each side of the head and thorax. The wing cases are darker brown, and dorsally the abdomen bears 2 rows of bold triangular processes that join on posterior segments to form forward-pointing chevrons. Immature stages of *B. epithore* are most similar to other *Boloria* spp. The larvae of *B. astarte*, *B. freija*, and *B. chariclea* are usually considerably darker, and those of *B. selene* have long horns over the head. Mature *B. bellona* larvae are browner than *B. epithore* and have a distinctive black and white side stripe. Our images appear to be the first published of immature stages of *B. epithore*.

Discussion

Eggs are easily obtained from gravid females, and larvae grow quickly, overwintering in L4. Rearing larvae under long or increasing daylengths may avert winter diapause. Mortality during winter diapause was considerable in our rearings, sometimes complete. High relative humidity appears to be important for survival of diapausing larvae; however, excessive humidity causes mold problems. Dormant L4 should be aged for ~10 days prior to overwintering. Postdiapause development is enhanced by prolonged overwintering (at least 6 months). Larvae are difficult to find as they feed nocturnally on low-growing violets under other vegetation. Late-season (high-elevation) larvae tend to be present after most violet hosts have senesced, an apparent maladaptive strategy unless other hosts are also used. Research is needed on the identity and impact of natural enemies influencing the abundance and distribution of this species.

Freija Fritillary   *Boloria freija* (Thunberg)

Adult Biology

Boloria freija is a circumpolar species found in most of Canada, extending to the Great Lakes and the N Rockies (WY, CO). In Cascadia it occurs in BC and Okanogan Co., WA in willow bogs and forest clearings above 5,000 ft. The single brood flies from June to July. Males patrol for females, flying erratically and close to the ground, sometimes even under low logs. Females are elusive but generally stay closer to host plants. Of 40 individuals seen near Keremeos, BC on June 15–16, only 6 were female. Both sexes visit flowers, including Bog Rosemary, buttercups, asters, and yarrow as well as mud. Larval hosts include *Vaccinium caespitosum* (Dwarf Bilberry), *V. uliginosum* (Northern Bilberry), *V. oxycoccus* (Wild Cranberry), *Arctostaphylos uva-ursi* (Kinnikinnick), *Empetrum nigrum* (Black Crowberry), *Rubus chamaemorus* (Cloudberry), and *Rhododendron aureum* (Rhododendron) (Pyle, 1981; Scott, 1986a; Nielsen, 1999; Douglas and Douglas, 2005). In Cascadia, *V. caespitosum* and *A. uva-ursi* appear to be the major hosts. Eggs are laid singly or in small clusters on or near hosts (Scott, 1992; Douglas and Douglas, 2005). We observed oviposition on the edge of a *V. caespitosum* leaf in Okanogan Co.

Immature Stage Biology

We reared this species once from 6 females obtained on June 16 near Keremeos, BC. Confined in plastic and muslin cylinders (13 cm dia × 8 cm tall) with cut *Vaccinium parvifolium* (Red Huckleberry) and *A. uva-ursi*, females laid ~100 eggs in 2 days. Eggs hatched in 6 days at 20–22 °C and L1 were provided with *Vaccinium scoparium* (Grouseberry) or *V. caespitosum*, reaching L2 after 4 days (July 1). L2 and L3 occupied ~8 days each with L4 reached by July 16. Feeding continued for 10–14 days then ceased, with larvae measuring ~23–25 mm. Thereafter larvae

rested on the host or inert surfaces, not seeking refuge. Larvae shrank to 20–23 mm, and on Aug 10, 9 larvae were placed at 5 °C for overwintering, followed by a second group of 22 on Aug 17. The second cohort was examined 35 days later, and only 4 larvae were alive. These were exposed to 25 °C and long days (~15 hr), but all died after 3–61 days without feeding. One larva from the first cohort was exposed to warm conditions after 128 days but died after 71 days without feeding. The first cohort was examined after 215 days (Mar 13); the 5 remaining larvae were exposed to 19–23 °C and supplied with new and old growth of *A. uva-ursi*. The larvae did not feed except for 1, which nibbled on a new growth leaf. After 4–6 days the larvae molted to L5 that did not feed; after 1–3 days (depending on temperature: 22–28 °C) they pupated. Adults eclosed after 10–12 days at 19–21 °C. Scott (1986a) and Pyle (2002) stated *B. freija* overwinters as a mature L4, while Nielsen (1999) and Douglas and Douglas (2005) suggested overwintering may occur as a pupa. Larvae fed on young leaves at first, then older leaves, eating from the edges. Larvae are solitary with no nests and most feeding is nocturnal, with larvae resting exposed on the host during the day. Larvae are active when disturbed and rapidly move away from any threat. L2–L5 possess an eversible ventral gland on segment 1 (James, 2008a) that is likely defensive. Prepupal larvae wandered before pupating from cage tops. Larval defense appears based on chemical protection (ventral gland, droplets on setae in L1), physical protection (spines), and evasive activity.

Description of Immature Stages

The shiny greenish white egg has ~35–45 shallow vertical ribs that amalgamate at the apex. L1 has a shiny black setaceous head and is inconspicuously banded, gray and dark brown, becoming purplish with maturity. There are 8–10 black bullae across each segment, each bearing a long dark seta with a droplet at the tip. L2 is black, developing gray areas with maturity. Six prominent conical black spines traverse each segment, each bearing ~12–14 setae. Mature L2 have an indistinct pale spiracular stripe and the prolegs are white. L3 is black with black spines, a shiny black head, and whitish tan prolegs. L4 is similar, unicolored black with distinct tan prolegs. Numerous tiny black bullae (~200 per segment), each bearing a short seta, give the body a textured appearance. L5 is black with greatly reduced green spines giving the appearance of a green-dotted black larva with tiny light-colored bullae. Some individuals have pale dorsolateral stripes. The head and spiracles

Adult male

Egg @ 0.9 mm

Pupa @ 13 mm

L1 @ 3 mm

L2 @ 6 mm

L3 @ 15 mm

L4 @ 22 mm

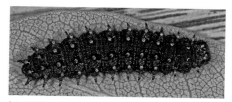

L5 @ 20 mm

L5 @ 21 mm

Freija Fritillary

Boloria freija (Thunberg)

are black. The pupa is black mottled with gray, brown, and white on the abdomen. There are 5 pairs of tubercles on the abdomen, the posterior 3 with connecting ridges. The black L4 and green-dotted black L5 of *B. freija* are unique in our fauna and should not be confused with other *Boloria* or *Argynnis* spp. Larvae of the related *Vaccinium*-feeding *B. chariclea* are closest in appearance, but mature larvae are speckled with white dots and have dull orange spines. An image of an apparent L4 is shown in Nielsen (1999) and appears similar to ours.

Discussion

Our rearing appears to be the first documented in North America. Eggs were readily obtained, and larvae fed well on *V. scoparium* and *V. caespitosum*, completing development to L4 rapidly under mild temperatures. Larvae are black in all instars, optimizing available warming. Overwintering occurred as mature L4. The absence of feeding in postdiapause L4 and a brief L5 is highly unusual and possibly unique. If post-hibernation feeding does not occur in wild populations, the need for L5 is puzzling; perhaps the species is evolving toward 4 instars? Attempts to break dormancy early were unsuccessful, with L4 remaining alive for lengthy periods without pupation. Between 4 and 7 months of cold temperatures appear necessary to break diapause; thus shorter winters may reduce overwintering survival. Studies on the overwintering and postwintering behavior of *B. freija* are needed to better understand the ecology of this species, which may be vulnerable to global climate change.

Astarte Fritillary   *Boloria astarte* (E. Doubleday)

Adult Biology

Boloria astarte is widely distributed in S BC, rare in WA and absent from OR. In WA it occurs in a few scattered colonies in the N Cascades (Okanogan and Chelan Cos.). Beyond Cascadia the range includes AK, YT, AB, and MT. A true alpine specialist flying above 7,000 ft during mid-July–mid-Aug, *B. astarte* frequents rocky ridges, slides, and steep scree slopes where the host plant grows. Males hill-top or patrol in search of females, often flying diagonally up and down slopes. Females eclose later than males and are more sedentary, basking on boulders or visiting nectar sources, especially *Arnica* spp. *Saxifraga bronchialis* (Spotted Saxifrage) is the only known host plant in Cascadia, although *S. tricuspidata* (Three-toothed Saxifrage) is used in N BC (Guppy and Shepard, 2001). *Boloria astarte* is biennial, requiring 2 years for larval development and flying on even-numbered years in Cascadia; Guppy and Shepard (2001) reported similar flights in the Coast Range of BC, with flights every year (2 separate biennial populations) in the Rockies. Oviposition occurs in mid–late July, with eggs laid on or near the host plant (Scott, 1986a). Captive females laid ~85% of their eggs on inert surfaces, the remainder well hidden on the undersides of host leaves very close to or touching the ground.

Immature Stage Biology

We reared this species once from 3 females collected July 21 at Slate Peak, Okanogan Co., WA. Of 150 eggs laid, 80% hatched after 8 days with eggshells not consumed. Larvae were reared in 2 cohorts on *Saxifraga bronchialis* at 20–26 °C and natural daylengths. Pyle (2002) reported that *B. astarte* overwinters as L1 the first winter and L4 the second winter; however, in our rearing, larvae developed from L1–L4 in the first season, taking ~28 days; many (~30%) of the larvae did not develop beyond L2–L3 and died before overwintering. In mid-Sep, L4 stopped feeding and sought shelter under paper toweling. Twelve L4 were overwintered at 5 °C from Sep 17. Small groups (1–3) were removed from overwintering on Dec 16, Jan 3, Feb 9, and March 13 and exposed to warm temperatures. The Dec group exposed to 19–21 °C and natural daylengths remained dormant, but the Jan group (*n* = 2) started feeding after 9 days at 28 °C in continuous lighting and molted to L5 2 days later. L5 fed avidly and pupated on the host plant ~20 days after exposure to warm temperatures. One adult eclosed ~12 days later. The larvae removed from overwintering in Feb died; those in March molted to L5, but after extended periods of feeding, wandering, and resting, they also died without pupating. Our rearing suggests that in some instances larvae may develop to adulthood in 1 year if they avoid early-instar dormancy, which, contrary to Guppy and Shepard (2001), does not appear to be obligate in at least a portion of the population. Larvae refused to feed on a number of plants, including *Viola glabella* (Stream Violet), *Arctostaphylos uva-ursi* (Kinnikinnick), and *Vaccinium caespitosum* (Dwarf Bilberry), confirming *Saxifraga* spp. as the sole host. Larvae tolerated each other well in crowded conditions, with no evidence of cannibalism. The larvae fed first on terminal leaf clusters, then the middle parts of leaves several tiers below the tip. Protection is based on camouflage, concealment, and likely chemical defense from setal droplets in L1 and a ventral gland on segment 1 in L2–L5. No nests are constructed, and there are 5 instars.

Description of Immature Stages

The pale yellow-orangish egg is densely covered in ~40 coarse vertical ribs at midheight, merging to ~20 near the apex. L1 is pale tan-brown with a shiny black setaceous head, persisting until pupation. Each segment has a transverse row of about 10 low, round, brown bullae, each bearing a medium-length brown seta with a tiny droplet at the tip. L2 is purplish gray and the bullae are tall conical grayish spines, each bearing 8–10 dark setae. There are 3 dark broken dorsal lines and an irregular dark line laterally. In L3 the dark lines expand, forming a mottled blackish ground color, while the pale markings constrict and form complex contrasting patterns. L4 is black with contrasting gray-white dorsal markings in complex but regular patterns of dashes and Vs. All spines and setae are black, the bases of the 2 dorsal rows circled

Adult

Egg @ 1.1 mm

Pupa @ 15 mm

L1 @ 2.2 mm

L2 @ 4.3 mm

L3 @ 10 mm

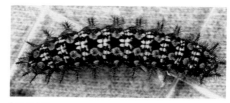

L4 @ 20 mm

L5 @ 25 mm

in yellow-gold late in the instar. L5 is similar, with the bases of all spines encircled in gold, contrasting strongly against the black body color. Each spine bears ~25 black setae. White markings are also strongly contrasting, and dorsally they meet at segment joints to form patterns resembling caricatures of white butterflies. White spots pepper the body laterally and the spiracles are black, narrowly encircled in white. The pupa is shiny dark brown on the wing cases and mottled dark and light brown elsewhere. There is a fairly deep depression behind the thorax, and the abdomen bears rows of low rounded bullae. Most other *Boloria* spp. larvae are lighter; one exception is *B. freija* in which the larvae are even darker; however, the patterns are different. No other *Boloria* spp. use *Saxifraga* spp. as hosts. The gold spine bases and prominent white markings of *B. astarte* are unique and readily identify mature larvae. Our images appear to be the first published of *B. astarte* immature stages.

Discussion

Immature stages of *B. astarte* are difficult to find, and females are elusive and difficult to catch. Females appear to oviposit for ~1 week so timing is important, as is exposure to full sun. Additional rearing is needed to determine the life history of this species, particularly concerning overwintering and the environmental cues determining biennialism. Nothing is known of the natural enemies affecting populations of *B. astarte*, nor of climatic tolerances. The restricted alpine habitat of *B. astarte* suggests vulnerability to climatic warming (Miller and Hammond, 2007), but further research is needed. Warmer summers might enable an annual rather than a biennial life cycle, but warmer winters might increase the mortality of larvae.

Arctic Fritillary *Boloria chariclea* (Schneider)

Adult Biology

The Arctic Fritillary is Holarctic, occurring from Siberia to Scandinavia to the North American Arctic. In North America it is found in most of Canada, MN, and NH in E US and AK through the Rockies S to NM in the W. In Cascadia it occurs in BC and the Olympic and Cascade Mts of WA. *Boloria chariclea* flies in a single brood in moist mountain meadows and alpine tundra (4,000–8,000 ft), mid-Jun–Sep, depending on elevation and seasonal conditions. The life cycle is annual or biennial, according to latitude and/or elevation (Pyle, 1981; Scott, 1986a, 1992), but Pyle (2002) considered it annual in Cascadia. It is locally common and often flies with *B. epithore*. Both sexes visit flowers, especially asters and goldenrod. Males patrol for females usually during the middle of the day, when courtship and mating most frequently occur. Scott (1986a) observed a female that appeared to be releasing pheromone from 2 everted red glands at the end of her abdomen. An unusually wide range of host plants have been recorded including *Viola* spp. (violets), *Salix* spp. (willows), *Vaccinium* spp. (blueberries), *Polygonum bistortoides* (Western Bistort), and *Dryas integrifolia* (Entireleaf Mountain Avens) (Pyle, 1981; Scott, 1986a; Guppy and Shepard, 2001). Eggs are laid singly on the undersides of host-plant leaves or in the vicinity of hosts. Pyle (2002) observed oviposition on *Leutkea* sp. (partridgefoot) in the Olympics, and we observed oviposition on *Vaccinium deliciosum* in S BC and the Cascades. At Chinook Pass (WA) we observed oviposition on *Lupinus* sp. Scott (1992) listed a large number of ovipositions and plant association records from CO and showed that *Vaccinium* was by far the most favored plant genus.

Immature Stage Biology

We partially reared this species from gravid females several times. Five females obtained Aug 13 at Twin Lks, Whatcom Co., WA produced 40 eggs; 25 larvae overwintered, 10 surviving to spring at 4 °C. Placed at 20–22 °C on April 23, they began feeding within 1 day on *Viola glabella*. Two survived to 9 mm (L3) after 20 days, but then ceased feeding and slowly died. One female collected Aug 12 at Slate Peak, Okanogan Co., WA produced 30 eggs, most of which hatched. Following overwintering, small groups of larvae were moved to 22 °C on 5 occasions between April 7 and June 11 and provided with 7 different food plants, including Red Huckleberry, Scouler's Willow, Pacific Willow, commercial blueberry, Stream Violet, Bitter Dock, and Sheep Sorrel. In spite of these many variations, the larvae failed to feed and all died in L1. Females obtained on July 27, 29, 30, Aug 6, 7 from Bear Creek Mt (Yakima Co., WA), Chinook Pass, and Apex Mt (BC) laid 50–400 eggs on each occasion. Confined in plastic boxes with muslin lids (30 × 23 × 12 cm) with senescing host plants (*Viola* spp., *Vaccinium* spp.) and exposed to sunshine at 22–28 °C, they laid eggs on leaves, twigs, and inert surfaces. Eggs hatched in 6–10 days at 22–27 °C; larvae consumed their eggshells but fed minimally or not at all; L1 sought shelter in curled leaves and under debris, and hibernated. Dormant L1 held at 21–27 °C, 30–40% RH for 4–6 weeks desiccated and suffered >90% mortality (2 cohorts). Dormant L1 transferred to winter conditions (5 °C, 75–85% RH) within 7–10 days of hatching survived better, with <50% mortality in 3 cohorts. After 89–192 days, L1 were transferred to 21–27 °C, 16 or 24 hrs of daylight, and provided with *Viola labradorica* (Labrador Violet), *Viola odorata* (Sweet Violet), *Vaccinium parvifolium* (Red Huckleberry), or *Vaccinium scoparium* (Grouseberry). Two cohorts (Apex Mt, Chinook Pass) fed well on both *Vaccinium* spp. and poorly on *Viola*, while the third (Apex Mt) fed well on *Viola* and poorly on *Vaccinium*. Feeding commenced after 3–5 days, and L4 was reached after ~30 days. In some instances L4 stopped feeding and reentered diapause. Mortality was high, especially in L1, and no larvae reached L5. Larvae were reared mostly on leaf platforms on moist cotton wool and were prone to drowning and disease. Larvae reared under drier conditions suffered from desiccation. Protection appears to be based on concealment (larvae appear to be nocturnal), a ventral gland, and spines to dissuade predators. There are 5 instars, and no nests are constructed.

Adult

Egg @ 0.8 mm

L1 @ 2 mm

L2 @ 5 mm

L3 @ 8 mm

L4 @ 12 mm

L4 @ 13 mm

Description of Immature Stages

The pale yellowish white egg, orange at maturity, is broadly oval, wider at the base, with ~30–35 vertical ribs merging to ~20 at the apex and terminating around the micropyle. L1 is brownish gray with a small sclerotized black collar. The shiny black head bears setae and remains this way until pupation. There are 8–10 shiny dark tubercles on each segment, each bearing a single long pale seta, sometimes with a droplet at the tip. L2 is blackish gray with a broad middorsal pale stripe. On each segment there are 6 conical black bullae, each with 10–15 spines. In some individuals there are indistinct lateral pale markings. L3 is similar but blacker and with more profuse spines. There is an indistinct middorsal pale stripe bisected by a black line. In some individuals there are pale areas laterally. The true legs are black and the prolegs are white. L4 is darker and spinier with some orangish and white freckles laterally, particularly at bullae bases, and has indistinct middorsal striping. The head is distinctly bifurcated and brownish dorsally. Mature larvae were described by Scott (1986a) and Douglas and Douglas (2005) as gray with black dorsal and lateral stripes with orange spines, the first subdorsal pair longer and yellow. Immature stages of other *Boloria* spp. are similar; however, larvae of *B. chariclea* are generally darker and less prominently marked than other *Boloria* spp. Our images are the first published of immature stages of this species in North America.

Discussion

Although eggs are readily obtained from gravid females and unfed L1 overwinter well, rearing early instars of this species is challenging. L1 are very susceptible to disease, and it is difficult to obtain development to L2. Later instars are hardier. Populations of *B. chariclea* appear to differ in host-plant preferences. Scott (1992) showed that *Vaccinium* was chosen most often by ovipositing females in CO; however, several of our larval cohorts preferred *Viola* to *Vaccinium*. Host-plant choice, larval utilization of hosts, and incidence of biennialism in Cascadia are fertile fields for study.

Family Nymphalidae
Subfamily Nymphalinae (part)
CHECKERSPOTS and *CRESCENTS*

Checkerspots and crescents are represented in Cascadia by 10 species in 3 genera. They are medium-sized orange and black butterflies found in habitats ranging from low shrub-steppe and coastal lowlands to alpine meadows. Males and females are similar in most species, and males eclose a week or so before females. Some species are widespread and can be abundant (*Euphydryas* spp.) while others are uncommon and localized (e.g., *Phyciodes pallida*). All nectar readily on flowers, and males are often seen in mudding congregations.

Eggs are laid in close-packed masses of 10–300, usually on the undersides of leaves. The rounded barrel-shaped green or yellow eggs are shiny with subtle vertical ridges. The spiny larvae are dark with bright yellow or orange aposematic markings when mature but drab gray/black in early instars, living in messy silk communal nests for protection. Mature larvae are solitary, coiling and falling to the ground if disturbed. There are usually 5 instars, but larvae may go through additional instars if there is more than 1 diapause. Pupae hang and are either tan-gray or white with black and orange markings. The Mylitta Crescent is multibrooded and the Northern Crescent may have 2 generations, but all other species have a single annual generation.

Larval hosts include asters (Compositae), figworts (Scrophulariaceae), and plantains (Plantaginaceae). *Chlosyne* checkerspots use only asters or rabbitbrush, while crescents use asters or thistles. *Euphydryas* checkerspots use a variety of figworts, including penstemon, paintbrush, snowberry, honeysuckle, and monkeyflower, as well as plantains. Many of these hosts contain iridoid alkaloids that are sequestered by aposematic larvae, making them unpalatable to vertebrate predators.

Overwintering occurs as partly grown larvae, often in communal nests, and some species remain partially active, feeding opportunistically as conditions permit. Some species are able to reenter diapause if spring conditions are unfavorable, extending life spans to 2–3 years. Postdiapause larvae sometimes switch host plants and spend much time basking in sunshine to hasten development.

CHECKERSPOTS and CRESCENTS

Hoffmann's Checkerspot   *Chlosyne hoffmanni* (Behr)

Adult Biology

Chlosyne hoffmanni ranges in a narrow band along the Cascades from S BC to OR. Beyond Cascadia the range extends into central CA in a narrow band along the Sierra Mts. The single flight is early May–mid-Aug. In Cascadia *C. hoffmanni* is found from 1,000 to 7,000 ft (Pyle, 2002; Warren, 2005), although most populations occur below 4,000 ft. Larval hosts are asters, including *Eucephalus ledophyllus* (Cascades Aster), *Eurybia conspicua* (Showy Aster), and probably others; Scott (1986a) reported that *Eucephalus breweri* (Brewer's Aster) is used in CA. It is a montane butterfly, preferring mountain meadows, logging roadsides, riparian zones, and seeps. Males frequently mud-puddle at wet spots along unpaved roads or along stream margins, often in the company of blues and *Euphydryas* spp. Both sexes visit flowers, and males patrol corridors in search of mates. After mating, typically in July, females lay large masses of eggs on the undersides of aster leaves, usually at midheight on the host, with a preference for plants in partial shade. Eggs are placed in neat rows crowded together. We found egg masses of 82 and 108 eggs; Newcomer (1967) determined that masses averaged 72 eggs (range 25–179) in 12 masses found in Bear Canyon, Yakima Co., WA. A female observed in Reecer Canyon., Kittitas Co., WA, took ~20 min to oviposit on a small aster, body inverted and with outspread wings; Newcomer (1967) observed a female lay a mass of 82 eggs in 7 min.

Immature Stage Biology

We partially reared this species 10 times from collected eggs or larvae. On July 7 and July 12 eggs were collected at Reecer Canyon, and on Aug 4 and Aug 11, L2 were collected at the same locality (all different years). Overwintered larvae were collected 6 times between June 2 and June 10 (different years) at Liberty, Kittitas Co., and on Aug 28 more than 100 nests with diapausing larvae were found at Reecer Canyon. In

several cases the larvae were reared through to adults. Eggs hatched 11–16 days post oviposition at 20–22 °C. Development from L1 to L3 took ~16 days, after which the larvae ceased growing but continued feeding, then entered diapause. Postdiapause development from L3 to pupation took ~14 days, and adults eclosed ~15 days later. Newcomer (1967) reported similar development times. The eggs hatch en masse, and L1–L2 are gregarious, feeding communally in large silk nests enveloping the top 4–8 inches of the host. Leaves are skeletonized, and the group moves after a leaf has been largely consumed. When disturbed, larvae coil into a ball and tumble to the ground. Diapausing larvae congregate in silk nests surrounding dead leaves. In the spring, larvae disperse to only 1 or 2 per plant, in small individual silked leaf nests. Late instars feed on upper surfaces of upper leaves of spring asters, sometimes openly, their black color contrasting with the green leaves. Asters with rectangular feeding holes in the youngest unopened apical leaves may betray the presence of a larva. Protection includes concealment (in nests or under leaves), tumbling to the ground when disturbed, defensive spines, and camouflage. There are 5 instars.

Description of Immature Stages

The eggs are pale greenish yellow with 18–20 indistinct vertical ribs on a shiny surface. L1 is whitish with dorsal, lateral, and ventrolateral rows of long pale setae, becoming tan with a brown side stripe. The head is shiny black and bifurcated. L2 is reddish brown with a broad lateral cream stripe. Three dorsal rows of elongated spines have black setae. The head is black, this persisting until pupation, with a shiny black collar and a small sclerotized plate on the posterior segment. Below the lateral cream stripe is a narrower brown band. L3 is black with small pale speckles and 5 dorsal rows of conical black spines bearing numerous black setae. Ventrolaterally there are 2 grayish stripes with an additional row of tall dark spines with pale setae. L4 is black dorsally with a broad brownish ventrolateral stripe bordered above by a broken pale line; there are numerous distinct whitish spots dorsally and low laterally. The black spiracles are distinct, outlined in white. In L5 the central dorsal spiny bullae are surrounded at the base with dull orange rings; a dorsal black line is fairly prominent, and white spotting is strong and contrasting. The ventrolateral spines are orange on the outer halves. The pupa is smooth with no setae; the head and thorax are pale lavender, the abdomen is blackish with distinct white spots, and there are 2

Adult

Eggs @
0.8 mm

L1 @ 2.6 mm

Larval nest (pre-winter)   Pupa @ 13 mm

L2 @ 5 mm

L3 @ 6 mm (pre-winter)

L4 @ 12 mm

L5 @ 20 mm

Hoffmann's Checkerspot *Chlosyne hoffmanni* (Behr)

lateral orange lines and another dorsally. There are 3 rows of purplish rounded bullae on each side and another dorsally, all outlined at the base with either white or orange rings; the wing cases are light brown. Similar species include *Phyciodes* spp. in which the larvae are brown rather than black and the pupae less ornate. The larvae of *C. palla* and *C. acastus* are similar, but in L4 and L5 their orange markings are considerably stronger and more contrasting. *Chlosyne hoffmanni* and *C. acastus* do not use the same host plants. North of central OR the ranges of *C. hoffmanni* and *C. palla* overlap only in Klickitat Co., WA. Our images appear to be the first published of the immature stages of *C. hoffmanni*. Newcomer (1967) provided a detailed description of larvae from Bear Canyon, similar to ours except that his L5 had no orange markings.

Discussion

Eggs and larvae of *C. hoffmanni* are not difficult to find in areas where adults are common, and they can be found in the same aster patches year after year. Larvae may be parasitized, with about 25% of a cohort of 52 we collected yielding wasp pupae in L3, killing the hosts. Postdiapause larvae rest exposed on host plants, probably to hasten development by raising body temperature as *Euphydryas editha* is known to do. Very little research has been conducted on *C. hoffmanni*, so little is known of the natural factors controlling distribution and abundance. The environmental cues controlling diapause are unknown and deserve study. Miller and Hammond (2007) found that *C. hoffmanni* benefits from infrequent controlled fires and forest thinning to create open habitat.

Sagebrush Checkerspot   *Chlosyne acastus* (W. H. Edwards)

Adult Biology

The Sagebrush Checkerspot is an arid lowland species ranging from S central Canada, MT, WY, and NB S to NV, NM, and E CA. In Cascadia it is found in basin areas of E OR and WA. *Chlosyne acastus* flies in a single brood from early April to early July in WA and mid-April to mid-June in OR. At Schnebley Coulee (1,750 ft in E Kittitas Co., WA), adults fly from about April 20 to May 10, matching growth of the host plant, *Erigeron linearis* (Linear-leaf Daisy). At Red Mt (1,400 ft, Benton Co., WA), where *Chrysothamnus viscidiflorus* (Green Rabbitbrush) is the larval host, adults fly early May–early June. Males eclose a few days earlier than females, and females do not mate until ~7 days after eclosion. In cloudy conditions and at dusk, both sexes often roost in loose communal groups on shrubs, especially in canyon bottoms. Males establish territories on paths and trails, challenging interlopers and pursuing females. In captivity, females oviposited 10–12 days post mating, after most males had died. Eggs are laid in clusters directly on the host plant or on inert surfaces close to the plant. Additional species of herb and shrub Compositae are likely used as hosts (Scott 1986a). Egg masses on *C. viscidiflorus* are usually large (100–150 eggs) and are generally placed on the undersides of leaves at the bases of plants. In captivity, smaller clusters of eggs are laid on the undersides of *E. linearis* leaves, on basal flower stalks, or on adjacent inert surfaces.

Immature Stage Biology

We collected and reared overwintered larvae on a number of occasions, and obtained and reared eggs from females on several other occasions. Eggs hatch after 6 days at 25 °C. L1 feed on *E. linearis* or *C. viscidiflorus* leaves and are gregarious. L1 and L2 occupy ~6 days each at 22–26 °C. In a captive Red Mt cohort, development slowed in L3, with larvae

Adult female

ultimately leaving the host plant seeking refuge at the base of the cage ~30 days post hatch (June 15). These larvae entered diapause and oversummered, then overwintered at the base of the cage. A cohort from Schnebley Coulee continued to feed and develop rapidly in captivity, reaching L4 ~19 days post egg-hatch; 6 days later they molted to L5 before entering diapause. At the same location, L3 and L4 are found in spring, so the natural overwintering stage is apparently L3. At Red Mt, L3 climb up *C. viscidiflorus* in mid-March and begin feeding as soon as new growth occurs. Molts to L4 and L5 took place after ~7 and 13 days, respectively, at 18–21 °C. Pupation occurred 11 days later, with first adults eclosing after an additional 18 days (April 28). Schnebley Coulee pupae produced adults after 7–11 days. Postdiapause larvae are solitary, feeding diurnally and resting openly on outermost branches of *C. viscidiflorus* or on exposed surfaces of *E. linearis*. Both host plants have numerous narrow leaves, rendering it difficult to recognize larval feeding damage. Larvae construct only light nests in L1 and L2, with delicate webbing to support their aggregations. Protection is based on gregariousness and sheltering in L1 and L2, whereas later instars appear to depend on spines and a ventral gland on the first segment that likely emits defensive chemicals.

Description of Immature Stages

The shiny green eggs bear ~25 indistinct longitudinal ribs and are laid in jumbled clusters. L1 is greenish with a shiny black head; each segment has a row of black bullae, each bearing a simple seta. There are dark sclerotized plates behind the head and on the terminal segment. The head in all instars is setaceous, becoming increasingly so with maturity. In L2 broad dark bands develop laterally, and the bullae grow much longer, each with ~8 simple setae. L3 is black with densely setaceous bullae, appearing fuzzy. In L4 most segments bear lateral and dorsal bright yellow-orange elongate spots, the black body bears numerous white spots, and each spiracle is encircled with a thin white ring. In L5 the pattern is similar except that the orange spots are paler in some populations, elongated into an almost continuous lateral stripe. The dorsal yellow-orange spots disappear in some individuals. The pupa is variegated mottled brown, white, and gray with irregular small black spots and dorsal rows of orange bullae. *Chlosyne palla* larvae are very similar; both species are variable, likely precluding reliable separation. *Chlosyne hoffmanni* larvae tend to be black in L4 and L5, except for a ventrolateral

Eggs @ 0.8 mm

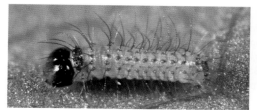

L1 @ 1.4 mm

L2 @ 3 mm

L3 @ 5 mm

L4 @ 10 mm

L5 @ 23 mm

Pupa @ 14 mm

Sagebrush Checkerspot Chlosyne acastus (W. H. Edwards)

gray stripe, and have very reduced orange spots. *Chlosyne palla* and *C. hoffmanni* utilize tall asters, not the hosts used by *C. acastus*. Some *Phyciodes* spp. are also similar but utilize different hosts. The larvae of *Euphydryas* are similar but are not found on asters and have distinct pupae. As *C. palla* and *C. acastus* were previously considered conspecific (Warren, 2005), it is difficult to determine which published images, if any, depict *C. acastus*.

Discussion

Chlosyne acastus may be common, even abundant in some habitats. At such locations larvae can often be found; for example, up to 30 larvae (L4–L5) on each *C. viscidiflorus* bush were found on Red Mt on April 26; larvae were sunning themselves on bare branches. Four weeks later a substantial adult population (~500 adults counted in an hour) was present at the same site. A few weeks later (early July), a search of basal branches and foliage of *C. viscidiflorus* revealed egg batches and early instars. It can be difficult to obtain gravid females for oviposition, as males and females have partially offset flight periods. Females flying late in the flight period when few or no males are present should be good candidates for obtaining viable eggs. It is likely important to provide the local host plant, which may vary with locality. Research is needed on the full range of hosts used by *C. acastus* in Cascadia. The developmental ecology of this species would provide an interesting research project. The 2 populations we studied utilized different hosts and developed at different rates and different times.

Northern Checkerspot   *Chlosyne palla* (Boisduval)

Adult Biology

Chlosyne palla and *C. acastus* are sometimes treated as a single species. Warren (2005) separated them on the basis of dimorphism in adult *C. acastus*, largely nonsynchronous flight periods, different habitats, and different host plants. The Northern Checkerspot ranges from SE BC, NE WA, and the E slopes of the Cascades to SE WA (Blue Mts). It is patchily distributed throughout montane OR and ID into N CA (Warren, 2005; Shapiro and Manolis, 2007). The flight period of the single annual generation in WA and BC is mid-June–mid-July, but earlier flights occur in OR. Preferred habitats are mid–high-elevation forested openings, roadsides, gullies, and creeks. Both sexes visit a wide variety of flowers, and males visit scat and ashes. Males fly for about a week before the females. The only confirmed host in Cascadia is *Symphyotrichum campestre* (Western Meadow Aster) (Pyle, 2002; Warren, 2005). *Erigeron speciosus* and *Solidago californica* were reported as hosts in CO and CA, respectively (Scott, 1992; Emmel and Emmel, 1973). We obtained oviposition on *Eucephalus ledophyllus* (Cascade Aster) and *Symphyotrichum foliaceum* (Leafybract Aster).

Immature Stage Biology

We reared this species twice, from eggs obtained from captive females from Okanogan Co., WA and Plumas Co., CA. Eggs were laid in masses of varying size (mean of 89 eggs (SD 24), range 17–160), on leaf undersides. The WA females, obtained in early July, produced 3 large egg masses (~300 eggs). Infertility and predation resulted in ~40 larvae reared to diapause. The CA females obtained in June produced 5 egg batches; approximately 100 larvae were reared, with 75% developing to adults by early Aug and the remainder entering diapause in L3. Eggs hatched in unison after 5–9 days at 20–26 °C. Initial development was rapid, the larvae reaching L3 just

9 days post egg-hatch at 20–26 °C. The WA larvae entered diapause as mature L3 by Aug 11, 24 days post oviposition. One batch of larvae was removed from overwintering on March 10 and reared at 15–21 °C, natural daylengths; L4 was reached in 10 and L5 in 17 days. Pupation occurred after 27 days, and 2 adults eclosed 9 days later. The second cohort was removed from overwintering on April 18 and reared at 18–25 °C under natural daylengths. Feeding was limited and development slow, larvae reaching L4 in 21 days, then becoming dormant (May 12) under shelter. On Aug 13 the larvae were returned to cold storage (5 °C). Batches of larvae (6, 6) were returned to summerlike conditions (27 °C, continuous lighting) after 10 and 24 weeks and molted to L5 after 15–16 days; however, they then reentered dormancy and died after 3–4 weeks. In contrast, pupation of direct-developing CA larvae occurred 28 days after egg-hatch, with adults eclosing 6 days later (20–27 °C). Clearly, the environmental cues determining larval diapause in *C. palla* are complex, and these observations indicate the species may be capable of delaying larval development. In early instars several asters and *Chrysothamnus viscidiflorus* (Green Rabbitbrush) were tried as host plants, but only *E. ledophyllus* and *S. foliaceum* were eaten. A commercial aster (*Aster novae-angliae*) and goldenrod (*Solidago* sp.) were also accepted by L5. Feeding was entirely on host plant leaves, which were skeletonized by feeding from the undersides, leaving them blackened and draped in abandoned frass, shed skins, and light webbing. Larvae are gregarious while feeding and enter diapause in tight groups, typically within curled dead leaves. During overwintering the larvae shrank considerably, but there were few losses. One L5 at 20 mm molted to L6. In later instars larvae were less gregarious. Distinct nests were observed only during diapause, although light silk webbing was present in L1 and L2. Feeding was mostly nocturnal, the larvae resting on the leaf undersides diurnally. The survival strategy includes concealment (in silk nests), protection from spines, and an escape response of dropping to the ground when disturbed.

Description of Immature Stages

The clustered eggs are pale green with ~18 low vertical ribs. L1 is green anteriorly and light brown behind. The head is shiny black with short setae; the body has moderately long brown setae arising from brown spots arranged in 1 transverse row per segment. L2 is chestnut brown with a broad grayish longitudinal stripe through the spiracles. Bullae are tall and pinkish or black, each with 4–6 dark setae. L3 is dark, each

Adult

Eggs @ 0.5 mm

Pupa @ 14 mm

L1 @ 1.5 mm

L2 @ 3.5 mm

L3 @ 7 mm

L4 @ 14 mm

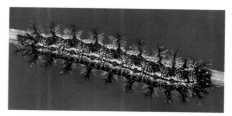

L5 @ 25 mm

segment with a dorsal yellow to orange-red saddle. The bullae have more setae. In L4 each bulla bears 20–25 setae, imparting a shaggy appearance to the larva. Each segment has 3 or more transverse rows of small white spots. Each dorsolateral bulla has an orange ring around its base, and there is a narrow white longitudinal line above the spiracles. In L5 these patterns continue, with white ventrolateral markings dominating below the spiracles. The pupa is dull purplish, sometimes with broad black swaths across the wing case and raised orange and black bullae along the abdomen. *Chlosyne acastus*, the most similar species, has L2–L5 that are usually blacker with reduced to absent yellow-orange markings. When intraspecific variation is taken into account, larvae and pupae of these species may not be reliably separable. There are few published images of the immature stages of *C. palla* or *C. acastus*. Allen et al. (2005) provide images similar to ours of both species, but showing more prominent dorsal orange markings.

Discussion

As with all checkerspot species, it is sometimes difficult to obtain gravid females. It is also difficult to find eggs or larvae. Once fertile eggs are obtained, larvae are relatively easy to rear. On one occasion we observed predation of an entire cohort of ~100 captive L1 larvae by a single lacewing larva. The *C. palla–C. acastus* complex is still poorly understood, and studies on the biology of immature stages may be valuable in resolving taxonomic issues. Studies on the environmental cues inducing larval diapause will help in understanding population dynamics and voltinism in this species. Additional information on Cascadia host plants is needed.

Northern Checkerspot *Chlosyne palla* (Boisduval)

Northern Crescent *Phyciodes cocyta* (Cramer)

Adult Biology

Phyciodes cocyta is part of the "Pearl Crescent complex" and was formerly included in *P. tharos* (which occurs over most of E and central US) or *P. selenis*. *Phyciodes cocyta* occurs through most of Canada, extending S into the Appalachians, Rockies, and E Cascades. In Cascadia it occurs in BC, E WA, ID, and NE OR, occupying grasslands, meadows, and forest glades from about 200 ft (Columbia Gorge) to >6,000 ft (Blue Mts, Wallowas). Populations are usually localized. There are 2 generations at lower elevations, with adults flying in May–June, then mid-Aug–Sep. At higher elevations, *P. cocyta* is univoltine, flying in July–Aug. Males patrol with a semigliding flight just above the ground in search of females, most often in the vicinity of host plants. Both sexes nectar at flowers, including thistles, dogbane, clover, milkweed, and asters, and males also visit mud. Larval hosts are various asters including *Symphyotrichum frondosum* (Alkali Aster), *S. laeve* (Smooth Aster), *S. foliaceum* (Leafybract Aster), *Eucephalus ledophyllus* (Cascade Aster), and *Eurybia glauca* (Gray Aster) (Warren, 2005). Eggs are laid in large single-, double-, or triple-layer clusters on the undersides of host leaves.

Immature Stage Biology

We reared or partially reared *P. cocyta* several times. Two females obtained from Cummings Creek, near the Tucannon R, Columbia Co., WA on June 4 were confined with 2 potted commercial asters (*Aster frikartii*, *A. alpinus*) in a bucket with a muslin lid (30 cm tall × 30 cm dia.). Held in sunshine at 23–28 °C, they laid 4 egg batches (egg numbers: 104, 198, 90, 9) during 3–5 days confinement. Three females collected from the same location on June 18 (different year), confined with *A. alpinus* in a plastic cylinder with a muslin lid (26 cm tall × 8 cm dia.) laid 1 egg batch (*n*

= 120) after 48 hrs. Eggs hatched in 7–8 days at 21–26 °C, with L1 exiting through the top, not consuming the shell. Complete egg-hatch did not occur in any batch, with 20–80% apparently infertile. L1 immediately fed on *A. alpinus* but not on *A. frikartii*. The June 4 cohort developed poorly, only a few reached L2, and all died by early July. L1 in the June 18 cohort were transferred from *A. alpinus* to *Erigeron peregrinus* (Peregrine Fleabane) within 12 hrs of hatching. L1–L5 fed well on this native and the related *E. subtrinervis* (Three-nerved Daisy), but failed to feed on a number of other *Erigeron* spp., including *E. speciosus* (Showy Fleabane). Development from L1 to pupation (July 28) took 26 days (21–26 °C, natural daylength), with 4, 8, 4, 4, and 6 days spent in L1–L5, respectively. Adults eclosed 8–9 days later. L1 did not produce webbing and fed gregariously on the undersides of leaves, creating holes on young leaves, "windowpanes" on older, coarser leaves. L3–L5 were less communal and tended to rest on the undersides of leaves, feeding from leaf edges. Prepupal larvae were purplish and wandered. Overwintering is reported to occur in L3 (Scott, 1986a; Pyle, 2002). Defense is based on gregarious behavior in early instars and concealment and crypsis later.

Description of Immature Stages

The egg is light green with 20–22 indistinct ribs running from above the base to the micropyle. L1 is distinctly segmented, has a shiny, slightly bifurcated black head, and is whitish yellow becoming green anteriorly. Long, simple brown setae arise in 6 longitudinal rows, together with less distinct, shorter setae. L2 is whitish green becoming dark brown-black with paler markings. There are 7 rows of prominent pale brown bullae, each bearing 10–22 dark setae, most profuse dorsally, giving the larva a spiny appearance. Laterally there is a subspiracular pale stripe, and the head is shiny black with 2 indistinct white patches dorsally. L3 is similar, mostly mottled black with an additional pale stripe dorsolaterally. Bullae are light brown with bases encircled with orange, and the posterior segment has a dark spot middorsally. The head is shiny black with white markings dorsally. L4 is black peppered with small white dots and has dorsolateral and subspiracular pale stripes. The bullae are black with orange bases and the spiracles are black encircled with white. The head of L4 is shiny black with dorsal, lateral, and central white dashes. L5 is similar, with distinct subspiracular and dorsolateral white stripes. The orange bases of bullae are prominent, giving the larva a distinct black,

Adult

Eggs @ 0.5 mm

L1 @ 2.0 mm

L2 @ 4 mm

L3 @ 6 mm

L4 @ 10 mm

L5 @ 22 mm

Pupa @ 13 mm

white, and orange appearance. The pupa is whitish brown, mottled with dark brown-black vermiform markings. There are 4 rows of 3 orange tubercles dorsally on the abdomen and a vague dark stripe laterally. This species could be confused with other *Phyciodes* spp., especially *P. pulchella*, which has a similar host range. Mature larvae of the latter species are paler, with less distinct orange body markings and white head markings than *P. cocyta*. Our images appear to be the first published of the immature stages of *P. cocyta*. Allen et al. (2005) state that the larvae of *P. tharos* and *P. selenis* (= *P. cocyta*) are identical; however, their image of a *P. tharos* larva as well as one in Wagner (2005) lack the vivid white dorsolateral stripe and orange-based bullae seen in ours. Larvae and pupae of *P. tharos* from IL, which we reared separately, are very similar to those of *P. cocyta*, as are those of *P. batesii* (Allen, 1997; Allen et al., 2005).

Discussion

Oviposition is relatively easy to obtain, even on commercial asters, but native hosts are required for successful rearing of larvae. Finding larvae has proven difficult. The full host range remains to be determined; our rearing experience suggests that some (but not all) *Erigeron* spp. are acceptable hosts. Rearing larvae during mid-June–mid-July results in a second brood of adults; rearing after mid-July with declining daylength likely would result in overwintering larvae. The environmental cues regulating diapause and overwintering need study. The impact of natural enemies on populations is unknown and also needs study.

Northern Crescent *Phyciodes cocyta* (Cramer)

Field Crescent *Phyciodes pulchella* (Boisduval)

Adult Biology

The Field Crescent ranges from AK to CA, AZ, NM, and Mexico. In BC and WA it occurs largely at mid–high montane elevations with only a patchy distribution W of the Cascades. In OR it is widespread, occurring at low elevations as well as in montane areas (Warren, 2005). *Phyciodes pulchella* flies in a single brood from early May to late Sep in meadows, prairies, and forested foothill and mountain terrain. The long flight period results from elevational differences and, according to Pyle (2002), from pupae eclosing over an extended period of time. In N CA there are 2–3 broods (Shapiro and Manolis, 2007). Males patrol in sunny areas (open fields, roadsides) and visit scat and mud. Both sexes visit a variety of flowers, including thistles, dogbane, milkweed, and mints. Females lay their eggs in masses on the undersides of leaves of asters. A female may lay multiple egg masses, each with 20–100 eggs, on leaves of various sizes and often in neat parallel rows. Scott (1992, 2006) reported 30–100 eggs in masses, the largest mass with 235 eggs. He also reported that masses are usually placed on aster seedlings, sometimes only an inch above the ground. In Cascadia the reported host plants include *Eurybia conspicua* (Showy Aster) and *Eucephalus ledophyllus* (Cascades Aster). Other asters have been reported as hosts in CA, and it is likely that other asters and fleabanes are also used in Cascadia.

Immature Stage Biology

We reared this species 3 times from gravid females obtained from Kittitas (Aug 12), Columbia (June 17), and Yakima (July 15) Cos., WA. Eggs were laid in 2–4 masses per female on the undersides of large basal leaves of live Cascade Asters (Kittitas Co. female) and an Old World aster cultivar, *Aster alpinus* (females from Columbia and Yakima Cos.) and hatched in

6–9 days. Yakima Co. L1 were transferred to *Erigeron subtrinervis* (Three-nerve Fleabane) after hatching; the other cohorts were reared on oviposition hosts. L1 gathered into groups, constructing communal silk nests and completely covering themselves. Leaves within the nests became black as a result of feeding damage. Six days after hatching, L2 migrated to adjacent fresh leaves. At this point the Columbia Co. larvae ceased feeding and died, apparently because the Old World aster was unacceptable. The Kittitas and Yakima Co. larvae continued to construct extensive nests of thick webbing under which they lived and fed. By ~2 weeks post egg-hatch, ~80 L2 had spread through several aster plants 50 cm tall, skeletonizing all leaves. Larvae appeared to feed diurnally; however, Scott (2006) reported nocturnal feeding. Late L2 communal groups dispersed into several smaller groups, each constructing silk nests. The nests typically measured 2–5 cm and comprised 1 or 2 skeletonized leaves tied together or to the plant stalk with heavy silk webbing. If nests were opened, larvae coiled and dropped to the ground. Feeding and development slowed in late L2. By ~36–40 days post egg-hatch, most larvae were L3 and had entered diapause. Diapausing larvae silked small nests (3–8 individuals) mostly in single, tightly coiled dead leaves. Kittitas Co. larvae were overwintered on senesced potted hosts in ambient outdoor conditions; only 4 survived to early spring. During March and April, larvae fed lightly but indiscriminately on hosts including *Taraxacum officinale* (Dandelion). Larvae were supplied with host asters in early May and growth thereafter was rapid, maturing to L5 and pupation within 7 days. The Yakima Co. larvae were overwintered at 5 °C for 137 days before exposure to 20–25 °C and 15 hr light. Overwintering survival was 60%, and feeding on 2 unknown *Erigeron* spp. commenced after ~10 days, with molting to L4 26 days after removal from overwintering. L5 and pupation occurred 11 and 20 days later, respectively. Adults eclosed after 33–40 days at ~22 °C. Postdiapause larvae rested preferentially on brown leaves. There are 5 instars; however, Scott (2006) reported 6 instars occur in CO. Protection is based on gregariousness (in early instars), spines, camouflage, and a ventral gland on the first segment that emits chemicals likely to be defensive.

Description of Immature Stages

The eggs are pale green, sculptured with ~20 indistinct rounded vertical ribs. The greenish cream L1 has a shiny black head and a distinctly segmented body. Sparse, long blond setae adorn the body and

Adults

Field Crescent

Phyciodes pulchella (Boisduval)

Eggs @ 0.5 mm

Pupa @ 12 mm

L1 @ 2 mm

L2 @ 4 mm

L3 @ 6 mm

L4 @ 11 mm

L5 @ 25 mm

head. In L2 there are 10 prominent conical white bullae on each segment, each bearing 6–8 pale setae. The head is shiny black. L2 is greenish cream with a dark green ring at each segment joint, and there are 2 distinct brown dorsolateral stripes. L3 is dark brown with 2 narrow black dorsolateral stripes. There are many small pale gray spots, and the bullae bear many setae. There is a distinct white longitudinal stripe laterally below the spiracles, and a less distinct one passing through them. L4 is similar except the body is almost black, mottled orange-brown between the 2 lateral white stripes. L5 is dark brown with complex black and orange-brown markings and prominent gray speckling. The pupa is light brown with darker brown mottling and streaks. This species is similar to other *Phyciodes* spp., particularly *P. cocyta*, which uses similar or the same hosts. Mature larvae of *P. cocyta* are brighter orange overall, with white markings on the head. Larval images similar to ours are published in Allen et al. (2005, as *P. campestris*), Guppy and Shepard (2001, as *P. pratensis*), and Miller and Hammond (2003).

Discussion

Achieving captive oviposition can be difficult in this species. Many apparently aged and mated females lay infertile eggs. Once fertile eggs are obtained, larvae are relatively easy to rear, although overwintering can be difficult, with significant mortality from desiccation or mold. Overwintering larvae in situ on potted hosts may produce the best results. Opportunistic feeding on non-hosts may be an important adaptation for overwintering survival. The full range of hosts used in Cascadia needs research. At least 3 species of *Erigeron* were acceptable to larvae in our rearings. The biotic and abiotic factors influencing population densities and distribution are unknown and deserve study.

Pale Crescent   *Phyciodes pallida* (W. H. Edwards)

Adult Biology

The Pale Crescent occupies the arid interior W ranging from S BC to AZ and CO. *Phyciodes pallida* is rare in Cascadia, occurring in small shrub-steppe colonies along rivers E of the Cascades in N WA, SE WA, and NC OR. The single annual brood flies from early May to early July, and larval hosts are native thistles with *Cirsium undulatum* (Wavyleaf Thistle) the preferred species. We observed courtship on May 26 near Oroville, WA, with males perching conspicuously and flying out to inspect all passersby in search of mates. The females remained hidden low in the vegetation near the host plants, nectaring on low flowers. The females visit first-year *C. undulatum* plants, ovipositing on the underside of midheight leaves about a third of a leaf length from the tip. Mature (flowering) thistles are avoided. A single plant may have >1 egg mass.

Immature Stage Biology

We reared this species once from collected eggs and a gravid female. On May 26, at the Similkameen R, Okanogan Co., WA, we collected several egg masses (mean of 89 eggs/mass, range 54–120, *n* = 10) on *C. undulatum*, rearing them in independent cohorts (Seattle/Yakima). In addition we obtained oviposition from a female confined with cut *C. undulatum*. A few egg masses had some nonviable eggs, but ~200 larvae were obtained. Most of the larvae developed to L3 or L4 and entered diapause in early to mid-July; however, 33 in the Seattle rearings developed to maturity, pupated (over a period of >10 days), and produced a second generation of adults in mid-July. All the Yakima-reared larvae entered diapause in L4. Eggs hatched en masse after 8–9 days, and

nondiapausing larvae developed to L2, L3, L4, L5, and pupae at 7, 15, ~19, 23, and 34 days post egg-hatch. Adults eclosed 52 days after egg-hatch. Yakima-reared larvae developed to L2, L3, and L4 at 6, 12, and 18 days after egg-hatch. The cohort from the captive female (egg-hatch ~7 days later than collected eggs) lagged in development during L3, taking up to 20 days in this instar. Diapause-destined L4 showed minimal growth compared with postdiapause L4. Larvae entering diapause ceased feeding and sought refuge in curled leaves or dark corners. In mid-Aug after a month of inactivity, larvae were overwintered (5 °C in 24 hr dark). After 9 weeks, 12 larvae were transferred to summerlike conditions (27 °C in 24 hr light) with thistles. Feeding commenced in 5–7 days with molting to L5 after another 7 days. Pupation occurred when larvae reached 30 mm, 26 days after removal from overwintering. Interestingly, pupation in the nondiapausing cohort occurred when L5s reached only 22 mm. Pupae from diapausing larvae averaged 16.5 mm, compared to 13 mm in the nondiapausing group. L1–L2 live communally in very loosely woven silk nests, usually in a fold of the host leaf. *Cirsium arvense* (Canada Thistle) was readily accepted as a surrogate host. L1–L2 bored into leaves, where they remained concealed while feeding between the 2 membranous surface leaf layers. Later instars remained on the exterior surface. Each larva was tethered to the nest or a leaf with a silk strand. Feeding produces skeletonized blackened leaves with much frass. Larvae coil and fall to the ground when disturbed. Larvae in small numbers placed on a thistle did poorly, faring much better in large groups. The survival strategy includes group behavior (synchronous reactions to disturbances), physical protection (silken webs), concealment (inside or under leaves), avoidance (dropping when disturbed), and camouflage (cryptic markings).

Description of Immature Stages

The eggs are greenish yellow becoming black prior to hatching, with 20–21 vertical ribs. L1 is brownish cream becoming yellowish green, with 1 dorsal and 3 lateral rows of setae. The head is shiny black throughout development. L2 is darker brown with spines, each with multiple setae. There are vague middorsal and lateral dark stripes. In L3 the dorsal spines are dark brown cones with many setae; lateral spines and setae are blond. There is a broad brown dorsolateral stripe that persists to L5. Early in L4 the larva is extremely spinose. The dorsum has much orange at spine bases, and the body is pale below the

Adult

Pupa @ 13 mm

Eggs @ 0.6 mm (fresh, green, mature, black)

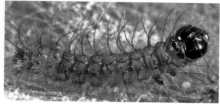

L1 @ 2 mm

L2 @ 2.6 mm

L3 @ 4.5 mm

L4 @ 8 mm

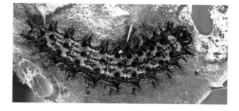

L5 @ 25 mm

spiracles. In L5 the setaceous spines are reduced, dorsal and ventrolateral spines have pronounced orange bases, the dark body has numerous irregular small white spots, and there are several broken longitudinal white stripes. The black head has 2 small orange patches dorsally, most obvious in postdiapause larvae. The pupa is brown on the wing cases and head, grading to light orange or brick red on the abdomen. *Phyciodes mylitta* has smaller, darker larvae with less dorsal orange. The larvae of *P. pulchella* lack dorsal orange coloration. L5 of *P. cocyta* is most similar but has less orange dorsally, more distinct unbroken longitudinal white stripes, and distinct white markings on the head. Woodward (2005) and Guppy and Shepard (2001) showed images of eggs, L5, and pupa, all agreeing closely with ours. Scott (1986a) and Pyle (2002) reported that this species makes no nests; however, we observed loose silk nests. Scott (1992) reported L3 as the overwintering stage in CO; we found that L3 or L4 may overwinter.

Discussion

Our experience suggests that obtaining oviposition from gravid females may be fairly easy, and finding eggs and larvae in the wild on *C. undulatum* is possible in the few sites where the species is found. This species is easily reared on Canada Thistle. Little is known about the biology and ecology of this rare species in Cascadia, including why populations fluctuate and are apparently declining, and why it rarely uses thistle species other than *C. undulatum,* despite their acceptability to larvae. The environmental cues controlling voltinism need elucidation. The finding of a significant number of nondiapausing larvae in our rearing suggests that a second generation may sometimes occur in Cascadia. The smaller size of larvae and pupae that produce a second generation suggests some trade-offs are involved in this strategy.

Pale Crescent *Phyciodes pallida* (W. H. Edwards)

Mylitta Crescent   *Phyciodes mylitta* (W.H. Edwards)

Adult Biology

The Mylitta Crescent occurs throughout W North America from S BC to Mexico, E to AZ, CO, UT, and MT. It is found throughout Cascadia in a variety of habitats, including weedy fields, vineyards, shrub-steppe, forest trails, meadows, marshes, and many other kinds of open space from sea level to mid-elevations. One of our earliest spring butterflies, *P. mylitta*, may be seen from late March to mid-Oct producing 2–3 generations. According to Scott (1986a), males perch all day in canyons or along creeks to await females, but in Cascadia they are more frequently seen patrolling territories along trails, roads, and watercourses. Both sexes nectar avidly on a wide variety of flowers, including thistles, goldenrod, asters, rabbitbrush, clover, and Pearly Everlasting; males also visit scat and mud. Larval hosts are thistles, including *Cirsium arvense* (Canada Thistle), *C. vulgare* (Bull Thistle), *C. undulatum* (Wavyleaf Thistle), *C. remotifolium* (Fewleaf Thistle), *C. edule* (Edible Thistle), *C. callilepsis* (Fingerbract Thistle), and *C. hookerianum* (Hooker's Thistle) (Pyle, 2002; Warren, 2005). They also reported *Centaurea diffusa* (Diffuse Knapweed), *C. solstitialis* (Yellow Star-thistle), and *Silybum marianum* (Milk Thistle) as possible hosts, and Scott (1986a) reported *Mimulus guttatus* (Common Western Monkeyflower) as a host. Eggs are laid in large clusters on the undersides of host leaves.

Immature Stage Biology

We reared *P. mylitta* several times from collected larvae or gravid females. Larvae collected June, July, Aug, and Sep in 4 WA counties represented 2–3 different broods; all were reared to adulthood. Two females obtained from near the Snake R, N of Pasco, Franklin Co., WA, on March 25, confined with potted *C. arvense* in a plastic bucket with a muslin lid (30 cm tall × 30 cm dia.) and held outdoors in sunshine (13–18 °C) laid 6 batches of eggs during 3 days. Two females obtained from Umatilla Co., OR on Sep 27 and confined with *C. arvense* in a plastic cylinder with muslin ends (13 cm tall × 8 cm dia.) laid a single batch of eggs on the side of the cylinder. The number of eggs in the batches ranged 40–270 (mean of 142). Eggs hatched in 8–11 days, and development from L1 to pupation on *C. arvense* took 25 days at 20–25 °C, with 3, 7, 8, 4, and 3 days spent in L1–L5, respectively. At 28 °C, development from L2 to pupation took 15 days. Adults eclosed after 8–10 days at 20–25 °C and 44 days at 15–20 °C. L1 feed gregariously on leaf surfaces, removing the green epidermis and leaving an opaque "window pane," but not making complete holes. Webbing produced by L1 and L2 retains frass, resulting in untidy nests. L3 are moderately gregarious, but L4 and L5 are more solitary; all instars feed openly. Overwintering is reported to occur as a "half-grown" larva (Pyle, 2002) or as L4 (Guppy and Shepard, 2001). We observed overwintering L3 on lower leaves of *C. arvense* at Yakima, WA, and also in King Co., WA. Defense is based on gregarious behavior, rapid coiling, falling, and dispersing as escape responses, and protective webbing in early instars. Later instars are spiny, produce regurgitant when disturbed, and have a ventral gland on segment 1 that may secrete predator-deterring chemicals. Pupation in captivity occurs on hosts or on cage sides. High-density rearing produces larvae that pupate at a smaller size, resulting in undersized adults.

Description of Immature Stages

The egg is whitish green becoming gray-black before hatching. There are 18–20 indistinct ribs running from midheight upward, terminating around the micropyle. L1 is whitish tan becoming tan-green and has a shiny black head. Long black setae arise in rows. Late L1 have a segmented appearance and are whitish dorsally with dorsal rows of black dots-dashes and tan markings dorsolaterally. L2 is tan-brown with rows of black markings middorsally and dorsolaterally. There are 7 rows of prominent brown-black bullae, each bearing 6–8 brown setae. L3 is gray and brown becoming black with a subspiracular white stripe. Bullae in the stripe are orange based; elsewhere they are black. Bullae bear 20–30 black spinelike setae, giving the larva a bristly appearance. The head is shiny black with 2 small white markings dorsally. L4 is dark brown-black dorsally with small white dots-dashes and a lateral pale band with orange-based yellowish bullae. In some individuals the dorsal white dots and dashes coalesce in L4 and L5 to form alternating white and black stripes. L5 is similar, with enlarged bullae and spines, some with orange bases, reduced white, and increased orange-brown markings on the pale lateral band. The black shiny head is marked with 3 white dashes dorsally, extending downward for about one-third the head capsule. The pupa is

Adult

Eggs @ 0.4 mm

L1 @ 1.2 mm

Pupa @ 10 mm

L2 @ 4 mm

L3 @ 5 mm (early)

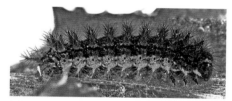

L4 @ 11 mm

L5 @ 14 mm

roughly textured and varies from tan-brown to gray, mottled with white and black, with orange tubercles dorsally. This species could be confused with other *Phyciodes* spp., although only *P. pallida* feeds on thistles and usually only on the native *C. undulatum*. Mature larvae of *P. pallida* are lighter than *P. mylitta* and have orange markings dorsally. Mature larvae illustrated in Guppy and Shepard (2001) and Allen et al. (2005) are similar to ours. Guppy and Shepard (2001) showed a pupa and Neill (2007) illustrated a nest of early instars on thistle.

Discussion

This is a common and easily reared species with gravid females ovipositing readily in captivity, rendering it an ideal model for laboratory-based studies on butterfly biology. Larvae can be reared in high densities without disease being a significant problem, although resulting adults are often small. The overwintering biology of this species deserves study, particularly with regard to physiology and phenology. Larvae appear to enter dormancy late in the year and resume feeding opportunistically in milder coastal locations, suggesting that diapause dynamics may be flexible; true diapause may only occur at cold, interior sites. Comparison of the biology and ecology of this species and that of the closely related thistle-feeding *P. pallida* may illuminate the reasons for the great difference in distribution and abundance of these 2 butterflies.

Mylitta Crescent *Phyciodes mylitta* (W. H. Edwards)

Anicia Checkerspot   *Euphydryas anicia* (E. Doubleday)

Adult Biology

The Anicia Checkerspot occurs throughout most of Cascadia E of the Cascades but is absent from the driest parts of the Columbia Basin. Beyond Cascadia it ranges from AK to Mexico, E to WY, but is absent W of the Cascade-Sierra Mts. *Euphydryas anicia* is locally common, occupying a wide range of habitats, including shrub-steppe, high grasslands, lithosols, conifer forest edges, canyons, alpine meadows, and mountain summits (Pyle, 2002). Single brooded, *E. anicia* flies mid-April–mid Aug, a little earlier in OR. Larval hosts include *Penstemon* (beardtongue), *Castilleja* (paintbrush), and in at least 1 area, *Symphoricarpos* (snowberry). Pyle (2002) listed 9 species of *Penstemon*, also *Castilleja thompsoni* (Thompson's Paintbrush) and *Collinsia sparsiflora* (Few-flowered Blue-eyed Mary). *Collinsia* appears to be used only in the spring by overwintered larvae. Both sexes visit flowers, and males mud-puddle, hilltop, and cruise corridors searching for mates. Eggs are laid in masses on the undersides of leaves.

Immature Stage Biology

We partially reared this species several times. Overwintered larvae (L3–L5) were found as early as March 20 at 3,400 ft in Kittitas Co., WA, and as late as June 26 at 5,700 ft at Mt Chumstick, Chelan Co., WA. Pupae were found in June twice, and eggs were obtained Aug 1 from gravid females from Slate Peak (7,400 ft), Okanogan Co., WA. On Aug 12 near Moses Meadows (3,700 ft), Okanogan Co., we collected eggs, L1, and L2 from separate nests. Eggs hatch after ~12 days and develop to L3 in ~2 weeks at 20–22 °C. Thereafter, they feed slowly without growing, gradually entering diapause. The rate of postdiapause development was not recorded, but adults eclosed 12 days after pupation. L1–L2 are gregarious, constructing web nests in which they feed. When food is exhausted the nests are expanded, abandoned, or reconstructed on fresh leaves. Larvae consume leaves, which are skeletonized; abandoned nests are blackened with frass and plant remains, surrounded with bare stems. Our larvae overwintered in L3, also reported by Guppy and Shepard (2001). Guppy and Shepard (2001) also reported that larvae overwintered in nests at Keremeos, BC, but Pyle (2002) stated that diapause occurs under rocks or litter. Pyle (2002) also reported that larvae may diapause more than 1 year if conditions are very cold or very dry. This species may also reenter diapause under certain environmental conditions as *E. colon* does. Postdiapause larvae are solitary, do not build nests, and feed openly on host plants. There are 5 instars, but extra instars may occur if there is >1 period of dormancy. Protection is based on concealment during L1–L3. Later instars have spines and are likely protected by chemicals (iridoid glycosides) sequestered from host plants. All instars react to disturbance by squirming and falling to the ground.

Description of Immature Stages

Eggs are shiny, ovate-conical, yellow turning black, and bear ~20 vertical ribs. L1 is pale yellow with greenish infusion dorsally, with 4–5 longitudinal rows of dark brown bullae laterally, each bearing a simple long white seta. There is a black collar and the head is shiny black with setae, remaining thus to pupation. The bullae in L2 are dark with multiple pale setae. The body is brown, with rows of large orange spots becoming white anteriorly. L3 is darker, almost black dorsally with a row of orange conical bullae except on the first 3 segments. All other bullae are black, although those laterally have orange around the bases. Setae are numerous on the bullae and most are pale. L4 and L5 are extremely variable, some mostly black with white markings and others mostly white with black markings. The contrasting colors are in longitudinal rows in either case. There is a thin black dorsal line surrounded by white. Conical bullae in dorsal, lateral, and ventrolateral rows are black, or orange with black tips. Each has a large orange spot at the base and bears numerous setae, dark dorsally, lighter laterally. Dark and pale forms are common in all populations, although the palest larvae are more prevalent in N WA and S BC. Pyle (2002) reported that larvae and adults are variable but independently so. The white pupa lacks setae and has contrasting black and orange markings. The abdomen has multiple rows of low orange bullae and multiple black spots, and the wing case is white with large black markings. The immature

Adult

Eggs @ 0.7 mm (fresh, *below*, mature, *above*)

L1 @ 1.8 mm

L2 @ 4.1 mm

L3 @ 8 mm

L4 @ 17 mm

L5 @ 26 mm (light form)

L5 @ 29 mm (dark form)

Pupa @ 14 mm

Anicia Checkerspot *Euphydryas anicia* (E. Doubleday)

stages of *E. colon* are virtually indistinguishable from those of *E. anicia*, although we have not seen larvae as pale as the palest *E. anicia*. The dorsal spines on L3–L5 of *E. editha* are typically all orange, whereas in *E. anicia* they are heavily black tipped, but the larvae are very similar. Guppy and Shepard (2001) published photos of L5 and a pupa similar to ours, with their larva like our palest variant. Neill (2001), Allen et al. (2005), and Neill (2007) show larvae similar to our darkest variants.

Discussion

Larvae of *E. anicia* are readily found in appropriate habitats and easily reared, particularly postdiapause. Host-plant selection may be driven by availability as well as by preference; within a relatively small area near Mt Chumstick, we found late instars on penstemons and paintbrush on the same day. Near Moses Meadows, we found *Symphoricarpos* heavily used where neither penstemons nor paintbrush grow. In Kittitas Co., postdiapause larvae are found on *Collinsia* in April, before other hosts have begun spring growth. In captivity, food sources can be successfully switched, and nonnative *Plantago* is readily accepted. Diapause dynamics and associated environmental cues deserve study. Parasitized pupae are sometimes encountered, but research is needed to identify natural enemies and their impact on populations. Research is also needed on the likely distastefulness of larvae caused by sequestration of iridoid glycosides from host plants (Kuussaari et al., 2004).

Snowberry Checkerspot *Euphydryas colon* (W. H. Edwards)

Adult Biology

Euphydryas colon was until recently considered a subspecies of *E. chalcedona* (Pelham, 2008). Most range maps combine *E. colon* and *E. chalcedona* (Guppy and Shepard, 2001; Pyle, 2002), and some older maps (Scott, 1986a) include *E. anicia*. The range in Cascadia includes SW BC, the Olympic, S Cascade, and Blue Mts of WA, NE WA, the ID panhandle, and much of OR. Beyond Cascadia *E. colon* occurs in CA and NV. *Euphydryas colon* flies from early May to mid-Aug in a single brood, depending on elevation, and occupies mountain, foothill, and upper shrub-steppe habitats, most commonly along open forest margins, in meadows, and along unsurfaced forest roadsides. There are numerous reported larval host plants; Pyle (2002) listed *Symphoricarpos albus* (Common Snowberry), *S. oreophilus* (Mountain S.), *S. mollis* (Creeping S.), *Penstemon fruticosus* (Shrubby Penstemon), *P. rupicola* (Rock P.), *P. davidsonii* (Davidson's P.), *P. procerus* (Small-flowered P.), and *Verbascum thapsus* (Common Mullein). Other hosts include *Penstemon subserratus* (Fine-toothed P.), *Lonicera* spp. (honeysuckle), *Mimulus* spp. (monkeyflower), *Castilleja* spp. (paintbrush), *Antirrhinum* spp. (snapdragon), and *Plantago major* (Common Plantain). Adults nectar readily on flowers, and males mud-puddle, hill-top, and cruise corridors in search of females. On July 12 at Reecer Creek, Kittitas Co., WA, we observed a female ovipositing on the lower surface of an uppermost leaf of *P. rupicola*. We have found egg masses on *Symphoricarpos* on lower leaf surfaces.

Immature Stage Biology

We reared this species several times from gravid females and field-collected eggs and larvae. Females obtained from the Tucannon R (Columbia Co., WA) on June 19, 21, and 23 (different years) laid egg batches (50–200 eggs/batch) on *S. albus* and *Penstemon serrulatus* (Cascades P.), hatching after 8–17 days at 20–25 °C. L1–L3 are gregarious, consuming leaves in messy silk nests, expanding or moving nests when food becomes exhausted. L3 was reached after 21–40 days, when larvae entered dormancy within nests on upper parts of the host plant (early Aug). In some rearings, most larvae overwintered as L2, abandoning nests and moving into detritus at the base of the plant. We overwintered L3 outdoors, then exposed separate cohorts to cool, short days (15–21 °C and 12 hr light) or warm, long days (25 °C and 24 hr light) on Jan 13 and Feb 6, respectively. Provided *P. serrulatus* or *P. major*, larvae commenced feeding in 2–13 days, and developed to L4 in 7–23 days. Under these and a variety of other postdiapause conditions, larvae reentered dormancy 1–3 additional times and developed 6 and possibly 7 instars before reaching adulthood. These observations suggest that larval diapause and instar number are extremely flexible and likely influenced by temperature, daylength, food quality, and possibly moisture availability. Young larvae are found from mid-July to Oct at mid-elevations such as Reecer Creek, Kittitas Co., WA, whereas overwintered larvae are found mid-May–mid-June, to early July at high elevations. Postdiapause larvae are solitary and feed openly, wandering prior to pupation. Field-collected postdiapause larvae develop in ~20+ days from first feeding to pupation, adults eclosing 10–20 days later. Protection is based on concealment prior to overwintering, with spring larvae depending on physical (spines) and chemical (iridoid glycosides sequestered from hosts, ventral gland?) defenses.

Description of Immature Stages

The yellow eggs turn orange-brown and have 18–20 vertical ribs. L1 is yellowish cream with a shiny black head. There are 4 rows of brown bullae laterally; all bullae and the head bear long white setae. L2 is purplish brown with black conical spines, each bearing a black seta at the tip. There is an additional middorsal row of setaceous spines, the base of each surrounded by a large orange spot. In L3, grayish stripes appear dorsally and laterally, and the lateral spines also have orange spots at their bases. L4 is similar, but the pale stripes are white, contrasting with the black ground color. The large branching setae impart a spiny appearance. The spines are shorter in L5, and the dorsal and lateral orange spots are contrasting. The pupa lacks setae and is white with contrasting black spots and several rows of low orange bullae on the abdomen. The immature stages of *E. colon* are very similar to those of other *Euphydryas* spp. The dorsal

Adult

Eggs @ 0.9 mm

Pupa @ 15 mm

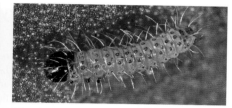

L1 @ 2.2 mm

L2 @ 4 mm

L3 @ 9 mm

L4 @ 15 mm

L5 @ 29 mm

L6 @ 25 mm

spines on L3–L5 of *E. editha* are orange, but they are usually black with orange bases in *E. colon* and *E. anicia*. *Euphydryas colon* and *E. anicia* are highly variable and do not appear to be reliably separable. Miller and Hammond (2003) picture a larva of *E. chalcedona* that differs in its black-tipped orange dorsal spines. An *E. chalcedona* larva from CA in Allen et al. (2005) is starkly black and white, unlike ours, but their other image from WY is like ours. Pupae in Pyle (2002) and Neill (2007) are similar to ours.

Discussion

Immature stages, especially larvae, are easily found, and gravid females readily oviposit in captivity. Larvae will feed on many penstemon species, also on exotic plantains. Overwintering larvae are prone to desiccation and mold. The biology of this species with respect to surviving adverse conditions is remarkable and deserves much more study. The dynamics of larval diapause are complex and flexible, allowing at least 3–4 periods of dormancy and 6–7 instars in 1 generation under laboratory conditions. The identity and role of environmental cues in diapause induction and termination need elucidation. The ability of larvae of some checkerspot species to reenter diapause "several times" was mentioned by Kuussaari et al. (2004), but detailed studies are lacking. We observed disease and parasitism in mature larvae, but research on the identity and impact of natural enemies is needed. The importance of iridoid glycosides sequestered from host plants for unpalatability and defense of *E. colon* also needs research.

Snowberry Checkerspot *Euphydryas colon* (W. H. Edwards)

Edith's Checkerspot   *Euphydryas editha* (Boisduval)

Adult Biology

Euphydryas editha occurs on both sides of the Cascades and in the Olympic and Blue Mts. Beyond Cascadia it occurs from S BC and AB to Baja, MT, WY, and CO. The single brood is sedentary (Ehrlich, 1961) and flies mid-April–mid-Aug, from near sea level to >8,000 ft elevation. Habitats include high ridges, coastal prairies, open woodlands, and alpine areas. Warren (2005) suggested that *E. editha* may be the most intensively studied noneconomic insect in North America, listing 32 studies and papers, summarized in Ehrlich and Hanski (2004). Many hosts are used in Cascadia, including paintbrushes (*Castilleja levisecta, C. parviflora, C. miniata, C. suksdorfii, C. thompsonii*), plantains (*Plantago lanceolata, P. maritima, P. major, P. macrocarpa*), *Collinsia* spp. (blue-eyed mary), *Penstemon* spp. (beardtongue), *Pedicularis* spp. (lousewort), and *Orthocarpos* spp. (owl clover) (Warren, 2005; Pyle, 2002). Scott (1986a) added *Mimulus* spp. (monkey flower) and *Lonicera* spp. (honeysuckle). Pre- and postdiapause larvae may use different hosts. Males perch on ridges or patrol for females. Females generally mate only once (Scott, 1986a) and lay up to 1,200 eggs in clusters of 20–350 on the undersides of leaves or on inflorescences.

Immature Stage Biology

We reared or partially reared this species 5 times. On May 11 a female from Ellensburg, Kittitas Co., WA, oviposited, producing 2 L4 that entered diapause but failed to overwinter. On June 27, 7 L4–L5 collected from *Castilleja* at Quartz Mt, Kittitas Co., were reared on cut paintbrush and produced 1 adult; 2 larvae were parasitized. On July 20 a female from Quartz Mt produced 40 eggs that were reared to L2. On July 23 and 29, females from Slate Peak, Okanogan Co., WA, and Bear Creek Mt, Yakima Co., WA oviposited, and larvae were fed on *Castilleja* and *P. lanceolata*,

respectively. The Slate Peak cohort overwintered as L4 and the Bear Creek Mt cohort as L3, producing adults the following spring. Eggs hatched in 8–9 days; development from L1 to L4 took 30 days at 22–27 °C. Young larvae are gregarious, living in communal silk nests on the undersides of leaves that are skeletonized or consumed; flowers and stems may also be eaten (Scott, 1986a). Most diapausing larvae in the Bear Creek Mt cohort sheltered in curled leaves in which they cocooned themselves with silk. Postdiapause development was variable. The Slate Peak cohort exposed to warm (~20–22 °C) conditions in Jan and fed on young potted *Castilleja* reached L5 in <2 weeks, pupating a few days later. Bear Creek Mt larvae took 18 days to break diapause at 15–20 °C and 10 hr light, but only 5 days under continuous lighting. One larva in the continuous lighting group pupated 22 days later, but remaining larvae in both cohorts reentered dormancy 12–14 days after exposure to warm conditions. Singer and Erhlich (1979) reported that postdiapause larvae reenter diapause under conditions of food stress, the likely trigger in our *P. lanceolata* rearing. Postdiapause larvae disperse widely, up to 10 m/day, and seek warm open slopes, pupating up to 2 weeks earlier than larvae remaining on cool slopes (Weiss et al., 1987). In E WA prediapause larvae may feed on *Castilleja* then switch to *Collinsia* postdiapause. There are usually 5 instars, although CA populations have 6 or more (Singer et al., 1994), and populations reentering dormancy may have extra instars. Protection is based on concealment and chemicals provided by iridoid glycosides sequestered from host plants (Kuussaari et al., 2004).

Description of Immature Stages

The eggs are yellow with ~20 faint vertical ribs. L1 is dull yellow-orange. The head is shiny black with setae, this persisting to pupation. Each segment has 8–10 bullae, each with a long pale seta. L2 is dark brown with black spines. A middorsal row of yellow spots, 1 per segment on segments 4–11, develops late in L2. L3 is black with grayish lateral and brown ventrolateral broken stripes. The conical spines bear black setae. There is a distinct dorsal row of bright orange, black-tipped spines, 1 per segment, on segments 4–11, also a thin black dorsal line. L4 is black with white speckles, and has larger orange spines. A bright orange patch surrounds the base of each orange spine in L5, and an orange crescent is present laterally above each spiracle. There is a distinctive broad black side stripe between the dorsal and lateral orange spots. The tips of the lateral and ventrolateral spines are orange. The

Adult female

Eggs @ 0.8 mm

Pupa @ 15 mm

L1 @ 2 mm

L2 @ 4 and 5 mm

L3 @ 10 mm

L4 @ 13 mm

L5 @ 23 mm

L5 @ 23 mm

Edith's Checkerspot *Euphydryas editha* (Boisduval)

pupa is white; on the abdomen there are 6 rows of yellow bullae with black edges. The wing cases are white with strong black markings. The most similar species are *Euphydryas anicia* and *E. colon*. L4 and L5 of *E. editha* are usually darker and have a broader, blacker side stripe. The orange bases of the middorsal spines in *E. editha* are larger than those of the other species. Miller and Hammond (2007) show a mature larva similar to ours, while Allen et al. (2005) show a black and white larva from WY.

Discussion
In all stages, *E. editha* occurs earlier than other *Euphydryas* spp., although there is some overlap. Overwintered larvae are found on host plants shortly after snowmelt, and eggs are readily obtained from captive gravid females. Larvae are not difficult to rear, but postdiapause larvae require sunny, warm conditions (body temp of 30–35 °C for optimal growth) and young, succulent host plants to prevent reentering dormancy. Most of the biological and ecological research on *E. editha* has been conducted on CA populations, and comparative studies in Cascadia are needed.

Family Nymphalidae (part)
Subfamilies Nymphalinae (part), Limenitidinae, and Danainae

OTHER BRUSHFOOTS, ADMIRALS, and MONARCH

This section, a composite of three subfamilies, includes 17 of our best known and most colorful larger butterflies, the admirals, ladies, commas, tortoiseshells, and Monarch. Adults are robust, strong fliers, long lived (up to 10 months), and often strikingly patterned. Some are common, but most are rarely abundant. Painted Ladies and California Tortoiseshells are exceptions, occurring in vast numbers when migration from the south is optimal. All are attracted to nectar, and many visit mud, carrion, dung, rotting fruit, and sap.

Eggs are green, tan, or yellow and barrel shaped, with strong, widely spaced vertical ribs; admirals have rounded cone-shaped eggs with hexagonal surface patterns and numerous hairlike spines. Eggs are usually laid singly, although 4 species oviposit in clusters, and the Satyr Comma may stack its eggs vertically.

Brushfoot larvae have large, well-developed branched spines and vary considerably in color and pattern. Some are cryptically patterned, blending in with their surroundings. Mature larvae of admirals are bizarre, bearing sparse but very large spines or horns, and may adopt threatening postures; early instars mimic bird droppings. The Monarch larva is banded in yellow, white, and black, advertising its distastefulness, and has a pair of horns at each end. Development of nymphalid larvae is usually rapid. Host plants are varied, including trees like willow, birch, poplar, cherry, and maple, as well as nettles, currants, *Ceanothus*, rose, *Spirea*, pussytoes, sage, thistles, mallows, Ocean Spray, serviceberry, chinquapin, and oak.

Pupae hang, and most have a complex shape with spines, angular processes, and a constriction behind the thorax. Many bear pearly or metallic reflective conical processes. Pupae of the admirals have large dorsal humps, and the Monarch pupa is green with reflective gold spots. Most species overwinter as adults; the several species unable to survive our winters must migrate southward each year and return the following season. Admirals overwinter as partly grown larvae in rolled-leaf hibernaculae.

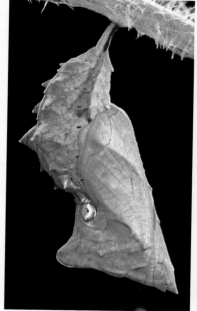

OTHER BRUSHFOOTS, ADMIRALS, and MONARCH

Satyr Comma   *Polygonia satyrus* (W. H. Edwards)

Adult Biology

The Satyr Comma occurs in most of Canada, the Great Lake states, and all of W US. It is common in Cascadia in a variety of habitats, including deciduous woodlands and moist areas from sea level to ~7,000 ft, and is often seen in disturbed and urban settings. The flight period is mid-March–Sep, but as the adult overwinters it may be seen as early as Jan and as late as Dec (Warren, 2005). Males are territorial, perching in sunny spots along forest trails, chasing off interlopers, and pursuing females. Males visit scat and mud, and both sexes visit thistles, willows, lilacs, overripe fruit, and sap flows. The primary hosts are nettles, particularly *Urtica dioica* (Stinging Nettle); *Urtica holosericea* (Hoary Nettle) is used in N CA (Shapiro and Manolis, 2007). Pyle (2002) also reported usage of *Humulus* (hops), but we have not seen it on commercial hops in the Yakima Valley. Remington (1952) reported oviposition on *Salix drummondiana* (Drummond's Willow), and Scott (1986a) recorded Salicaceae as a host for an AZ population. The female lands on and inspects a series of plants, and if a suitable plant is found, she hangs by the edge of a leaf, bending her abdomen and laying 1 or more eggs on the underside. The eggs are often stacked in a column up to 6 or 7 tall, hanging vertically under a nettle leaf. *Polygonia satyrus* is bivoltine in Cascadia, possibly triple brooded in N CA, although probably univoltine in high-elevation Cascadia habitats.

Immature Stage Biology

We reared this species several times from field-collected larvae and gravid females. L3–L5 were collected on 4 occasions (10 individuals) and reared

to adults. Eggs are easily overlooked as their hanging green strings are very well camouflaged against the nettle leaves; they are typically placed on larger leaves on the lower part of a plant. A female obtained May 31 at Mission Canyon, Chelan Co., WA produced eggs that were reared to L4. Females obtained from the Umatilla NF, Columbia Co., WA on April 20 and April 28, caged outdoors over potted nettles in dappled sunlight (20–25 °C), oviposited in 7–14 days. Eggs hatch after 5–7 days, and at 19–25 °C with increasing daylength, development is rapid, with instars lasting 3–4 days each and pupation occurring 23 days post oviposition. Adults eclose after 8–9 days. New-generation adults were produced in the first week of June. Larvae developing under declining daylengths produce adults that enter reproductive diapause and overwinter. L1–L2 may be gregarious or solitary, resting in the open on the underside of a nettle leaf. Individual nests are constructed by folding the edges of a nettle leaf together, usually downward, upper side out. The nest is silked loosely together, leaving the ends partially open with the larva visible inside. The larvae feed either from the leaf edge or at midleaf, making deep jagged holes that do not cross leaf veins. Larvae of *P. satyrus* may share a nettle plant with *Vanessa atalanta* or *Aglais milberti* larvae. Early L3 (*n* = 12) confined with *Salix scouleriana* (Scouler's Willow) failed to feed on this potential host and died. In contrast, another group of 10 early L2 that were transferred from nettle to hops readily accepted this host and developed to adulthood. The survival strategy of *P. satyrus* larvae is primarily concealment, and there are 5 instars.

Description of Immature Stages

Polygonia satyrus is the only anglewing in Cascadia that stacks its eggs in strings. The egg is light green with ~10 raised white ribs patterned laterally with numerous tiny septae. L1 has a shiny black head and is variegated brown, white, and black. Long dark simple setae arise from dark bullae; shorter setae are on the head. In L2 the bullae become exaggerated into tall pointed cones, mostly black anteriorly and alternating white and black behind. Each cone has multiple dark setae on the sides and a single white-based, black-tipped seta at the tip. L3 is black; the black spiracles are circled in white and are underlain by a mottled white stripe. Two or 3 transverse triplets of conical bullae are white, while the other bullae are all or mostly black. Two horns behind the head are prominent, shiny black with branched spines. In L4, broad white saddles appear across several abdominal

Adult male

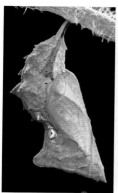

Pupa @ 20 mm   **Pupa @ 21 mm**

Eggs @ 0.8 mm (fresh, *left*, mature, *right*)

L1 @ 3 mm

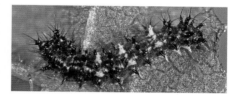

L2 @ 5 mm

L3 @ 10 mm

L4 @ 15 mm

L5 @ 25 mm

segments and coalesce to a continuous dorsal white patch. L5 is almost entirely white dorsally, as are most of the dorsal and ventrolateral bullae and their spines. The anterior horns are small, black, and antlerlike. Black chevrons interrupt the white dorsum, 1 per segment. The pupa is yellowish gold to brown with 3 pairs of prominent solder-silver dots dorsally. The larvae of *P. satyrus* are similar to those of other *Polygonia* spp. In later instars, other *Polygonia* spp. have varying amounts of yellow or orange dorsally, especially on the bullae and spines. *Polygonia satyrus* is the only anglewing with a striking, virtually all-white dorsum in L5. There are many published images of larvae and pupae. Among recent images are larvae in Guppy and Shepard (2001) and in Pyle (2002), which are similar to ours. Allen et al. (2005) show an AZ larva in which the dorsum is only partially white and the sides have significant orange markings. A pupa in Guppy and Shepard (2001) is darker than ours, but their image of eggs is very similar to ours.

Discussion

Polygonia satyrus is easily reared. Partially grown larvae can be common in Stinging Nettle patches and are found by searching for drooping tied leaf nests. Obtaining eggs from females is more challenging but can be achieved on potted nettles; females will not oviposit on wilting plants. Gravid females are best sought in May–June. This is our best known anglewing; however, our knowledge of its biology and ecology is still fragmentary. Research needs include the effect of environmental cues on adult diapause and voltinism, identification and impact of natural enemies on population regulation, and host-plant choice and larval acceptance.

Green Comma   *Polygonia faunus* (W. H. Edwards)

Adult Biology

The Green Comma ranges from AK to CA and along the Rockies to NM. It also occurs across S Canada and in some E states. It occupies most of Cascadia except drier basin areas. *Polygonia faunus* flies in a single brood, favoring moist woodlands in montane areas in S BC, along the Cascades, and in SE WA and NE OR. The flight period extends from March to Sep, but adults may be seen earlier or later in favorable weather. Adults overwinter and feed on animal scat, rotting fruit, and tree sap and occasionally nectar on flowers. In the spring, adults can often be seen near willow trees; males perch conspicuously and fly out to challenge all passersby in search of mates, returning to the same perch. During April–early June, eggs are laid singly on stems, terminal buds, or catkins of host plants. One female was observed in Whatcom Co., WA visiting several branches of *Salix sitchensis* (Sitka Willow) before placing a single egg on a small woody stem ~6 inches from the tip. Several trees and shrubs are utilized as larval hosts, including *Salix* (willow), *Alnus* (alder), *Betula* (birch), and *Rhododendron*. Scott (1992) reported that *P. faunus* also uses *Ribes* (currant) in CO, and Janz et al. (2001) reported that *Urtica dioica* (Stinging Nettle) could be used for rearing. Willows appear to be the most commonly used hosts in WA, with *Salix sitchensis* a favored species. New-generation adults eclose in late July or Aug and enter reproductive diapause. During late summer and autumn they feed heavily on rotting fruit and sap to build fat reserves for overwintering.

Immature Stage Biology

We reared this species several times. One female obtained in Kittitas Co., WA in early May laid 7 eggs on *Salix sitchensis*. Females obtained from Umatilla

Adult male

NF, Columbia Co., WA on May 31 laid ~15 eggs on *Salix hookeriana* (Hooker's Willow). Eggs were laid singly, placed on willow catkins or leaves. Hatching occurred after 4–6 days and larvae were reared on *S. scouleriana* and *S. hookeriana*. L1–L2 did not wander, staying on a single leaf until it was substantially eaten, each larva tethering itself with a silk strand. When moved, the larvae were observed to cling quickly to any available leaf or stem. Larvae ate leaves from the edges, producing jagged-edged margins. Larvae in the Columbia Co. cohort began dying from a viral disease in late L1, with all dead 10 days post egg-hatch. By ~16 days post hatch, feeding by L4 in the Kittitas Co. cohort turned the willow leaves into jagged shapes, leaving only the central stems and tattered remains. Larvae generally rested on the undersides of leaves in a "cane" position, with the head and thorax turned back parallel to the abdomen. The larvae did not construct nests and were observed to twitch in unison when disturbed. Pupation began 32 days post egg-hatch and eclosion commenced 14 days later. Larvae were found on a number of occasions on *Salix sitchensis*. In one instance a field-collected L2 pupated 16 days later, a faster growth rate than that of the brood described above. Protection is derived from camouflage and spines and there are 5 instars.

Description of Immature Stages

The egg is green with ~12 raised septate ridges running vertically up the sides, converging around the micropyle. L1 is green anteriorly grading to orange posteriorly with long unbranched setae and has a shiny black setaceous head. Late L1 is brown, 5 segments with contrasting paired white patches dorsally, each patch bearing 1–2 contrasting dark bullae. In L2 the simple setae are replaced with branched spines, in alternating yellow-green or black dorsal sets, and the larva is complexly patterned in dusky lavender and black. L3 is similarly patterned, but the dorsal spines are much larger and more branched, and are bright yellow on the thorax and straw yellow posteriorly. A pair of black branched horns adorns the head. In L4 the pattern is similar, with the branched spines more boldly colored, bright yellow-orange on the anterior segments and cream posteriorly. L5 is strongly patterned, the dorsal spines bright orange on the thorax and bright white on posterior segments. Laterally the pattern is orange and black with small white branched lateral spines. In L3–L5 each segment has 3–4 contrasting white rings encircling the body between the clusters of spines. A pair of spiny black branched horns is prominent on the head in L3–L5.

Egg @ 0.9 mm

Eggs @ 1.0 mm

L1 @ 3 mm

L2 @ 7 mm

Pupa @ 22 mm

L3 @ 13 mm

L4 @ 20 mm

L5 @ 28 mm

The purplish tan pupa has a complex pattern of spines, ridges, and reflective metallic cones. All *Polygonia* spp. larvae are fairly similar; however, the larvae of *P. satyrus* are generally black and white (no yellow or orange) and are found only on *Urtica* (nettle). Early instars of *P. gracilis* are black with white spines; L4–L5 are marked in burnt orange. *Polygonia gracilis* and *P. oreas* feed on *Ribes* (currant). Images of larvae shown in Guppy and Shepard (2001), Miller (1995), and Miller and Hammond (2003) agree well with ours.

Discussion

Gravid females will oviposit in captivity on *S. sitchensis* and *S. hookeriana,* and likely other *Salix* spp. as well. Mating occurs in early spring, and females obtained in May are generally ready to oviposit. A reasonably large cage with a potted host plant is required for oviposition; wilting hosts are not utilized. Larvae can be found by searching the undersides of *S. sitchensis* leaves along road shoulders or other corridors where the species flies. Branches protruding into the corridor seem to be preferred, particularly at heights of 4–6 ft. Overhead backlit leaves can be scanned from below, searching for larval silhouettes, and ragged leaf edges may indicate larval browsing. Near Snoqualmie Pass, WA, larvae were found between June 18 (L2) and Aug 16 (L5). The larvae are hardy and can be reared fairly easily.

Green Comma *Polygonia faunus* (W. H. Edwards)

Hoary Comma   *Polygonia gracilis* (Grote & Robinson)

Adult Biology

The Hoary Comma is a W North American species ranging from AK to S CA and NM. It is also found throughout most of Canada. In Cascadia it is found in all montane areas but is largely absent from coastal and arid basin areas. *Polygonia gracilis* is locally common in mountain and foothill habitats (streams, trails, meadows) usually above 3,000 ft. Adults have a long flight period, from mid-March to early Oct, and adults overwinter so may be seen at any time, depending on seasonal conditions. This is one of our longest-lived butterflies, some living for up to a year. Scott (1986a) suggested that adults may fly to higher elevations in summer, returning downslope in the autumn. Males perch on bushes or shrubs awaiting females. Males visit animal scat and mud, and both sexes visit flowers (e.g., asters, rabbitbrush, mints), rotting fruit, and sap. The larval hosts are mostly currants, including *Ribes viscosissimum* (Sticky Currant), *R. lacustre* (Swamp Gooseberry), *R. cereum* (Squaw Currant), *R. inerme* (Whitestem Gooseberry), and *R. aureum* (Golden Currant). *Rhododendron albiflorum* (Cascade Azalea) and *Ulmus* (elm) are also reported hosts. *Ribes viscosissimum* and *R. lacustre* appear to be the most commonly used hosts in mountainous areas, whereas *R. cereum* is more likely used in the shrub-steppe. We observed oviposition behavior in late May in Whatcom Co., WA (1,800 ft); a female visited multiple leaves of a currant shrub, frequently bending her abdomen under the leaves, but without ovipositing. Eggs are laid singly or in groups of 3 or 4 on the undersides of host-plant leaves. *Polygonia gracilis* appears to be single brooded in Cascadia, although Pyle (2002) suggests a second generation may occur. Scott (1986a) suggested 3 flights occur in lowland CA.

Immature Stage Biology

We reared this species several times from gravid females and field-collected larvae. Three females obtained from Columbia Co., WA on May 28 held in an outdoor cage June 2–11 over potted *R. cereum* laid ~60 eggs in 7 days. Many were lost due to nocturnal predation by earwigs and ants. Eggs hatched after 4–5 days at 21–27 °C and larvae were reared on *R. cereum*. Nine females from Columbia Co., WA caged over *Ribes divaricatum* (Straggly Gooseberry) on June 2 (different year) produced 9 eggs after 3 days. One female from Okanogan Co., WA on June 18 laid 7 eggs on *R. divaricatum*. Larvae were reared on *R. cereum* or *R. divaricatum*. Development from oviposition to pupation took 30 days, each stage averaging 5 days. The pupal period was 9 days, giving a total development from oviposition to eclosion of 39–45 days. A commercial variety of gooseberry was also accepted by larvae. The larvae rest on stems or leaves of the host, often but not always on the underside, where they are hidden from aerial predators. The larvae do not construct shelters but feed openly, eating jagged holes from leaf edges. In later instars, larvae disperse, with only 1 or 2 per shrub. Survival is based on concealment and camouflage, but some of the hosts (as well as the larvae) are spiny, providing additional protection from predators. Nylin et al. (2001) found experimentally that larval coloration in the similarly marked *Polygonia album* is to some degree aposematic. There are 5 instars.

Description of Immature Stages

The eggs are semiglossy green, with 9 narrow pale ribs. L1 is cinnamon brown with white saddles on half of the segments and has a shiny black head bearing dark setae. Black bullae each bear a single long black seta. L2 is darker, and the black bullae are spine-bearing cones. Three segments have white saddles, and there is a dark brown dorsal line. L3 is mostly black anteriorly and dark brown posteriorly. There are 4 contrasting white dorsal saddles with white bases. A dorsal dark line is bordered on each side by white marks. The head is shiny black with 2 conical "horns." In L4, the anterior 3 segments have mustard or bright orange saddles and bullae; the remaining segments have alternating dark and yellow bullae. The spines are black dorsally but mostly orange ventrolaterally. Dorsally, posterior segments are whitish. Laterally there are 2 wavy orange lines. L5 is similar, with

Adult

Hoary Comma *Polygonia gracilis* (Grote & Robinson)

Egg @ 0.9 mm

Pupa @ 20 mm

L1 @ 4 mm

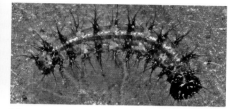

L2 @ 6 mm

L3 @ 13 mm (premolt)

L4 @ 17 mm

L5 @ 23 mm

L5 @ 33 mm

the anterior region mostly black with orange or mustard spines and the posterior area black with nearly solid white frosting dorsally. The black sides have rusty orange wavy lines resembling links in a chain. Some larvae become bright rusty orange anteriorly prior to pupation. The pupa varies from tan with dark brown mottling to dark gray-brown with black markings, with 2 rows of orange-tipped dorsal bullae and 3 pairs of reflective silver cones. The larvae of *Polygonia satyrus* are darker in most instars, lacking orange coloration. *Polygonia faunus* is lighter than *P. gracilis*, and L5 have less black. *Polygonia oreas* is most similar but with less white dorsally. Guppy and Shepard (2001) show a larva (as *P. zephyrus*) and pupa similar to ours, and Allen et al. (2005) illustrate 2 larvae similar to prepupal L5.

Discussion

Larvae can be found fairly easily and are easily reared on a number of *Ribes* spp. Obtaining eggs from gravid females is more challenging, as the adults are long lived and oviposit only briefly in mid–late spring. The type of cage and environmental conditions to which the females are exposed are important factors in obtaining oviposition. We had good results with a steel-framed muslin bag cage over potted *R. cereum* held outdoors in dappled sunlight at 22–28 °C. Larvae can be found by searching the undersides of leaves of Sticky Currant, paying attention to plants with feeding damage. Swamp Gooseberry is more difficult to search as the leaves are small and very spiny. Research is needed to explain why this species undergoes severe population fluctuations. Natural enemies may be responsible, but little is known about the predators and parasitoids attacking this species. It is unknown why some individuals turn bright orange prior to pupation while others do not. Environmental cues determining reproductive diapause in adults need study to help clarify the number of generations in Cascadia.

Oreas Comma   *Polygonia oreas silenus* (W. H. Edwards)

Adult Biology

Polygonia oreas is a NW species occurring from S BC and S AB S to N CA, UT, WY, and CO. It is local and generally rare in Cascadia, where it is found in moist forested mountains and foothills on both sides of the Cascade Mts (ssp. *silenus*) and similar habitats in SE BC, NE WA, and the Blue Mts of WA and OR (ssp. *threatfuli*). Records in OR extend from sea level to ~5,000 ft (Warren, 2005). Flight records extend from late Jan to mid-Sep in OR (Warren, 2005), with few reports during June. Larval hosts are *Ribes* spp. (currants and gooseberries) growing in forest clearings near stands of conifers. *Ribes divaricatum* (Straggly Gooseberry) and *R. lacustre* (Swamp Gooseberry) are the principal host plants in Cascadia; Pyle (2002) also reports *R. sativum* (red currant cultivar), and Scott (1986a) reported *R. inerme* (Whitestem Gooseberry) and *R. leptanthum* (Trumpet Gooseberry) from CO. Flowers are occasionally visited, but rotting fruit, scat, carrion and flowing sap are preferred. Males perch on host plants awaiting females, and pairs encircle each other, flying up into trees during courtship or when disturbed. Females carefully inspect hosts, landing on several leaves before laying a single egg on a leaf underside. *Polygonia oreas* is single brooded in WA and BC; however Dornfield (1980) and Warren (2005) report 2 broods of *P. o. silenus* in the Siskiyous and the N coast range of OR.

Immature Stage Biology

We reared single individuals of *P. o. silenus* on 2 occasions. On May 30 along the Nooksack R in Whatcom Co., WA, we discovered an L1 adjacent to an eggshell, on the underside of a *R. lacustre* leaf; the larva was reared to an adult. On May 19 (different year) we collected 1 egg on *R. divaricatum* at Reecer Canyon, Kittitas Co., WA, which was also reared to adulthood. Both larvae grew rapidly, developing from L1 to pupation in 16–19 days, producing adults 28 days

after egg-hatch. When not feeding, the larvae rested, usually in a position resembling a J, on the underside of a host-plant leaf. Both larvae were reared at 20–22 °C under natural daylengths, the Nooksack larva on *R. lacustre* and the Reecer larva on *R. divaricatum*. The early instars fed by nibbling at the leaf edges, ceasing feeding for about 1 day prior to each molt. From L2 onward the larvae folded a leaf downward, upper side out, tying the edges together with silk to make a nest in which they rested. Some of the later nests involved multiple leaves; pupation occurred within the final shelter for 1 larva, and hanging in the open for the other. Most feeding was nocturnal, when the larvae left their nests to feed, and leaves were consumed throughout development by eating jagged sections from the edges. Protection is based on concealment in leaf nests, nocturnal foraging, and profuse spines.

Description of Immature Stages

The egg is green with numerous white spots and has 10 raised vertical ribs with numerous small horizontal septae. L1 is greenish anteriorly, whitish posteriorly, and segments 4, 6, 8, 10, and 11 are red-brown dorsally. Each segment has a row of large brown spots, each giving rise to a single long dark seta, and the head is shiny black. L2 is mottled cream and brown, with pointed bullae alternating in pairs between black and creamy yellow on segments 5–9, each with several black setae. L3 is black with 3–4 narrow contrasting pale bands encircling the body on the rear half of each segment; these persist through L5. The spines are large, bright orange, and strongly branched, imparting a bristly appearance, and there is a pair of branched brown horns at the back of the head. L4 is similar except the white bands dominate dorsally, resulting in a lighter appearance. L5 is very dark ventrally, yellow-orange dorsally, and the orange branched spines are

Adult

Eggs @ 0.8 mm

L1 @ 2.8 mm

Pupa @ 20 mm

L2 @ 5.0 mm

L3 @ 10 mm

L4 @ 17 mm

L5 @ 30 mm

Oreas Comma *Polygonia oreas silenus* (W. H. Edwards)

proportionally smaller. The body is black laterally with a diamond-chain pattern of narrow orange lines. The pupa is variegated in brown tones and with several reflective pearly spots dorsally. The Whatcom Co. larva was darker during L3–L5, with smaller and less-branched orange spines, and the pupa was reddish brown. Other *Polygonia* spp. have similar larvae; L5 of *P. faunus* are mostly white posteriorly and mostly yellow anteriorly. L5 of *P. satyrus* are almost entirely white dorsally. The larvae of *P. gracilis* are most similar but have heavier white coloration on the posterior half of the dorsum. An L5 pictured in Neill (2007) is darker and has smaller and paler branched spines than ours. Images of an L5 and pupa from OR in Guppy and Shepard (2001) are similar to ours.

Discussion

This species is difficult to find as either adults or immature stages. Considerable persistence examining host-plant leaves where the species flies may produce occasional eggs or larvae. We obtained limited oviposition from female *P. o. threatfuli* caged over a potted *Ribes* host after ~2 weeks. Little is known about the variability of this species or about factors controlling its distribution. Larvae in particular may exhibit great variability in coloration and spine development, as indicated by our 2 individuals of *P. o. silenus*. The immature stages of *P. o. threatfuli* have not been described or illustrated. Natural mortality factors are unknown and deserve study. In some locations, cattle and sheep grazing has adversely affected gooseberry plants, reducing the hosts available for *P. oreas*.

Compton Tortoiseshell   *Nymphalis l-album* (Esper)

Adult Biology

The Compton Tortoiseshell is a Holarctic species with a North American range centered on the US-Canada border region. In Cascadia it is found locally in BC, N WA, SE WA, and NE OR. It is usually rare in Cascadia, but outbreak years occur occasionally. The flight period is late March–Sep and adults overwinter. Preferred habitats are open areas in forests and woodlands, including watercourses, trails and roads. *Nymphalis l-album* is single brooded throughout its range. The larval hosts are *Betula papyrifera* (Paper Birch) and *B. occidentalis* (Water Birch), although *Salix* (willow), *Populus* (aspen), and others have been reported (Scott, 1986a). Males perch conspicuously awaiting passing females along forest edges. Several males pursue females along corridors into forested areas. We observed oviposition behavior on April 15 in Okanogan Co., WA. Several adults were present in a fairly dense forest of Western Larch and Douglas Fir, with small (20–30 ft) birch trees in the shaded understory. Adults perched high in the trees, but occasionally a female would drop to ground level; one landed ~6 inches from the tip of a *B. papyrifera* branch ~4 ft above ground level. She briefly examined the branch then flew to 2 or 3 similar sites, frequently touching her abdomen to the branches. Oviposition was not observed; however, when similar birch branches were searched, 6 egg clusters were located. All were 4–5 ft above the ground and positioned on the terminal 0.5–0.75 inch of a branch tip bearing a green, unopened leaf bud. All egg clusters were on the underside of the outermost twig of a primary branch, never on side branches or shorter twigs with buds. The eggs were usually placed in parallel rows, but sometimes in irregular masses. The 6 egg masses each contained 17–36 eggs, averaging 26. Oviposition continued for ~4 weeks, judging from the presence of newly hatched L1 5 weeks later.

Immature Stage Biology

We reared this species twice, splitting stock between the authors for 4 rearings. Six egg clusters collected April 15 from Okanogan Co. comprised 160 eggs in 2 cohorts of ~80 each. Five groups of L1 collected May 25 comprised 135 larvae. In all rearings we encountered serious problems with disease; from 160 eggs, many larvae reached L4 and L5, but only 2 pupated and only 1 eclosed. All of the 135 larvae of the second batch died before reaching L4. On July 5 we collected an L5 and reared it to adult. The disease was virulent and widespread, affecting all larvae despite collection on different occasions from different trees. Larvae were reared in multiple, separate cages under clean conditions. Eggs hatched in unison after ~9 days. Larval development occupied 35 days at 21–25 °C with 4, 5, 4, 5, and 17 days spent in L1–L5. An adult eclosed after 16 days. The eggshells were left uneaten except for an exit hole in the top. L1 congregated in parallel alignment under leaves. The larvae were highly gregarious, moving, "thrashing" (when disturbed), and feeding in unison on leaves from the edges. Each larva was attached to the plant with a single silk strand. By L2, larvae fed in advancing lines, consuming entire leaves. *Betula occidentalis* was provided as food from L3 onward to 1 cohort and readily accepted by the larvae. In L4, larvae dispersed and began feeding individually. Disease appeared in this instar, and within a few days more than 50% of each cohort had died or were dying. Characteristic symptoms included cessation of feeding, loss of turgor, and an inability to grasp with the thoracic legs, leaving larvae hanging by the prolegs. These symptoms indicate a viral disease, most likely a nuclear polyhedrosis virus (NPV). Except for some light silk webbing in L1–L2, no nests or shelters are constructed; there are 5 instars. Protection is based on oral regurgitation, gregariousness, synchronous "thrashing," spines, and a ventral gland on the first segment, which likely emits defensive chemicals.

Description of Immature Stages

The egg is pale blue becoming tan, with 10–12 vertical ridges terminating near the micropyle. L1 is light yellow-brown with black dorsal tubercles bearing long black setae. L2 is dark brown with small white patches and black bullae, each bearing several setae. The shiny black head has 2 black tubercle horns that persist to pupation. There are 2 cream stripes ventrolaterally. L3 is darker with parallel broken white lines and stiff branching spines. In late L3 sclerotization of the posterior abdominal segment is apparent, creating the

Adult

Eggs @ 0.9 mm

L1 @ 2.5 mm

L2 @ 7 mm

L3 @ 15 mm

L4 @ 21 mm

L5 @ 45 mm

appearance of a head at both ends of the body. L4 is similar, with taller and more branched spines. The black head is dotted with very small white bullae each bearing a short seta. L5 is similar except the parallel lines are less prominent. An L5 collected from Okanogan Co. was pale green with black spines. The pupa is pale green turning tan, with several silvered conical processes. *Nymphalis antiopa* (Mourning Cloak) is the most closely related species but easily separable in all instars. Wagner (2005) illustrated a green L5 from E US. Allen (1997) and Allen et al. (2005) illustrate black larvae, and Guppy and Shepard (2001) illustrate the egg, L1, L5, and pupa, all similar to ours.

Discussion

We failed to obtain eggs from gravid females in 2 attempts; birches may need to be in bud but not leafed out to stimulate oviposition. Eggs and larvae can be found in April and May in known habitats. A major research need is identification and epidemiology of the virulent disease that apparently affects larval populations. Dramatic population fluctuations are common in this species, and it is likely that disease is a major contributing factor. Epidemiology of the disease in our rearing suggests it was present either on or in the eggs. Transovarial transmission of NPV has been reported for some moth species (Fuxa et al., 1999), and it is possible that the disease in our larvae originated from the parent females. Information on seasonal biology would determine whether adults aestivate or migrate after eclosion, and what environmental cues mediate reproductive diapause, overwintering, and oviposition.

Pupa @ 23 mm

Compton Tortoiseshell *Nymphalis l-album* (Esper)

California Tortoiseshell   *Nymphalis californica* (Boisduval)

Adult Biology

Occurring from S BC to S CA and E to the Rockies, the California Tortoiseshell has evolved reproductive and migratory strategies that optimize the use of host plants and climate to maximize population potential. Shapiro and Manolis (2007) summarized this biology for N CA, where hibernated adults produce a local spring generation of adults that migrate N in late May to breed in the Cascades. Hibernation also occurs in coastal and lowland Cascadia, with individuals active on mild days during Jan–Feb (Pyle, 2002; Warren, 2005). During March–June adults move higher, seeking fresh growth of the larval hosts, *Ceanothus velutinus* (Mountain Balm), *C. integerrimus* (Deerbrush), or *C. sanguineus* (Redstem Ceanothus) for oviposition. This brood emerges in early July, migrates N, then flies S-SW from mid-Aug to late Sep. During Sep 24–26, 30–60 per hr were seen in Mt Rainier NP flying down N–S oriented valleys. Migrating *N. californica* laden with yellow fat are in reproductive diapause (James, unpubl.). There is 1 generation in Cascadia; adults are long lived and may be seen anywhere in any month. Many kinds of flowers such as *Monardella odoratissima* (Coyote Mint) are visited, and prior to autumn migration, montane populations congregate on conifers, feeding on sap and honeydew. Eggs are laid in clusters of up to 250 on upper or lower leaf surfaces (Reinhard, 1981; Scott, 2006; James, unpubl.).

Immature Stage Biology

We reared or partially reared this species several times. Two females obtained from Columbia and Yakima Cos., WA, on April 20 and 24 were confined with potted *C. sanguineus* in a plastic bucket (28 cm

dia × 31 cm tall) with a muslin lid and held at 25 °C (continuous light) for 3 weeks. No eggs were produced until the cage was placed outdoors in sunshine on May 13. Five eggs were laid on the underside of a single leaf within 24 hrs. Six females from Columbia Co. on June 27 oviposited within 1 hr when confined outdoors with *Ceanothus thyrsiflorus* (Blueblossom Ceanothus). Four batches (each containing ~250 eggs) were laid in 2 days. Eggs hatched in 4–5 days at 22–27 °C, L1 partially eating shells then feeding on new leaves (making holes, chewing edges) of *C. sanguineus* or *C. velutinus*. Development was rapid at 21–25 °C, with pupation 25–27 days after egg-hatch and 5, 3, 4, 5, and 10 days spent in L1, L2, L3, L4, and L5, respectively. Adults eclosed after 6–11 days, 5 weeks after oviposition. We found and reared larvae on many occasions, especially from Kittitas and Chelan Cos., WA. Larvae do not make nests, although L1 and L2 use silk to cover and join leaves together. L1–L3 fed only on young foliage. Larvae fed and rested openly on leaves, gregarious during L1–L3, dispersing in L4–L5. Protection is based on aggregation, intimidation, confusion, and spines. When disturbed, L1–L3 jerk heads in unison to intimidate natural enemies. The heavily sclerotized posterior segments in L3 and L4 resemble the head capsule, giving the larvae a 2-headed appearance, which may divert the attention of birds (James, 2008b). A ventral gland on segment 1 during L2–L5 produces secretions that may deter predators, and the greater development of spines in L4–L5 affords increased protection. There are 5 instars.

Description of Immature Stages

The shiny, light green egg has 8 whitish ridges that terminate near the micropyle. L1 is whitish green with brown markings, including a broken middorsal line, and has a shiny black setaceous head. There are 12 rows of small black bullae, each with a long black seta. L2 is dark brown to black, whitish dorsally with a middorsal line of black dashes and pale lateral patches. Black setae are shorter with an extra row middorsally. L3 is black with a broad white area dorsally, bisected by a black stripe; there is a yellow seta-bearing bulla on each segment. Other bullae are black laterally, orange dorsolaterally. L4 has large orange spines and numerous white spots on a black body, with 2 white stripes dorsally and long white setae on the head and body. The posterior 2 segments are heavily sclerotized. L5 is black with profuse white setae on the body and head. White dorsal lines are lacking and spines are mostly black. The middorsal bullae are prominent

Adult

Eggs @ 0.9 mm

L2 @ 8 mm

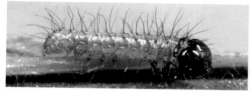

L1 @ 1.5 mm

L3 @ 15 mm

and yellow with an orange forked seta. The pupa varies from tan to blue-gray with 2 pairs of large white spots above the thorax and a dorsal row of black spots. The pale tan abdomen has irregular black spotting and dark, orange-tipped tubercles. The closely related *N. antiopa* and *N. l-album* differ significantly and should not be confused. Images of *N. californica* immature stages appear in Guppy and Shepard (2001), Pyle (2002), Miller and Hammond (2003), Allen et al. (2005), and Neill (2007). Most agree with ours, although in some the pale markings are more yellow.

L4 @ 20 mm

Discussion

Obtaining eggs from gravid females is achievable, and finding larvae on *Ceanothus* bushes during late spring in the Cascades is easy, particularly in outbreak years. Larvae are easily reared on *Ceanothus* and develop rapidly under warm temperatures. Larval biology is likely to play a major role in the population dynamics of this species and should be researched. Reports of acres of defoliated host plants and "millions of shimmering pupae" suggest that larval survival can be excellent. We have little knowledge of the regulating factors controlling populations and how they operate. The natural enemies of *N. californica* have not been identified, although we witnessed destruction of entire egg masses by a stink bug (Pentatomidae) within an oviposition cage. The importance of "double headedness," head-jerking, and defense secretions in protecting larvae from predators would make an interesting study. Population crashes of *Polygonia* spp. in Cascadia often occur simultaneously with population explosions of *N. californica*.

L5 @ 45 mm

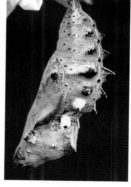

Pupa @ 23 mm

California Tortoiseshell   *Nymphalis californica* (Boisduval)

Mourning Cloak   *Nymphalis antiopa* (L.)

Adult Biology

The Mourning Cloak ranges from AK to Venezuela as well as across much of N Europe and Asia. It is common and widespread throughout Cascadia except for the maritime NW coast of WA; however, it is subject to significant population fluctuations, with great abundance often followed by scarcity. It is univoltine in Cascadia and may be seen on the wing from Feb to Oct. A large number of larval hosts are used, including *Salix exigua* (Coyote Willow), *S. scouleriana* (Scouler's Willow), and *S. sitchensis* (Sitka Willow). Poplar, aspen, cottonwood, elm, rose, hackberry, hawthorn, birch, spirea, alder, maple, apple, and others have also been recorded as hosts (Scott 1986a). Oviposition was observed in early April at low elevations in E WA, and larvae were found as late as Aug 23 at higher elevations. On April 6 a female was observed ovipositing in Kittitas Co., WA on budding *Salix exigua*. The female fluttered among the willows 4–5 min before choosing a stem in the middle of the grove, ~4 ft above the ground. She landed head down 4 inches from a branch tip and very slowly crawled down the stem, probing with her abdomen. After ~15 min she was 7 inches from the tip and began ovipositing with wings spread open. With antennae fixed about 60 degrees above her body, she laid a batch of 180 eggs in diagonal rows encircling the 2 mm dia. branch. Newly eclosed adults are seen briefly in June before aestivating in trees or old buildings. Migration may precede aestivation; Cech and Tudor (2005) and Shapiro (1986) suggested that N CA *N. antiopa* undergo regular elevational movements. In Cascadia, activity resumes in Sep for a period of heavy feeding and building of fat reserves before the adult overwinters.

Immature Stage Biology

We reared this species several times from field-collected eggs and larvae and gravid females. Eggs (180) were obtained on April 6 from Kittitas Co.

Adult

Eggs were also obtained from single caged females collected from Yakima Co., WA on May 9 and 16 (different years). Both oviposited in muslin cages over the nonnative *Salix purpurea*. Larvae reared on this plant died, only 3 reaching L2. Oviposition occurred at temperatures of 18–26 °C, with 165 eggs produced in 6 unequal batches, 1 with 44 eggs encircling a 20-mm dia. branch, 2 other batches encircling 2 mm dia. twigs. Egg-hatch occurred after 5–9 days, with 88% of May 9 eggs failing to hatch. Virtually all (98%) of the Kittitas Co. eggs hatched after 10 days, L1 leaving shells uneaten except for an exit hole at the top. Larvae were reared on *S. scouleriana* and developed rapidly, averaging ~3 days per instar, pupating ~16 days after egg-hatch. Adults eclosed 12 days later on May 13, 37 days post oviposition. Larvae are highly gregarious, feeding and moving in groups throughout development if food plant allows. The larvae surround themselves with strands of silk in a loose nest during L1–L3, each larva tied to the group with a single strand. Larvae react in unison when disturbed, rearing their heads and waving them about. Consumed leaves are left enshrouded in a black mass of abandoned silk and frass. A large group of larvae can denude a small willow tree. Mature larvae leave the host before pupating and wander on paths or roads. Protection is based on aggregations, synchronous head jerking, spines, and likely emission of repellent chemicals from a ventral gland on the first segment. Mature larvae have bright orange or red patches that may be warning coloration signifying toxicity to predators. There are 5 instars.

Description of Immature Stages

The egg is orange-brown and has 8 low ribs terminating near the micropyle. L1 is variable orange-brown with a shiny black head and short black setae arising from black bullae. L2 is burnt orange-brown with a darker dorsal stripe, low orange and black bullae with pale setae, and lighter stripes ventrolaterally. The dark posterior segment is more prominent as a result of sclerotization increasing in later instars. James (2008b) hypothesized that posterior segment mimicking of the head is an adaptive feature of some nymphalid species, aimed to confuse visually orienting predators into attacking the "wrong" end, thus providing attacked larvae with an opportunity to invoke other defenses like thrashing in unison and expelling ventral gland contents. L3 is darker, with each bulla supporting a blond seta at the tip. The head is black with short setae, and most segments have a pair of dull orange dorsal spots. Bullae in L4 are elongated with indistinct

Eggs @ 1.0 mm

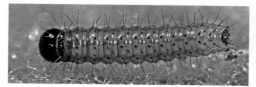

L1 @ 3 mm

L2 @ 7 mm

L3 @ 15 mm

L4 @ 25 mm

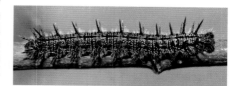

L5 @ 50 mm

setae. Laterally there are complex white markings, prolegs are bright orange, and dorsal orange spots are prominent. L5 is black with numerous tiny white spots in broken transverse lines. The dorsal spots are bold and may be orange or red. Setae are short, white, and numerous, imparting a shaggy appearance. The pupa is stocky, drab purplish gray with orange-tipped dorsal spines. L5 are distinctive with bright orange or red dorsal spots and should not be confused with other nymphalid species. Pyle (2002) and Guppy and Shepard (2001) illustrated the eggs, mature larva, and pupa, agreeing well with ours.

Discussion

It can be difficult to obtain eggs from captive females as they are long lived and oviposition occurs during a relatively short period in spring. If potted willows are provided in a muslin cage along with nectar sources, eggs may be obtained. Searching for eggs in willow thickets is tedious and usually unproductive. It may be productive to watch adults flying near willows, particularly Coyote Willow, where females may be observed ovipositing. Partly grown larvae are often encountered, and mature larvae can be found by searching for severely damaged host willow trees along roads and trails or in thickets. Larvae are easily reared, but crowding can cause virus outbreaks in late instars. Our knowledge of natural enemies of this species and their impact on populations is limited; research on these and the defensive strategies of *N. antiopa* would improve our understanding of population dynamics. Research on the physiology of the 2 apparent diapauses (summer and winter) in this species and associated behavior and environmental cues is needed.

Pupa @ 27 mm

Mourning Cloak *Nymphalis antiopa* (L.)

Milbert's Tortoiseshell *Aglais milberti* (Godart)

Adult Biology

Aglais milberti ranges throughout most of North America except the far N and far S. It occurs throughout Cascadia and is found in a variety of habitats, including parks, gardens, shrub-steppe canyons, watercourses, montane meadows, and open woodland. In most of Cascadia it is double brooded. Adults hibernate in tree holes, under bark, in hollow logs, etc., and are among the first butterflies to appear in spring. Adults were seen flying on Feb 17 in Waterworks Canyon, Yakima Co., WA. Overwintered butterflies mate and lay eggs in lowland areas in April, and first-generation adults emerge in late May–early June; the second generation flies in July–Aug. Some first-generation adults move to higher elevations where they appear to oversummer, nectaring on phlox, asters, Coyote Mint, and thistles in mountain meadows. During autumn they descend to lower elevations, feeding on late-blooming rabbitbrush, rotting fruit, and sap, then hibernate. These individuals may live for 10–11 months. Resident populations at higher elevations (e.g., Blue Mts) are univoltine, with a single larval generation in July–Aug. *Urtica dioica* (Stinging Nettle) is the only recorded larval host in Cascadia, but other *Urtica* spp. have been recorded elsewhere (Scott, 1986a). Eggs are laid in large, untidy masses of 20–900 eggs, often piled on top of each other, usually on the underside of a terminal nettle leaf.

Immature Stage Biology

We reared or partially reared *Aglais milberti* on several occasions. On Aug 3, a nettle patch near Mt Misery (6,336 ft) in Umatilla NF, Garfield Co., WA, yielded 15 L5 that pupated within 72 hrs. Pupae at 25–27 °C produced adults after 7 days. One female obtained from Waterworks Canyon April 6 was confined in a steel-framed muslin cage (54 × 32 × 20 cm) placed over nettle plants outdoors at Benton City, WA. One egg batch (~80 eggs) was laid on April 21 underneath a midsized leaf. Held at 18–25 °C, eggs hatched on April 27 and L1 fed communally on *U. dioica*, skeletonizing and webbing leaves. Frass was retained on the webbing. Webbing increased in L2, providing support and access for larvae between leaves and shoots. L3 used little webbing but remained gregarious. L4 were mostly solitary and formed folded leaf nests tied with silk, sometimes with >1 individual in a nest. In L5 there was a tendency to leave nests and feed openly. Prepupal larvae wander, and most pupae are formed away from the host. Development is rapid, pupation occurring 21 days post egg-hatch, after 4, 3, 4, 3, and 7 days spent in L1–L5, respectively. Adults eclosed 33 days post oviposition. We found larvae on many occasions in Kittitas, Yakima, Chelan, King, Clallam, and Skamania Cos., WA, and in some favored nettle patches they occur predictably from year to year. Within a large nettle patch it is not uncommon to find all instars present, attesting to the staggered nature of oviposition in this species. Protection is based on aggregation in early instars, then concealment in leaf nests. Mature larvae are spiny and have a ventral gland on segment 1 that likely secretes defensive chemicals when a larva is disturbed. There are 5 instars, and adults overwinter.

Description of Immature Stages

The shiny pale yellow-green egg has 9 vertical ridges that terminate near the micropyle. L1 is creamy yellow with a shiny black head and intermittent reddish brown dorsolateral stripes. There is a middorsal line of brown dashes, a dorsal black collar on segment 1, and a small dark sclerotized area on the posterior segment. Small black bullae sparsely cover the body, each sporting a single long black seta. L2 is similar, with wider dark dorsolateral stripes. Laterally, L2 is yellowish white with a wavy dark subspiracular stripe. L3 is black dorsally with well-developed black bullae and spines. Ventrolaterally it is yellowish with a wavy black stripe along the spiracles. The shiny black head is covered sparsely with short setae and the posterior segment is sclerotized dorsally. L4 is black peppered with yellowish white spots, each sporting a short white seta. Spines are black, each with a long black seta. Laterally, there are 2 intermittent wavy yellow stripes bordering the spiracles, and the prolegs are white. L5 is similar but with more profuse white spotting and development of medium-length white setae, particularly ventrolaterally. The ventrolateral stripes are creamy yellow and prominent, bordering black spiracles encircled in white. There is a sometimes indistinct middorsal black stripe, and the

Adult

Eggs @ 0.75 mm

L1 @ 2.5 mm

Pupa @ 18 mm

L2 @ 5 mm

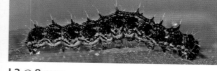

L3 @ 8 mm

L4 @ 13 mm

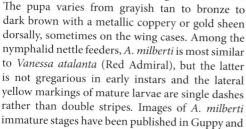

L5 @ 25 mm

L5 @ 26 mm

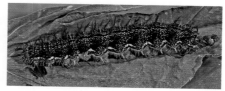

Milbert's Tortoiseshell *Aglais milberti* (Godart)

head is black with medium-length white setae. The pupa varies from grayish tan to bronze to dark brown with a metallic coppery or gold sheen dorsally, sometimes on the wing cases. Among the nymphalid nettle feeders, *A. milberti* is most similar to *Vanessa atalanta* (Red Admiral), but the latter is not gregarious in early instars and the lateral yellow markings of mature larvae are single dashes rather than double stripes. Images of *A. milberti* immature stages have been published in Guppy and Shepard (2001), Pyle (2002), Allen (1997), Allen et al. (2005), Wagner (2005), Woodward (2005), and Neill (2007). All agree closely with ours.

Discussion

Larvae are readily found on nettles. The early instars form distinctive webbed aggregations and often defoliate hosts; the blackened defoliated nettles make a good field mark. Guppy and Shepard (2001) reported substantial parasitism of field-collected larvae in BC. Minute Pirate Bugs (*Orius* spp.) are common residents of nettle patches and likely prey on early instars. The environmental cues dictating reproductive or nonreproductive behavior in newly eclosed adults are unknown, as are the details of migration and aestivation. Elucidation of these factors would help explain why this species is single or multibrooded in different, often nearby locations. This is an easily reared species and makes a good subject for laboratory studies.

American Lady   *Vanessa virginiensis* (Drury)

Adult Biology

Vanessa virginiensis is common over most of the US, less so in the W, and seen only occasionally in Cascadia. It is an unlikely permanent resident in Cascadia but can turn up anywhere as a migrant from the S. It is rarely seen in the spring, with most records during July–Oct, and adults are not likely able to survive winters in most of Cascadia. Multivoltine in the S, it is capable of 1 or 2 broods here depending on how early it arrives. Adults fly rapidly with an erratic and dodging flight usually just a few feet above the ground as they go from flower to flower. Favored nectar sources include dogbane, clovers, thistles, milkweed, and rabbitbrush, and adults also visit sap, mud puddles, and carrion. Males perch, often on the ground, and fly out to investigate passersby. Scott (1986a) records ~30 larval hosts, mostly Asteraceae, particularly *Anaphalis margaritacea* (Pearly Everlasting), *Gnaphalium* spp. (cudweeds), and *Antennaria* spp. (pussytoes). The exotic bedding plant *Gazania* is sometimes used (Shapiro and Manolis, 2007). Eggs are laid singly or in small groups of 2–6 in captivity, mostly on the undersides of host leaves. On *A. margaritacea*, eggs are inserted into leaf pubescence and nearly disappear from view.

Immature Stage Biology

We reared this species once from a female caught by Jasmine Vanessa James (5 years old) by the Feather R, Plumas Co., N CA, on Oct 13. Held in a plastic box (30 x 20 x 8 cm) with 1 muslin side, the female laid 65 eggs on *A. margaritacea* during 2 days in partial sunshine at 18–22 °C. Larvae were reared on *A. margaritacea* and *Antennaria luzuloides* (Woodrush Pussytoes) in 2 groups divided between the authors. Eggs hatched after 6–7 days at 18–25 °C, shells were partially eaten, and L1 excavated lightly silked nests under leaf pubescence; most L1 stayed between leaf membranes feeding on the inside layers, producing clear "windowpane" areas. L2 moved to the outer leaf surface and continuing through L4 formed increasingly complex nests by silking leaves together and using pieces of pubescence or dry chaff from leaves as reinforcements. Large amounts of silk were produced by L2–L4, forming mats or thick messy webs over the nests. Feeding occurred inside and outside nests, mostly nocturnally. L5 largely abandoned nests and rested exposed on leaves and stalks. At 25–28 °C, 3 days were spent in each of L1 and L2, 4 days each in L3 and L4, and 6 days in L5; at cooler temperatures of 20–22 °C, all stages took longer, ~25 days from egg-hatch to pupation. Other potential hosts, *Urtica dioica* (Stinging Nettle) and *Cirsium arvense* (Canada Thistle), were also supplied; the latter was accepted by starved L5, enabling pupation. Pupation occurred mostly on the host, and this stage lasted 5–7 days, for a total developmental period of 31–34 days at 25–28 °C. In most of its range *V. virginiensis* has no dormant overwintering stage; however, adults (Scott, 1986a) or pupae (Douglas and Douglas, 2005) may overwinter in N areas. Larval defense is based on concealment in nests during L1–L4, but this strategy is abandoned in L5, when protection may accrue from spines and bold warning colors and patterning. L2–L5 have a ventral gland on segment 1 that likely has a defensive function. Larvae are solitary in the wild and nests may be inconspicuous.

Description of Immature Stages

The egg is dull mustard yellow, darkening prior to hatching. There are 13–16 narrow raised ribs. L1 is gray-green dorsally tending to tan laterally. Six black bullae occur on each segment, giving rise to single, long setae. L2 is mottled black, brown, orange, and white. Black bullae are enlarged, have orange bases, and each sports 4–6 short setae and 1 moderate-length spine. Paired white spots occur dorsolaterally on segments 5–11. L3 is black with more prominent orange-based black bullae. The overall appearance of L4 is of a spiny black-and-white banded caterpillar with prominent orange and white spots. White intersegmental areas comprise 5 to 6 indistinct white bands on a black background. Late in L4 and in L5, dorsal bullae bases become red and expand. In L5 the intersegmental white areas expand and become creamy, or yellowish in some individuals, contrasting further with the black bands and red dots. The head is

Adult

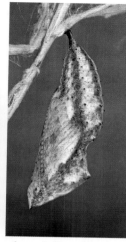

Pupa @ 23 mm (green form, *left*) and 24 mm (gray form, *right*)

L5 @ 36 mm (head)   **Egg @ 0.5 mm (fresh, *left*, mature, *right*)**

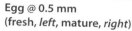

L1 @ 1.5 mm

L2 @ 5 mm

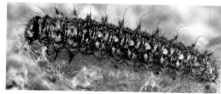

L3 @ 7 mm

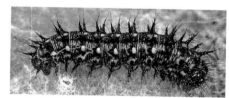

L4 @ 15 mm

L5 @ 32 mm

L5 @ 37 mm

American Lady   *Vanessa virginiensis* (Drury)

black and bears numerous long white setae. Larvae of *V. virginiensis* are unlikely to be confused with any other species. The pupa is more streamlined than other vanessid pupae and occurs in 2 color forms. The commonest form is whitish gray with 4 dark stripes running lengthwise along spiracles and bullae, both of which are orange. The rarer form is gold dorsally with metallic green wing covers. In our rearings, gold-green pupae were formed mostly when pupation occurred on green plants. Many images of *V. virginiensis* larvae have been published, including those in Pyle (1981, 2002), Wagner (2005), Allen et al. (2005), Cech and Tudor (2005), and Minno et al. (2005). All are similar to ours, but some show a greater area of white-yellow banding and reduced black coloration (e.g., Allen et al., 2005). Pupae were illustrated by Pyle (1981), Scott (1986a), and Allen (1997); all showed gray-form pupae, similar to ours.

Discussion

Gravid females readily oviposit in captivity and larvae are easy to rear. Much is known about this common species in E US, but little is known of its biology in the W. Its abundance in NE states is not matched in Cascadia and the reasons for this are unclear. Research on migration, overwintering biology, and behavior of W *V. virginiensis* is needed. Mortality from predators, parasitoids, and abiotic factors (drought, heat) may be greater in the W. Acceptance of *C. arvense* as a host by larvae suggests the host range may be wider than realized. The genetic and/or environmental basis of the 2 pupal color forms would make an interesting study.

West Coast Lady   *Vanessa annabella* (W. D. Field)

Adult Biology

Ranging throughout the W US, *Vanessa annabella* is an irregular breeding visitor to Cascadia, potentially occurring in all habitats from mountains to backyards. It is the lady most often seen in early spring and late autumn, with records extending from early March to mid-Nov. Ovipositing females have been seen as late as Oct 25 in Benton Co., WA. One to 3 generations may be produced, depending on how early arrivals appear from the S or whether overwintering has occurred locally. Adults are probably unable to overwinter in most of E Cascadia, and there may be some inconspicuous S migration in autumn. One adult found on Nov 12 at Benton City sheltering inside a plant pot was unable to fly (10 °C) and was placed in an unheated garage for overwintering. Temperatures during Dec fell to -10 °C and the butterfly died. N spring movement is better defined and often occurs within the mass migrations of *Vanessa cardui* (Painted Lady). Males are territorial, defend perches, and live up to 28 days in summer and several months in winter (Shapiro and Manolis, 2007). Flowers, including asters, dogbane, and mints, are visited frequently as well as animal scat. Scott (1986a) records ~30 larval hosts, but Malvaceae (mallows) and *Urtica dioica* (Stinging Nettle) are mostly used in Cascadia (Pyle, 2002; Warren, 2005). Eggs are laid singly, usually on the upper surfaces of leaves.

Immature Stage Biology

We reared this species 3 times from late-instar larvae collected on nettles in N CA (April) and SE WA (August) and from eggs laid by 3 females obtained from Benton Co., WA, Oct 17–23. Held in a steel-framed muslin cage (30 × 30 × 30 cm) at 20–22 °C (natural daylength) in sunshine, they laid 70–80 eggs on *Malva neglecta* (Cheeseweed) during 14 days. Eggs hatched after 8 days, and larvae fed immediately on

Adult male

M. neglecta. Larvae rasped leaf surfaces, producing clear "windowpane" areas in L1 and holes in L2. Feeding was on both sides of the leaves, and larvae produced small amounts of silk to cover themselves, sometimes forming loosely silked nests between leaf veins, especially near petioles. L3–L5 fed mostly at leaf edges and formed loose, untidy leaf shelters. Feeding occurred inside and outside nests at any time of day or night. Pupation occurred mostly on plant stems and under leaves. Development from egg-hatch to pupa took only 15 days, with 2, 3, 4, 3, and 3 days spent in L1–L5, respectively. Adults emerged after 5–6 days, 28–29 days after oviposition, similar to that reported by Dimock (1978). Pupae held at 13–15 °C took 4 weeks to produce adults. Pupae held outdoors in Dec with temperatures down to -10 °C failed to survive, indicating pupae are unlikely to overwinter successfully in E WA. Larval defense is based on concealment in nests; L2–L5 also possess a ventral gland on segment 1 that likely secretes antipredator chemicals. Larvae are solitary but tolerate each other well in captivity.

Description of Immature Stages

The green egg has 12 prominent white ribs; 10–14 were recorded by Dimock (1978). L1 is purplish brown with a black head, becoming dark reddish brown with dorsal white markings. There is a black collar, and each segment has 6 dark bullae with a long black seta arising from each. Late in L1, cream markings occur on segments 5, 7, and 9. L2 is dark brown to black with dorsal creamy yellow markings on segments 5, 7, and 9. Two indistinct creamy yellow lines also occur dorsally. L3 is black with creamy yellow dorsal markings and prominent black or yellow spines with multiple setae. The head is black and densely covered with setae. L4 is black, speckled with yellow and white dashes and dots, and has 2 broken yellow dorsal lines. As L4 matures, areas of reddish orange develop. The body and black head are densely covered with setae, giving the larva a shaggy appearance, which persists into L5. L5 is a mosaic of coloration that is variable even within a cohort (Dimock, 1978). Black, gray, yellow, white, orange, and red are the principal colors, with black marked or speckled with orange-red commonest. The black spines are long and prominent. The pupa is gray to light brown with inconspicuous black markings, especially on the abdomen. It has a rough, granular appearance and there are 4–6 white nonmetallic cones dorsally behind the thorax, the lowest pair the largest. Dorsally there are 5 rows of orange bullae on the abdomen. Larvae of *V. annabella*

Egg @ 0.75 mm

L1 @ 3 mm

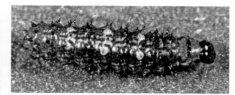

L2 @ 5 mm (late)

L3 @ 8 mm

L4 @ 15 mm

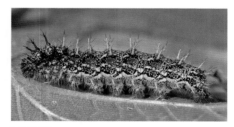

L5 @ 28 mm

Pupa @ 18 mm

could be confused with *V. cardui*. Both species exhibit wide variation in coloration; however, pupae of the 2 species differ more reliably. Images of mature larvae were published in Guppy and Shepard (2001) and Allen et al. (2005), both similar to ours, although the prominent black spines seen in our larvae are lacking in both. The larva in Guppy and Shepard (2001) is lighter colored with white spines, but their pupa is identical to ours. Dimock (1978) published black-and-white images of all stages of *V. annabella* and provided detailed descriptions and notes on natural history.

Discussion

Gravid females readily oviposit in captivity, and larvae are easy to rear on mallows and nettles. Little is known about the biology of *V. annabella*. The variability in coloration of mature larvae has been noted previously and may have an environmental component. Dimock (1978) suggested that lighter morphs occur under brighter rearing conditions and the influence of environment on coloration would make an interesting study. *Vanessa annabella* depends on immigration from the S but may overwinter in S OR and perhaps W OR and W WA. Further studies on low temperatures and survival of immature stages and adults of *V. annabella* would provide an indication of areas in which overwintering may be possible. Moderate and short-lived subfreezing temperatures may permit survival, and *V. annabella* could increase its abundance in Cascadia under a warming climate. The impact of predators and parasitoids on this species in Cascadia is unknown, although Dimock (1978) reported frequent parasitism by tachinid flies in CA. Minute Pirate Bugs (*Orius* spp.) are common residents of mallow and nettles in E Cascadia, and early instars were preyed upon by this predator in our rearing.

West Coast Lady *Vanessa annabella* (W. D. Field)

Painted Lady   *Vanessa cardui* (L.)

Adult Biology

Ranging over much of the planet, the Painted Lady is the world's most cosmopolitan butterfly. In Cascadia, adults occur from early April to mid-Nov and can be found in virtually every habitat. *Vanessa cardui* is unable to overwinter in Cascadia, spending the cool season in S CA, AZ, and Mexico. When winter conditions are favorable for host-plant growth in the desert SW, vast populations develop, with huge migrations of sometimes millions of individuals heading N, attracting public attention. These migrations sometimes result in very large numbers reaching Cascadia in May and June. In such years up to 2 generations are produced in Cascadia and larvae can be found on nearly every thistle plant. Moderate populations occur in Cascadia in most years, but the species may be absent when migrations are small. An amazing variety of plants are utilized as larval hosts, including thistles, legumes, borages, mallows, lupines, sages, nettles, and soybeans. Thistles, especially the introduced *Cirsium arvense* (Canada Thistle), are the most favored hosts. Females usually lay 1 egg per host plant, except when populations overwhelm host-plant availability. The S return migration in autumn is much smaller but can be observed along the SW ocean coast of WA, when E winds concentrate migrants near the ocean.

Immature Stage Biology

We reared this species several times from gravid females and field-collected larvae. Eggs obtained from females collected in Benton Co., WA on May 1 hatched in 6 days at 21–26 °C. The inconspicuous eggs are laid singly on the host plant; however, in captivity females will oviposit many eggs on single leaves of hosts, usually on upper surfaces. Development on *C. arvense* was rapid at 21–26 °C, taking 21 days from egg-hatch to pupation, with 2–3 days spent in early instars (L1–L3) and 7 days each in L4 and L5. Adults eclosed ~10 days later. Larvae feed on leaves, and all instars build protective silk web nests incorporating leaves curled or silked together during L3–L5. L1 feed and rest on upper leaf surfaces, covering themselves with a few strands of silk. Nests are more complex as larvae mature, with only 1 larva per nest and usually only 1 nest per host plant. The nest completely encloses the larva that feeds within it. When the leaf is consumed, the larva moves to another location and builds a larger nest. A considerable amount of frass collects in the bottom of nests, which are up to 3 inches across. Nests are easily seen in thistle patches and remain intact for several weeks after the larvae have left. Larvae usually wander before pupating, but pupae can occasionally be found on the host plant. Immature stages do not diapause, but development during declining late summer–autumn photoperiods produces adults in reproductive diapause, as indicated by endocrine studies (Herman and Dallman, 1981); this is linked to the S migratory movements sometimes seen at this time (Pyle, 2002). Protection is based on concealment, webbing, spines, and a ventral gland on segment 1 that likely emits defensive chemicals. There are 5 instars.

Description of Immature Stages

The pale bluish green egg has 18–20 vertical ribs. L1 is brown with a purplish tinge and has 8 rows of small dark bullae, each giving rise to a single unbranched seta. L2 is dark brown-black and the bullae are black except for 3 or 4 yellow dorsolateral pairs. L3 is brown-black, with 5 pairs of yellow dorsolateral bullae. There are 2 pale middorsal stripes; 3 or 4 yellow rings encircle the body between spines, and bullae bear multiple setae. L4 is darker and all bullae have branching setae. Most of the paired dorsolateral bullae are partially yellow to orange, and a dorsal yellow spot occurs between some pairs. In L5 the dorsolateral bullae may be bright orange, contrasting with the black body, or they may be yellow with reddening on adjacent areas. The branched spines are drab yellow with black tips, the body is speckled with numerous small pale spots, and the black head has dense long unbranched setae. Larvae are extremely variable, especially in L4 and L5, some with a much lighter hoary appearance, and others with very distinct rings between bullae. Pupae vary from almost white to tan to copper-gold with a metallic sheen, all peppered with tiny black spots. Short conical spines on the dorsal surface reflect a bright gold metallic luster. Larvae and pupae could be confused with those of *V. annabella*, which occur

Adult

Painted Lady *Vanessa cardui* (L.)

Egg @ 0.8 mm (fresh, *left*, maturing, *right*)

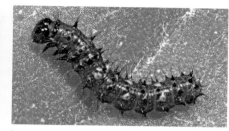

L2 @ 4 mm

L1 @ 2 mm

L3 @ 9 mm

L4 @ 17 mm

L5 @ 32 mm

on some of the same hosts. Early instars are very similar, and late instars of both species are highly variable, making positive identification difficult, although *V. cardui* generally have more yellow coloration. There are many photos of the larvae and pupae in the literature showing variation but generally agreeing with ours.

Discussion

Mating and oviposition can be obtained in small cages. We used 30-cm dia. plastic buckets with potted thistles or mallows, enclosed by a muslin lid, to obtain oviposition. The hardy larvae are easily reared. Commercial rearing kits, used in classrooms, are readily available with a formulated agar-based diet for larval food; resulting adults routinely mate and produce a second generation of eggs. During outbreak years, larvae are easy to find; any patch of thistle will typically have larvae. Variability in larval and pupal coloration is substantial and possibly a consequence of the wide range of hosts utilized. *Vanessa cardui* preferred to oviposit on *C. arvense*, when given a choice between this and *Urtica dioica* (Janz, 2005). Reared under cool temperatures and short daylengths, or warm temperatures and long daylengths, the Australian *Vanessa kershawi* develops faster, producing smaller, migration-adapted adults (James, 1987). Early-spring and late-summer populations of *V. cardui* in W US may display a similar strategy, with early-season colonizers in Cascadia often undersized.

Pupa @ 19 mm

Red Admiral   *Vanessa atalanta* (L.)

Adult Biology

The Red Admiral is one of the northern hemisphere's best known and most charismatic butterflies. It is found throughout subarctic North America and Eurasia, occurring in many habitats, including urban settings, gardens, woodlands, meadows, orchards, and riparian areas. In most of Cascadia, *Vanessa atalanta* is a breeding immigrant from the south, flying in 2–3 broods from late March to late Oct, from sea level to >7,000 ft. A dispersed return autumn migration may occur, but in mild winters, particularly in coastal areas, some adults may overwinter successfully (Warren, 2005). Males are pugnacious and territorial, perching on hilltops, bushes, or the ground, flying out to intercept intruders. Many kinds of flowers are visited, including buddleia, thistles, milkweed, and lilac, but both sexes are also partial to rotting fruit, sap flows, dung, and carrion. The primary larval host in Cascadia is *Urtica dioica* (Stinging Nettle). Other hosts include *Urtica urens* (Dwarf Nettle), *Parietaria* spp. (pellitory), *Soleirolia soleirolii* (Baby's Tears), *Boehmeria cylindrica* (False Nettle), *Pipturus albidus* (Mamaki), *Pilea* spp. (clearweeds), *Laportea canadensis* (Wood Nettle), and *Humulus lupulus* (Hop) (Scott, 1986a; Shapiro and Manolis, 2007). Eggs are laid singly on undersides of leaves, typically on a leaf vein.

Immature Stage Biology

We reared or partially reared this species several times. Adults obtained from wild-collected L4–L5 in King Co., WA, mated in captivity (tent-style cage under artificial lighting with nettles) and produced many eggs during Sep 12–18 (Martha Robinson, pers. comm). Martha and her children's butterfly

group (INSTARS) kindly provided eggs that began hatching on Sep 19; L1 immediately fed on *U. dioica* leaves. Held at 22–25 °C, larvae developed rapidly, with pupation 22 days post egg-hatch; larvae spent 5, 4, 4, 4, and 5 days in L1–L5, respectively. Adults eclosed after 14 days at 17–21 °C. A subcohort of 50 L3 was switched from nettle to *H. lupulus* but did not feed and died within a few days. On at least 9 other occasions, larvae were collected from Kittitas, King, Skamania, Pierce, and Chelan Cos., WA, and reared on *U. dioica* with similar development. L1–L2 rasped leaf surfaces, creating holes. Feeding occurred on both sides of leaves, and larvae produced small amounts of silk to cover themselves, forming loosely silked nests on leaf veins, especially near petioles. L3–L5 fed at leaf edges and formed shelters by folding a leaf over or silking a group of leaves together. Larvae often fold nettle leaves downward into a tent and feed from within (Minno et al., 2005). In captivity, feeding occurred inside and outside nests at any time of day or night. Pupation was mostly on plant stems and under leaves, sometimes drawn together with silk to create an "umbrella." Defense is based on concealment in nests, but L2–L5 possess a ventral gland on segment 1 that likely secretes antipredator chemicals. Larvae are solitary but tolerate each other well in captivity.

Description of Immature Stages

The green egg has 8 white ribs. L1 is whitish green with sparse black setae becoming brown with white markings. L2 is dark brown with 6 rows of black bullae, each bearing multiple short setae and 1 long black spine. L3 is black with prominent bullae and longer setae and spines. There is an indistinct subspiracular row of white dashes, and the setaceous head is shiny black. L4 is similar, with well-developed spines, prominent subspiracular creamy white dashes, and white spots peppering the body. In some individuals the ground color is variegated reddish brown rather than black. L5 is black peppered with white dots and short white setae. The spines (black or pale) are prominent, as are the creamy white subspiracular dashes. The prolegs are brown and the head is black with white dots and short pale setae. Mature larvae vary considerably in ground color, from black to greenish gray or brown, or even whitish. The pupa is also variable, ranging from gray-brown to whitish, with 4–5 pairs of metallic gold dots and flecks dorsally and 5–6 pairs of orange-red–tipped tubercles on the abdomen. The larvae and pupae are distinctive and separable from other common nettle feeders like *V. annabella*, *Polygonia satyrus*, and *Aglais milberti*.

Adult

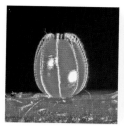

Egg @ 0.8 mm (fresh)

Egg @ 0.7 mm (mature)

L1 @ 2.5 mm

L2 @ 5 mm

Pupa @ 19 mm

L3 @ 10 mm

Many images of *V. atalanta* larvae and pupae have been published, including those in Allen (1997), Guppy and Shepard (2001), Miller and Hammond (2003), Wagner (2005), Woodward (2005), Allen et al. (2005), and Neill (2007).

Discussion

Vanessa atalanta is easily reared on nettles, and mating and oviposition can be obtained in captivity. Larvae can be found in nettle patches by looking for drooping leaf nests. Despite being recorded as a host, *H. lupulus* was refused by L3 in our rearing. Scott (2006) reported a female displaying pre-oviposition behavior on *H. lupulus*, but no eggs were laid. Eight years of regular leaf sampling of Yakima Valley, WA Hops yielded no evidence of use by *V. atalanta* (James, unpubl.). Overwintering of adults in Cascadia may occur more frequently with climate warming, and the possibility of pupae overwintering as reported by some authors, e.g., Allen (1997), should be examined. Environmental cues determining migratory behavior in adults need study. Parasitism of larvae and pupae occurs commonly in the UK but has not been reported or studied in Cascadia.

L4 @ 15 mm

L5 @ 29 mm

L5 @ 29 mm

Common Buckeye   *Junonia coenia* Hubner

Adult Biology

Junonia coenia is a common resident of S US, extending N each summer as far as S Canada. In Cascadia it occurs in much of OR and has been recorded within 30 miles of the WA border. Pyle (2002) considered *J. coenia* to be a year-round resident on the S OR coast, but most of the individuals seen in OR likely originate from N CA, where it is present for most of the year (Shapiro and Manolis, 2007). It frequents disturbed, weedy habitats along roadsides, watercourses, and other open areas. In Cascadia it may be seen from May to Oct, and 2 broods are likely in S OR. The Buckeye visits a great variety of flowers (e.g., clovers, asters, milkweed) and usually perches on the ground, often in hot, dusty places like gravel roads and bare earth. It is a rapid flier and difficult to approach. Mating usually occurs in late morning or early afternoon (Scott, 1975), when optimal conditions (32–34 °C and high-intensity sunshine) (McDonald and Nijhout, 1996), most often occur. Eggs are laid, usually singly, on lower surfaces of host leaves or on terminal shoots. Larval hosts are species of Plantaginaceae (plantains), Scrophulariaceae (figworts, penstemons), Acanthaceae (acanthus), and Verbenaceae (verbena). Bowers (1984) listed 43 recorded hosts. All these plants contain iridoid glycosides, which function as larval feeding stimulants (Bowers, 1984).

Immature Stage Biology

We reared *J. coenia* once from 5 females collected at Suisun Marsh, Solano Co., N CA, on Oct 11–12. Confined in a plastic and muslin cage (30 × 30 × 30 cm) with cut *Plantago lanceolata* (Narrow-leaved Plantain) and *Antirrhinum majus* (Snapdragon), they laid >200 eggs over 4 weeks on both hosts. Oviposition occurred only in sunshine and at temperatures >15 °C. The eggs were split between the authors and reared separately. Eggs in 1 cohort hatched in 3 days at

28 °C. Shells were consumed before L1 began feeding on *P. lanceolata* and *A. majus*. L1–L3 fed exposed on the upper surface of leaves, producing clear patches or spots. Development from egg-hatch to L5 took 9 days at 28 °C, with a further 6 days as L5. In the other cohort the larvae were reared at cooler temperatures (18–21 °C); eggs hatched in 7 days and development from egg-hatch to pupation took 24 days. The cool cohort was reared on *A. majus* then switched to *P. lanceolata* in L5; the larvae accepted the host switch with reluctance. L4 and L5 feed openly on host plants and consume leaves from the edges. L2–L5 possess a ventral gland on segment 1 (likely emits defensive secretions) and are prone to much oral regurgitation caused by interactions with other larvae. The larvae often wandered off the food plants during the day, then returned at night to feed. Larvae reacted sharply when touched by another larva; however, no cannibalism or disease was observed despite overcrowding. Pupation occurred on host plants, sometimes under lightly silked leaf shelters, or on inert surfaces.

Description of Immature Stages

The egg is light green with 12–13 narrow raised whitish ribs. L1 is yellow-orange becoming dark brown with indistinct white spots. On each segment, 6 dark brown-black bullae each sport a single long black seta, and several setae adorn the shiny black head. L2 is dark brown-black except for the sides of segment 1, which are bright orange. There is an indistinct middorsal darker line. The bullae are black except for the lowest row on each side of the body, which have orange bases. All bullae sport 4–6 short setae and a single long spine. Indistinct white spots appear late in the instar. L3 is black with 2 broken white lines dorsally and white spotty markings laterally between the 2 rows of bullae. The orange bases of the lower row expand, and some bases of the upper row are orange. Dorsal bullae are black, enlarged, and spinelike. Paired black bullae with setae occur on the upper part of the head capsule, with smaller white bullae scattered lower on the head. Bullae on the first segment are bright orange on each side. L4 is black and similarly marked, but the dorsal bullae and spines show an iridescent blue sheen in bright light. The lateral white markings are 2 longitudinal stripes, and orange surrounds the lateral spine bases or extends between the 2 lateral bullae on each segment. A pair of small black bullae with setae is present on each proleg. The head is black with upper paired black bullae and spines, scattered prominent white bullae, and a central orange spot. L5 is black dorsally with 2

Adult female

Egg @ 0.8 mm

L5 @ 30 mm (head)

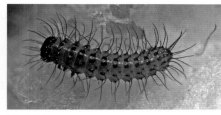

L1 @ 2.5 mm

L2 @ 6 mm

Pupa @ 17 mm

L3 @ 8 mm

L4 @ 17 mm

L5 @ 38 mm

L5 @ 38 mm

orange stripes broken into small spots. Many tiny white spots are present and the dorsal bullae and spines show increased iridescence. Laterally, the orange and white markings are bold and the paired proleg bullae are orange and long. The head is bright orange (upper) and black (lower). The pupa is gray-black on the wing cases and mottled brown, black, orange, blue, and white on the abdomen and thorax. There are 5 Y-shaped markings in a line on each wing case. Mature larvae were illustrated by Pyle (1981), Scott (1986a), Allen (1997), Allen et al. (2005), Wagner (2005), Cech and Tudor (2005), Minno et al. (2005), and Spencer (2006), and all are very similar to ours, as are the pupae illustrated by Scott (1986a) and Allen (1997).

Discussion

Gravid females oviposit readily in captivity; obtaining successful matings is more difficult but achievable, even in small cages (10 × 10 × 10 cm) as long as the butterflies are in direct sunlight (McDonald and Nijhout, 1996). This species has been reared often and is ideal for laboratory-based studies on butterfly biology (e.g., Bowers, 1984; Smith, 1991). Developmental rate and larval duration and size are strongly affected by photoperiod in the related Australian species *Junonia villida* (James, 1987), and this may also be the case for *J. coenia*.

Common Buckeye *Junonia coenia* Hubner

Lorquin's Admiral   *Limenitis lorquini* Boisduval

Adult Biology

Lorquin's Admiral is a W North American species ranging from BC to S CA and inland to W MT and NV. It is a common species in Cascadia in many habitats (rivers, roadsides, parks, gardens, canyons), and is found throughout except for a coastal fringe in NW WA and the driest parts of the Columbia Basin. It flies from early May to early Oct and is usually double brooded in Cascadia, triple brooded in N CA. Recent studies suggested *L. lorquini* is a mimic of the distasteful California Sister (*Adelpha californica*) (Prudic et al., 2002). Males are territorial, perching along natural corridors, rivers, trails, or roadsides, flying out to vigorously challenge passersby in search of mates. Both sexes visit flowers, including thistles, dogbane, milkweed, asters, and mints. Groups of males can sometimes be found puddling or feeding on animal scat. Mated females often chain-visit host plants; when a suitable site is found, a single egg is laid on the upper surface of a leaf tip. Lorquin's Admiral uses a variety of larval hosts, including *Salix* (willow), *Populus* (poplar), *Prunus* (cherry), *Malus* (apple), *Prunus virginiana* (Chokecherry), *Spiraea douglasii* (Douglas Spiraea), *Holodiscus discolor* (Ocean Spray), *Amelanchier alnifolia* (Serviceberry), and *Ceanothus velutinus* (Mountain Balm). Additionally, an Okanogan Co., WA female was observed ovipositing in mid-July on *Pachistima myrsinites* (Mountain

Boxwood), but the resulting larva would not feed on this plant.

Immature Stage Biology

We reared this species several times from field-collected eggs and larvae and from gravid females. Two worn females obtained near the Tucannon R in Columbia Co., WA, on July 11 were confined with potted Hooker's Willow (*Salix hookeriana*) in a muslin cage and held in a sunny room at ~26 °C. Eggs ($n = 6$) were laid within a few hours, but the females expired after a day. Eggs hatched in 5–6 days at 23–28 °C. L1 did not feed on *S. hookeriana* leaves, wandered off, and were transferred to *H. discolor*, which they readily accepted. Larvae were reared at 23–28 °C under 18 hr daylength and development was rapid, taking about 5 days for each instar and 6 days as pupae, for a total developmental period of 34 days. Reared on *Salix scouleriana* (Scouler's Willow), larvae took 37 days to develop. L1 fed on the tips of leaves, eating either side of the midrib, creating a "pier" that was extended by adding frass pellets to the tip. L1 and L2 rested on these piers, where they blended very well and looked like dead vegetation. L4 and L5 rest for extended periods on the upper sides of host leaves, out of direct sunlight, usually on a very thin silk mat. The larval silhouette is sometimes visible through the leaf from below when backlit by the sky. In the second generation, larvae overwinter in L2. The diapausing larva rolls a host-plant leaf into a hibernaculum, tying it with silk and binding it to a twig. Larval defense appears to be based on a combination of camouflage (white saddle "bird dropping" markings, resting on piers) and intimidation. When disturbed, L4–L5 rear up anteriorly, waving their "horns." They also often rest with posterior segments held at 45 degrees, breaking up the caterpillar outline. There are 5 instars.

Description of Immature Stages

The egg is light silvery green and pincushion-like, with a pattern of numerous polygons formed by the intersections of raised ridges; at each angle around the polygon, a narrow spike protrudes radially outward. L1 is olive green changing to tan with black splotchy markings and indistinct white spots. The head is brown and bifurcated, and the body is covered with numerous irregular bumps and nodes; there are sparse short stubbly setae around the ventral periphery. L2 is black with irregular brown markings, small white spots, and short bristled tubercles. The head is dark and bears numerous brown and black nodes. A contrasting white "bird dropping" saddle develops

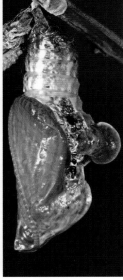

Adult

Pupa @ 23 mm

late in L2. In L3 the ground color is black but is masked by pairs of short brown bristled bullae on most segments, reduced and white on the saddle. On the second segment there is a pair of large brown-and-black horns. The spiracles vary in size, largest on anterior and posterior segments. The same pattern occurs in L4, with the head, horns, and ground color reddish brown. The spiracles are of equal size and dark, outlined with thin white rings. The head progressively flattens during development and bears upper protuberances in L4 and L5. L5 has a significant reduction of bristled bullae, the white saddle area is larger, and the ground color is rusty orange anteriorly. Anterior segments are cream colored and the horns are longer. Striking, regularly spaced, tiny bright blue dots occur on most brown areas of the body. The pupa is mostly white with brown wing cases and a large brown rounded dorsal knob. In Cascadia only the Viceroy, *L. archippus,* can be confused with *L. lorquini.* The final instars of both species are similar, reddish brown with tiny bright blue spots and a white saddle, although *L. archippus* appears more "bristly" as a result of greater retention of bristly bullae. The saddle in *L. lorquini* is more extensive, running anteriorly and posteriorly on both sides, whereas in *L. archippus* it extends posteriorly only. The head of *L. lorquini* is dark brown while that of *L. archippus* is orange. The earlier stages show greater differences: *L. archippus* eggs are light tan and L1 are orange; L2–L4 are generally darker than those of *L. lorquini.* Mature larvae of *L. lorquini* were illustrated by Miller and Hammond (2003) and Woodward (2005); however, their larvae are olive green. Allen et al. (2005) and Guppy and Shepard (2001) show mature larvae that are reddish, as well as an egg and pupa similar to ours.

Discussion

Many plants are potential hosts, but not all may be acceptable to different populations. Ovipositing females can be followed in June and July and their eggs collected. Confining mature gravid females with potted host plants can be successful, and a single egg is often sufficient to rear this species. Females obtained early in the flight period are often unmated. Partially grown larvae can be obtained by searching for eaten leaves and/or silhouettes in bush or tree canopies. Studies on the influence of environmental cues on voltinism would clarify why this species is variably single or double brooded. Little is known of the biotic and abiotic factors regulating population densities of this widespread species.

Egg @ 1.1 mm

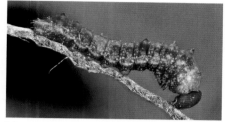

L1 @ 3.5 mm

L2 @ 8 mm

L3 @ 12 mm

L4 @ 18 mm

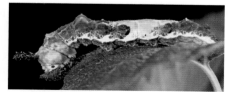

L5 @ 30 mm

Lorquin's Admiral *Limenitis lorquini* Boisduval

Viceroy *Limenitis archippus* (Cramer)

Adult Biology

Limenitis archippus is widely distributed in North America from Canada to S Mexico. In Cascadia it is locally distributed E of the Cascades, mostly along major rivers. It occurs in small populations in low-elevation riparian habitats with abundant willows. It is on the wing from mid-April to late Sep and there are 2 generations. The Viceroy may be confused with the Monarch, which it mimics. Prudic et al. (2007) showed that *L. archippus* sequesters phenolic glycosides from willows, making it toxic like its model. Adults are attracted to willow sap and visit a number of flowers including thistles and milkweed. Nelson (2003) showed puddling was an important component of adult behavior. Scott (1986a) listed more than 25 larval hosts, mostly species of *Salix* (willow) and *Populus* (poplar) but also *Malus* (apple) and *Prunus* (plum). A confirmed host in OR and WA is *Salix exigua* (Coyote Willow) (Warren, 2005). Eggs are laid singly at the upper side tips or edges of host leaves. Females at Horn Rapids, Benton Co., WA, during early afternoon on May 23 chose *S. exigua* leaves 2 m high for oviposition.

Immature Stage Biology

We reared this species twice from females collected at Horn Rapids on May 23 and July 27. In May, 1 female was held in a steel-framed muslin cage (55 × 30 × 20 cm) with potted Purple Willow (*Salix purpurea*). This European cultivar was used because potted *S. exigua* was not available and cut *S. exigua* quickly wilts. Oviposition occurred in partial sunshine at 20–22 °C with 7 eggs laid over 5 days. Eggs hatched after 6 days at 18–25 °C and larvae fed mostly at night on *S. purpurea*. L1 fed on leaf edges at the tip of a leaf, leaving an exposed vein (pier) on which they constructed a silk mat and rested. Frass pellets were added to the tip of the midrib, extending the "pier." A ball of leaf and frass was attached with silk to the base of the vein. L2 and L3 occurred after 3 and 5 days, respectively.

Thereafter, growth slowed, and all but 1 larva died before L4, 9 days post egg-hatch. The single L4 died shortly after molting. In July, 4 females caged with potted *S. purpurea* produced ~60 eggs over 5 days. Half the larvae were reared under indoor ambient conditions (20–27 °C, natural daylength) during Aug and half at 28 °C and continuous lighting. Apple foliage was provided as food initially but replaced with *S. exigua* in the ambient group after larvae showed a reluctance to feed and 80% mortality had occurred. By L2 only 1 larva remained in the 28 °C cohort, still feeding on apple, and 9 in the ambient cohort. The single larva reached L5 14 days after egg-hatch and pupated 10 days later. Pupation of the ambient-held larvae occurred on foliage 4 days later, and adults emerged Sep 8–12, giving a total development period of 40 days. Overwintering of *L. archippus* occurs as a diapausing L3 in a hibernaculum constructed from a host-plant leaf (Scott, 1986a). Clark and Platt (1969) showed that L2 exposed to daylengths less than 14 hrs entered diapause. In our ambient rearing, L2 exposed to daylengths of ~14.2–14.5 hrs did not enter diapause, indicating a third generation may occur in Cascadia. Larval defense is based on camouflage ("bird dropping" markings, resting on "pier") and intimidation (waving horns).

Description of Immature Stages

The egg is light tan and pincushion-like, with a pattern of numerous polygons formed by the intersections of raised ridges. L1 is orange–light brown with indistinct white spots. Each segment carries 6 bullae, some darker and enlarged dorsally on segments 2, 3, 5, 10, and 11. The head is orange and bifurcated. L2 is dark chocolate brown–black with bristly white bullae, giving the appearance of profuse white spotting on the body and black head. A white "saddle" occurs on segments 7–9. L3 is similar; the enlarged dorsal bullae on segment 2 are hornlike, tipped with a rosette of smaller orange bullae. L4 is bristly black becoming smoother and dark brown. The horns are large (2–3 mm), and enlarged dorsal bullae sport distinct rosettes of orange tubercles. Irregular tiny bright blue spots cover the body. L5 is smooth and reddish brown with a white saddle, white posterior ventral markings, tiny blue spots, and long spiked horns (6–8mm). The head is orange and flat with knobby protuberances. The pupa is mostly white on the abdomen and brown to black on the wing cases. There is a large dorsal brown-black knob, and the overall appearance is shiny-wet. In Cascadia only *L. lorquini* could be confused; minor differences are

Adult

Egg @ 1.2 mm

L5 @ 35 mm (head)

L1 @ 5 mm

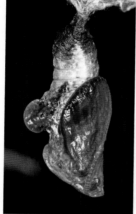

Pupa @ 14 mm

L2 @ 8 mm

L3 @ 11 mm

L4 @ 21 mm

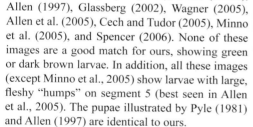

discussed in the *L. lorquini* account. Mature larvae were illustrated by Pyle (1981), Scott (1986a), Allen (1997), Glassberg (2002), Wagner (2005), Allen et al. (2005), Cech and Tudor (2005), Minno et al. (2005), and Spencer (2006). None of these images are a good match for ours, showing green or dark brown larvae. In addition, all these images (except Minno et al., 2005) show larvae with large, fleshy "humps" on segment 5 (best seen in Allen et al., 2005). The pupae illustrated by Pyle (1981) and Allen (1997) are identical to ours.

L5 @ 35 mm

Discussion

Females oviposited readily on a European willow, but larvae failed to survive on it and did not fare well on apple. Research on host use in Cascadia is needed. Variation in coloration and morphology of larvae deserves study to determine the genetic and/or environmental factors involved. The Viceroy appears well defended, yet in Cascadia it is relatively uncommon and absent from many seemingly suitable habitats. Research on the impact of natural enemies and abiotic factors might provide some insights. Nelson (2003) showed that *L. archippus* needs a variety of resources found only in well-functioning riparian ecosystems. The ecosystems in Cascadia may not fulfill the resource requirements of *L. archippus*.

Viceroy *Limenitis archippus* (Cramer)

California Sister   *Adelpha californica* (Butler)

Adult Biology

Adelpha californica occurs infrequently in SW WA (Pyle, 2002); however, it regularly occurs in S and W OR. Common in the Siskiyous, it can be abundant along the edges of the Willamette Valley (Warren, 2005). Elsewhere it ranges into the SW states and Mexico. Adult *A. californica* and *Limenitis lorquini* resemble each other; *A. californica* is moderately unpalatable to captive birds, whereas *L. lorquini* is palatable, suggesting it mimics *A. californica* (Prudic et al., 2002). Male *A. californica* are territorial, perching and patrolling along riparian corridors, rivers, and roads near oaks. Populations often appear to be male dominated as a result either of protandry or of females "lying low." Both sexes visit mud, flowers, fermenting fruit, and dung, and 2–3 broods may be produced from March to Nov. Larval hosts are various oaks, including the evergreen *Quercus chrysolepis* (Golden Oak) and *Quercus agrifolia* (Coast Live Oak). The preferred host in OR is *Quercus garryana* (Garry Oak) (Warren, 2005), and *Chrysolepis chrysophylla* (Golden Chinquapin) is also used. Eggs are laid singly, mostly on upper leaf edges, often at the tip, with no apparent preference for young leaves.

Immature Stage Biology

We reared *A. californica* from 3 females collected near the Feather R, Plumas Co., N CA, on Oct 13. Confined with a potted *Q. garryana* in a dome-shaped muslin cage (60 × 60 × 60 cm), 13 eggs were laid during 8 days. The cage was held outdoors at 15–20 °C. Eggs were placed at the ends of leaf veins or midribs and hatched after 12 days at 19–25 °C. The upper one-third of the egg was opened like a lid, and the larva exited without eating the shell. L1 immediately fed at the leaf edge, forming a "pier" from the vein or midrib upon which it rested when not feeding. Frass was used to extend the pier by silking pellets together, such that 35–40% of the pier length (up to 25 mm) comprised frass; larvae invariably rested on the frass portion. Excess frass was stored in a lightly silked "bag" constructed at the base of the leaf, hanging from the stem. Most larvae used a single pier during L1, and piers were used until L3. Development from L1 to pupation took 36 days at 20–30 °C, with 10, 5, 8, and 13 days in L1, L2, L3, and L4, respectively. From L2 onward a serpentine posture was adopted at rest, with head and posterior segments raised. L4 were aggressive and attempted to bite when disturbed. L2–L4 possess a ventral gland on segment 1, likely for defense. Overwintering occurs as a partially grown larva, but details of the instar and behavior have not been published. Our larvae had 4 instars; Aiello (1984), Otero and Aiello (1996), and Freitas (2006) reported 5 or 6 instars for 14 other *Adelpha* spp. Pupation occurs on branches or twigs of the host plant, and the hanging pupa often twists, remaining in a "bent" position for hours or days. Larval defense strategies include camouflage, pier construction, aggression, spines/horns, unpalatability, and ventral gland secretions (James, 2009a).

Description of Immature Stages

The egg is green and pincushion-like. Its surface is covered with hexagonal pits; at each angle around the hexagons, a narrow spike protrudes radially outward. L1 is yellow-green with indistinct white spotting and a large orange-brown head. On segments 1, 2, 3, 5, and 11, a pair of dorsal enlarged brown tubercles are present. L2 is tan-olive-brown with a profusion of raised white spots over the body and head. The dorsal tubercles on segments 2, 5, and 11 are dark brown, spiny and enlarged; the other tubercles are reduced. There is an indistinct white sublateral line, and the tan head has 4 dark stripes. L3 is darker reddish brown, particularly dorsally, with profuse white spotting; the enlarged spiny dorsal tubercles are hornlike, measuring 2–3 mm. The head has multiple black-tipped white spines laterally, and a larger pair of black spines at the top. Early L4 is similar but by mid-instar is bright green dorsally with many tiny white spots. The spiny horns are long (4–6 mm), orange-red, and largest on segments 2, 3, 5, and 11. The spiny head is dark purplish brown with a pair of enlarged, black-tipped bullae resembling eyes. Prior to pupation, the larva yellows, becoming light orange-brown as a

Adult male

Egg @ 2 mm   L4 @ 30 mm (head)

L2 @ 9 mm (on frass pier)

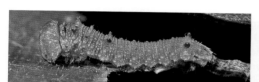

L1 @ 4 mm

L3 @ 20 mm

L4 @ 30 mm

California Sister *Adelpha californica* (Butler)

prepupa. The pupa is whitish tan tending to gray dorsally with gold and brown spots and dashes dorsally and a large dorsal abdominal projection. The body tapers toward the cremaster, and the posterior wing margins protrude like keels. There is a dark middorsal line and a reflective silver/sapphire butterfly-shaped motif on the dorsal thorax. The immature stages are unlikely to be confused with any other species in Cascadia. Images of late-instar larvae appear in Miller and Hammond (2003, 2007) and Allen et al. (2005) and are similar to ours.

Discussion

Finding eggs and larvae of *A. californica* can be challenging, although the search image of leaf/frass piers should be useful in locating early instars in high-abundance habitats. Miller and Hammond (2007) successfully collected larvae using a beating sheet. Females will oviposit in captivity but are sometimes hard to obtain among the often more numerous males. Instar number has not previously been reported for this species, but our observation of 4 is unusual given that 5–6 instars have been recorded for other *Adelpha* spp. It is possible that instar number varies with temperature and photoperiod. This is a species that may extend N with a warming climate; successful overwintering is likely to be the key for northward expansion. No information is available on the natural enemies of this species in Cascadia, and further studies are needed on its unpalatability (adults and larvae) and the chemistry involved. Research on the function and value of frass piers and frass storage is needed. Minno et al. (2005) speculated that ants find it difficult to locate L1 of *Limenitis* spp. when they are resting on frass piers.

L4 @ 35 mm

Prepupa   Pupae @ 28 mm

Monarch   *Danaus plexippus* (L.)

Adult Biology

Danaus plexippus is the archetypal butterfly, known throughout the world in books, in movies, on clothing and souvenirs, etc. Famous for its transcontinental migrations from Canada to overwintering grounds in Mexico and back (now a threatened phenomenon), the Monarch is a regular visitor to Cascadia, but in most years it is not common. Most of the monarchs seen in BC, WA, and OR likely originate from CA, where overwintering occurs on the S and central coast. When overwintering populations are large, and N movement is aided by good environmental conditions (e.g., in 2007), individuals reach OR in May and WA/BC in late May–early June. At this time the larval hosts, primarily *Asclepias speciosa* (Showy Milkweed) and *A. fascicularis* (Narrow-leaved Milkweed) grow rapidly, and the former in particular is common in much of E BC, E WA, and OR. *Danaus plexippus* occurs throughout OR and E of the Cascades in WA and S BC, usually near major rivers along which it migrates. One or 2 generations may be produced before adults begin moving S-SW in Sep–Oct. Pyle (1999, 2002) also observed autumn movement to the SE from S WA, suggesting that some of Cascadia's Monarchs migrate over the Rocky Mts to and from Mexico. Males patrol milkweed patches for mates and are often "violent" in their courtship/mating behavior, sometimes dragging females to the ground. Females lay eggs singly on the undersides of leaves, often near the petiole, or on terminal shoots.

Immature Stage Biology

We reared or partially reared this species several times. On June 4 a female was seen ovipositing on *A. speciosa* near Naches in Yakima Co., WA. Ten eggs were collected and hatched 6 days later. L5 was reached 12 days post egg-hatch at 20–28 °C, with 2, 4, 3, and 3 days spent in L1, L2, L3, and L4, respectively, but the cohort died in L5. On June 7 a female from Trinity Co., N CA, was confined with *A. speciosa* in a dome-shaped muslin cage (60 × 60 × 60 cm) held outdoors, and ~200 eggs were laid over 6 days. Egg-hatch occurred after 6 days, and larval development occupied 18 days at 20–28 °C, with pupation on July 2 and adult eclosure 9 days later. On July 23, 5 eggs were collected in Chicago, IL; 2 hatched and were reared with slightly faster development. Much is known about the biology of the immature stages of *D. plexippus*, with many published studies on ecology, development, behavior, and natural enemies. The unpalatability of larval and adult *D. plexippus* to birds and other vertebrates due to larval storage and concentration of cardenolides from milkweeds is a well-known factor in the species' success (Brower et al., 1972). The distastefulness of larvae is well advertised by striking, banded coloration, and there is evidence that larvae are also poisonous to some invertebrate predators (James, 2000). L1–L2 are inconspicuous, hiding in terminal shoots or under leaves, but most mortality occurs during these instars from ants, spiders, and other small predators (Zalucki and Kitching, 1982). Larvae feed on all parts of milkweed plants and L3–L5 rest openly, using warming from the sun to hasten development (Rawlins and Lederhouse, 1981; James, 1986).

Description of Immature Stages

The egg is whitish yellow with 18–20 vertical ribs. L1 is grayish white with ~10 short black setae on each segment. The head is shiny black, and there are pairs of dark brown bullae on segments 2 and 11. L2–L5 are colored and marked very similarly; each segment is transversely banded in black (center), white and yellow (edges of segment). Under cool temperatures, black bands enlarge at the expense of white bands, which may disappear, enabling larvae to attain higher body temperatures (James, 1986). In L2 the black bullae on segments 2 and 11 enlarge and many tiny black setae cover the body. In L2 and L3 the head is banded black and white. In L3 the bullae elongate to fleshy filaments, the anterior pair larger. In L4 and L5 additional black bands develop to a greater or lesser extent between the white and yellow bands. The head capsule bands are black and yellow, and the anterior filaments extend beyond the head for >5 mm. In most Cascadia larvae, black bands are wider than white ones. The cylindrical pupa is bright green, blue-green with maturity. A line of gold spots extends from near each eye to the back of the thorax. The pairs of gold spots on the thorax, near the eyes, and in midlateral position are usually the largest. On abdominal segment 3 there is a gold line, edged with broken

Adult female

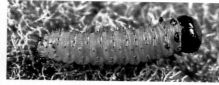

L1 @ 2.5 mm (early)

Egg @ 1.0 mm (fresh)

Egg @ 0.9 mm (maturing)

L1 @ 3.4 mm

Pupa @ 25 mm

L2 @ 7 mm

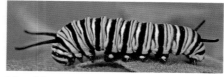

L3 @ 14 mm

L4 @ 25 mm

L5 @ 50 mm

black. The immature stages of *D. plexippus* are well known, and images abound in scientific, layman, and even children's books. Recent images of larvae and pupae were published in Guppy and Shepard (2001), Allen et al. (2005), and Neill (2007).

Discussion

Danaus plexippus is probably the best-known and most-researched butterfly species in the world. Despite this, there are still significant gaps in knowledge of its biology and ecology, particularly in Cascadia. The environmental cues used by *D. plexippus* in the W US determining when adults become nonreproductive and migrate S have not been elucidated as they have for Australian populations (James, 1993). The destination of autumn migrants from Cascadia needs research to see whether some of the population overwinters in Mexico. The impact of natural enemies in Cascadia is unknown. A larval monitoring program based at the University of Minnesota helps track annual population densities as well as aspects of biology and ecology, and participants in Cascadia are sought. *Danaus plexippus* is easily reared and often used as an educational tool in schools. There is immense public interest in this species, and the input from "citizen scientists" is being harnessed, particularly in E US, to increase our knowledge of this charismatic butterfly.

Monarch *Danaus plexippus* (L.)

Family Nymphalidae
Subfamily Satyrinae
SATYRS

There are 9 species of satyrs or "browns" in Cascadia. The sedentary adults are typically brown, tawny, gray, or dull orange, most with fine striations and eyespots on the wings. The wings have swollen veins at the bases which contain auditory sensors. Satyrs occupy a wide range of habitats from grassy lowlands and shrub-steppe to arctic-alpine tundra. All species develop on grasses, sedges, or rushes. Eggs are inflated-barrel shaped with vertical ribs that may be obscure (Ochre Ringlet) or prominent (arctics).

Eggs are laid singly on or near grasses and sedges, or dropped loose. Larvae are smooth, elongated, and tubular with short paired "tails," red in green species and tan in brown species. Development in annual and biennial species is very slow. Larvae are nocturnal, resting at the bases of grasses by day. They are green, brown, tan, or pinkish, with contrasting longitudinal stripes, providing excellent camouflage in a grassy habitat. Green larvae entering winter diapause turn brown. Small shiny eyes are prominent, usually black, but emerald green in *Cercyonis* spp. The head is granular in appearance, and setae on the body are mostly very short. There are usually 5 instars, although some species may have an extra instar under certain environmental conditions.

Pupae of most species hang, sometimes in a rudimentary shelter made from grass blades, and are compact with a squarish head and broadly pointed abdomen. Pupae of *Oeneis* spp. are formed loose on the ground or in a rudimentary cocoon.

The Ochre Ringlet is multiple brooded with a long flight period. Other species have a single generation that flies during summer. Females of 2 of the 3 wood nymph species spend a few weeks aestivating in midsummer before beginning oviposition. Overwintering occurs as unfed (wood nymphs) or as partly grown larvae, or as eggs (Vidler's Alpine). The 3 *Oeneis* spp. and likely Vidler's Alpine are biennial, requiring 2 winters to complete their life cycles.

Common Ringlet   *Coenonympha tullia* (Muller)

Adult Biology

Coenonympha tullia is the commonest satyr in Cascadia and a wide-ranging Holarctic species found across much of North America. It is found throughout Cascadia in low–high-elevation grassy habitats but is absent from some coastal areas. The Common Ringlet has a very long flight period, early April to late Oct, and is double brooded, triple brooded in some areas beyond Cascadia. Males patrol all day for females with a characteristic bouncing flight. Both sexes visit flowers, including dandelions, asters, and thistles. Grasses and sedges are the larval hosts for this species. Scott (2006) listed species of *Stipa*, *Bouteloua*, *Festuca*, *Bromus*, and *Carex* as hosts. He also reported observations of ovipositing females in CO, e.g., "female flew slowly then landed and crawled on litter and bent abdomen…then she flew 2 m to tiny *Stipa comata* clump and laid egg on dead leaf 4 cm above ground." Most ovipositions were on *Poa pratensis* (Kentucky Bluegrass), usually on the underside of a bent or horizontal dead grass blade, 3–6 cm above the ground (Scott, 2006).

Immature Stage Biology

We reared or partially reared this species several times from gravid females. Females obtained May 8 in Yakima Co., WA laid ~50 eggs which developed to pupation in 54 days (July 3) at 21–27 °C and natural daylengths. Females obtained May 10 in Kittitas Co., WA also produced second-brood adults. Eggs hatched after 6 days, and 8, 11, 12, and 25 days were spent in L1–L4, respectively, with pupation 62 days post oviposition (20–22 °C). A similar rate of development (60 days) was recorded for *C. tullia california* from N CA, reared at 20–28 °C during April–June. Eggs obtained from females June 23, Kittitas Co., hatched, but development was slow and L2 entered dormancy after ~57 days (late July). Eggs from Steens Mt, OR females on Aug 9 developed to pupation in 52 days at 21–27 °C and natural daylengths. These rearings suggest that direct development occurs during spring and early autumn but dormancy (aestivation) may intervene in summer. Populations of *C. t. california* in N CA are known to undergo summer dormancy as adults (Shapiro and Manolis, 2007). Eggs were laid on live grass or nearby inert surfaces. Larvae are extremely well camouflaged, green longitudinal stripes blending with grasses. A small proportion (~20%) of larvae in our E WA and CA rearings were brown throughout development and produced tan pupae. Nests are not constructed except for diapausing larvae, which build loose silken shelters. Scott (1986a) reported that diapausing larvae overwinter in a thick mat of dead grass. The survival strategy of *C. tullia* is based on camouflage, which this species has evolved to near perfection. There were 4 instars in all the cohorts we reared.

Description of Immature Stages

The egg is cream with faint vertical ribs. Distinctive reddish brown speckling and an equatorial brown ring develop with maturity. L1 is green anteriorly and yellowish posteriorly, with 7–8 narrow orange dorsolateral stripes. The head is dull brownish yellow with tiny bumps and a vertical median line. The tip of the abdomen terminates in a pair of "tails" that persist to pupation. Late L1 are greener and the stripes are whitish. L2 is green with a bluish cast. The tails are pale peach and a white lateral stripe is prominent. The bullae are white, giving a lightly speckled appearance to the body and head. In L3 the bold white lateral stripe is distinct and the ventrolateral sides below the stripe are bluish green while above the line the body is green. There are several obscure to distinct narrow pale white dorsolateral stripes along the body. A prominent middorsal dark green-black stripe is often present. L4 is a larger version of L3, and brown larvae are marked identically to green larvae. The pupa is yellow-green, often with several graceful contrasting curved dark brown streaks on and near the wing cases. These brown streaks are sometimes absent in E WA populations. Brown larvae produce tan-colored pupae, sometimes with dark brown streaks. The larvae of the wood nymphs (*Cercyonis* spp.) are similar to *C. tullia*; however, the brown-streaked pupa of *C. tullia* is unique. Among other satyrs the larvae of *Oeneis* spp. and *Erebia* spp. are brown, so may be confused with the brown morph of *C. tullia*. Some sulphurs (*Colias* spp.) have superficially similar green larvae; however,

Adult

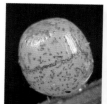

Egg @ 1.0 mm

Pupa @
10 mm (CA)

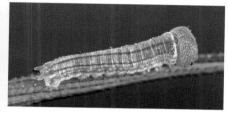

L1 @ 2.5 mm

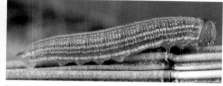

L2 @ 8 mm (late)

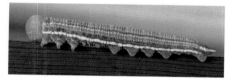

L3 @ 13 mm

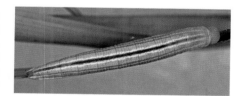

L4 @ 22 mm (ssp. *california*)

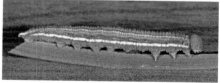

L4 @ 23 mm

they do not have the characteristic double tails of the satyrs, nor do they use grasses. Wagner (2005) and Allen et al. (2005) illustrated mature larvae, and Woodward (2005) and Guppy and Shepard (2001) each illustrated a mature larva and pupa; all these images agree closely with ours.

Discussion

Immature stages are extremely difficult to find; however, eggs can be readily obtained by caging gravid females or by watching females oviposit. Scott (2006) found it difficult to break diapause in overwintered larvae, and we had similar problems; however, rearing in spring or late summer usually avoids dormancy. Late spring–early summer larvae may aestivate and resume development in late summer, but this requires confirmation. Populations in N CA (*C. tullia california*) reproduce only in spring and autumn and undergo adult aestivation during summer (Shapiro and Manolis, 2007). It is possible that summer adult dormancy as well as larval dormancy also occur in inland populations in Cascadia. Autumn larvae enter dormancy, although it is unknown which instar overwinters in Cascadia. According to Scott (1986a), overwintering can occur in L1–L4. Further work is needed to determine what factors cause some populations to be single brooded and others to be bivoltine in Cascadia. Brown larvae accounted for ~10–20% of reared populations from E WA and N CA and may be related to the greater incidence of senescing grass in these areas. Only 4 instars occurred in our rearings, although 5 have been reported for some subspecies (Webster, 1998). It is possible that larvae with development interrupted by diapause have 5 instars, with direct-developing larvae having 4 instars, as reported for the related European species *Coenonympha pamphilus* (Garcia-Barros, 2006).

Common Ringlet *Coenonympha tullia* (Muller)

Common Wood Nymph   *Cercyonis pegala* (Fabricius)

Adult Biology

Cercyonis pegala is found throughout the US, N Mexico, and S Canada. It occurs throughout Cascadia except for most of the coast and NW counties of WA. A wide range of habitats up to ~7,000 ft are used, including prairies, meadows, roadsides, parks, forest openings, and dry gullies. A single brood flies from mid-May to late Sep. Males emerge first and live for 3–4 weeks; females live for >8 weeks. *Cercyonis pegala* is often scarce during the hottest periods in late July–early Aug. Females rest in the shade, aestivating in groups of 6–20, and do not feed or oviposit. In late Aug females resume activity and begin oviposition. Non-aestivating *C. pegala* visit a variety of flowers, including thistles, asters, blanket flowers, Oxeye Daisy, and teasels. Both sexes visit sap, and males visit dung, carrion, and mud. Various grasses have been recorded as larval hosts elsewhere, including *Poa pratensis* (Kentucky Bluegrass), *Danthonia spicata* (Poverty Oatgrass), *Tridens flavus* (Purpletop), *Avena fatua* (Wild Oat), *Festuca* spp., and *Andropogon* spp. (Scott, 1986a; Cech and Tudor, 2005; Allen, 1997; Guppy and Shepard, 2001). Eggs are laid singly on the host or nearby surfaces.

Immature Stage Biology

We reared or partially reared this species several times from gravid females. On July 11, 2 females from the Tucannon R, Columbia Co., WA were confined with *Setaria glauca* (Yellow Foxtail) in a plastic box with a muslin lid (36 × 23 × 12 cm) and held in sunshine at 22–26 °C under natural daylengths. The females did not feed or lay eggs for >2 weeks. Feeding recommenced and reproductive dormancy broke after 17 days, with ~50 eggs laid July 28–Aug 7. A second group of females from the same location on July 27 did not oviposit until Aug 20. Females captured and confined with grass on Aug 21 oviposited within 24 hrs, tucking eggs deep in a clump. Eggs hatched in 9–10 days, and unfed L1 diapaused. L1 were held at 18–26 °C until Oct, when overwintered at 5 °C. Eight L1 exposed to 20–22 °C on Jan 15 fed poorly, none surviving past L2. A cohort of 18 L1 was transferred to 25 °C and continuous lighting on Feb 23. Feeding commenced immediately and growth was rapid, with 9, 9, 5, 9, and 10 days spent in L1–L5, respectively. Adults eclosed 12 days later. On Aug 20, 3 females from Fish Lk, Chelan Co., WA, laid 40 eggs; unfed L1s were overwintered on Sep 2. Exposed to 18–20 °C on April 26 and reared on lawn grass, most larvae failed to feed and died. A few fed and pupated in ~80 days, with 26 days in L2 and ~13 days in each of the other instars. Larvae fed mostly at night on grass-blade edges, often spending the day at the base of the grass. No nests were made; survival is based on camouflage. There were 5 instars in our rearings and pupation occurred on the host, suspended from a bent-over stem or blade, sometimes encircled by silk strands. The unfed L1 overwinters at the base of host grasses, emerging to feed in March or April.

Description of Immature Stages

The egg is creamy white and finely reticulated, with 18–20 well-defined vertical ribs. The mature egg has distinctive reddish spots. L1 is light brown and purplish. Dorsolaterally there are 6 dark stripes and a reddish middorsal stripe. There are 6 longitudinal rows of backward-bent medium-length pale setae. Postdiapause L1 is green with dark stripes and a subspiracular white stripe. L2 is green with lines of white spots, indistinct dark stripes, and a faint subspiracular white stripe. There are 2 red-tipped tails on the posterior segment, and the head is green with white spots, both characters persisting until pupation. L3 is similar, grass green but lacking dark stripes. The subspiracular white stripe is distinct, and the body is clothed with white spots and short white setae, creating a whitish sheen. L4 has increased numbers of setae, giving a whitish green appearance with an indistinct middorsal dark stripe. L5 is yellowish green, densely clothed with white spots, and short white setae, with a dark middorsal stripe. There are 2 yellow stripes laterally, the lower one bolder. The pupa is grass green with vermiform darker green markings. The dorsal edge of the wing cases and lower edge of the pupa are lined in white and the cremaster is red. Larvae of *C. pegala* may be confused

Adult

Eggs @ 1.0 mm
(fresh, *above*,
maturing, *below*)

Pupa @ 15 mm

L1 @ 2.5 mm (prediapause)

L1 @ 3.5 mm (postdiapause)

L2 @ 6 mm

L3 @ 11 mm

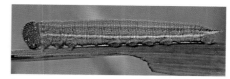

L4 @ 17 mm

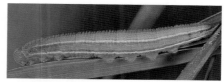

L5 @ 30 mm

L5 @ 25 mm
(head)

with other *Cercyonis* spp. and *Coenonympha tullia*; however, *C. sthenele* L5 have tan-green heads and those of *C. oetus* have more distinct striping. Many images have been published of mature *C. pegala* larvae, including those in Allen (1997), Guppy and Shepard (2001), Wagner (2005), Allen et al. (2005), and Minno et al. (2005); all agree closely with ours, as does the pupa in Allen (1997).

Discussion

Surprisingly little is known about the biology and ecology of this widespread and common species. It is easily reared (although overwintering can be difficult) and makes an excellent candidate for laboratory-based studies. Black and white pupal morphs, as described for *C. p. blanca* from NV (Emmel and Mattoon, 1972), should be looked for in Cascadia. Adult aestivation and summer reproductive dormancy need further study; it is one of only a few insect species with diapause in 2 of its life stages, similar to the Australian satyr *Heteronympha merope* (James, 1988, 1999). The occurrence of 5 instars is noteworthy; Emmel (1969) stated that *C. pegala* has 6 instars, although Emmel and Mattoon (1972) reported 5 instars for *C. p. blanca*. A study on instar number variability and its causes would be valuable. The larval host grasses used in Cascadia are currently unknown. Nothing is known about natural enemies and other mortality factors.

Common Wood Nymph *Cercyonis pegala* (Fabricius)

Great Basin Wood Nymph   *Cercyonis sthenele* (Boisduval)

Adult Biology

Cercyonis sthenele occurs in arid zones of the W US from S BC to N AZ, E to mid-CO. In Cascadia it is widespread and locally common E of the Cascades in BC, WA, and most of OR, extending into the Cascades and Siskiyous in S OR. Habitats include dry canyons, sagebrush rangelands, and juniper woodlands from near sea level to >5,000 ft. Adults fly in a single brood beginning mid-June (a month later than other *Cercyonis* spp.) until mid-Sep, according to elevation and seasonal conditions. Males emerge before females, which aestivate for a few weeks in midsummer, delaying oviposition until Aug. Males patrol for females, and both sexes visit a variety of flowers, including rabbitbrush, thistles, milkweed, yarrow, and sweetclover, sometimes in the company of the very similar *Cercyonis pegala* and *C. oetus*. Larval hosts are grasses; species used are undocumented, although *Poa* spp. and *Festuca* spp. were reported by Ferris and Brown (1981) and Miller and Hammond (2007), respectively. Eggs are laid singly at the bases of host plants or on nearby plants and are often deposited loose.

Immature Stage Biology

We reared or partially reared this species 3 times from gravid females. One female collected July 10, Grant Co., WA, aestivated for 2 weeks before oviposition. Sixty-four eggs were produced during 5 weeks, 2 hatching, 15 maturing (pink), and the remaining 47 collapsing (infertile). No eggs or larvae survived overwintering. Three females obtained on July 21 from Sun Lakes, Grant Co., were confined with *Setaria glauca* (Yellow Foxtail) in a plastic box with a muslin lid (36 × 23 × 12 cm) and held at 22–26 °C under natural daylengths.

The females were largely inactive, sought shelter, and did not lay eggs for 14 days. During the following 2 weeks, 138 eggs were laid, lightly glued on grass and inert surfaces at the bottom of the box. Five females obtained on July 24 from Columbia Co., WA behaved similarly and did not produce eggs for 20 days. Most eggs in both cohorts matured and appeared viable but failed to hatch. After 5–7 days a small number (~25) produced L1 which rested on dead grass and entered diapause without feeding. Eggs and L1 were overwintered at 5 °C on Sep 5. On Jan 3, 13 L1 and 14 eggs were transferred to 22–28 °C and 16 hr light, and L1 feeding commenced (on *S. glauca*) within 24 hrs. No additional eggs hatched. Development from L1 to adult eclosion took 110 days, with L1–L5 and pupa occupying 26, 10, 8, 12, 22, and 32 days, respectively. Larvae fed on blade edges, mostly at night, returning to the base of host grasses by day. No nests were made, and survival is based on camouflage. When larvae were reared together, cannibalism of an L3 by an L5 occurred. Pupation was on the hosts, suspended from a bent stem or blade sometimes encircled by silk. There are 5 instars, and unfed L1 overwinter under host grasses, emerging to feed in March–April.

Description of Immature Stages

The egg is creamy white with 16–18 vertical ribs. As the egg matures, distinctive red spots develop. L1 is tan to pink with 4 lateral and 1 middorsal red to dark brown stripes. There are 8 longitudinal rows of backward-bent white setae and the head is pink-tan with small dark spots. Postdiapause L1 is identical except green. L2 is grass green peppered with white spots, each bearing a single tiny white seta. There are 4–6 indistinct dark stripes, a prominent subspiracular white stripe, and 2 crimson-red–tipped "tails" which persist for the rest of development. The head is green with white spots. L3 is grass green with longer white setae, giving a slightly bristly appearance. There are a distinct middorsal and 2 dorsolateral dark stripes as well as a subspiracular white stripe. L4 is similar, but the subspiracular stripe is yellowish white and a pale dorsolateral stripe is present. L5 is yellowish green peppered with tiny white spots and setae and has prominent broad yellow-white stripes above and below the spiracles. There is a distinct broad middorsal black stripe bordered by white, and the head is greenish tan with white spots. The pupa is grass green with fine darker vermiculations. The dorsal edge of the wing case is lined in white and the cremaster is red. The larva of *Cercyonis oetus* is similar, but the stripes may be less prominent and the

Adult male

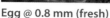

Egg @ 0.8 mm (fresh)   Egg @ 0.8 mm (mature)

L1 @ 2.5 (prediapause)

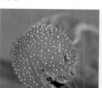

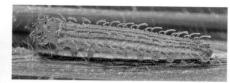

L1 @ 3 mm (postdiapause)

L4 @ 15 mm (head)

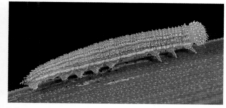

L2 @ 6 mm

Pupa @ 12 mm

L3 @ 11 mm

head has no tan overtones. The head of *C. pegala* is darker green, and L4 has indistinct middorsal dark and dorsolateral white stripes and a whitish sheen. Ours appear to be the first published color images of immature stages of *C. sthenele*. Black-and-white images appear in Emmel and Emmel (1973) and Sourakov (1995) and look similar to ours.

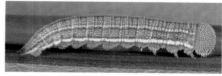

L4 @ 15 mm

Discussion

Our rearings support the hypothesis that adult female *C. sthenele* aestivate and are nonreproductive for a period in midsummer, similar to the closely related *C. pegala*. Females obtained in July were inactive and did not oviposit for 2–3 weeks. This may be a strategy to ensure that diapausing L1 are not exposed to the harshest conditions in midsummer, but aestival dormancy in *C. sthenele* and *C. pegala* needs further research. Many eggs in our rearings developed (as indicated by color change) then failed to hatch prior to or after overwintering. The possibility that the egg may also serve as the overwintering stage in *C. sthenele* (and *C. pegala*) under some conditions should be investigated. Post-overwintering larvae were easily reared on *S. glauca*, but the natural host grasses for *C. sthenele* remain unknown and need study. Natural mortality factors are also unknown.

L5 @ 22 mm

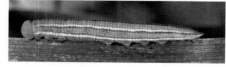

L5 @ 30 mm

Great Basin Wood Nymph   *Cercyonis sthenele* (Boisduval)

Dark Wood Nymph   *Cercyonis oetus* (Boisduval)

Adult Biology

Cercyonis oetus is widespread in W North America from S Canada to NV and E to CO. In Cascadia it is found E of the Cascades in various habitats including montane meadows, open woodland, shrub-steppe, canyons, dry hillsides, and roadsides. In many areas it is abundant, flying in a single brood from late May to mid-Aug (later farther S), according to elevation and seasonal conditions. Males emerge first, patrol for females, and sometimes visit mud. Both sexes commonly visit flowers, including thistles, milkweed, yarrow, buckwheats, rabbitbrush, and asters. Larval host plants are grasses, but the species are undocumented in Cascadia. Scott (1992) reported that *Poa pratensis* (Kentucky Bluegrass) and *Festuca idahoensis* (Idaho Fescue) are used in CO. Females do not oviposit until ~5 days old (Scott, 1986a) and eggs are laid singly on hosts or nearby plants and litter.

Immature Stage Biology

We reared or partially reared this species 5 times from gravid females. On 4 occasions females were collected between July 24 and Aug 3 from various localities in Kittitas Co., WA. In each instance good numbers of eggs were laid (>200 total) on inert cage surfaces, and larvae hatched after ~14 days. Eggshells were consumed, then L1 hibernated in groups under host plants. Most larvae overwintered but fed poorly on a variety of lawn grasses (*Festuca* sp., *Poa* sp.) in spring, with only a few reaching L2 before dying. Females obtained on July 16 from Pend Oreille Co., WA, confined in a plastic cylinder with muslin ends (13 cm dia × 8 cm tall) with *Elytrigia repens* (Quackgrass) laid ~50 eggs on the grass and inert surfaces. Eggs hatched in 10 days at 22–27 °C, and unfed L1 entered diapause, resting on dried grass and debris. L1 were held at 21–26 °C from early Aug until Sep 14 with little mortality, then transferred to 5 °C for overwintering.

After 7 weeks, 22 L1 were transferred to summerlike conditions (26 °C, continuous light) and commenced feeding on *E. repens* after 5 days. Overwintering mortality of L1 was negligible. Development was fairly rapid, with pupation after 55 days and 14, 8, 10, 8, and 15 days spent in L1–L5, respectively. Adults eclosed after 18 days. Larvae do not build nests and feed on the edges of grass blades in early instars, consuming entire blades in late instars. Larvae tolerate each other well and feed mostly at night, spending days resting low down on the host plant. Survival is based on camouflage, with larvae blending in very well with their grassy environment. Pupation occurs on grass stems or blades, adjacent blades sometimes pulled down with silk to form a slight protective "tent." Scott (1992) reported similar protection for *C. oetus* pupae in CO. There are 5 instars, and overwintering occurs at the bases of host grasses as unfed L1 that emerge to feed in March or April.

Description of Immature Stages

The finely reticulated oval egg is creamy white becoming lemony yellow with 20–22 vertical ribs. As the egg matures, distinctive red spots develop. L1 is light brown to orangish with 4 lateral and 1 middorsal dark brown stripes. White, backward-bent setae arise from black bullae in longitudinal rows. The head is dark brown with many darker markings and spots. Postdiapause L1 is identical except grayish or green. L2 is light green peppered with tiny white setae arising from black or white spots. There is a prominent white subspiracular stripe and indistinct lateral (3) and middorsal dark stripes. There are 2 crimson red-tipped "tails" on the posterior segment and the head is green with white spots, both features persisting for the rest of development. L3 is grass green and similarly marked. Transverse folds on each segment are more developed, giving the larva a more "wrinkled" appearance. L4 is yellowish green with a distinct middorsal dark green stripe and a prominent subspiracular white stripe. A less prominent white stripe is present dorsolaterally. L5 is yellowish green peppered with tiny white spots and setae and a prominent subspiracular yellow-white stripe. There is an indistinct middorsal dark stripe. The pupa is grass green with fine inconspicuous dark markings. The dorsal edge of the wing cases and lower edge of the pupa are lined in white, and the cremaster is red. The larva of *C. sthenele* is most similar, but the middorsal dark and dorsolateral white stripes are more prominent and the head is greenish tan. The mature larva of *C. pegala* has indistinct middorsal

Adult

Egg @ 1.0 mm (fresh, *left*, mature, *right*)

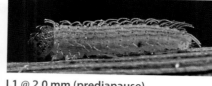

L1 @ 2.0 mm (prediapause)

Pupa @
12 mm

L2 @ 5 mm

L3 @ 8 mm

L4 @ 14 mm

L5 @ 23 mm

Dark Wood Nymph *Cercyonis oetus* (Boisduval)

dark and lateral white stripes and a whitish sheen. Our images appear to be the first published of immature stages of *Cercyonis oetus*.

Discussion

Adult *Cercyonis oetus* do not appear to aestivate as adults of the closely related *C. pegala* and *C. sthenele* do, and eggs are readily obtained from gravid females during most of the flight period. Oviposition will occur even in small containers and without sunshine as long as grass is available. Larvae of *C. oetus* are difficult to find, as is the case with most grass-feeding species, which feed primarily at night. Dormant L1 appear to be hardy and overwinter well in captivity. We overwintered L1 for only 7 weeks at 5 °C, but this was sufficient to break diapause and allow rapid development when subsequently exposed to warm temperatures, continuous lighting, and grasses with new growth. We successfully used *E. repens* as the larval host plant, but the range of grass species used by *C. oetus* in Cascadia needs to be determined. Pyle (1981, 2002) reported the occurrence of brown and black (with light stripes) pupae; we did not see these in our rearing, but they should be looked for. Brown larval morphs may also occur in *C. oetus* as they do in *Coenonympha tullia*. This is a very successful species in most parts of its range, but little is known of habitat and climate requirements. Scott (1992) reported parasitic wasps emerging from a pupa of *C. oetus*, but little else is known regarding natural enemies and other mortality factors.

Common Alpine   *Erebia epipsodea* Butler

Adult Biology

The Common Alpine ranges from AK through the W half of Canada, S along the Rockies to NM. In Cascadia it is found in SE BC, NE WA and the WA Cascades, SE WA, and NE OR. The single generation flies May–Aug in moist montane habitats, including meadows and low-elevation grasslands in BC (Guppy and Shepard, 2001). In WA and OR, *E. epipsodea* is found from 2,000 to 8,000 ft (Warren, 2005). Both sexes visit flowers, and males visit mud or dung. Males patrol circular routes (Brussard and Erhlich, 1970b) while females tend to remain hidden in grass, resting or ovipositing. Sunshine and 16–18 °C is necessary for flight and both sexes spend much time basking (Brussard and Erhlich, 1970b). Larval hosts are various grasses and sedges, although preferences have not been determined. Larvae have been reared on *Poa* spp. (bluegrass) (Brussard and Erhlich, 1970a; Scott, 1992; Pyle, 2002). Scott (1992) also witnessed oviposition on *Poa pratensis* (Kentucky Bluegrass) and thought this a likely host in CO. In the Blue Mts of SE WA, *E. epipsodea* appears to be associated with *Setaria* spp. (foxtail). Eggs are laid singly on grass or nearby surfaces, and females walk through grass selecting oviposition sites.

Immature Stage Biology

We partially reared this species several times. Females collected June 7, 21, and 23 (different years) in Okanogan Co., WA, laid 30–100 eggs, but larvae failed to survive past L1 or L2. Four females from near Sinlahekin R, Okanogan Co., WA, (May 27) produced 80 eggs, ~75% on live grass, the remainder on inert surfaces. Larvae were reared on live potted *P. pratensis* in a fabric sleeve at ~15–25 °C, natural daylength, and 7 reached L3 7 weeks post egg-hatch and entered diapause. They were overwintered at 4 °C on Sep 20, but none survived. Females (2–4) were obtained on May 17, June 4, 11, 12, and 19 (different years) from near the Tucannon R, Columbia Co., WA

and laid 5–170 eggs on each occasion. Five females from Apex Mt, BC, on July 27 and 1 female from Mt Hull, Okanogan Co. on June 23 laid 60 and 43 eggs, respectively. All females were confined in muslin-covered plastic boxes (31 × 12 × 23 cm) with cut grass *Elytrigia repens* (Quackgrass), *Poa annua* (Annual Bluegrass), *Cynodon dactylon* (Bermuda Grass) and *Setaria* sp., and held in sunshine at 22–28 °C. Eggs hatched in 8–10 days at 25–27 °C. Larvae were fed on the above grasses and completed L1 in 10 days. The Tucannon larvae spent 30–35 days in L2 before molting to L3 and entering diapause. The Apex Mt and Mt Hull larvae spent 10 days in L2 and 8 and 16 days in L3, respectively, before most larvae overwintered as early L4. All Tucannon cohorts died prior to or during overwintering (5 °C). One Mt Hull larva was exposed to warm conditions after 81 days but died quickly. Apex Mt larvae (*n* = 20) were exposed to 26 °C and continuous lighting after 68 or 80 days and immediately fed on *Setaria* sp. One larva molted to L5 after 11 days. After ~8 days, the L5 stopped feeding and reentered dormancy. Larvae feed nocturnally, spending the day resting at the bases of grasses. Our larvae diapaused as L3 or L4, with none completing development prior to overwintering as recorded by Scott (1992) for lab-reared larvae. Diapausing larvae seek refuge at the bases of grasses, resting on dead blades. There are 5 instars (although Lyman, 1896, recorded 4 instars in larvae that did not overwinter), nests are not constructed at any stage, and camouflage appears to be the primary defense. Spiders and predatory mites (Cheyletidae) killed many larvae when potted grass was used for rearing.

Description of Immature Stages

The egg is white with ~22–24 subdued vertical ribs. L1 is tan to pinkish with 3 brown lateral stripes and a dorsal stripe. Each segment has 8 dark bullae, each with a very small seta. L2 is dark greenish tan with alternating thin red-brown and thick white stripes. L3 is similar but greener, especially anteriorly. The white stripes are bolder and there are an increased number of pale setae. There are 2 indistinct "tails." The prolegs on the posterior segment are large. L4 is pinkish tan with a prominent middorsal black stripe. The less distinct lateral white stripes are bordered below in black. L5 is similar, with more distinct black edging to the lateral white stripes and a very prominent middorsal black stripe. Profuse short pale setae arise from tiny white bullae, giving an overall granulated appearance, and the tails are short. The setaceous head is greenish tan. *Erebia epipsodea* is similar to *E. vidleri*, although small

Adult

Egg @ 1.3 mm (fresh)   Egg @ 1.1 mm (mature)

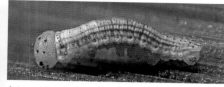

L1 @ 3.5 mm (prefeeding)

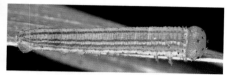

L1 @ 3.5 mm (after feeding commenced)

L2 @ 9 mm

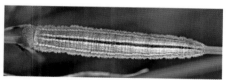

L3 @ 13 mm (late)

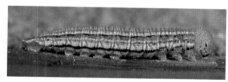

L4 @ 12 mm (early)

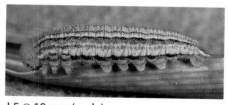

L5 @ 18 mm (early)

Common Alpine *Erebia epipsodea* Butler

differences occur in most stages; for example, the number of ribs in the egg differs and the L1 head of *E. vidleri* has numerous speckles whereas that of *E. epipsodea* has 6–8 dark spots. The larvae of *E. vidleri* have well-developed "tails" while those of *E. epipsodea* are short. Head color, setae, and prominence of the dorsal stripe differ in L5. Images of immature stages appear in Guppy and Shepard (2001) (egg and L1) and Allen et al. (2005) (L5). All appear similar to ours except for the larva in Allen et al. (2005), which is much darker.

Discussion

This is a difficult species to rear. Obtaining eggs from gravid females is relatively easy, but mortality of larvae is high during early instars and it is difficult to overwinter larvae. Moist conditions during rearing on live grass were highly detrimental to survival of early instars; however, high relative humidity is required for overwintering, while too much causes mold. We were unable to avoid diapause, but rearing early instars under increasing daylengths or continuous lighting might allow uninterrupted development. Larvae broke diapause readily on exposure to warm conditions and developed to L5 but then stopped feeding and reentered dormancy. Exposure of these larvae to continuous lighting may have caused the second dormancy by mimicking midsummer conditions. This suggests that overwintered larvae with development delayed by poor spring conditions may have the ability to overwinter a second time. Alternatively, this species may be biennial in Cascadia. Host-grass preferences are unknown and need study. Thermal requirements for *E. epipsodea*, which is not as alpine-limited as other *Erebia* spp., may be less restrictive but need study. Natural mortality factors, including the impact of predators and parasitoids, are unknown.

Vidler's Alpine   *Erebia vidleri* Elwes

Adult Biology

Erebia vidleri is a Pacific Northwest endemic, found only from central BC S to central WA. It occurs on both sides of the Cascades in WA, the Olympic Mts, and N into BC, but is absent from Vancouver Is. Flying in higher-elevation habitats from 3,500 to 8,300 ft (Pyle, 2002), it is found in moist alpine and subalpine flowery meadows, in forest openings, and near rockslides and seeps, late June–early Sep. Larval hosts are not reported but are likely various grasses and sedges; Guppy and Shepard (2001) report an association with *Calamagrostis rubescens* (Pinegrass). Adults nectar on arnica, yarrow, dandelion, and other subalpine composites (Pyle, 2002). Females lay eggs singly, gluing them to grasses or nearby inert surfaces. *Erebia vidleri* is reported to be single brooded (Guppy and Shepard, 2001); however, our rearings strongly suggest it is biennial. As *E. vidleri* is found in both odd and even years, this suggests that there are 2 temporally separated populations in Cascadia. Other biennial species of *Erebia* are known from BC (Guppy and Shepard, 2001).

Immature Stage Biology

We reared this species to L5 twice. Females obtained July 27 near Slate Peak, Okanogan Co., WA, produced 37 eggs during 10 days, gluing them to cage surfaces. The eggs were held at 20–25 °C for a month then stored at ~4 °C for overwintering. The eggs were moved to 18–20 °C, natural daylength, on April 3 and began hatching within 24 hrs. Larvae were reared on lawn grass (*Poa* sp.), which was readily accepted; development was very slow, taking 3 months to reach L5 (July 3) before the cohort died. On July 22 (different year, same locality) 3 females produced ~90 eggs, glued to live grass blades. After 2 weeks the eggs showed embryonic development and

Habitat: Slate Peak, Okanogan Co., WA

were overwintered at ~4 °C. Some were overwintered on potted fescue grass outdoors. The eggs were removed from overwintering in 4 time cohorts. On Feb 10, half a cohort of 21 hatched (<4 hrs) and the larvae fed lightly on *Poa* sp. and *Festuca* sp. at 20–22 °C. Development was slow, and all died within 5 weeks. On Feb 19, 30 eggs, overwintered outdoors, were exposed to 18–20 °C. Half hatched after 11 days, but feeding was poor and all died within 3 weeks. On March 22, 9 eggs were exposed to 20–22 °C and hatched within 2 hrs. Again, feeding was poor, and all died within 2 weeks. The final cohort (*n* = 13) of eggs was removed from overwintering on April 9; egg-hatch was very rapid, 3 hatching prior to removal from refrigeration. Larvae were reared at 20–22 °C on potted *Stipa tenuissima* (Mexican Feathergrass). Development was relatively slow; some reached L3 within a month and 3 attained L5 by early June. In mid-June feeding ceased and the larvae sought refuge under the cage liner, where they shrank and darkened. On June 30 the 3 survivors were overwintered again. Two larvae were exposed to room temperature (18–20 °C) in mid-Dec and provided with live *Festuca* sp.; they were active and fed minimally but died after several days. The final larva succumbed to mold during overwintering. L1 do not eat their eggshells. Larvae feed nocturnally, resting in shaded places at the base of the host plant during the day. Larvae nibble the tips of grass blades, often head-down as the blades bend under their weight. Grass blades may be consumed from the tip or scalloped from the side to the midrib. In later instars, blade tips are often cut off during feeding, falling to the ground. There are 5 instars, nests are not constructed, and camouflage appears to be the primary defense. Our rearings indicate that *E. vidleri* is biennial, with the first winter spent as a diapausing

Adult

Eggs @ 1.1 mm (fresh, *left*, mature, *right*)

L1 @ 2.2 mm

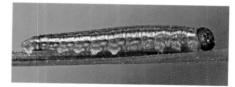

L2 @ 5 mm

L3 @ 10 mm

L4 @ 14 mm

L5 @ 21 mm

L5 @ 21 mm

egg and the second as a diapausing mature or nearly mature larva. Egg dormancy of 7–8 months appears necessary for development.

Description of Immature Stages

The egg is white with ~14 widely spaced subdued vertical ribs. L1 is bright pink with 3 red lateral stripes and 1 stronger dorsal stripe. Each segment has 8 dark bullae, each with a very small pale seta. The slightly bifurcated head bears short white setae and is red with numerous dark speckles. L2 is bronzy pink with alternating red-brown and white stripes. L3 is heavily striped in alternating brown, black, and white, and covered with bristly pale setae. Two strong "tails" persist from L3 to pupation. The head, tail, and legs are pale whitish, and a broad white lateral stripe is bordered below in black. The head bears numerous pale tubercles, and there are large white prolegs below the abdomen tip. The striping in L4 is less contrasting and the setae less bristly; however, the same pattern persists and the head is pale brown. L5 is pinkish with a purple-brown bifurcated head. Dark setae cover the body, with brown setae on the head. A distinct dorsal stripe is brown behind the head, becoming black posteriorly. The pupa is not known. *Erebia epipsodea* is similar to *E. vidleri*, although small differences occur in most stages, e.g., egg and L1. We are unaware of any published images or accounts of the immature stages of *E. vidleri*.

Discussion

The immature stages of *E. vidleri* need further study. We suggest that *E. vidleri* is biennial, but additional rearing and field observations are needed for confirmation. *Erebia vidleri* appears to be the only North American alpine known to overwinter as an egg (first winter). Study is needed to determine to what extent separate year populations exist in the same areas and how they are interrelated. Like all biennial species, *E. vidleri* is difficult to rear; moist conditions are needed for overwintering eggs as relative humidities <70% cause desiccation. Larvae are difficult to find and the host grasses are unknown. Nothing is known of population dynamics and the factors involved.

Vidler's Alpine *Erebia vidleri* Elwes

Great Arctic   *Oeneis nevadensis* (Felder & Felder)

Adult Biology

The Great Arctic is a strictly NW species, ranging from the BC coast range and Cascades S to N CA. In N WA it occurs E of the Cascades, but from central WA S through OR it is found on both sides of the Cascades and Siskiyous, also in outlier populations in OR. *Oeneis nevadensis* is generally common and flies in a single generation from early May to late Sep in forested areas, favoring meadow edges, forest roads, rocky hills, and forest openings from sea level to subalpine areas up to 8,000 ft in OR (Warren, 2005). It is biennial, flying in alternate, usually even-numbered years. Odd-year sightings occur, however, and established odd-year populations exist on Vancouver Is, interior BC, and some parts of OR. Larval host plants are not reported but are probably grasses (which larvae eat in captivity) and sedges. Little is known of the oviposition behavior, but Neill (2007) reported that females stick eggs to blades of grass or nearby structures close to the ground.

Immature Stage Biology

We partially reared this species several times, but none of our cohorts survived beyond L3. Keith Wolfe supplied images of *O. n. iduna* L5 from CA and contributed to this discussion. Females obtained from Derby Canyon, Chelan Co., Rimrock Lk, Yakima Co., and Liberty, Kittitas Co., WA on June 16, July 2, and July 2, respectively (different years), laid 45–70 eggs, mostly on inert surfaces. Larvae hatched in 12–13 days and were reared on *Poa* sp., reaching L2 and diapausing, but none survived overwintering. One female obtained from Rimrock Lk on June 24 laid 5 eggs in 3 weeks. Another female from the same location on July 21 laid ~80 eggs in 4 days. Eggs hatched in 11 days at 22–26 °C, and L1 fed on *Elytrigia*

Adult

repens (Quackgrass), molting to L2 after 28–30 days. Thereafter, feeding slowed and larvae were dormant by the end of Sep. L2 were overwintered at 5 °C for 58 days with 70% survival. Feeding resumed on *E. repens* within 24 hrs at 15–21 °C, 13 hr light. After 6 days larvae were transferred to 27 °C and continuous light, molting to L3 after 12–14 days. Feeding stopped after 16–18 days and larvae reentered dormancy. All larvae died during this second dormancy. L1 partially or completely consumed eggshells after hatching. Larval growth was slow and larvae were lethargic, spending considerable time resting on grass blades and brown vegetation, or beneath cage liners. Feeding was on grass tips or edges near the tips, and no nests were made. CA larvae were supplied with *Carex* and *Scirpus* sedges (Cyperaceae) and *Calamagrostis* and *Imperata* grasses (Poaceae); larvae fed on all plants and were reared on *Scirpus altrovirens* (Green Bulrush). Development of 1 individual from egg-hatch to L5 took 48 days, with ~12 days in each instar, but the larva died before pupating. Other larvae survived >106 days post egg-hatch, but none pupated and all died. Later instars mostly fed nocturnally, resting or hiding during the day at the base of the host plant, usually head downward. In our rearings the overwintering stage appeared to be L2; Scott (1986a) reported that hibernation occurs as L2–L3 the first winter and as L5 the second winter, pupation occurring the following spring. The CA cohort developed to L5 without diapause. Defense is based on concealment and camouflage.

Description of Immature Stages

The whitish egg is ornamented with ~25 intricately branched vertical white ribs. L1 is pinkish white, often greenish with feeding, with 5 brown to reddish pink stripes dorsally. There are 2 "tails" posteriorly, and the head is tan-pink. Eyes are shiny black and persist to pupation. L2 is variegated with stripes of cream, tan, whitish, red, and black, a dark brown-black lateral stripe being the most prominent, especially on the posterior half. The head is cream-tan and granular and has 6 vertical brown stripes. L3 is similar, the "tails" remaining prominent and the vertical head stripes faint. The black lateral stripe is bold, and below it are 2 thin red lines enclosing black spiracles. L5 (CA) is generally similar, except the colors are paler, grayish with a pink cast dorsally. There is a distinct narrow black dorsal stripe. A broad lateral stripe, blackish posteriorly but fading anteriorly, is bordered above and below with broad pale stripes. The grayish head bears short setae and has indistinct vertical dark

Egg @ 1.3 mm (fresh, *left*, mature, *right*)

L1 @ 3.5 mm

L1 @ 7.5 mm (late, premolt)

**L2 @ 9 mm
(head)**

L2 @ 9 mm

stripes. The spiracles are black. Below the "tails" a pair of large prolegs is prominent in all instars. The larvae of *Erebia* are similar but differentiated by their lack of head stripes, and eggs are less ornamented. *Oeneis chryxus* appears to have pinker post-L1 instars than *O. nevadensis*; the larvae of *O. melissa* are unknown. Allen et al. (2005) picture a CA larva more reddish than ours, and Neill (2007) shows an egg and L5 similar to ours.

Discussion

It is not difficult to obtain females or eggs of this species; however, it is difficult to successfully rear the larvae beyond the first diapause. Immature stages are exceedingly difficult to find. The life history of this species is imperfectly known, and it needs to be reared and studied through a normal 2-year cycle to pupation. The possibility of newly eclosed females entering reproductive dormancy should be investigated. Larvae will feed on a variety of grasses and sedges; however, natural host plants are unknown, and careful field observations of ovipositing females are needed. The environmental or genetic factors responsible for annual broods in some locations deserve study. Nothing is known of the natural enemies or other mortality factors regulating populations of this species.

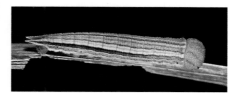

L3 @ 13 mm

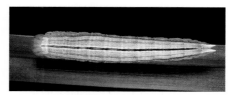

L5 @ 31 mm (photo K. Wolfe)

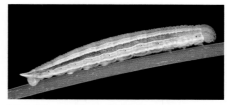

L5 @ 30 mm (photo K. Wolfe)

Great Arctic *Oeneis nevadensis* (Felder & Felder)

Chryxus Arctic   *Oeneis chryxus* (E. Doubleday)

Adult Biology

Oeneis chryxus ranges across North America from AK to Quebec and S to WI. In the W the range extends S through BC, WA, ID, and MT to NM, with outlying populations in CA. In Cascadia it occurs in SE BC and NE WA, with isolated populations in the Olympic Mts and S Cascades. It is found in dry meadows and grasslands, open pine forest, logging road shoulders, and upland sage bunchgrass habitats. *Oeneis chryxus valerata*, found in the Olympic Mts, is being considered for federal listing. Host plants include *Poa* (bluegrass) and possibly *Festuca idahoensis* (Idaho Fescue) (Pyle, 1981, 2002); in CA it is often associated with *Carex spectabilis* (Scott, 1986a). *Carex rossii* was found to be the most-used host in CO, and 4 other *Carex* spp. and *Poa pratensis* were also reported as hosts (Scott, 2006). The grasses *Danthonia spicata*, *Oryzopsis pungens*, and *Phalaris arundinacea* have been reported elsewhere (Scott, 1986a; Douglas and Douglas, 2005). The flight period is early June–mid-Aug (Pyle, 2002), and in Cascadia it is biennial. Flights in odd and even years represent 2 temporally separated populations. Adults sip mud and males perch all day, usually on hilltops or ridgetops to await females (Scott, 1986a); nectaring occurs on Pearly Everlasting, paintbrush, Showy Phlox, geranium, and Puccoon (Pyle, 2002). In CO, Scott (1992) found *O. chryxus* eggs on dead or living lower branches of Ponderosa Pine, with *C. rossii* growing below.

Immature Stage Biology

We partially reared this species 3 times; however, our cohorts failed to survive beyond L1. Discussion and images of *O. chryxus* beyond L1 are courtesy of Keith Wolfe, who reared CA populations. Females collected June 17 and 22 (different years) at Mt Hull, Okanogan Co., WA, laid 30 and ~100 eggs, respectively. Eggs

hatched in ~15 days and were fed *Poa* spp. at 20–22 °C, under natural daylengths. The June 17 cohort fed poorly and died, while larvae in the June 22 cohort fed for 13 days, then ceased; 14 individuals entered diapause as late L1 but died from mold during overwintering. Ten worn females collected from Apex Mt, BC on Aug 4 oviposited within hours when confined with grass. Eggs were laid mostly on grass but also on *Buddleia* flowers. Egg-hatch was over a relatively lengthy period beginning after 12 days, most hatching in 24 days (Sep 1). Feeding was poor and many L1 died, but 16 mature L1 were overwintered at 5 °C on Sep 18. Mortality was high (75%), and larvae were removed from overwintering after 89 (*n* = 3) and 107 (*n* = 1) days. All larvae failed to break dormancy and died after a few days. *Oeneis chryxus ivallda* females captured June 30, Mono Co., CA, confined with 3 species of Cyperaceae and Poaceae, laid eggs on inert surfaces. Eggs hatched after 13 days, and L1 were maintained under normal daylengths at 20–22 °C. Eggshells were partially consumed. Larvae were supplied with 5 potential hosts: 2 species of *Carex* (sedge), *Cyperus* (sedge), *Imperata*, and *Stenotaphrum* (both Poaceae). Larvae readily accepted all except *Stenotaphrum* and were reared on an unidentified *Carex* sp. from the collection locality. Larvae fed during the day or night and rested on blades near the base of the sedge. L1 fed from the edges of blades, while L4 and L5 ate blades starting slightly below the growing tip, leaving severed apexes scattered around the sedge base and numerous evenly clipped, slightly angled blades throughout the *Carex* clump. Larvae reached L5, with 1 pupating 88 days after egg-hatch. Pupation occurred unattached on the ground, the adult eclosing after 14 days. An *Oeneis chryxus stanislaus* female captured July 26 in Mono Co., CA, produced eggs that hatched after 10 days. Larvae were offered 5 sedges and 2 grasses; all the sedges were accepted. Larvae were reared on potted *Scirpus atrovirens*, also fresh cuttings of *Carex stricta* and *Carex spissa* under natural daylengths, developing to L5 in 63 days. The L5 died without pupating. Rearing *O. c. valerata* in WA, Pyle (2002) reported diapause in L1. His cohort was "overwintered" for only 2 weeks then reared to L4 on *Poa* when a second diapause occurred; the larvae died during the second overwintering. *Oeneis chryxus* appears to diapause as late L1 in the first winter, then as L4 or L5 in the second winter. In captivity in CA, *O. c. ivallda* developed directly to adulthood, and *O. c. stanislaus* grew to L5 without diapausing. There are 5 instars and no nests are constructed; protection is based on camouflage.

Adult

Eggs @ 1.6 mm

L1 @ 3.2 mm

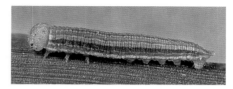

L1 @ 6 mm

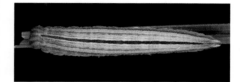

L5 @ 26 mm

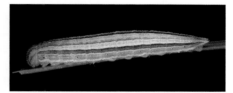

L5 @ 26 mm

Pupa @ 15 mm (photo K. Wolfe)

Chryxus Arctic *Oeneis chryxus* (E. Doubleday)

Description of Immature Stages

The egg is white ornamented with ~20 intricately branched, vertical white ribs separated by gray zigzag lines. L1 is pink or pinkish brown with darker magenta-pink longitudinal lines dorsally and dorsolaterally and a broad supraspiracular stripe. There are exaggerated pink prolegs below the posterior segment, which has paired pink "tails." Late L1 is darker with white subspiracular stripes. Some mature L1 are tan-brown with a green tinge anteriorly, and the lines and stripes are darker brown and white. In both forms dark eyes are present on the granulated pink-tan head. L5 is pinkish to tan or brownish, with fairly short paired tails. The bifurcated head bears 6 distinct brown to black vertical stripes and the head and body have irregular small untidy setae. There is a black dorsal stripe and another above the spiracles, plus a narrow dark dorsolateral line. The pupa is brown with extensive charcoal black on the head and wing cases. There are heavy black spots on the abdomen, sparse on the thorax. *Oeneis nevadensis* is very similar and may not be reliably separated. Pyle (2002) presented an image of a mature larva from WA that is similar but less pink than ours, and Emmel et al. (1992) published a black-and-white image of a mature larva from CO.

Discussion

Obtaining eggs of *O. chryxus* from gravid females is easily accomplished, and rearing L1 to overwintering is not difficult. Overwintering L1 appears to require a narrow range of relative humidity for optimal survival. RH below ~70–80% results in significant desiccation, and >90% sometimes causes mold. The biennial life history of *O. chryxus* presents significant challenges for rearing, and immature stages are rarely found in the wild. Very little is known of the life history, natural enemies, and other environmental factors influencing population dynamics. Miller and Hammond (2007) considered *O. chryxus* sensitive to climate change; thus it is important that the environmental requirements of this species be determined.

Family Hesperiidae
Subfamily Pyrginae
SPREADWING SKIPPERS

There are 11 spreadwing skippers in Cascadia. Colored in grays, browns, and blacks and often checkered or marked with white or yellow, many are mothlike, particularly the duskywings. Most rest with their wings outspread and flat, giving them their popular name. A wide range of habitats are occupied, from disturbed lowland sites to alpine tundra. About half are widespread and common, the remainder more localized.

Eggs are inflated-barrel shaped, many wider than high, often with very coarse vertical ribs in a variety of patterns. Usually green or white when first laid, some duskywing eggs mature to bright orange or red. Eggs are laid singly, usually on young leaves of host plants.

Larvae are elongate and cylindrical, mostly green or brown with wide black, brown, or orange heads. A cervical shield behind the head, often shiny and contrasting, is characteristic. Setae are short, often T- or Y-shaped. Head markings provide an excellent diagnostic feature for duskywing larvae. Premolt larvae withdraw their head from the old head capsule into the collar area, causing a characteristic temporary swelling. Larvae of all species construct shelters by folding leaves or parts of leaves with silk. Feces are usually propelled away from nests to put predators off the scent.

Pupae are formed horizontally, typically secured with posterior silk threads and a silk strand around the middle. Pupae have a broadly rounded head and a pointed posterior tip, some with dense setae, but other species are smooth.

Larval hosts are a wide variety of leafy plants, including *Lotus*, vetch, pea, willow, aspen, birch, oak, *Ceanothus*, lupine, Alfalfa, cinquefoil, strawberry, mallow, chenopods, and others. Most spreadwing skippers overwinter in the final instar or as a pupa. Those that overwinter as a mature larva usually pupate without feeding the following spring. All our spreadwings are single brooded except 3 which are multiple brooded. Single-brooded species may develop a second brood when reared in captivity.

Silver-spotted Skipper   *Epargyreus clarus* (Cramer)

Adult Biology

Epargyreus clarus occurs throughout most of the US and S Canada. In Cascadia it is found in SW BC, W of the Cascades in WA and OR, also disjunctively in E WA (S Columbia Basin, lower Blue Mts) and NE OR. It is our largest skipper and can be locally common in some areas. It is absent from higher elevations, preferring low–mid-elevation open flowery areas, scrub-heath, and riparian habitats. The single brood flies early May–late July. Flight is very fast; however, adults nectaring on dogbane, *Lotus*, and teasel are easily approached. Scott (1986a) listed >25 larval hosts. In Cascadia primary hosts are *Lotus crassifolius* (Big Deervetch) W of the Cascades, and *L. nevadensis* (Nevada Deervetch), *Robinia* (locust), and *Glycyrrhiza lepidota* (Wild Licorice) on the E side. Scott (2006) reported that *E. clarus* uses some of the largest legumes as host plants in CO, some tree-sized. In mid–late June near Hood Canal, Mason Co., WA, females fly among *L. crassifolius* plants, where they lay their eggs singly on small terminal leaves, usually the underside of the topmost leaf on the plant, typically 1 egg per plant.

Immature Stage Biology

We reared this species several times from field-collected eggs and larvae. On June 30, July 8, and July 8 (different years) on the Tahuya Peninsula, Mason Co., we found ~150 occupied larval shelters, containing L1–L4. The simultaneous presence of multiple instars indicates an oviposition period of >2 weeks. Larvae were collected and reared on *L. crassifolius* at 20–22 °C under natural daylengths. L4 and L5 refused *Lotus corniculatus*. Many pupated, and some eclosed to second-brood adults (50% in 1 cohort), the remainder overwintered. On hatching, L1 cuts a small terminal leaf twice, both cuts on the same side and perpendicular to the midrib, and folds the flap over the upper surface, silking it in place to form a shelter. See Neill (2007) for a sequence of images showing

this behavior. In L2–L3 the larva silks 2 leaves with flat sides together, and in L4–L5, 3 or more leaves are tied together. Larvae spend most of their time resting in the nest, typically with head recurved, parallel to the body. The larva leaves the nest to feed, mining leaf surfaces in L1, consuming entire leaves thereafter. The mature larva continues feeding lightly for about a week without further growth, then constructs a new, roomy nest of multiple leaves, turns pink, and pupates. The pupa is suspended horizontally from the ceiling of the shelter with 3 silk threads, 2 extending horizontally in opposite directions from the cremaster and another slung loosely under the middle of the pupa, tied to the ceiling on either side. Development from mid-L3 to pupation took 18 days; development from egg-hatch to pupation is estimated at ~30 days based on partial rearings and observations. Adults eclosed from nondiapause pupae in 13–14 days at 22–28 °C. There are 5 instars, all of which construct leaf nests, and the pupa overwinters. Protection is based on concealment, nocturnal feeding, frass ejection, a large red ventral gland on the first segment, and head twitching when disturbed. Larval coloration may be aposematic, warning predators of toxicity or distastefulness. Larvae are solitary, objecting to the presence of other larvae with violent head twitching; however, we did not witness cannibalism. Weiss (2003) showed that frass ejection from leaf shelters (up to 40 body lengths at speeds of 13 m per sec) serves to eliminate chemical cues for natural enemies.

Description of Immature Stages

The egg is green with a red top (Scott, 1986a; Emmel and Emmel, 1973), becoming orange, and has 15–16 vertical pale ribs. L1 is yellow-orange with a black head, sparsely covered with short Y-tipped setae. There is a constriction behind the head with a shiny black collar, a character that persists to pupation. L2 is greenish gold; the head is cleft dorsally, giving an inflated heart-shaped appearance that persists to pupation. L3 is green-gold banded transversely with olive-green spots and streaks. In L4 a transverse black bar and broken black lines occur on each segment. Bright orange false eyespots appear on the black head; the first segment (with ventral gland) is red-orange with a black collar dorsally, and the legs and prolegs are yellow. In premolt L4, the false eyespots temporarily disappear, reappearing in L5. L5 is similar, with more distinct black markings. The pupa is torpedo shaped, bluntly rounded at the head with a stemlike projection at the rear, cryptically patterned with vermiform markings in reddish brown. No other

Adult

Egg @ 1.0 mm (mature)

L5 @ 32 mm (head)

L1 @ 3.3 mm

Pupa @ 23 mm (suspended, viewed from below)

L2 @ 5.5 mm

L3 @ 10 mm

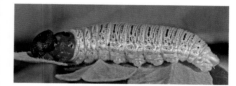

L4 @ 18 mm (premolt)

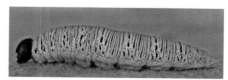

L5 @ 37 mm

Silver-spotted Skipper *Epargyreus clarus* (Cramer)

species in Cascadia has similar larvae. The skippers *Erynnis persius* and *Thorybes pylades* also use *Lotus* spp., but larvae lack false eyespots and the eggs are distinct. There are many published images of mature larvae, including those in Miller (1995), Allen (1997), Tveten and Tveten (1996), Wagner et al. (1997), Nielsen (1999), Miller and Hammond (2003), Wagner (2005), Allen et al. (2005), Minno et al. (2005), Spencer (2006), Neill (2007), and Betros (2008). Of these, 8 depict larvae from E US, 1 from mid-US, and 3 (plus ours) from the W. All the E larvae (except Nielsen, 1999) have reddish-brown heads, while those from mid-US and W US have black heads. The Wagner et al. (1997) larva appears to have a black head, but the text describes it as red-brown.

Discussion

The eggs and larvae of *E. clarus* are readily found in locations where the species is common. Studies are needed to determine whether the larvae of *E. clarus* are aposematic, sequestering toxins from host plants. Secretions from the large red ventral gland are likely defensive but have not been analyzed. Research is also needed to identify parasitoids and predators and their impact on Cascadia populations. Betros (2008) shows a robber fly feeding on an adult *E. clarus*. The full range of larval hosts is unknown in Cascadia and needs study. The existence of a partial second adult generation in laboratory cohorts suggests that this may also occur naturally under some conditions, as it does routinely in E and S US populations. The environmental cues determining diapausing or nondiapausing pupae need elucidation.

Northern Cloudywing   *Thorybes pylades* (Scudder)

Adult Biology

Thorybes pylades occurs over most of North America but is local and usually uncommon in Cascadia. It is found in S BC, along the slopes of the Cascades of WA and N OR, and in the Ochoco and Siskiyous Mts of E and S OR. Preferred habitats include upper shrub-steppe in E areas, and prairie remnants and heaths near the coast. It often occurs near forests or in recovering clear-cuts, also in riparian or other open evergreen forests. A single generation flies from late April to mid-July, depending on elevation. Both sexes nectar on flowers, including dogbane and vetches. Males visit mud and perch on shrubs near host plants, flying out to challenge passersby in search of mates. *Thorybes pylades* uses many legumes as larval hosts, including *Lathyrus* spp. (wild pea), *Lotus nevadensis* (Nevada Deervetch), *L. crassifolius* (Big Deervetch), *Trifolium* spp. (clover), *Astragalus* spp. (milkvetch), and *Vicia* spp. (vetch). Females lay eggs singly on terminal leaves. Scott (2006) found eggs on the undersides of *Lathyrus* leaves in CO.

Immature Stage Biology

We reared or partially reared this species 4 times from gravid females obtained from Satus Pass, Klickitat Co., WA. Seven eggs were obtained from a female on June 19, laid on inert surfaces near host leaves. One hatched after 8 days and was reared on leaves of a naturalized nonnative legume, *Lotus pedunculatus* (Big Bird's-foot Trefoil). Development was rapid, with 5, 9, 14, and 11 days spent in L1–L4, respectively. L5 fed and grew slowly, entering diapause after ~1 month (mid-Sep). It was overwintered in ambient conditions (unheated garage) and exposed to 15–20 °C under natural daylengths on Jan 10. Within a few days the larva began spinning silk, apparently preparing for pupation, but escaped. An L3 found on July 6 on *L. nevadensis* and held at 23–28 °C under natural

daylengths reached L5 in 13 days. After the larva was switched to *Lotus corniculatus* (Bird's-foot Trefoil), development slowed; after 4 weeks (Aug 16), the larva wandered and found refuge in a curled leaf. It spun a few protective silken strands, turned dark brown, and entered dormancy. On Aug 30 it was transferred to 5 °C for overwintering but died within a few weeks. Two females obtained on June 25 (different year) laid 9 eggs on *L. nevadensis,* hatching after 7 days. Reared at 20 °C and 16 hr daylength, they reached L3 in 18 days, but all were killed by Minute Pirate Bugs. Four females obtained during June 26–29 (different year) laid ~70 eggs on *L. nevadensis* and cage surfaces. Eggs hatched after 5 days at 22–27 °C and development was rapid, larvae reaching L5 in 25 days. L5 fed well initially then slowed, becoming dormant in lightly silked shelters under paper toweling after 4 weeks (Aug 28). Larvae were overwintered at 5 °C for 72 days, then exposed to 26 °C and continuous light. Half the cohort died during overwintering; the remainder pupated within their shelters after 7–10 days at 26 °C. One pupa was formed during overwintering. Adults eclosed after 12–14 days. L1 and L2 constructed simple nests between 2 terminal leaves, lightly bound with silk. L3 nests were constructed from 3 or 4 larger leaves stitched together with equally spaced silk ties. Larvae fed nocturnally and frass shooting was observed. L5 constructed well-camouflaged nests from leaves or under paper toweling, and overwintering occurred within the final nest. Pupation also occurred in this nest, with the pupa attached by 2 strands of silk, 1 attached to the cremaster and to the substrate on either side, the other passing over the thorax and fixed on either side, holding it in place. Protection is based on ventral gland secretions, camouflage, concealment, nocturnal feeding, and frass shooting to confuse predators and parasitoids. *Thorybes pylades* has 5 instars and the mature L5 overwinters.

Description of Immature Stages

The egg is glossy white with 12–14 vertical ribs. L1 is greenish yellow with short blond setae and a black collar bordered with purplish brown. The head is shiny black with an inverted-Y–shaped median cleavage. L2 is light golden brown becoming green with numerous pale yellowish spots. The head and body are covered with numerous tiny setae, and the collar is shiny black. L3 is variable, green with narrow yellow dorsolateral stripes, an indistinct spiracular stripe, and a dark green dorsal heart line. L4 is greenish brown or yellowish green with similar markings to L3. L5 varies from sandy tan peppered with tiny black-and-white

Adult

Egg @ 1.2 mm

L4 @ 15 mm (head)

L1 @ 3.0 mm

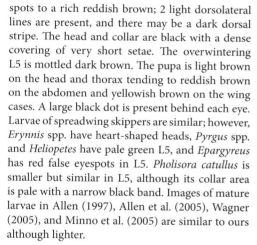

Pupa @ 20 mm

L2 @4 mm (early)

L3 @ 12.5 mm (late)

spots to a rich reddish brown; 2 light dorsolateral lines are present, and there may be a dark dorsal stripe. The head and collar are black with a dense covering of very short setae. The overwintering L5 is mottled dark brown. The pupa is light brown on the head and thorax tending to reddish brown on the abdomen and yellowish brown on the wing cases. A large black dot is present behind each eye. Larvae of spreadwing skippers are similar; however, *Erynnis* spp. have heart-shaped heads, *Pyrgus* spp. and *Heliopetes* have pale green L5, and *Epargyreus* has red false eyespots in L5. *Pholisora catullus* is smaller but similar in L5, although its collar area is pale with a narrow black band. Images of mature larvae in Allen (1997), Allen et al. (2005), Wagner (2005), and Minno et al. (2005) are similar to ours although lighter.

Discussion

Thorybes pylades is challenging to find and adults are fragile in captivity; however, eggs and larvae are relatively easy to rear. Gravid females are best sought in W WA (Mason Co.) in May and in E WA (Chelan or Klickitat Co.) in June. Larvae may be found by searching host plants, particularly terminal shoots, for tied nest shelters. Little is known about natural enemies or other aspects of ecology, including host-plant range and choice. Larval coloration is variable and deserves study, perhaps focusing on host-plant influence. The role of environmental cues on induction of diapause in L5 instead of direct development needs study, as does the chemistry and role of the ventral gland in defense.

L4 @ 18 mm

L5 @ 30 mm

Dreamy Duskywing   *Erynnis icelus* (Scudder & Burgess)

Adult Biology

Erynnis icelus is a skipper of montane areas, favoring forest clearings, roadsides, and waste areas in forested habitats, where it occurs singly or in small numbers in 1 brood from mid-April to late July. In Cascadia it occurs from sea level to near 7,000 ft along both sides of the Cascades, to the Puget Trough in W WA, and mostly W of the Cascades in OR. Beyond Cascadia, *E. icelus* occurs throughout S Canada and N US, extending S to GA in the E, and NM and CA in the W, but is absent from the S Great Plains and Gulf states. The favored larval hosts in Cascadia are *Salix* spp. (willow) with broad-leaved species preferred; *Populus* (poplar), *Betula* (birch), and *Robinia* (locust) are also reported hosts (Scott, 1986a). Both sexes visit flowers; males often imbibe at mud puddles and perch low along flight corridors, frequently on dead woody material, awaiting passing females. Females lay eggs singly on the undersides of leaves. We found eggs on a leaf 6 ft above ground on a 15 ft *Salix sitchensis* (Sitka Willow) in Okanogan Co., WA. We obtained captive oviposition on *S. scouleriana* (Scouler's Willow) while *S. lasiandra* (Pacific Willow) was refused. Scott (1992) observed oviposition on a seedling *Populus tremuloides* (Quaking Aspen), on a stem near the tip, in CO; he indicated that only seedling plants are used.

Immature Stage Biology

We reared or partially reared this species 3 times. On July 24, 2 eggs were collected in Okanogan Co., on *S. sitchensis* but were parasitized. On June 25, 1 female from Yakima Co., WA, confined in a small plastic cylinder (13 cm dia × 8 cm tall) with cut *S. scouleriana*, produced ~50 eggs in 5 days. Eggs hatched after 5 days (22–28 °C) and larval mortality was high, with few reaching L3, 20 days post egg-hatch; all larvae died by late July. On June 1, 2 females from open oak forest in Yakima Co. laid 3 eggs; 1 was reared to adulthood on *S. scouleriana*. Eggs hatched after 7 days and development was rapid, 2 reaching L5 in 19

days post egg-hatch. The mature larvae continued to feed for a few weeks before entering diapause within leaf nests; thereafter they were overwintered (~5 °C). On April 16 the larvae were exposed to 20–22 °C, and 1 pupated 19 days later without feeding; an adult eclosed 16 days later. L1–L2 feed on leaves from the middle, chewing small holes between veins. Leaf flap nests are constructed by cutting a leaf in 2 places, from the margin perpendicularly halfway to the midvein. The cut leaf segment is folded over and silked in place as a shelter. L3–L5 silk whole leaves together (as in a stack) and live between them, leaving the shelter nocturnally to consume leaves from the margins. Frass was scattered, indicating "frass shooting" occurs, a predator-avoidance strategy in which pellets are expelled away from the nest. Throughout development, larvae rested with head crooked back parallel to the body in a cane shape. In spring the mature larva is slow to pupate, shrinking substantially and appearing moribund before forming a pupa that is relatively small compared to the preshrunken L5. There are 5 instars and overwintering occurs as a mature L5. Protection is based on concealment in nests, frass shooting, and nocturnal feeding.

Description of Immature Stages

The egg is creamy yellow turning red within 3 days. Coarsely ornamented, it has 9–10 ribs. L1 is greenish yellow with a black head. Small, T-shaped white setae occur in 2 dorsolateral rows and straight white setae occur along the ventral periphery. Unlike other *Erynnis* spp., *E. icelus* has no black collar at any stage. L2 is green with an indistinct narrow dark stripe and numerous small T-shaped white setae. The posterior half of each segment has 4 lateral folds that persist to maturity. Numerous white spots form 2 lines laterally. The setaceous head is black. L3 is similar, with a narrow dark green dorsal stripe. The lateral white lines are more distinct, and the black head is broadly notched at the crown. L4 and L5 are similar with more distinct striping. The L5 head capsule is brown and heart shaped with a black lining around the margin; the "face" is marked with a broad black vertical mark from the dorsal notch downward, bifurcating into 2 diagonal lines. L5 is greenish infused with yellow near the head. The pupa is dark purplish brown, covered with numerous white speckles. The wing cases are lighter, and there is a thin white dorsolateral stripe. Short white setae occur on the head and body, concentrated across midsegments. Immature stages of *E. icelus* are similar to those of other *Erynnis* spp., but none occur on the same hosts as *E. icelus*. *Epargyreus*

Adult

Eggs @ 0.8 mm

Pupa @ 15 mm

L1 @ 3.5 mm

L2 @ 6 mm

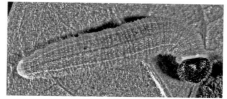

L3 @ 10 mm

L4 @ 15 mm

L5 @ 21 mm

clarus is the only Cascadia skipper found on an *E. icelus* host (*Robinia* sp.), but its larvae are distinct and easily separated. The egg, mature larva (head capsule), and pupa of *E. icelus* can be differentiated from those of other *Erynnis* spp. by comparing images. Photographs of mature larvae appear in Allen (1997), Wagner (2005), and Allen et al. (2005), all from the E US. These show lighter-colored larvae, and none have the distinctive head capsule markings shown in our images.

Discussion

Eggs are readily obtained from captive females. Finding eggs or larvae of this species is difficult. We found eggs on terminal leaves of willow branches protruding into open corridors, about head height. Larvae are fragile in early instars, but later instars are relatively easy to rear. The larval host range in Cascadia needs study; a purported host, *Betula* sp. (birch) was rejected by our larvae. Oaks are thought to be hosts in some parts of E US (Cech and Tudor, 2005). The red eggs resemble tiny galls commonly seen on *Salix* spp., thus may be gall mimics. Interestingly, newly hatched L1 are not red (nor are the eggshells). Very little is known about the biology and ecology of this species, including parasitoids, predators, or diseases. Abiotic factors regulating population densities and voltinism, are also unknown. The distinct difference in head capsule markings of Cascadia larvae and the head capsules of E US larvae warrants further investigation from a taxonomic perspective.

Dreamy Duskywing *Erynnis icelus* (Scudder & Burgess)

Propertius Duskywing   *Erynnis propertius* (Scudder & Burgess)

Adult Biology

Erynnis propertius ranges from S BC to N Baja California with a generally coastal distribution. In Cascadia, *E. propertius* is locally common, following the distribution of the major larval host *Quercus garryana* (Garry Oak) extending inland to the E slopes of the Cascades. Other oaks may be used in SW OR, including *Q. kelloggii* (California Black Oak), *Q. chrysolepis* (Canyon Live Oak), and *Q. vaccinifolia* (Huckleberry Oak). Open oak woodland is the primary habitat, with adults flying in glades, along roadsides and trails, and on nearby flowery hillsides. Males perch on low branches to await passing females and also hill-top. Males congregate on mud, and both sexes nectar on many flowers, including *Brodiaea* sp., dogbanes, vetches, and Desert Parsley. In Cascadia, *E. propertius* is usually univoltine, but a partial second brood sometimes occurs. At low–mid-elevations, adults appear in late March–early April and fly for 6–8 weeks, peaking in most locations during April–May. Stragglers (or second-generation adults) fly until the end of July. Eggs are laid singly on the upper and lower surfaces of young oak leaves and buds; females often select oak saplings or the lower branches of large trees.

Immature Stage Biology

We reared or partially reared this species several times. Two females from Bear Canyon, Yakima Co., WA on April 9 were confined with potted *Q. garryana* in a wood and muslin cage (90 × 90 × 90 cm) and held outdoors in sunshine at 20–25 °C. Eggs (~50) were laid within a few hrs on young leaves and opening buds and hatched after 7–8 days at 18–25 °C, with larvae exiting through the top of the egg. Development was rapid, with L5 reached after 25 days (4, 7, 4, and 10 days at L1–L4, respectively). Development slowed in L5, with all but 1 larva (*n* = 21) pupating after 35–83 days, producing adults in 10–14 days. The single exception overwintered as a

prepupa. Two females from Bear Canyon on April 27 (different year), confined with *Q. garryana* twigs in a plastic cylinder with muslin ends (13 cm dia. × 8 cm tall), produced ~50 eggs. Eggs were laid on unopened buds; development was similar to the first cohort, with L5 reached after 30 days at 18–25 °C. L5 duration was shorter, with the entire cohort (*n* = 25) pupating after 25–35 days and adults eclosing during mid–late July. Newly hatched L1 construct shelters, creating "leaf flaps" by cutting 2 parallel channels in a leaf then folding over and silking down the "flap." Larvae emerge to feed, primarily at night, skeletonizing the nearby leaf, and rest by day in the leaf flap shelter. From L2 onward, some time is spent resting exposed on upper leaf surfaces, but leaf shelters are still constructed and utilized. L4–L5 construct larger shelters, often silking down half a leaf or silking 2–3 leaves together. All instars propel feces out of their nests. L5 wander, and prepupal L5 construct a silked cocoon in a folded leaf. Under natural conditions, many oak leaves containing overwintering prepupae likely fall to the ground. Pupation occurs in early spring. Survival of larvae is based on concealment, although young larvae are vulnerable to small predators like Minute Pirate Bugs that can enter nests.

Description of Immature Stages

The pale yellowish green egg has 16–18 vertical ribs. L1 is tan-orange with an orange bifurcated head. White setae sparsely cover the body (~16–18 per segment). Dorsally and dorsolaterally the setae are Y shaped, while on the head and ventrolaterally they are simple. L2 is greenish tan changing to yellowish green with many small white spots (~100 per segment), each bearing a tiny white Y-shaped seta. There are 2 indistinct white stripes dorsolaterally and an indistinct middorsal dark stripe. The head is black with numerous short white setae and a pale collar. L3 is similar but with a dark brown-black head with 6 peripheral white markings. L4 is whitish green with better-defined dorsolateral white stripes and white spotting. The head is orange with indistinct brown and reddish markings. L5 is similar. In some individuals the reddish orange head is strongly marked with transverse and vertical black bands; others retain the head color and markings of L4. Variation in head coloration and markings may be characteristic of this species. The pupa is pale greenish white or tan, with many short pale setae, especially dorsally. Ventrally there are variable black markings on the wing cases and abdomen. Two black structures protrude from behind each eye. This is the only oak-feeding

Adult

Eggs @ 1.0 mm

L1 @ 2.1 mm

L2 @ 7 mm

L3 @ 12 mm

Pupae @ 20 mm

L4@ 18 mm

duskywing in Cascadia, reducing confusion with other Cascadia *Erynnis* spp., all of which have similar mature larvae. Images of mature larvae appear in Miller and Hammond (2003, 2007) and Neill (2007). All are similar to ours except for different head color and markings.

Discussion

This species can be reared from gravid females, which oviposit readily on potted or cut *Q. garryana*. The distinctive cutting and folding of oak leaves by early instars provides an excellent field mark by which larvae can be found. Captive rearing invariably results in the production of second-generation adults. The reasons for this are unclear but likely related to temperature, photoperiod, and perhaps relative humidity. Overwintering occurs as a prepupa within a cocoon and leaf shelter, but field details of this vulnerable period in the life cycle are lacking. Hellmann et al. (2008) showed that *E. propertius* population densities increased with latitude on Vancouver Is, BC and concluded that the distribution of oaks is more critical in determining northward expansion than climate. Conservation of the remaining patches of oak habitat is clearly critical to this species, particularly in WA.

L5 @ 25 mm

L4 @ 18 mm (head)   **L5 @ 25 mm (head)**

Propertius Duskywing   *Erynnis propertius* (Scudder & Burgess)

Pacuvius Duskywing *Erynnis pacuvius* (Lintner)

Adult Biology

Erynnis pacuvius occurs from S BC through E WA, OR, and CA to Baja, Mexico and E to AZ and CO. The flight period is short, ~3 weeks in most locations (Shapiro and Manolis, 2007), early June–mid-July in WA and May–early Aug in OR (Warren, 2005). Preferred habitat is the transition zone between upper shrub-steppe and lower open pine forest up to ~8,000 ft (Warren, 2005). *Erynnis pacuvius* frequents unsurfaced roads and scrub areas, sometimes visiting mud and often perching on dead woody material. Warren (2005) reported hill-topping behavior. Favored nectar sources include *Apocynum* (dogbane), *Asclepias* (milkweed), and *Ceanothus*. The only known host in Cascadia is *Ceanothus velutinus* (Mountain Balm), although association with *C. integerrimus* (Deerbrush) is reported in OR, and other *Ceanothus* spp. are used elsewhere (Scott, 1986a; Shapiro and Manolis, 2007). *Erynnis pacuvius* is never found far from its host plant. After mating, usually in early–mid-June, females lay eggs singly on upper surfaces of small, new-growth leaves, usually on the basal half of the leaf, often in the middle of a dense host shrub. Scott (2006) reported similar oviposition on *C. fendleri* in CO. The white egg contrasts with the dark green leaves of *Ceanothus*. *Erynnis pacuvius* is single brooded in Cascadia, but Scott (1986a) reports several broods in AZ and CA.

Immature Stage Biology

We reared or partially reared this species twice. On June 8, a single egg was collected in Chelan Co., WA on *C. velutinus*. Reared on *C. sanguineus* (Redstem Ceanothus), it reached L2 on June 17 but accidentally drowned 16 days later. On June 12 (different year) at the same locality a single egg was collected and hatched on June 14. The larva was reared at 20–22 °C on cut *C. velutinus*. Development from egg-hatch to pupation was 38 days, and a second brood adult eclosed 10 days later. L1–L2 ate small circular holes through young terminal leaves, the holes neatly arranged in a string. From L2, nests were made by folding a leaf with the upper surface on the outside, then silking the sides together with the ends partly open. The larva remained in the nest during the day, usually in a cane-shaped position, with the head bent back parallel to the body. Feeding was mostly nocturnal, and frass was typically left in a pile near the nest, suggesting limited wandering. Pupation occurred within the final instar nest. Larvae are likely solitary (evidenced by widely separated eggs), and there are 5 instars. Protection is provided by concealment in nests and nocturnal activity. Our second brood adult eclosure was likely an artifact of captive rearing, as related *Erynnis* spp. are known to overwinter as L5, pupating in the spring. Scott (2006) reported L5 as the overwintering stage in CO.

Description of Immature Stages

The egg is pale greenish maturing to rich orange (although Scott 1986a, 2006, reported no color change in CO). There are 19 strong vertical ribs, each bearing a translucent fringe, terminating in a circle around the micropyle. L1 is golden yellow with a bifurcated black head and black collar. There are 2 irregular dorsolateral rows of T-shaped white setae, and simple white setae occur on the head and around the ventral periphery. L2 is similar, although the setae are proportionally smaller and more numerous. There is an indistinct greenish dorsal stripe. Late L2 is dull greenish with 2 narrow white dorsolateral lines; the dorsal stripe is dark green. L4 is uniformly blue, paler behind the head, and covered with numerous small white spots; the posterior half of each segment has 4 transverse folds. Tiny pale setae are numerous, and there is a distinct white dorsolateral line. The head is dark reddish brown with blackish patches; the crown is broadly and deeply notched, producing truncated horns. The collar is thick and white. L5 is similar although more greenish blue, the head is richer reddish brown, and there is a distinct narrow dark bluish dorsal stripe. The pupa is shiny and dark brown. The abdomen is banded with 3 distinct broad white rings. Contrasting light lines follow the veins on the wing cases, and there are 2 distinct black false eyespots. Other *Erynnis* spp. are similar but do not use *Ceanothus* as a host. Late instars of *E. propertius*, *E. icelus*, and *E. persius* all differ in head markings and pupae, with *E. persius* the most similar. The head capsule of *E. icelus* is darker (still black in L4), and the markings are different; also the pupa is darker and lacks the distinct white rings. In *E. persius*, head

Adult

Eggs @ 0.9 mm

L1 @ 2.4 mm

L4 @ 15 mm (head)   Pupa @ 19 mm (head)

L2 @ 5 mm

Pupa @ 19 mm

L2 @ 8 mm (late)

L4 @ 15 mm

L5 @ 22 mm

capsules and pupae are fairly easily separated. Allen et al. (2005) published an image of a mature larva very similar to ours.

Discussion

It is a challenge to obtain gravid females of this uncommon skipper, and it is also challenging to find eggs or larvae. Scott (2006) obtained oviposition and larvae by placing females in a net bag over *Ceanothus fendleri* in CO. A diligent search of small new-growth terminal *C. velutinus* leaves in mid-June in areas where this species occurs may yield eggs. *Ceanothus sanguineus* was accepted as a host in one of our rearings and may be an additional host in Cascadia. Further study is required on the host range. Nothing is known of parasitoids, predators, diseases, and other mortality factors. The ecological preferences and constraints on this species are unknown and clearly go beyond the presence of *Ceanothus* spp., which are widely distributed and common in Cascadia. The environmental cues determining voltinism are unknown but likely related to temperature and daylength.

Persius Duskywing   *Erynnis persius* (Scudder)

Adult Biology

Erynnis persius is one of our commonest spreadwing skippers, widely distributed in Cascadia from S BC through OR, absent only from the most arid regions and some areas W of the Cascades. Beyond Cascadia the range includes the North American W from AK to the Mexican border and E through the Rockies, plus NE US. The flight period is mid-April–late July in WA and OR. Several larval hosts in the pea family are used, including *Lotus crassifolius* (Big Deervetch), *L. nevadensis* (Nevada Deervetch), *L. unifoliolatus* (Spanish Clover), *Thermopsis montana* (Goldenbanner), *Lupinus latifolius* (Broadleaf Lupine), *L. sericeus* (Silky Lupine), and *Astragalus* spp. (milkvetch). Additionally, we have obtained oviposition on *Lotus pedicularis* (Big Bird's-foot Trefoil) and *Trifolium* sp. (clover). Preferred habitats include forest clearings, roadsides, woody clear-cuts, and heath scrublands from near sea level to over 9,000 ft (Warren, 2005). Males hill-top, also wait on low perches along flight corridors for prospective mates. Both sexes visit a variety of flowers, including lupine and phlox, and males visit mud. Eggs are laid singly on young terminal host leaves or unopened leaf buds, typically in May and June near sea level, later at higher elevations. In WA there is a single brood, but according to Warren (2005), late records in OR indicate a partial second brood, and 3–4 generations occur in CA (Shapiro and Manolis, 2007).

Immature Stage Biology

We reared or partially reared this species 3 times. On July 14 a worn female obtained at 5,700 ft in Kittitas Co., WA laid 15 eggs on *Lotus pedicularis*; only 1 was fertile and it died in L1. On July 7 and July 8 (different years), 3 early-instar larvae were collected in Mason Co., WA, on *L. crassifolius* and reared on this plant to pupation and adulthood. Early instars cut a *Lotus* host leaf in 2 parallel incisions perpendicular to the central vein, folding over the resulting flap and silking it in place to form a nest. Older larvae made nests by silking *Lotus* leaves together, upper surfaces in contact. The nests become more complex in later instars, constructed of several leaves silked together. Larvae leave nests to feed nocturnally, eating only leaves. In a separate rearing, eggs hatched in ~8 days, and development from L2 to diapausing L5 was rapid, taking only ~14 days. The onset of diapause occurred rapidly, with feeding ceasing and larvae becoming dormant. L5 remained in the leaf nest for overwintering. Two L5 were overwintered at ~5 °C, with 1 surviving. On April 1 it was exposed to 18–20 °C; it appeared moribund initially but then pupated after 6 days without feeding. An adult eclosed 13 days later. The larvae of *E. persius* are solitary, and defense is based on concealment in nests, nocturnal feeding, and frass shooting to confuse natural enemies. Additionally the nests are strongly tied with silk to prevent easy access by predators. There are 5 instars and overwintering occurs as a fully fed L5.

Description of Immature Stages

The egg is pale green with a pinkish cast. There are 15 vertical ribs terminating in a circle around the micropyle. L1 is yellow-tan with a shiny black head and black collar; there is 1 dorsolateral and 1 ventrolateral row of T-shaped white setae laterally, and simple white setae around the ventral periphery. L2 is variable, bluish and dull yellow with numerous T-shaped white setae on the body and head. Late L2 and L3 are similar, becoming more yellowish and the black collar becoming inconspicuous. The posterior half of each segment has 4 lateral folds. In L4 the numerous white setae are proportionally shorter, and there is an indistinct dorsolateral white line and a thin dark green dorsal line. The segment folds are more distinct, and the collar is yellow-green. The head is broadly notched dorsally and solid black. L5 is yellowish green and similar; the head is deeply and broadly notched, creating a rough heart shape. The head is medium brown with black peripheral edging, and the "face" is conspicuously marked with a black transverse patch. Prior to diapause the larva becomes more yellow. The pupa is brown and shiny; the wing cases are dark brown with pale vein lines, the abdomen lighter brown. The spiracles are dark brown, and the cremaster is dull black. The immature stages of *E. persius* are similar to those of other *Erynnis* spp.; however, these do not use legumes as hosts. The L5 head capsules of *E. icelus* and *E. pacuvius* lack a bold black horizontal marking, and the pupae differ. The

Adult

Egg @ 0.7 mm   Egg @ 0.8 mm

L1 @ 1.8 mm

Pupa @ 19 mm

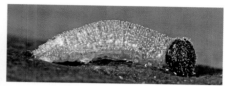

L2 @ 5 mm

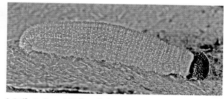

L2 (late) or L3 (early) @ 8 mm

head of *E. propertius* is most similar; however, it is a much brighter reddish brown, the larva is larger, and the pupa is blackened ventrally. *Epargyreus clarus* and *T. pylades* may be found on the same host plants. The eggs of both species are larger than those of *E. persius* and have more ribs. The larvae of *E. clarus* are unmistakable, and those of *T. pylades*, while similar, always have black heads; both are readily distinguished by comparing images. There appear to be no published photographs of the immature stages of *E. persius*.

Discussion

Eggs can be obtained from gravid females caged with an appropriate host. Immature stages can be found by searching *Lotus* terminal leaves for nests or eggs in June, particularly in heath-scrub habitats such as those on the Tahuya Peninsula, Mason Co., WA. Other skipper eggs and larvae occur on the same host plants (and they also build similar leaf nests), so care should be taken in identification. A variety of lycaenid species use the same host plants, but these are readily separable. Little is known of the ecology of this species, including natural mortality factors. The possibility of a partial second brood in WA exists, but research is needed to confirm this and to provide information on the environmental cues responsible for diapause or direct development in L5.

L4 @ 15 mm

L5 @ 20 mm

L5 @ 25 mm

Persius Duskywing   *Erynnis persius* (Scudder)

Grizzled Skipper   *Pyrgus centaureae* (Rambur)

Adult Biology

Pyrgus centaureae is rare in Cascadia, known only from BC and the Cascades of N WA. It has a Holarctic distribution, occurring in boreal regions of North America and Eurasia. In North America, scattered populations occur from Labrador to Cascadia and AK and S in the Rocky Mts to NM. In Cascadia it occurs above timberline on mountain tops and talus slopes, in moist meadows, and in other remote alpine habitats, rarely in large numbers. The single generation flies from early June to early Aug. Males patrol for females, and their rapid, low, bouncing, winding flight is difficult to follow. Females can sometimes be observed basking on rocks or nectaring from yellow asters. Oviposition has been reported elsewhere on cinquefoils (*Potentilla* spp.) and wild strawberries (*Fragaria* spp.), with eggs laid singly on the undersides of leaves. In BC and WA the specific host plants are unknown; however, *P. centaureae* flies where woody alpine cinquefoils are common. Scott (1992) reported preoviposition behavior in CO on *Potentilla diversifolia* (Diverse-leaved Cinquefoil) and oviposition on nearby *Vaccinium caespitosum*. Scott (2006) reported 4 ovipositions on *V. caespitosum* and confirmed it as a host for early instars. In Europe *Rubus chamaemorus* is used (Scott, 1986a).

Immature Stage Biology

This species has not been reared previously in Cascadia (Miller and Hammond, 2007). We reared *P. centaureae* once, from a female collected at Slate Peak, N of Winthrop, WA on Aug 6. The female was caged with *Potentilla fruticosa* (Shrubby Cinquefoil), *Rubus discolor* (Himalayan Blackberry), and *Fragaria virginiana* (Mountain Strawberry) in a plastic box (28 × 14 × 7 cm) with 1 muslin side. Held at 21–28 °C in sunshine, the female oviposited on *P. fruticosa* within a few hours. During 9 days, 57 eggs were laid,

all on *P. fruticosa*. Egg-hatch began after 9 days; L1 chews a hole at the top of the egg through which it emerges, leaving the rest of the shell uneaten. Larvae supplied with *F. virginiana* refused to feed and died. Larvae supplied with *P. fruticosa* immediately fed and reached L2 in 5 days at 21–28 °C. Only ~15 eggs hatched, the remainder shriveling. L1 constructed lightly silked folded leaf nests in which they remained except when feeding. Larvae fed on adjacent leaves, mostly at night, and always returned to their shelters. L3 were provided with *F. virginiana* as an alternative host, which they used. L4 fed mostly on *F. virginiana* and L5 were provided only with this host. Nests were used throughout larval development, new ones formed when larvae became too large. Frass was ejected by larvae from the nests. Development from L2 to L5 took 10–12 days in each instar at 21–28 °C. Growth was slower in L5, occupying 27–28 days before pupation in the final larval nest in late Oct. Scott (1986a) considered *P. centaureae* in Cascadia to be biennial. Guppy and Shepard (2001) thought this unlikely. Our rearing showed full larval development occurred in 8 weeks at 21–28 °C. Ambient temperatures in BC and N WA alpine areas during the Aug–Oct growth period would clearly be much lower, although exposure to sunshine would accelerate development. The slow development we observed in L5 suggests overwintering could occur in this or an earlier instar, as reported in MI by Nielsen (1999). Cannibalism occurred in our rearing, and we also had problems with predation by Minute Pirate Bugs (*Orius tristicolor*), which were able to access larval nests and kill L1–L3.

Description of Immature Stages

The egg is white to pale green and there are 14–16 slightly wavy raised ribs with cross-latticing. L1 is light tan to olive becoming pale orange-brown with small white blotches. There is an indistinct dark middorsal line and the head is shiny black with a black collar. Medium-length white setae sparsely cover the lower sides of the body. Dorsolateral setae are shorter, arise from small white bullae, and are forked. L2 is brown with a middorsal dark line and peppered with tiny white dots (bullae) bearing short white knobbed setae. L3 is similar, although the ground color is darker and white bullae more numerous and distinct. The head is black and densely clothed in pale setae. The black collar on segment 1 is larger, and the posterior segment is dark. The appearance of L4 and L5 is similar to that of L3, with an increased number of white spots, some without setae. Late in L4 and

Adult female

Egg @ 0.7 mm

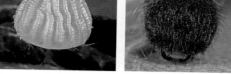

L5 @ 18 mm (head)

L1 @ 2.5 mm

Pupa @ 17 mm

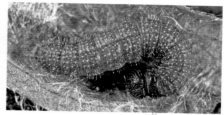

L2 @ 6 mm

L3 @ 8 mm

L4 @ 15 mm

L5 @ 25 mm

in L5, some coagulation of spots is evident and intersegmental folds are apparent. In L5 the first segment collar is brown and the posterior segment is orange, as are the spiracles. Late in L5 the ground color becomes more orange-brown and there are 2 indistinct dark dorsal lines. The pupa is dark brown speckled with black and covered with a powdery white bloom, particularly on the wing covers and thorax. Larvae of other *Pyrgus* spp. in Cascadia are generally green and occur mostly in nonalpine habitats. Images of *P. centaureae* larvae have been published in Nielsen (1999), Allen (1997), and Allen et al. (2005). Two of these are of the E US subspecies (*wyandot*) and are pale green. Larvae in Allen et al. (2005) and Nielsen (1999), both from MI, are pinkish brown, more comparable to our images. The pupa illustrated in Allen (1997) is similar to ours.

Discussion

Obtaining eggs from a gravid female was readily achieved, although many were not viable. Finding gravid females is more difficult. Cech and Tudor (2005) discuss the difficulties in finding and observing *P. centaureae* and considered the species to have "legendary evasive abilities." Much remains to be learned about the biology of this vulnerable skipper in Cascadia, including its ability to adapt to likely warming of its alpine habitat. Whether this species is able to complete development in a single season or is biennial, as Scott (1986a) suggests, needs study. The hosts used in Cascadia also require study, particularly whether *Vaccinium* spp. are used, as reported by Scott (2006) in CO.

Grizzled Skipper *Pyrgus centaureae* (Rambur)

Two-banded Checkered Skipper   *Pyrgus ruralis* (Boisduval)

Adult Biology

The Two-banded Checkered Skipper is a W North American species ranging from S BC to S CA, inland to MT and CO. It occurs throughout most of Cascadia except for dry basins and the coastal fringe of WA. *Pyrgus ruralis* flies from late March to mid-July and is one of the earliest butterflies to appear at lower elevations, but it is a midsummer species at high elevations. Often common, this single-brooded skipper favors meadows and clearings, power line rights-of-way, and roadsides in moist forested areas. Males visit mud and patrol and sometimes perch on the ground for females. Both sexes visit a wide variety of flowers, including dandelions and strawberries. Mated females oviposit a single egg on the underside of a host-plant leaf. Known host plants are members of the rose and mallow families, including *Fragaria* (strawberry), *Potentilla* (cinquefoil), and *Sidalcea* (mallow), also probably *Geum* (avens) and possibly *Horkelia* (horkelia). We report acceptance of the ericaceous plant *Vaccinium scoparium* (Grouseberry) as an additional larval host.

Immature Stage Biology

We reared this species several times from gravid females, including cohorts from a mid-elevation (2,100 ft) population in Whatcom Co., WA (W cohort) and a high-elevation (6,000 ft) population at Bear Creek Mt, Yakima Co., WA (Y cohort). The W cohort was reared on *Fragaria* sp. (wild strawberry) and the Y cohort on a commercial cultivar of *Potentilla fruticosa* (Bush Cinquefoil). A female from W laid eggs on June 1 on the undersides of leaves. Females from Y on July 19 also laid >30 eggs on leaf undersides. Eggs hatched in 3–4 days at 22–25 °C and L1 left eggshells uneaten, producing silk webbing on leaf undersides. L1 rested under silked areas, feeding beyond the outer edge by chewing multiple small holes halfway through the leaf, producing a green refuge surrounded by a ring of brown feeding marks. L2 fed similarly, expanding the

shelter in size. By late L2, larvae began pulling leaves together, folding them down and creating leaf nests. Folded leaf nests were used during the remainder of development, with each new nest larger and more complex, some folded on 3 axes like a pyramid. From mid-L3, larvae skeletonized leaves, mostly within the shelters, or consumed leaves from the edges. Larvae in the Y cohort also fed on *Potentilla* flowers and on leaves of *Vaccinium scoparium* (Grouseberry), an unrecorded host. Nests on this host were constructed from groups of 4–6 stems with multiple leaves. Larvae in the W cohort did not tolerate each other well in later instars, the smaller ones driven from the host plant and sometimes cannibalized. In contrast, >20 Y cohort larvae were reared in a small cage with no evidence of cannibalism. In L4–L5 additional leaves were used in nest construction, producing untidy silked structures with 3–4 leaves. When at rest, larvae adopted a "cane" posture with the head parallel to the body (see L4 image). Development from L1 to pupation (July 26) of the W cohort occupied 55 days at 20–22 °C. Development of the Y cohort took 32 days at 23–27 °C, with 4, 6, 6, 5, and 11 days spent in L1–L5, respectively. Pupation occurred in late Aug, generally in heavily silked final-instar nests, with the pupa weakly attached at the posterior end. Pupae were overwintered at 4–5 °C. After 62 days, pupae were transferred to 15–21 °C and adults eclosed 56 days later. Larval defense is based on camouflage and concealment in nests. There are 5 instars.

Description of Immature Stages

Eggs from the W female were white, turning gray at maturity, while eggs from Y females were light green when laid. There are ~22 vertical ribs, crossed with many smaller but distinct horizontal ribs. L1 varied from bluish gray (W) to yellow-green (Y) with shiny black heads, a purplish brown neck area, and a shiny black collar. Setae are sparse, stout, and white and confined to the head, collar, and ventral margins. L2 is yellowish green, chestnut brown near the head, with indistinct white spotting. L3 is similar, with a dark green middorsal stripe through which pulsations of the circulatory system can be seen. The black head is covered with fine white unbranched setae. L4 is variable, pale to dark green with numerous short pale setae and a more distinct dark green middorsal stripe. L5 is similarly green with several pale longitudinal stripes laterally. Setae are numerous and short, creating a slightly fuzzy appearance. The setaceous head is black with a black collar anteriorly edged in white. The pupa is chestnut brown on the abdomen

Adult

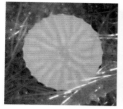

Egg @ 0.7 mm (fresh)

Egg @ 0.8 mm (mature)

L1 @ 3 mm (premolt)

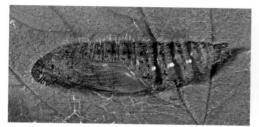

Pupa @ 15 mm

L2 @ 5 mm

L3 @ 10 mm

L4 @ 15 mm

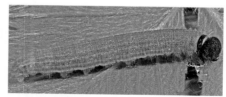

L5 @ 21 mm

and gray on the thorax and wing case and covered with fairly long pale setae. There are 4 constricted lateral bands on the abdomen and dark spots on most segments. Other species using some of the same host plants with similar larvae include *Pyrgus communis* and *Heliopetes ericetorum*. The immature stages of *P. centaureae* are tan to brown, rather than green, and occur only in alpine habitats. Surprisingly, our images appear to be the first published of this species.

Discussion

Eggs of *P. ruralis* can be readily obtained by caging gravid females with host plants. Eggs and larvae can be found by inspecting the undersides of host-plant leaves or by watching for oviposition, leaf nests, or skeletonized leaves. Our experience with 2 geographically distinct populations suggests that significant variation in coloration and morphology occurs in the immature stages of *P. ruralis*. Some of this may be due to different host plants, but it is likely that there are inherent differences between low–mid- and high-elevation populations of *P. ruralis*. Warren (2005) discussed likely differences between lower- and higher-elevation populations of *P. ruralis* in OR, and subspeciation is possible. Acceptability of and development on *V. scoparium* by larvae of *P. ruralis* is surprising and deserves further study.

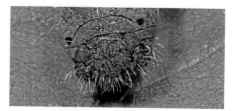

Pupa@ 15 mm (head)

Two-banded Checkered Skipper *Pyrgus ruralis* (Boisduval)

Common Checkered Skipper *Pyrgus communis* (Grote)

Adult Biology

Pyrgus communis is common in Cascadia, flying from early April to mid-Oct in most of S BC, all of WA E of the Cascades, and throughout most of OR. It is common in parks and gardens in the Columbia Basin of WA. Beyond Cascadia it occurs throughout most of the US and S Canada. A wide variety of open sunny habitats is utilized, but it is seldom present in more than nominal numbers at any one site. Males can often be seen mudding at damp spots along unsurfaced roads, sometimes in small groups, particularly in semiarid areas where host plants are common. Both sexes readily nectar, and males perch on low objects awaiting passing females. The larval hosts include a variety of mallows such as *Iliamna rivularis* (Streambank Globemallow), *Sidalcea oregana* (Oregon Checkermallow), *Malva parviflora* (Cheeseweed), *Malva neglecta* (Dwarf Mallow), *Sida hederacea* (Alkali Mallow), *Sphaeralcea munroana* (Munro's Globemallow), and *Malvella leprosa* (also called Alkali Mallow). Nonnative Hollyhock may be used in gardens; Pyle (2002) adds *Malva rotundifolia* (Dwarf Mallow) and Warren (2005) lists additional native mallows used in OR. On the E flank of the WA Cascades, *I. rivularis* is a highly favored host. Females lay eggs singly, but numbers are often found on the undersides of host leaves, mostly near the leaf margin; we have observed up to 10 eggs on a single *Iliamna* leaf. Scott (1992) found eggs on both surfaces of mallow leaves in CO. Adult populations in WA peak in June and Aug (Pyle, 2002), indicating 2 broods. Warren (2005) reported the species is "at least" double brooded in some areas of OR, but central Cascade populations above 4,000 ft are single brooded.

Immature Stage Biology

We reared this species several times from gravid females and field-collected eggs and larvae. Eggs were obtained from females collected in Benton Co., WA and Yakima Co., WA on May 15 and Sep 8, respectively, and reared to adulthood. On May 22, 8 eggs were collected from Swakane Canyon, Chelan Co., WA and reared to L5, but they died before pupation. On June 12, eggs and early instars were collected from the same location and reared to adulthood. On June 20, 20 eggs and several L1 were collected from Swakane and reared to adulthood. In all cases larvae were reared at 20–25 °C using *I. rivularis* or *M. neglecta* as the host. Eggs hatched in 3–5 days, and development took 25–37 days from egg-hatch to pupation; adults eclosed 11–12 days later. The developmental rate of autumn larvae is highly variable. L1 leave eggshells uneaten and begin feeding on a leaf underside. L1–L2 graze a layer half through in a patch, which turns dark. The larvae weave a small silk web over the grazed patch, remaining below it for protection. As they grow larger, they fold leaves, upper side out, silking the edges together to make shelters lined with silk on the inside. Progressively larger shelters are used until pupation; the larvae stay in nests during the day, venturing out to feed at night. Only host-plant leaves are eaten, later instars consuming the entire leaf from the edges. Cannibalism may occur under high population densities and declining food supply. Widely scattered frass indicates frass shooting, a strategy to lead predators away from larvae. The larvae rest with their heads curved back parallel to the body in a cane shape. L5 weave an extensive but open area of webbing, then pupate under it; this may occur within a nest but is more often on an inert surface. In the final brood, L4 or L5 overwinter, pupating the following spring, sometimes without feeding. From a group of 78 L4 and L5 overwintering outdoors in Yakima Co. (temperatures down to -25 °C), only 1 survived. There are 5 instars, and protection is based on concealment in nests, frass shooting, a ventral neck gland (emission of defense chemicals), and nocturnal activity.

Description of Immature Stages

The egg is greenish becoming white, with ~20 vertical ribs. L1 is pinkish amber with a shiny black head and collar. There are several rows of dorsal and lateral light brown Y-shaped setae. The setae are more numerous in L2 and the black collar is surrounded with a dark magenta area; the yellowish body is infused dorsally with olive green. L3 is darker green, especially dorsally; there are 3 lateral folds in the posterior half of each segment, and there is a narrow dark green dorsal stripe. The head and collar are black and there are numerous white speckles on the body. L4 is bright green with numerous pale setae. Indistinct darker green stripes are present laterally. L5 is yellow-green with a black head and collar. There are several narrow darker green stripes dorsally and laterally. The pupa is pale tan with greenish wing cases and is speckled in

Adult

Egg @ 0.7 mm (*above*) and 0.5 mm   Pupa @ 16 mm (head)

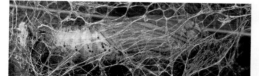

Pupa @ 16 mm (in nest, *above*)

L1 @ 1.5 mm

L2 @ 3 mm

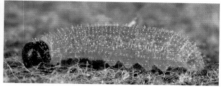

L3 @ 6 mm

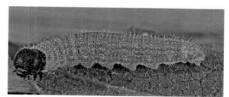

L4 @ 13 mm

L5 @ 23 mm

Common Checkered Skipper *Pyrgus communis* (Grote)

black. The species most similar are other *Pyrgus*; *P. centaureae* larvae are brown with longer setae on the head. The larvae of *P. ruralis* are similar, but the pupa is dark brown. *Heliopetes ericetorum* larvae are also similar but usually browner or whiter with simple-tipped setae. Images of mature larvae in Miller (1995) and Miller and Hammond (2003) are darker than ours but similar. Those in Tveten and Tveten (1996), Allen (1997), Wagner (2005), Allen et al. (2005), and Neill (2007) are similar. The pupa in Allen (1997) is similar but lighter.

Discussion

Eggs and larvae are relatively easy to find in the proper habitat; larval nests are fairly obvious on mallows, and gravid females oviposit readily in captivity. The larvae are easily reared if fresh mallow plants are available. Some moths utilize the same mallows, and their eggs can be very similar to *P. communis* but are typically more finely ribbed. Little is known of the parasitoids, predators, or diseases of *P. communis* or of the ecological factors influencing abundance and distribution. Studies are needed on overwintering, including confirmation of the overwintering instar(s). Chemistry and function of the ventral gland on segment 1 would make an interesting study.

Northern White-Skipper   *Heliopetes ericetorum* (Boisduval)

Adult Biology

Heliopetes ericetorum is a fast-flying white skipper of arid habitats, found in foothills, uncultivated waste areas, and coniferous forests just above the shrub-steppe belt. The range in Cascadia includes the E slopes of the Cascades below timberline, extending into the Columbia Basin from mid WA to mid OR, also along the Snake R, into the Blue Mts and NE OR. Beyond Cascadia, it ranges E to the Rocky Mts and S to Mexico. The flight period is early May–late Sep, with 2 or 3 broods in Cascadia. Scott (1986a) reported "several flights" in the S extending into Dec in places. Both sexes visit flowers; males are often seen mud-puddling in mixed species groups and patrolling flight corridors for females. Larval hosts are mallows, including *Iliamna rivularis* (Streambank Globemallow) and *Malva sylvestris* (Common Mallow); Pyle (2002) listed *Malvella leprosa* (Alkali Mallow), *Malva parviflora* (Cheeseweed), and *Althaea rosea* (Hollyhock), and Warren (2005) added *Sphaeralcea munroana* (Munro's Globemallow). Scott (1986a) lists 14 additional mallows used beyond Cascadia. Females lay eggs singly on the undersides of terminal leaves.

Immature Stage Biology

We reared or partially reared this species several times from field-collected larvae and gravid females. On April 21, L2 were collected near Rocky Reach Dam, Chelan Co., WA, and reared to adults. On May 22 and 28, L4 and L5 were collected and reared to adults. On July 18, an L3 was collected and reared to L4. These were all collected in the same area but in different years. The larvae were reared on *I. rivularis*, the host on which the larvae were found, at 20–22 °C. On June 10, a female obtained in Benton Co., WA laid ~25 eggs on *Malva neglecta* (Dwarf Mallow). Development

from oviposition to pupation was about 48–49 days at 22–28 °C with 7, 6, 6, 10, 7, and 14 days spent as egg and L1–L5, respectively. Prepupal L5 were quiescent for a week or more before pupation, and the pupal period was lengthy (18–28 days) despite warm conditions. L1–L2 hide in the furls of young terminal leaves of *I. rivularis* and construct folded leaf nests silked together at the edges. On *M. neglecta* L1 rest under lightly silked nests constructed between leaf ribs. L1–L2 feed halfway through leaves, producing "windowpanes," then produce round holes late in L2. Nests become larger and involve more leaves as larvae grow, and pupation occurs in the final nest. Feeding is mostly nocturnal, and larvae shoot frass away from themselves. Feeding occurs only on host leaves, and there are 5 instars. We did not overwinter this species, but Allen et al. (2005) indicate that overwintering occurs as a "partially grown caterpillar," likely L5. The survival strategy includes concealment in nests, nocturnal feeding, and frass shooting to avoid or confuse enemies, and strongly silked leaf nests deter predators. Predation of L1–L3 by Minute Pirate Bugs, common residents of mallows, occurred in some of our rearings.

Description of Immature Stages

The egg is white, delicately and uniquely ornamented with numerous vertical and horizontal ridges. There are numerous tiny spikes pointing radially outward, 1 located at each intersection in the latticework. L1 is dull whitish yellow with numerous short white simple setae and a shiny black head and collar. L2 is whitish green, and the body is covered with a regular pattern of round white spots, 4–5 rows of spots per segment. L3 is similar except there are more rows of smaller pale spots on the segments. The white setae are more numerous on the head and body, imparting a shaggy appearance. Indistinct pale longitudinal lines are present, particularly on the posterior half. L4 is pale gray-green; pale setae are numerous but proportionally short. There may be a distinct blue dorsal line, and 2 indistinct blue lateral lines. L5 is variegated pale bluish green and pale yellow; there are numerous white speckles on the body, and the pale setae are numerous and long enough to appear shaggy. The black head is densely clothed in setae, imparting a hoary appearance. The collar is white with small brown speckles. Prepupal larvae are golden brown; Allen et al. (2005) reported that late-season larvae are pink tinged. The pupa is reddish brown to purplish with heavy irregular black spots on the abdomen and numerous dark speckles elsewhere. The pupa is

Adult male

Egg @ 1.0 mm

Pupa @ 19 mm (head)

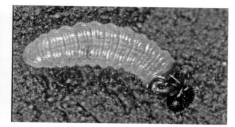

L1 @ 3 mm (late)

Pupa @ 21 mm

L2 @ 3.5 mm

densely covered with long pale setae. The larvae of *H. ericetorum* are similar to those of *Pyrgus* spp. Pale coloration and an overall hoary appearance help distinguish *H. ericetorum,* and the spiny egg is unique. The reddish purple pupa with large dark spots is also characteristic. Care is needed to separate larvae from those of *Pyrgus communis,* as both are often found on the same mallow host plants; however, *P. communis* larvae bear strongly Y-tipped setae whereas those of *H. ericetorum* are simple. Allen et al. (2005) published a photograph of a mature UT larva very similar to ours.

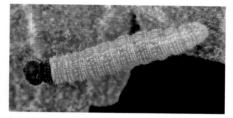

L3 @ 9 mm

Discussion

Larvae of *H. ericetorum* are fairly easily found, using folded leaf nests and leaf damage on the upper third of plants as field marks, and this is an easily reared species. Larvae can be fed cut native or introduced mallows or transferred to cultivated mallows such as Hollyhock. Favored habitats in WA include arid canyon mouths along the Columbia and Yakima rivers near Wenatchee, Yakima, and the Tri-Cities. Many aspects of biology and ecology need research, including the possibility of aestival dormancy in prepupal L5 and/or pupae and determination of the overwintering stage. Little is known of natural enemies or other factors affecting population dynamics. Shapiro and Manolis (2007) suggest that *H. ericetorum* is a seasonal migrant in N CA and evidence for migration should be looked for in Cascadia.

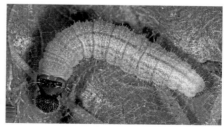

L4 @ 15 mm (late)

L5 @ 23 mm

Northern White-Skipper   *Heliopetes ericetorum* (Boisduval)

Common Sootywing   *Pholisora catullus* (Fabricius)

Adult Biology

Pholisora catullus is a common but inconspicuous small skipper, found in desert sagebrush and riparian areas as well as in gardens and vacant lots and along roadsides where host plants grow. It ranges E of the Cascades, from S BC through WA into N OR. Beyond Cascadia it occurs in much of the US, S Canada, and Mexico. The flight period is early April–early Sep with 2 generations. In the far S, *P. catullus* flies year-round with multiple broods. Larval hosts include a variety of chenopods and amaranths such as *Chenopodium album* (Lambsquarters), *C. rubrum* (Red Goosefoot), *C. fremontii* (Fremont's Goosefoot), *Salsola kali* (Russian Thistle), *Amaranthus graecizans* (Prostrate Pigweed), *A. retroflexus* (Rough Pigweed), and *Atriplex rosea* (Red Orache). Numerous other related species are reportedly used elsewhere (Scott, 1986a). Many of the hosts are hardy, nonnative Eurasian weeds typically found in waste areas and disturbed roadsides. Both sexes visit flowers, including clovers, mallows, and dandelions. Males mud-puddle, fly low to the ground with a rapid bouncing flight, or perch on the ground or low vegetation, challenging passersby in search of females. Females lay eggs singly; on *C. album* and *A. retroflexus*; oviposition occurs on the upper surface of an older midsized leaf. The tiny pink-reddish eggs are usually laid on veins, blend well, and are easily overlooked. The females usually lay 1 egg per leaf, but we have found up to 3. Allen (1997) reported in WV that the female deposits an egg quickly, then darts off in search of another plant.

Immature Stage Biology

We reared or partially reared this species several times from field-collected larvae and gravid females. On June 16, 6 L2–L5 were obtained from *C. album* in Swakane Canyon, Chelan Co., WA, and reared to pupae. On June 20, 5 L2–L5 were reared to adults. On June 25, 15 eggs were reared to L2, all cohorts from the same host and locality. One female collected near Benton City, Benton Co., WA on May 27 caged with *A. retroflexus* laid ~50 eggs during 4 days. Eggs hatched in 5–6 days at 24–28 °C; development from L1 to pupation took 22 days, with 4 days in each instar except L1 (6 days). Adults eclosed after 7–8 days. In preferred habitats larvae can be common, resting in leaf nests on the upper new-growth leaves of hosts like Lambsquarters. Leaves near nests typically show considerable feeding damage. Multiple immature stages are often found simultaneously, indicating that oviposition occurs over an extended period. L1–L2 feed at the tops of plants on young leaves, resting on upper surface midribs with a few silk strands for cover. L2 build nests by cutting a leaf inward in 2 places, then folding the loose flap over and silking it in place along the margin. L3–L5 bend entire leaves inward toward the midvein, fastening the edges together with silk. Larvae feed nocturnally on leaf edges away from the shelter, and frass shooting, especially in the later instars, occurs. Overwintering occurs as a mature larva, according to Scott (1986a), Wagner (2005), and Cech and Tudor (2005), with hibernation in the final nest and pupation in spring. There are 5 instars, and protection is based on concealment, frass shooting, nocturnal feeding, and strongly tied leaf nests.

Description of Immature Stages

The eggs are pinkish red and uniquely ornamented. There are 8 swollen vertical rounded ribs, bold, serrated, and extended at the top, terminating around a deep micropyle, fading downward and becoming obscure by midheight. There are ~22 encircling cross ribs, fairly bold at the top, but increasingly fine and closely spaced toward the base. L1 is yellowish orange with a shiny black head and collar. There are 3 rows of knobbed dark setae laterally, plus additional simple setae on the head. L2 is yellow-green with ~24–26 white spots per segment. The head and collar are black, the latter edged anteriorly with white; these characters persist until pupation. L3 is similar, with more numerous short knobbed setae and white spots. There are 4 distinct folds on the posterior half of each segment. L4 is medium green posteriorly, yellower anteriorly, and the setae on the head and body are short and numerous, imparting a fuzzy appearance. The black collar is cloven into 2 halves, starkly contrasting with the underlying white color. L5 is similar, medium to dark green with numerous white spots. There is an indistinct middorsal dark stripe. The pupa is black, but a waxy pruinose surface layer imparts a purplish gray appearance. Viewed anteriorly, the pupa has a "baby seal head" appearance with a black spot above

Adult

Eggs @ 0.6 mm

L1 @ 1.5 mm

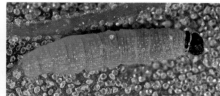

L2 @ 4.5 mm

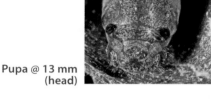

Pupa @ 13 mm (head)

L3 @ 7 mm

L3 @ 10 mm (in tied-leaf nest)

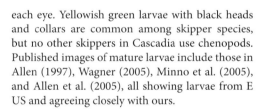

Pupa @ 14 mm

L4 @ 13 mm

L5 @ 19 mm

each eye. Yellowish green larvae with black heads and collars are common among skipper species, but no other skippers in Cascadia use chenopods. Published images of mature larvae include those in Allen (1997), Wagner (2005), Minno et al. (2005), and Allen et al. (2005), all showing larvae from E US and agreeing closely with ours.

Discussion

All immature stages of this species are easily found; eggs can also be obtained from gravid females held with host plants. The folded leaf nests provide a good search pattern in the field, and larvae are easily reared, provided a fresh supply of the host plant is available. Roadside spraying to control Eurasian weeds often adversely impacts *P. catullus* populations locally. Little is known of the biology, ecology, and population dynamics of this species, including the identity and impact of natural enemies. Shapiro and Manolis (2007) report that *P. catullus* populations are declining in N CA; this does not appear to be the case in Cascadia, but an understanding of its ecology here may help determine vulnerabilities.

Family Hesperiidae
Subfamily Hesperiinae
MONOCOT (GRASS) SKIPPERS

Seventeen species of grass skippers occur in Cascadia. We add *Carterocephalus palaemon* to this group because of its use of grasses; it is the sole representative of the subfamily Heteropterinae in Cascadia. The grass skippers are named for their larval hosts, grasses or sedges. Grass skippers rest in 2 characteristic poses, with their wings directly over the back, or held in 2 planes, the hindwings horizontal and the forewings diagonally upward. Adults visit flowers and fly with a rapid, darting flight; mudding and hill-topping are common among males.

Eggs are generally laid singly on or near the host plant. Grass skipper eggs are remarkably uniform, white or pale green, smooth and rounded-dome shaped, except for *Thymelicus*, which produces cough drop–shaped eggs in rows. All are unmarked except *Euphyes*, which has a red "bull's-eye."

Larvae are elongate and cylindrical, green or brown, some with longitudinal striping and many with black heads. Most species have a constricted collar behind the head, with exceptions in *Thymelicus* and *Oarisma*. Larvae tend to be proportionally large in comparison to the adults. Most larvae appear hairless or have very short setae, and many have a few longer setae at the posterior end. Larval facial patterns may be useful in identification. Most species construct silk and grass-blade nests.

Pupae are formed mostly horizontally, usually on the ground and most within nests; a few species have girdle threads in addition to a posterior anchorage. Pupae are elongate, most with bluntly rounded heads and pointed abdomens. *Thymelicus* and *Oarisma* have pointed projections at the head end. Some species have smooth, hairless pupae; others are densely setaceous.

Most grass skippers are single brooded, but 2 have multiple broods. Some species aestivate as mature larvae, delaying eclosion until optimal environmental conditions occur. A variety of overwintering strategies are used, roughly a third of our species hibernating as eggs, another third as pupae, and the remainder as partly grown larvae.

Arctic Skipper   *Carterocephalus palaemon* (Pallas)

Adult Biology

Carterocephalus palaemon is an uncommon skipper found in the mountainous parts of WA and OR, including the Cascades and Olympics and NE WA. Beyond Cascadia it is found in AK and much of Canada, extending S to CA and the CO Rockies, also into the upper Midwest and New England. This circumpolar species is also found in Europe and E across N Eurasia to the Pacific. It occurs in moist woodland mountain habitats from near sea level to ~7,000 ft (Warren, 2005), favoring forest clearings along riparian corridors. In Cascadia, populations are small and isolated and fluctuate significantly. Adults nectar readily on flowers and live for up to 3 weeks (Ravenscroft, 1992). Males perch ~3–4 ft above ground level, darting out in search of mates. Females oviposit singly on grasses. The single generation flies from mid-May to late July. Larval hosts are poorly documented but include *Calamagrostis purpurascens* (Purple Reedgrass), *Bromus* spp., and probably other grasses (Scott, 1986a; Warren, 2005).

Immature Stage Biology

We reared or partially reared this species 4 times. On June 10, 3 females collected near Liberty, Kittitas Co., WA, produced 7 eggs. Two developed to L5 but died during overwintering. Females obtained June 18 (same locality, different year) and July 23 from Slate Peak, Okanogan Co., WA produced 19 eggs that hatched, but all larvae died in L1–L2. Eggs hatched in 7 days, and development from L1 to L5 occupied 44 days. L5 remain active for ~3 weeks before entering diapause. Early instars pull together the edges of grass blades, forming hollow open-ended tubular nests, tying them with silk cross-ties. Feeding larvae often clip grass blades, and the outer 2–3-inch tips fall.

Stitched grass-blade nest of L2

Larger larvae require larger grass blades to construct nests; if only smaller blades are available, they may tie several together. On Oct 15 in Tumwater Canyon, Chelan Co., WA, a diapausing L5 was found in a rolled leaf nest on *Phalaris arundinacea* (Reed Canary Grass). The upright nest was on the upper part of the plant, near a blade tip. The nest was tied in a manner similar to that of *Ochlodes* spp. except that it had only a few silk cross-ties; it was fully silk lined on the inside, and there was no wax. After overwintering, on Feb 9 the larva was exposed to 20–22 °C; it pupated 5 days later without feeding, and the adult eclosed after 12 days. The pupa did not hang but was attached to nearby grass with a silk strand at the posterior tip. By contrast, Frohawk (1892) described pupation on a grass stem or blade with a silk girdle and cremaster, in a tentlike structure formed of 6 blade tips. In Scotland, later instars are reported to abandon nests prior to overwintering (Ravenscroft, 1992). *Carterocephalus palaemon* has 5 instars, and protection is based on concealment in nests, flinging frass, and nocturnal feeding to avoid predators.

Description of Immature Stages

The egg is white and smooth but faintly marked with obscure divots. L1 is white with a shiny black head and collar. Laterally there are 4 rows of short, straight, dark setae. L2 is dull yellowish with green infusion laterally and short pale setae. There is 1 dorsolateral and 1 lateral white stripe, and dorsally there is a narrow green stripe. There is a conspicuous vertical dividing line on the head, bifurcating below the midpoint. L3 is green and similar to L2. The head is black, and the black collar is narrow. L4 is paler except for 3 narrow dorsal stripes; the head is gray and light brown. There is a broad dorsal white stripe and a narrow one dorsolaterally. In L5 the head is light cinnamon brown and strongly divided vertically. The dark green body stripes are paler, and there is a bold white stripe laterally bordered on both sides by green. The pupa is pale white and gray on the ventral abdomen and strongly striped in brown and white dorsally. The wing cases are pale, but the veins, antennae, and

Adult

Pupae @ 17 mm (ventral, *above*, lateral, *below*)

Egg @ 0.8 mm

L1 @ 2.0 mm

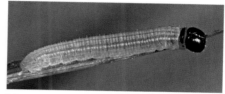

L2 @ 5 mm

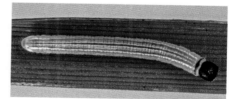

L3 @ 10 mm

L4 @ 20 mm (late)

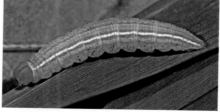

L5 @22mm

proboscis are strongly outlined in magenta. There is a sharply pointed projection anteriorly, and the posterior end terminates in 2 pointed "tails" joined together with a translucent membrane. The larvae in all instars are slender, elongate, and striped and can be confused with only a few other skippers. *Thymelicus lineola* has highly distinctive eggs and pupae, and L4–L5 have striped heads. In *Oarisma garita* the head capsule is greenish in all instars. The larvae of *Amblyscirtes vialis* differ in color in all instars, and the pupae lack projections. The head capsule of *Euphyes vestris* larvae is distinctly marked, and the pupae are black. There are few published images of American *C. palaemon*; Allen et al. (2005) show a mature NH larva similar to ours except that the back of the head appears angled (ours is rounded) and the head is green (ours is cinnamon brown). Some Web sites have images of European larvae and pupae that appear very similar to ours, and the detailed descriptions by Frohawk (1892) of immature stages from the UK also correspond well with our observations.

Discussion

Females fed poorly in captivity, and we found it difficult to obtain eggs. Losses were high during rearing, especially in the early instars. Very little is known about this species in the US, including host grasses, natural enemies, and other factors affecting abundance and distribution. In contrast, much is known about its ecology in the UK, where it is a threatened species (Ravenscroft, 1992). In some parts of its circumpolar range, *C. palaemon* populations have declined in recent decades (e.g., extirpated from England in 1976), and it appears vulnerable to habitat disturbance and/or climate change. In Okanogan Co., good survival of *C. palaemon* populations occurred following a severe wildfire that was detrimental to other butterflies.

Garita Skipperling   *Oarisma garita* (Reakirt)

Adult Biology

Widely distributed in W North America from S Canada through MT and ND to Mexico, the range of *Oarisma garita* encompasses NE WA, SE BC, N ID, and NE OR. It is found mostly in montane and transition grasslands, in wetter pastures, and along watercourses and can be locally common. Recently it has extended its range and habitat to include drier grasslands and disturbed habitats such as mowed lawns and roadsides (Pyle, 2002; Warren, 2005). A single generation flies from early June to early Aug, peaking in July. Host plants are bunchgrasses, hay grasses, and sedges. Scott (1992) reported oviposition in CO on 12 species in the genera *Bromus, Bouteloua, Sitanion, Stipa, Agropyron, Poa, Muhlenbergia, Agrostis, Koeleria*, and *Carex* and considered *O. garita* to be the most polyphagous skipper. Males patrol, flying over short grass or through longer grass (Warren, 2005). Both sexes visit flowers of clovers, asters, and thistles, and it is one of the few butterflies to nectar on lupine. The flight period lasts for 3–4 weeks, with males eclosing first. In the first 7–10 days, males outnumbered females by ~10:1 in Tiger Meadows, Pend Oreille Co., WA, with females dominating later in the flight period. Eggs are laid singly on host grasses, nearby vegetation, or inert surfaces.

Immature Stage Biology

We reared or partially reared this species several times from gravid females. Seventeen females obtained from Tiger Meadows on June 26 and confined at 22–26 °C with cut *Bromus carinatus* (California Brome) in a plastic container with a muslin lid (30 × 20 × 10 cm), produced ~500 eggs over 9 days. Twenty females from the same location on July 2 (different year) produced a similar number of eggs. Most oviposition occurred in late afternoon, and eggs were glued to grass stems and blades as well as to inert surfaces. Eggs hatched after 7–9 days; newly hatched larvae fed immediately and were strongly phototactic, congregating on the lighted side of the cage. The June 26 cohort was reared at 22–26 °C under natural daylengths on *B. carinatus, Setaria glauca* (Yellow Foxtail), *Lolium multiflorum* (Annual Ryegrass), and *Poa annua* (Annual Bluegrass), and development from L1 to L4 took ~50 days, with 9, 8, and 33 days spent in L1–L3, respectively. L4 developed to full size (~23 mm), entered dormancy, and were overwintered Sep 10 at 5 °C, but all larvae died during overwintering. The July 2 cohort was reared under constant illumination on *Cynodon dactylon* (Bermuda Grass), and development from L1 to pupation (4 instars) took 43 days, with 8, 12, 8, and 15 days spent in L1–L4, respectively. Adults eclosed in late Aug after 7 days. Larvae fed openly on grass blades and pupated on grass stems and blades as well as on the bottom and top of the cage, with the pupa attached by cremaster and silk girdle. No nests were made, and protection is based on the striped green larvae being cryptic in grass. Later instars possess a small ventral neck gland that may have a defensive function, but they lack wax glands used to protect pupae in many other skippers. Larvae had only 4 instars in our rearing, and overwintering occurs as a mature L4 (Scott, 1986a; Guppy and Shepard, 2001).

Description of Immature Stages

The smooth egg is creamy white maturing to dark bluish green. L1 is creamy white sparsely clothed with tiny black setae (~10/segment), becoming greenish white. The head is bifurcated, light brown with darker markings frontally. L2 is green with multiple white stripes. A broad dark green dorsal stripe is bisected by a narrow white line. Five narrow white stripes occur on either side, and there is a broad subspiracular white stripe. The body and tan head are clothed with short black setae (~60/segment). L3 is green with some infusion of yellow and is identically marked to L2. The head is light green, and there are 5–6 transverse folds posteriorly on each segment. The posterior segment has a "flap" dorsally, which is lifted when excreting (see image). L4 is whitish green with a green head and similar striping. The tan-colored pulsating heart is visible within the broad dorsal stripe on segments 9 and 10. A subspiracular white stripe is broad and prominent. There are 5 transverse folds posteriorly on each segment. The pupa is whitish green to green with white stripes on the thorax, abdomen, and wing cases. The abdomen is tinged with yellow and the head horn is pinkish tan. Larvae are superficially similar to those

Adult male

Eggs @ 0.8 mm

L4 @ 21 mm (head)

L1 @ 2.5 mm

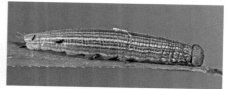

L2 @ 4.5 mm

L2 @ 8 mm

Pupa @ 15 mm

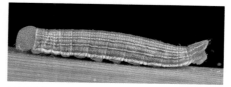

L3 @ 15 mm

L4 @ 21 mm

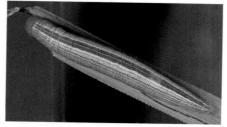

L4 @ 21 mm

of some satyrs, including *Coenonympha tullia* and *Cercyonis* spp.; however, *O. garita* lacks the "tails" found in satyr larvae. Among the skippers, only *Thymelicus lineola* and *Carterocephalus palaemon* have immature stages likely to be confused with those of *O. garita*. Both species have green-and-white striped larvae but lack a bold dorsal dark stripe and subspiracular white stripe and have either brown or striped heads. The pupa of *T. lineola* is very similar to that of *O. garita*, but the pupa of *C. palaemon* is not green. Allen et al. (2005) illustrated brown and green forms of *O. garita* larvae, the latter resembling ours except for a brown head.

Discussion

Gravid females oviposit readily in captivity and larvae are reared on a variety of grasses. Care must be taken to ensure early instars are not held in disease-promoting moist conditions. Rearing under long daylengths ensures full development, but declining daylengths during July–Aug produce overwintering L4. Research is needed to determine whether postwintering larvae pupate without feeding or molt to L5. We did not encounter the brown form of *O. garita* larvae described by Allen et al. (2005), and studies are needed to see whether it occurs in Cascadia. The remarkable similarity of *O. garita* larvae to some satyr larvae deserves study, as does the impact of natural factors on the distribution and population dynamics of this species.

European Skipperling   *Thymelicus lineola* (Ochsenheimer)

Adult Biology

An Old World species, *Thymelicus lineola* was found in Ontario, Canada around 1910, and in BC in 1960 (Guppy and Shepard, 2001). In 2002 *T. lineola* was discovered near Blaine, WA by Thea and Robert Pyle, and is now found in much of NW Whatcom Co., WA. One of us (DJ) found 3 males and 3 females flying at Tiger Meadows, Pend Oreille Co. in NE WA on July 22, 2009. Adults are similar to *Oarisma garita*, although the immature stages differ. The flight period in our area is based on few records, but adults are found throughout July. It is sometimes common where it has established. Males perch on tall grasses and fly out to challenge passersby, while females frequent lower vegetation, basking and seeking nectar, especially on *Lotus pedunculatus*. Eggs are laid in strings of up to 20, end-to-end but not quite touching, delicately glued to the concave side of a grass leaf. Food plants in WA include *Phalaris arundinacea* (Reed Canary Grass), *Phleum pratense* (Timothy Hay), and *Dactylis glomerata* (Orchard Grass), the latter used only sparingly. Allen (1997) reported *Holcus lanatus* (Velvetgrass) as a host; however, we tried unsuccessfully to rear larvae on this grass. *Thymelicus lineola* is single brooded, and it has reached pest status on hay and pasture grasses in some parts of Canada and E US, where it can be exceedingly abundant (Pengelly, 1960; Cech and Tudor, 2005).

Immature Stage Biology

We partially reared this species from eggs laid by females obtained on July 22 and July 25 from Pend Oreille Co. and Lk Terrell nr Ferndale, Whatcom Co., respectively. Females were caged with *Cynodon dactylon* (Bermuda Grass) and *P. pratense*, 100 and 600 eggs laid, respectively. Embryonic L1 were visible in eggs after ~3 weeks, then placed at 4–5 °C for overwintering. In mid-March, ~100 eggs were exposed to 20–22 °C and natural daylengths, and some began hatching within 2 days. Hatching was extended over a protracted period (50 days). Larvae left eggshells uneaten except for an escape hole; they were placed on *P. pratense*, but only a few fed. Most larvae appeared to be dormant and all eventually died. Some larvae that initially fed stopped after a few days and became dormant. Remaining eggs were removed from overwintering in 2 batches on April 22 and May 17; these eggs hatched faster, mostly within 4 days. Rearing temperature was the same as previously, but natural daylengths longer, suggesting a role of photoperiod in egg-hatch. These larvae fared better, but only a few survived to L3. On May 30, 16 larvae were found on Reed Canary Grass at Lk Terrell, most L3–early L4. The grass was 12–14 inches tall, and the larvae had constructed nests in the upper third of plants. A single grass leaf is pulled into a neat tube and stitched together with tidy silk cross-ties. Nests are very similar to those constructed by *Ochlodes sylvanoides*, which were found cohabiting the same plants, so larvae had to be examined to distinguish them. *Ochlodes* larvae have a black head and are easily distinguished after L2. The invariable presence of *T. lineola* larvae within tube nests during the day suggests they are nocturnal feeders. These larvae were reared on potted *P. arundinacea*, but most failed to survive beyond L5. A further 12 L4 and L5 larvae were obtained on June 11, most surviving to pupation, 10 reaching adulthood. Development from egg-hatch to pupation appears to be ~50 days, and the adult ecloses 7 days later. Larval feeding is evidenced by grass blades eaten in from 1 side, usually near the tip, and the tubular larval nests are readily found. The larvae pupate in the tied leaf nests. There are 5 instars, and the larvae are solitary. Concealment appears to be the primary survival strategy.

Description of Immature Stages

The pale yellow-green eggs are unique, oval in side view with broadly concave sides, laid in strings. L1 is creamy tan, with a shiny black head and collar. Each segment is folded into irregular wrinkles, and setae are almost lacking. L2 is slender and cream colored with 3 dorsal green stripes; the head is shiny black and bears stubbly setae. Fine green spots speckle the body. L3 is olive green; the head capsule is bifurcated and purplish brown. Setae at the head and tail are short, white, and fairly conspicuous. L4 is lighter green, and the head capsule is light brown with 1 median and 2 lateral vertical dark brown stripes. Numerous black speckles cover the body, each with a tiny dark seta. L5 is similar except the ground color is blue-green and the head capsule is whitish with dark vertical stripes.

Adult male

European Skipperling *Thymelicus lineola* (Ochsenheimer)

Eggs @ 1.0 mm

L1 @ 1.6 mm

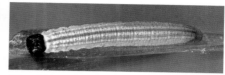

L2 @ 4.5 mm

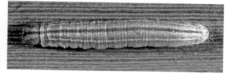

L3 @ 7 mm

L5 @ 18 mm (head)

L4 @ 12.5 mm

L5 @ 18 mm

Body segments are strongly creased with lateral folds, and small spots are restricted anteriorly and posteriorly. The pupa is elongate and green, with a long flesh-colored pointed rostrum. The eggs and pupae of this species should not be confused, and following L2 the larvae are easily identified. *Carterocephalus palaemon* larvae are fairly similar but have bold white striping and the head capsules lack vertical stripes; *Oarisma garita* has a green head and lateral white stripes. The mature larva is pictured by Layberry et al. (1998), Wagner (2005), and Allen et al. (2005), and a mature larva and pupa by Allen (1997); all images closely match ours.

Discussion

This species is not easily reared from overwintered eggs or early instars. The majority of larvae that hatched appeared to be dormant, or reentered dormancy after a short period of feeding. Prolonged termination of egg diapause in *T. lineola* was described by McNeil and Fields (1985). Further studies on the overwintering biology of this species are needed to determine optimal post-overwintering environmental conditions. *Thymelicus lineola* populations are expanding southward in W WA, and its expansion there and in NE WA should be monitored. The potential impact on commercial crops of Timothy grass in Cascadia needs study as the species is pestiferous in some locations. Competition between *T. lineola* and native skippers such as *O. garita* and *Ochlodes sylvanoides* needs study, as well as the impact of endemic natural enemies. The flight period and food plants of *T. lineola* also need study.

Pupa @ 17 mm (lateral, *above*, dorsal, *below*)

Juba Skipper   *Hesperia juba* (Scudder)

Adult Biology

Hesperia juba ranges throughout most of W US. In Cascadia it is widely distributed, occurring nearly everywhere except coastal areas N of S OR. Adults fly from mid-April to early Oct and occur in a wide variety of habitats, but they are most often seen in E xeric grassland/steppe areas. They are also common at high elevations (up to 9,500 ft), particularly in late spring–midsummer, with apparent upslope movement during spring (Warren, 2005). Columbia Basin, WA populations emerge in April and disappear within 3 weeks. Populations of medium-worn individuals appear in late May–early June in high-elevation areas in the Cascades and Blue Mts. A second generation of adults emerges in lowland areas in mid–late Aug and are often seen nectaring on rabbitbrush. Berkhousen and Shapiro (1994) suggested that montane *H. juba* in CA is univoltine and adults overwinter. Males are territorial and perch on the ground or low vegetation. Both sexes frequent flowers, and males sometimes mud-puddle. Larval hosts are various grasses, including *Bromus*, *Poa*, *Deschampsia*, and *Stipa* (Scott, 1986a). Eggs are laid singly on dead grass inflorescences (Scott, 1992), at the bases of grasses, sometimes on soil or other substrates.

Immature Stage Biology

We reared or partially reared this species several times. Gravid females obtained from Columbia Co., WA (May 21), Franklin Co., WA (Oct 10), and Yakima Co., WA (Sep 13) laid eggs on potted or cut *Setaria glauca* (Yellow Foxtail) within 24–48 hrs. Eggs laid in May produced autumn-flying adults whose progeny formed the spring adult generation. May-

and Sep-laid eggs hatched after 10–11 days outdoors. In contrast, most eggs laid in Oct diapaused and did not hatch until the following Feb or March (over a period of 6 weeks) and contained fully developed embryos during winter. One Oct egg hatched after 2 weeks and overwintered as L1 after briefly feeding. Larvae from Sep-laid eggs developed rapidly, with L2 reached 7 days after hatching; thereafter, feeding (mostly nocturnal) and development slowed, and by the end of Oct, dormant mature L2 rested in silked larval shelters. Temperatures in early Nov fell to -10 °C, and substantial mortality occurred. Two surviving L2 (from ~100) were transferred to 15–22 °C in early Dec and developed to adults in 51 days, with 2, 10, 11, and 12 days spent in L2, L3, L4, and L5, respectively. May-laid eggs hatched in 7–10 days at 21–26 °C, and larvae developed rapidly at 20–30 °C and natural daylengths, with 7, 10, 10, and 10 days spent in L1, L2, L3, and L4, respectively. L5 entered dormancy (aestivation) in early–mid-July, resting in shelters and not feeding. Dormancy lasted 31 days and pupation occurred Aug 10–15, adults eclosing Aug 25–31. All instars silked grass blades together, with shelters more complex in each successive instar. L1 wove a few silken strands into a vague "nest," while prepupal L5 constructed untidy but tightly woven shelter tubes. Pupation occurred within a silken cocoon lined with flocculent material produced by ventral abdominal glands. Our rearings suggest that *H. juba* overwinters in Cascadia either as fully developed eggs or L1 if oviposited in mid-Oct, or as L2 if eggs are laid in mid-Sep. Eggs laid in early Sep may overwinter as L3. Rapid development by overwintered L2 when transferred to warm conditions confirms Scott's (1992) assumption that sufficient time is available to produce adults in April–May. Larval defense is based on concealment, although in our rearings, predation by the Minute Pirate Bug (*Orius tristicolor*) was a significant problem, with these small (0.5–2.0 mm) predators able to access larval shelters.

Description of Immature Stages

The creamy white, smooth egg develops a pinkish tint after 5–7 days. L1 is cream colored, sparsely dotted with tiny black setae. The head is glossy black, and there is a dorsal, anteriorly white-margined black collar on segment 1 that persists through development. L2 is similar, greenish white with an increased number of tiny black setae, giving a peppered appearance. L3 is dark brown with a black head, indistinct dorsal and lateral gray areas, and numerous tiny black setae. L4 is initially similar to L3, with an indistinct black dorsal

Adult

Eggs @ 1.3 mm

L5 @ 24 mm (head)

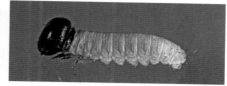

L1 @ 2.6 mm

stripe, becoming orange-brown and developing transverse folds on each segment. The textured black head has pale vertical parallel stripes with a pale inverted V at their base. L5 is dark orangish brown peppered with tiny black setae and has 6 well-defined transverse folds on the posterior half of each segment. The head markings are more pronounced. Ventrally there are 2 pairs of patches of a white flocculent secretion between abdominal segments 7–8 and 8–9. The pupa is brown and black, darkest on the thorax and head, with a slight waxy bloom. The abdomen is light brown ventrally with intersegmental light brown-orange banding. The larvae of *H. juba* may be confused with those of other *Hesperia* spp.; however, mature larvae of *H. colorado* and *H. comma manitoba* have brown rather than black heads. The mature larva of *H. nevada* is very similar to that of *H. juba*, but early instars are green. An image of L5 from CA appears in Allen et al. (2005) and is similar to ours, but lighter colored.

L2 @ 8 mm

L3 @ 15 mm

Discussion

Our rearings confirm that *H. juba* is bivoltine and demonstrate that overwintering can occur as diapausing eggs or early instars in Cascadia (see James, 2009b, for more details). In addition, we showed that a period of aestival dormancy occurs in mature larvae of the spring (but not autumn) generation, ensuring that the second adult generation does not fly before mid–late Aug. In most lowland *H. juba* habitats, larval hosts senesce in midsummer. Delaying the flight period until late summer–autumn increases the potential of females to find green hosts. Research is needed to determine whether a different strategy is used in less xeric environments. Two species of grass, *S. glauca* and *Elytrigia repens* (Quackgrass), were successfully used in our rearings. Work is needed to identify the grass species favored by *H. juba* in Cascadia. MacNeill (1964) reported that *H. juba* may go through 5 or 6 instars. We saw only 5, but an additional instar may occur under certain conditions. Winter mortality of larvae was significant in our rearings; a study of the thermal tolerances of *H. juba* larvae would be of interest.

L4 @ 20 mm

L5 @ 24 mm

Pupa @ 18 mm

Juba Skipper

Hesperia juba (Scudder)

Common Branded Skipper   *Hesperia comma manitoba* (Scudder)

Adult Biology

Hesperia comma manitoba is part of the *Hesperia comma* complex, with many described species occurring across boreal North America and Eurasia (Forister et al., 2004). In Cascadia, *H. c. manitoba* is confined to high-elevation areas of S BC extending into the N Cascade and Olympic Mts of WA. Alpine meadows and mountaintops are preferred habitats, and a single brood flies from June to Aug. Males are territorial and commonly hill-top, and both sexes visit flowers. Populations do not appear to be very dispersive, similar to the finding for *H. comma* in the UK, where 90% of individuals moved less than 100 m in patches of suitable habitat (Hill et al., 1996). *Festuca ovina* (Sheep's Fescue) is the only host recorded for *H. c. manitoba* (Pyle, 2002). The endangered UK subspecies of *H. comma* is restricted to breeding on *F. ovina* (Davies et al., 2005). Eggs are laid singly on grass blades and on nearby surfaces. Female *H. comma* in the UK show a high degree of selectivity when ovipositing (Thomas et al., 1986).

Immature Stage Biology

We reared or partially reared this species 3 times. Four gravid females obtained on July 29 from Apex Mt in S BC were confined in a small plastic cylinder (13 cm dia × 8 cm tall) with muslin ends, and ~70 eggs were laid on *Setaria glauca* (Yellow Foxtail) and inert surfaces during 8 days at 25–28 °C and partial sunshine. After 14–21 days, eggs had dark micropyles, indicating mature embryonic L1 were present, and on Aug 30, the eggs were placed at 5 °C for overwintering. Eggs were removed from overwintering on Jan 4 and 25 and held at 25 °C under 12 hr light. Eggs hatched within 48 hrs, with the larvae eating the tops of the eggs, leaving the remainder intact. The larvae immediately commenced feeding on *S. glauca*, eating blade edges.

Adult

Development from L1 to L6 took 60 days at 20–26 °C and 16 hr light, with 10–19 days spent in each instar. L6 fed for a few days then entered dormancy and remained immobile and nonfeeding for about 80 days. Total development period from egg-hatch to pupation occupied 143 days, with eclosion after a further 15 days. On Aug 7, a female from Parachute Meadows, Okanogan Co., WA, produced 2 eggs. One egg hatched, and the larva fed, then diapaused in L1 but died before spring. The other egg overwintered and hatched the following spring, developing to L2 before dying. On Aug 7 (different year), a female from Slate Peak, Okanogan Co., produced a single egg. It was overwintered 3 weeks later. The following spring it hatched 3 days after exposure to warm conditions but later died. Leaf nests are used throughout development, with complexity increasing in each successive instar. L1 form slightly rolled blade shelters secured with a few strands of silk. Late instars form messy but secure nests from multiple blades. Frass is generally stored within shelters. Pupation occurred in a silken cocoon within a final shelter, lined with flocculent material produced by abdominal glands. Larval defense is based on concealment, and there were 6 instars in our rearing. The overwintering stage is the egg (usually) or L1.

Description of Immature Stages

The creamy white egg is textured with hundreds of tiny polygons and develops a pinkish tint after 5–7 days. L1 is orange-brown, becoming greenish tan sparsely bristled with tiny black setae. L2 is greenish yellow and the head is shiny black. A black collar anteriorly edged in white persists throughout development. L3 is similar, with increased green-tan coloration and increased numbers of tiny black setae bristling the body, increasing further at each molt. L4 is orangish brown, some individuals dark brown. The head is dark brown-black with 2 vertical pale orange stripes and an inverted V at the base. L5 is similarly colored, tending darker brown with 5–6 transverse ridges on the posterior half of each segment. There is an indistinct middorsal dark stripe. L6 is dark brown with a reddish purple cast and an indistinct middorsal dark stripe. The brown head is marked as in L4–L5. The pupa is dark olive green on the head, thorax, and wing cases. The abdomen is greenish white, flecked with irregular brown markings, and there is a dark stripe dorsally. Immature stages of *H. c. manitoba* may be confused with those of the closely related *H. colorado*. Larvae of the latter are very similar, although lack the middorsal dark stripe seen in mature *H. c. manitoba*

Egg @ 0.9 mm

L6 @ 32 mm (head)

L1 @ 2.5 mm

Pupa @ 18 mm

L2 @ 7 mm

L3 @ 11 mm

larvae. Pupae of *H. colorado* have a W mark on the thorax and lack a dorsal stripe. Mature larvae of *H. juba* and *H. nevada* have black rather than brown heads, less complex pale markings on the head capsule, and are generally darker. A mature larva of *H. comma* (*laurentina*?) from Quebec is shown in Allen et al. (2005) and appears similar to ours.

L4 @ 15 mm

Discussion

Hesperia spp. including *H. c. manitoba* are easily reared. Gravid females confined with grass produce eggs that usually overwinter. Final-instar dormancy occurs in *H. c. manitoba*, similar to that seen in *H. colorado* and *H. juba*, extending larval development to at least 3–4 months, even under warm temperatures (James, 2009b). The function of delayed summer development is unclear but likely related to synchronization of adult eclosion with optimal conditions for survival and reproduction. Research on host-plant choice and use is needed. *Hesperia comma* in the UK is restricted to *Festuca ovina*, which to date is the only grass recorded as a host for *H. c. manitoba*. Other aspects of the biology of this species would also reward study, including possible variation in instar number, incidence of L1 overwintering, impact of natural enemies on populations, and identification of critical habitat features similar to that reported for *H. comma* in the UK (Thomas et al., 1986).

L5 @ 20 mm

L6 @ 32 mm

Common Branded Skipper

Hesperia comma manitoba (Scudder)

Western Branded Skipper   *Hesperia colorado* (Scudder)

Adult Biology

Hesperia colorado is part of the *Hesperia comma* complex, which occurs across boreal North America and Eurasia (Forister et al., 2004). In areas E of the Cascades in WA and OR, *H. colorado* is widespread (Pyle, 2002; Warren, 2005), occurring commonly in a wide range of habitats, including grasslands, dry steppe, and alpine meadows. A single brood in Cascadia flies from early May to late Sep, depending on elevation and seasonal conditions. Males visit mud and are territorial, perching or hill-topping. Both sexes visit nectar. Larval hosts are not reported locally but in CO are various grasses, including *Bromus*, *Bouteloua*, *Andropogon*, and *Lolium* (Scott, 1992). Eggs are laid singly at the bases of host plants and nearby surfaces.

Immature Stage Biology

We reared this species twice from gravid females obtained from Yakima Co., WA (June 29), and Wallowa Co., OR (Aug 20). Females in plastic boxes (30 × 23 × 8 cm) with muslin lids laid eggs on potted or cut *Setaria glauca* (Yellow Foxtail) within 24–48 hrs. The majority of eggs in both cohorts held outdoors remained dormant until Jan–March; however, about 25% of Yakima Co. eggs hatched after 45–65 days during late Aug and Sep into dormant L1 that did not feed, but rasped grass-blade surfaces and imbibed released water droplets. They also produced slight webbing to cover themselves. Overwintering eggs contained embryonic L1, as indicated by dark micropyles and verified by dissection. A few larvae in both cohorts hatched under outdoor ambient conditions during Jan–Feb. Most (~95%) eggs placed in summerlike conditions (25 °C and continuous lighting) during Jan–Feb failed to hatch but remained viable, suggesting that holding conditions were suboptimal. Eighteen larvae were reared, 8 of which overwintered as eggs and 10 as dormant L1. Initially, development of larvae in both cohorts was rapid,

feeding on *S. glauca* and *Elytrigia repens* (Quackgrass), and L5 was reached 36–39 days after exposure to summerlike conditions. Larvae in the Wallowa Co. cohort were L5 for 27 days before pupating; adults eclosed 13 days later. Larvae in the Yakima Co. cohort spent only 12 days as L5 but entered L6, which persisted for 46 days before pupation; adults eclosed 18 days later. Thus, both cohorts had dormant final instars with development to adulthood delayed. All instars silked grass blades together to form tubular nests, with construction more complex in each instar. On one occasion a late instar removed paper toweling used as a jar plug and constructed a shelter from it. Final instars constructed untidy but tightly woven nests. Pupation occurred in a silken cocoon within a final nest. Some prepupal wandering was observed. The cocoon was liberally decorated with flocculent material produced by ventral abdominal glands. Larval defense is based on concealment, although in our rearings predation by the Minute Pirate Bug (*Orius tristicolor*) was a significant problem, with these small (0.5–2.0 mm) predators readily accessing larval shelters.

Description of Immature Stages

The creamy white, textured egg develops a pinkish tint after 5–7 days. There is a distinct lip at the egg base. L1 is pale yellowish green (nondormant) or uniformly yellow-orange in dormant, overwintering individuals. The head is shiny black with a dorsal black collar on the first segment. L2 is pale yellowish green with increased speckling of tiny black setae. The head is shiny black, and the black collar has a white anterior edge that persists throughout development. L3 is similar but greener, with a further increase in tiny black setae. L4 is whitish to tan with profuse speckling of tiny black setae. There is a pair of small black dots dorsally on each segment. The dark brown head is textured with distinct, pale vertical parallel stripes and an inverted V at the base. L5 is olive brown to gray with pale patches on the brown head, in addition to the stripes. Five transverse ridges occur on the posterior half of each segment. L6 is similar, with increased light brown-orange replacing the gray-olive cast. The head is light brown with the same lighter-colored vertical stripe pattern. The pupa is yellowish tan tending to green on the thorax, with a slight waxy bloom, particularly on the wing covers. On the thorax there are 2 or 3 dark wavy lines; 1 approximates a W shape. Brown-black dots and dashes sparsely cover the abdomen. The larva of *H. colorado* may be confused with other *Hesperia* spp. Larvae of the closely related

Adult

Egg @ 0.8 mm

L6 @ 30 mm (head)

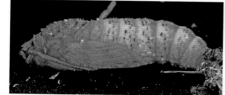

Pupa @ 22 mm

L1 @ 3.5 mm

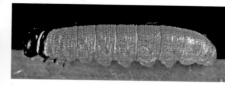

L2 @ 7 mm

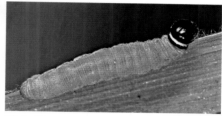

L3 @ 10 mm

L4 @ 16 mm

L5 @ 22 mm

L6 @ 30 mm

H. comma manitoba are very similar but have an inconspicuous dark dorsal stripe in the final instar. Pupae lack the thorax W mark and also have a dark dorsal stripe. Mature larvae of *H. juba* and *H. nevada* have black rather than brown heads and are generally darker colored. Our images appear to be the first published of immature stages of this species, except for those in James (2009b).

Discussion

Our rearing of *H. colorado* identified some important aspects of biology not previously recognized (see James, 2009b, for further information). Overwintering of species in the *H. comma* complex is reported to occur in the egg stage over most of its temperate range (Hardy, 1954; Scott, 1986a; Allen et al., 2005), although older larvae and pupae may overwinter in the Arctic, where it is biennial (Scott, 1986a). In our rearing, most individuals overwintered as eggs, but a significant number in the Yakima Co. cohort overwintered as L1. A second dormancy or summer diapause was seen in final-instar larvae. During this apparent aestivation, no feeding occurred. The function of delayed development in the final instar is unclear but likely a hedge against unfavorable conditions. Variation in final-instar dormancy length might explain the protracted flight period of this single-brooded species. Other aspects of the biology of this species would reward study, including the impact of natural enemies on populations and larval host preferences.

Western Branded Skipper   *Hesperia colorado* (Scudder)

Nevada Skipper   *Hesperia nevada* (Scudder)

Adult Biology

Hesperia nevada occurs in most W Rocky Mountain states but is localized in Cascadia, found on xeric, shrub-steppe ridges and plateaus in S central BC, central WA and NE, and S central OR. A single generation flies early May–late July depending on latitude and seasonal conditions. Populations in Yakima and Kittitas Cos., WA, on Manashtash and Umtanum ridges and nearby high steppe areas, fly late May–late June and can be locally common. Under cool, partially cloudy conditions, adults are thermally constrained and easily captured. In warm sunshine they are very difficult to pursue and capture. Males perch on rocks and along ridge or fence lines, darting out to challenge intruders. Females tend to occur a little below the highest points, resting in grass or nectaring. Larval hosts are grasses, including *Achnatherum* (*Stipa*), *Sitanion* (*Elymus*), *Festuca, Koelera*, and *Bouteloua* (Scott, 1986a, 1992; Warren, 2005); details of host choice and use in Cascadia are lacking. A female observed at Umtanum Ridge showed oviposition behavior on June 3 at the base of *Achnatherum occidentalis* (Western Needlegrass). Eggs are laid singly on the underside of low grass blades (Scott, 1992).

Immature Stage Biology

We reared this species twice from females from Umtanum Ridge, Yakima Co., WA. Females obtained June 3–5, confined in a small plastic cylinder (13 cm dia × 8 cm tall) with muslin ends, laid 63 eggs singly or in pairs on cut *A. occidentalis* and *Elymus elymoides* (Squirreltail) over 5 days. On the second occasion, females collected from Umtanum Ridge on June 3 (different year), confined in a plastic cylinder (13 cm dia × 24 cm tall), laid ~50 eggs on potted *A. occidentalis* during 4 days. On both occasions, most eggs were laid on grass stems and inflorescences. Eggs hatched in 9 days at 18–25 °C, and L1 wandered before settling and feeding on blade edges. Many

L1 appeared to move toward light, to developing or senesced inflorescensces of *A. occidentalis* and *E. elymoides*, where feeding began. These hosts were used by L1 in both cohorts, with *Cynodon dactylon* (Bermuda Grass) also supplied for L2–L4 (cohort 1) and *Bromus carinatus* (California Brome) and *Setaria glauca* (Yellow Foxtail) provided for L3–L5 in cohort 2. L1 and L2 produced slight silking when resting, but no nests were formed, with much movement between feeding sites. L1 lasted 8–9 days, L2, 11–13 days, and L3, 23–29 days at 18–27 °C under natural daylengths; 2 larvae reached L4 in the first cohort (Aug 4) and 7 in the second (July 27). Feeding was reduced in L3 and very limited in early L4, indicating diapause onset. Tubular nests were constructed from grass blades, stalks, inflorescences, and silk during L3–L4, always on the ground. L4 remained in nests, only rarely emerging to feed at night. L4 in the first cohort were transferred to pre-overwintering conditions (12 °C) on Aug 18; both larvae died in early Sep. Cohort 2 larvae (4 L3, 3 L4) were overwintered (5 °C) on Aug 13. The cohort was removed Oct 24–30 and placed at 27 °C under continuous lighting. Feeding began within 48 hrs, new nests were built, and pupation began within 23 days. L5 remained largely hidden in nests on the ground. Pupation occurred within a cocoon constructed in the final nest from silk and flocculent material produced by ventral glands. L4 and L5 did not tolerate each other well, and some cannibalism occurred. Mortalities to L1–L3 occurred in our rearings from the Minute Pirate Bug (*Orius tristicolor*). Larval defense in *H. nevada* appears to be based largely on concealment, but early instars appear to be poorly protected. We observed 5 instars, and L3 or L4 overwintered.

Description of Immature Stages

The pure white, weakly reticulated egg becomes greenish after 24–48 hrs. A tiny spike is often seen protruding from the egg of this and some other *Hesperia* spp. (see image). L1 is white and has a shiny chestnut brown head. The body is sparsely covered (10–12/segment) with short dark brown setae, and there is a dorsal collar on segment 1 banded white anteriorly, black posteriorly, and brown in the middle. L2 is whitish green, fairly translucent, and speckled with many (~180/segment) tiny black setae, which increase in number at each successive molt. The collar, banded black-and-white, persists throughout development, and the black head has 2 vertical white stripes extending from the top to about midway down the "face." L3 is similar but darker green, becoming

Adult

Eggs @ 1.1 mm   **L5 @ 25 mm (head)**

L1 @ 4 mm

L2 @ 7 mm

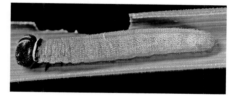

L3 @ 12 mm

L4 @ 16 mm

L5 @ 25 mm

Pupa @ 20 mm

brownish with maturity. There is an indistinct middorsal dark green stripe. The head is black, and the white vertical stripes meet a weakly defined inverted V at their base. L4 is dark reddish brown with a dark middorsal stripe. Each segment has 5–6 transverse folds on the posterior half, persisting to L5. The head is black with well-defined white stripes and orange inverted V. L5 is dark purplish brown with a similar head. Each segment has 5–6 transverse folds on the posterior half, and there is a dark middorsal stripe. Ventrally there are 2 pairs of patches of a white flocculent secretion between abdominal segments 7–8 and 8–9. This secretion adheres strongly and appears to be more profuse in *H. nevada* than in other Cascadia *Hesperia* spp. The pupa has a dark brown thorax and head and black wing cases. The abdomen is light brown with dark markings. Mature larvae of *H. colorado* and *H. comma* have brown rather than black heads. The mature larva of *H. juba* is very similar, but early instars are brown. All Cascadia *Hesperia* spp. have similar head capsule markings, but only *H. nevada* develops them in L2. Our images appear to be the first published of immature stages of this species.

Discussion

MacNeill (1964) provided a detailed description of immature stages of *H. nevada* with L2–L3 brown rather than green, as in our rearing. Variation in larval coloration may occur across the extensive range of *H. nevada* as appears to occur in other *Hesperia* spp. The function of the tiny spike seen on the eggs of this and other *Hesperia* spp. is unknown and deserves study. Our rearing indicates overwintering occurs as L3–L4, enabling the relatively early flight period. The larval host preferences of this species in Cascadia require determination. This species appears to be quite specific in its habitat requirements. Little is known of natural mortality factors or their importance.

Sachem *Atalopedes campestris* (Boisduval)

Adult Biology

The Sachem ranges throughout S US and Mexico, expanding and contracting its range annually. It is a relatively recent addition to Cascadia, expanding N during the past 4 decades. Prior to the mid-1960s, *A. campestris* was a rare summer migrant (Guppy and Shepard, 2001), but since 1967 has colonized W OR, including the Willamette Valley (Dornfield, 1980; Warren, 2005). In 1990, Pyle (2002) found it on the WA side of the Columbia R and it has since spread to the Columbia Basin and along the Snake R into Wallowa Co., OR and W ID. It has also reached Yakima, WA, with the current range edge S of Ellensburg (Crozier, 2004). Crozier (2003, 2004) presented experimental evidence that a 3 °C increase in minimum winter temperatures in E WA during the past 50 years facilitated the expansion. Most frequently found in parks, gardens, pastures, and roadsides, it also occurs in open woodlands and prairies (Warren, 2005). There are 2–3 broods, flying mid-May–late Oct with successive generations increasingly abundant. Males perch in sunshine on the ground or low vegetation. Adults visit flowers, including thistle, buddleia, marigold, milkweed, and aster. Larval hosts are grasses, with *Poa pratensis* (Kentucky Bluegrass), *Cynodon dactylon* (Bermuda Grass), *Digitaria* spp. (crabgrass), *Eleusine indica* (Indian Goosegrass), *Festuca rubra* (Red Fescue), and *Stenotaphrum secundatum* (St. Augustine Grass) recorded (Scott, 1986a; Pyle, 2002). Warren and Roberts (1956) reported economic damage to *C. dactylon* in AK by *A. campestris* larvae. Eggs are laid singly on grass blades (Allen, 1997).

Immature Stage Biology

Females collected May 23 at Richland, Benton Co., WA, produced 25 eggs, hatching in 8–10 days. Larvae were reared on lawn grass (*Poa* sp.) at 20–22 °C, and

development from hatch to pupation took 30 days. Nine pupated and eclosed to adults by July 20. The fall (presumably third) brood was reared from females obtained Oct 15 (Benton City, Benton Co., WA) and confined with *P. pratensis* in a plastic box with a muslin lid (28 × 16 × 7 cm). More than 200 eggs were laid during 6 days on inflorescences and stems, hatching in 7 days at 21–25 °C. L1 were divided into 2 groups; 1 was held outdoors in a shaded location, the other transferred to 28 °C with continuous lighting. The outdoor larvae fed little and were still L1 at the end of Nov. The temperature fell to -14 °C and was below 0 °C for a week in Dec, killing all larvae. At 28 °C, development from L1 to pupation took 45 days. L5 and pupae held at 15–21 °C occupied 56 and 33 days, respectively. As a subtropical species, *A. campestris* has no stage with hibernal diapause, but likely overwinters in Cascadia as larvae. Crozier (2004) suggested that most larvae in Benton Co., WA overwinter as L3. Crozier (2003) showed that the lethal temperature for L3 and L5 was -5 to -6 °C; high mortality also occurred under constant 0 °C conditions. Nevertheless, observations at Benton City showed that *A. campestris* survived the winters of 2001–2005, when minimum temperatures fell as low as -25 °C and temperatures remained below 0 °C for extended periods. Adults regularly reappeared in early–mid-May. All instars silked grass blades together, with shelters more complex in each successive instar. L1 simply wove a few silken strands into a vague "nest," whereas prepupal L5 constructed untidy but tightly woven silk-lined shelter tubes. Nests are usually at ground level, where they escape destruction by lawn mowers. Pupation occurred on the ground within a silken cocoon lined with flocculent material produced by ventral abdominal glands. Larval defense is based on concealment, although in our rearings predation by the Minute Pirate Bug (*Orius tristicolor*) was a significant problem, with these small (0.5–2.0 mm) predators readily accessing larval shelters.

Description of Immature Stages

The unsculptured egg is creamy white, developing a pinkish gray tint. L1 is grayish white; the head and collar are shiny black and remain so to pupation. The body and head are sparsely dotted (~20/segment) with tiny black setae; 3 long posterior setae project backward. L2 is yellowish green with an indistinct middorsal dark stripe clothed with many tiny dark setae (~65/segment). The collar has an anterior white margin. L3 is similar, yellowish green with tiny black setae (~250/segment) and an indistinct middorsal

Adult

Egg @ 0.9 mm

L1 @ 2.0 mm

Pupa @ 19 mm

L2 @ 6 mm

L3 @ 9 mm

dark stripe. L4 is olive brown with coalescing tiny black spots and setae. Each segment has 5 transverse folds posteriorly, and there is a distinct dark brown middorsal stripe. L5 is similar but brown, occasionally gray. The pupa is dark brown-black dorsally and on the head, thorax, and wing cases, with a slight powdery bloom. Ventrally, the abdomen is lighter with black spots. The larva of *A. campestris* may be confused with other grass-feeding skippers, particularly *Hesperia* and *Polites* spp. *Polites sabuleti*, *P. sonora*, *H. juba*, and *O. sylvanoides* may co-occur with *A. campestris*; late-instar *H. juba* and *P. sabuleti* have pale-colored stripes on the head, and the head of mature *O. sylvanoides* is brown. Larvae of *P. sonora* are most similar but are generally darker, as is the pupa. Images of *A. campestris* larvae appear in Allen (1997), Wagner (2005), Allen et al. (2005), and Minno et al. (2005); all are similar to ours. Allen (1997) shows a pupa that appears to be lighter and has more powdery bloom than ours.

L4 @ 16 mm

Discussion

The Sachem is easily reared, is relatively fast developing, and has no obligatory dormancy stage. The N and E range expansion of this species should be monitored. Continued residence of *A. campestris* in the Columbia Basin, despite frequent occurrence of cold temperatures during 2001–2005 shown to be lethal by Crozier (2003), suggests significant variation in tolerance within populations or behavior that ameliorates the effect of low temperatures. The impact of natural enemies on *A. campestris* is unknown, but larval parasitism by wasps was noted by Warren and Roberts (1956).

L5 @ 30 mm

Peck's Skipper   *Polites peckius* (W. Kirby)

Adult Biology

Polites peckius is primarily an E US species with a distribution that reaches the E edge of Cascadia. The range includes most of Canada and the US except the far W and S. In Cascadia, *P. peckius* occurs in SE BC, NE WA, N ID, and NE OR, a single brood flying mid June–late July. There are 2–3 broods in S areas (Scott, 1986a; Layberry et al., 1998). The habitat in E US includes meadows, power-line cuts, suburban lawns, and roadsides (Allen et al., 2005). In the NW US it is restricted to undisturbed habitats (Pyle, 2002), mostly moist grassy meadows, often near water and along riparian corridors. In OR populations occur at 3,000–5,000 ft (Warren, 2005). Adults visit flowers, including *Leucanthemum vulgare* (Oxeye Daisy) (Pyle, 2002), *Vicia* spp. (vetch), *Prunella vulgaris* (Self-heal), and *Trifolium* spp. (clover). Larval hosts are grasses, including *Leersia oryzoides* (Rice Cutgrass) in Manitoba and E US, and Scott (1992) reported *Distichlis spicata* (Saltgrass), *Poa pratensis* (Kentucky Bluegrass), and probably *Bromus inermis* (Smooth Brome) in CO. Adult males perch conspicuously on tall grasses, while females fly low to the ground seeking nectar or oviposition sites. Eggs have no glue and are dropped into the grass (Scott, 1992).

Immature Stage Biology

We partially reared this species several times, and females laid infertile eggs on 3 occasions. On July 1, 3 females from Tiger Meadows, Pend Oreille Co., WA produced 27 eggs that were dropped into grass. Eggs began hatching at ~9 days, and larvae were placed on cut lawn grass (*Poa* and *Festuca* spp.) at 20–22 °C. The larvae fed poorly and did not survive to L2. On July 10 (different year), 4 females from the same location produced 75 eggs that began hatching after ~8 days. Most eggs hatched, and larvae were reared outdoors on live potted *P. pratensis*, at ~15–25 °C. They grew slowly, 48 reaching L3 in ~40 days, and continued to feed sluggishly for another month, gradually entering diapause. None of the larvae survived overwintering. L1 did not consume eggshells. L1–L2 curled up in tied, bundled grass nests with a thin silk sheath inside. The vertical nests were constructed low on the grass, many almost touching the soil. Feeding larvae strongly preferred the sunny side of the plant; frass sprinkled throughout the enclosure suggested ejection. As larvae approached diapause, new nests, oriented horizontally, were constructed. The bundled grass-blade nests were tied with silk strands and lined with tightly woven silk; they were placed mostly at the base of grass, some in the upper layers of the soil. Scott (1986a) mentions the presence of nests, also the apparent absence of them (Scott, 1992). Scott (1986a) reported that larvae and pupae hibernate, while Guppy and Shepard (2001) reported that the mature larva is the presumed overwintering stage. Guppy and Shepard (2001) reported that pupation occurs inside a loose cocoon constructed from a bent grass blade. There are 5 instars and protection is based on concealment, frass ejection to misdirect predators, and probably nocturnal feeding.

Description of Immature Stages

The greenish egg is smooth; Scott (1986a) reported that it matures to irregular reddish mottling. L1 is light tan with numerous dull reddish freckles; each segment has several small dark bullae, each with a single long seta. There is a distinct black collar; the bifurcated head is shiny black and bears short setae. L2 is darker brown with numerous small black bullae and short brown setae. The black collar has separated spots at the lateral tips and bears numerous setae. The diapausing L3 is darker, blackish with small white vermiform markings, and bears numerous setae. The black collar is bordered anteriorly with a contrasting

Adult

Adult male

white line, and the head is black and densely covered with short setae; the true legs and prolegs are dark. Scott (1992) described the mature larva as dark brown with a blackish heart-line along the abdomen, the posterior segment bearing 2 black U's side by side, the open ends facing anteriorly. He added that there is a black collar and the head is black with a brown stripe and a brown crescent in front of the eyes. Wagner (2005) described these head markings as "indistinct white" and also reported that wax glands are absent. Scott (1992) described the pupa as brown-black with 2 blue-gray spots on the vertex, and with the wing cases and legs blue-gray on black. The larvae of *P. mystic* and *P. sonora* are lighter colored and those of *P. mardon* are greener, but the larvae of *P. themistocles* and *P. sabuleti* are similar. The larvae of *Hesperia* spp. are also similar but generally greener or lighter than *P. peckius*. The "facial" head patterns are important in grass skipper identification; however, published images (lateral views) suggest L5 "facial" markings in *P. peckius* are not very distinct (Wagner, 2005; Allen et al., 2005). In this respect, *P. peckius* appears to closely resemble *P. mystic* and *P. sonora*. Scott (1992) suggested that *P. peckius* is closely related to *P. mystic* and *P. sonora* because all lay unattached eggs, the pupae are blackish, and the pupal proboscis is longer than the cremaster. Published images of mature larvae of *P. peckius* are found in Wagner (2005) and Allen et al. (2005), and Allen (1997) illustrated both L5 and the pupa.

Discussion

This species is easily reared in regions where it is double or triple brooded; however, in Cascadia it is restricted in distribution and fairly uncommon, making it difficult to obtain. Gravid females produce eggs fairly well in captivity, although many may be infertile. Larvae are best reared in relatively dry conditions and fail to survive if kept too moist; however, overwintering larvae can succumb to desiccation. Larvae are difficult to find as they make nests deep in the grass, at or slightly below ground level. This habit has allowed *P. peckius* to include suburban lawns among its habitats in E US, safe from most lawn mowers. While this species is common and fairly well documented elsewhere, little is known of its life history in Cascadia or of natural enemies and other factors affecting its distribution and population dynamics. It is unknown why *P. peckius* has not colonized seemingly suitable habitats in W Cascadia and CA. Some of the best *P. peckius* habitat in NE WA. (e.g., Tiger Meadows) has recently been invaded by *Thymelicus lineola* (European Skipperling), and it remains to be seen whether competition will adversely impact its populations there.

Egg @ 0.7 mm

L1 @ 2.0 mm

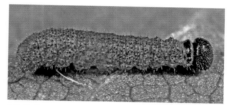

L2 @ 4 mm

L3 @ 7 mm

L2 tied grass nest

Peck's Skipper *Polites peckius* (W. Kirby)

Sandhill Skipper   *Polites sabuleti* (Boisduval)

Adult Biology

Polites sabuleti is found in W North America from S BC to Baja California and across the Great Basin to AZ and CO. In Cascadia it occurs E of the Cascades in WA and OR in a variety of habitats, including lawns, parks, alkali flats, marsh edges, dry hillsides, and, in S OR, higher-elevation meadows. It is common in the Columbia Basin of WA, occurring mostly on lawns and in parks in cities and towns (Newcomer, 1966) and on dry hillsides near rivers. At lower elevations, *P. sabuleti* is double brooded, flying in May and June with a larger brood in Aug–Sep. At higher elevations it is single brooded, flying from July to Sep. Males visit mud, are territorial, and guard perches from near ground level. Both sexes visit flowers, including thistles, asters, rabbitbrush, clover, dandelions, and Alfalfa. Larval hosts are various grasses, including *Distichlis spicata* (Saltgrass), *Cynodon dactylon* (Bermuda Grass), *Poa pratensis* (Kentucky Bluegrass), and *Festuca idahoensis* (Idaho Fescue) (Scott, 1986a; Warren, 2005). Scott (1992) additionally reported oviposition on *Hordeum jubatum* (Foxtail Barley) and *Puccinellia distans* (Weeping Alkali Grass) in CO, and found that grazing does not adversely affect abundance and may be beneficial. Eggs are laid singly on hosts, nearby plants, or soil (Scott, 1986a).

Immature Stage Biology

We reared this species twice from gravid females. One fall brood female obtained from Hat Rock State Park, Umatilla Co., OR, on Sep 27 confined with *Setaria glauca* (Yellow Foxtail) in a plastic cylinder (13 cm dia × 8 cm tall) with muslin ends laid 40 eggs during 7 days at 18–25 °C. Eggs hatched after 9 days at 15–21 °C, and L1 were held outdoors from Oct 7 and fed on *S. glauca*. Development was slow, and the cohort was transferred to 28 °C and continuous light on Oct 25. Within days some of the cohort molted to L2, and development was rapid, with 11, 8, 5, and 5 days spent as L2–L5, respectively. Pupae were held at 15–21 °C

for 20 days then transferred to 27 °C, where eclosion began after 4 days. Many (~50%) larvae did not develop rapidly when transferred to warm conditions in late Oct, languishing and dying, suggesting they had entered diapause. On June 1, in Richland, Benton Co., WA, 12 spring-brood females were collected and produced ~100 eggs, singly on inert surfaces, none on the provided grass. Egg-hatch began after 6 days, and larvae developed very slowly through 4 instars, feeding on cut lawn grass at 20–22 °C. Mortality was high, and feeding limited, indicating larvae had entered summer dormancy or aestivation. Pupation occurred after 63 days, adults eclosing 98 days post egg-hatch (Sep 16). According to Newcomer (1966) and Dornfield (1980), this species overwinters as pupae; however, Scott (1992) wrote that "larvae must hibernate about half grown" in CO. Larvae feed on blade edges and build silk-tied nests. L1 and L2 rolled blade edges to form lightly silked tubular nests in the fall brood; however, in spring, nests were not built until L2. L1 wandered off the grass and aggregated under cover. The nests of L3–L5 are clusters of grass blades tied together in untidy, loosely constructed bundles, and larvae spend much time out of them. Frass accumulates in the nests. Late instars readily drop and curl up when disturbed, more so than in other *Polites* spp. Prepupal larvae leave nests and construct loose silk "cocoons" on inert surfaces and pupate within them, vertically or horizontally. Survival appears to be based on concealment, particularly in the early instars.

Description of Immature Stages

The greenish white egg is covered with numerous shallow divots. L1 is yellowish green, sparsely covered with short black setae. The head is shiny black with a dark brown collar. L2 is brown, finely speckled with white dots and tiny black setae. There is a dark middorsal stripe and the head and collar are black. L3 is grayish green peppered with tiny black setae, and the black collar is edged in white anteriorly. The posterior segment has a sclerotized black dorsal patch. L4 is orangish to dark brown with a distinct middorsal dark stripe and indistinct dorsolateral stripes. The black head is marked with 2 pale stripes that run from the top to the mandibles and less prominent pale marks ventrolaterally. L5 is dark brown with paler areas and has a distinct black middorsal stripe and indistinct dorsolateral and ventrolateral dark stripes. Each segment has 5 transverse folds posteriorly. The head is marked as in L4, and the posterior segment is marked dorsally in black and white. Secretions from ventral glands between segments 7–8 and 8–9 are

Adult

Egg @ 0.9 mm

L5 @ 27 mm (head)

L1 @ 3 mm

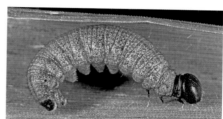

Pupa @ 15 mm

L2 @ 7 mm

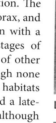

minor compared with those of other *Polites* spp. and not used greatly in cocoon construction. The pupa is dark green-brown on the head, thorax, and wing cases. The abdomen is yellowish tan with a middorsal dark stripe. The immature stages of *P. sabuleti* could be confused with those of other *Polites* spp., particularly *P. mardon*, although none coexist commonly in the same regions and habitats as *P. sabuleti*. Allen et al. (2005) illustrated a late-instar larva from UT similar to ours, although more reddish brown.

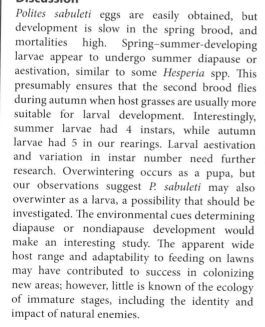

L3 @ 14 mm

Discussion

Polites sabuleti eggs are easily obtained, but development is slow in the spring brood, and mortalities high. Spring–summer-developing larvae appear to undergo summer diapause or aestivation, similar to some *Hesperia* spp. This presumably ensures that the second brood flies during autumn when host grasses are usually more suitable for larval development. Interestingly, summer larvae had 4 instars, while autumn larvae had 5 in our rearings. Larval aestivation and variation in instar number need further research. Overwintering occurs as a pupa, but our observations suggest *P. sabuleti* may also overwinter as a larva, a possibility that should be investigated. The environmental cues determining diapause or nondiapause development would make an interesting study. The apparent wide host range and adaptability to feeding on lawns may have contributed to success in colonizing new areas; however, little is known of the ecology of immature stages, including the identity and impact of natural enemies.

L4 @ 18 mm

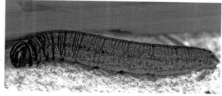

L5 @ 27 mm

Mardon Skipper   *Polites mardon* (W. H. Edwards)

Adult Biology

The world range of *Polites mardon* is restricted to a small number of sites in Thurston, Yakima, Skamania, and Klickitat Cos. in WA, Jackson and Klamath Cos. in OR, and Siskiyou Co., CA. It is listed as a federal candidate endangered species and is state-listed as endangered in WA. On the E slopes of the Cascades, *P. mardon* occurs in meadow habitats at elevations of 2,000–6,000 ft. In Thurston Co. it occurs in prairies of the S Puget Trough, and in S OR in wet meadows at about 5,000 ft. Most colonies consist of <50 individuals, although some colonies number >200 individuals (Jepsen et al., 2007). One generation flies from early May (Puget Prairies) to June–July (Cascade sites), and the flight period is 2–4 weeks. Adults fly low visiting flowers, including dandelions, vetch, asters, wallflower, and violet. Males are territorial, perching on low vegetation or rocks. Larval hosts are native bunchgrasses and sedges, including *Festuca idahoensis* (Idaho Fescue), *F. rubra* (Red Fescue), *Danthonia californica* (California Oatgrass), and *Carex* spp. (Pyle, 2002; Warren, 2005; Beyer and Black, 2007). Beyer and Schultz (2010) listed 22 grasses and sedges on which oviposition was observed in the S Cascades of WA. Eggs are laid singly at the bases of host plants.

Immature Stage Biology

We reared this species twice under permit from the WA Dept of Fish and Wildlife (WDFW). On June 29, 2 females from near Mt Adams produced 21 eggs. Most hatched, with 4 surviving to late L1, feeding on live lawn grass before dying. On June 9, 3 females from Scatter Creek, Thurston Co., produced many eggs; larvae were reared at 20–22 °C on cut lawn grass. Larvae developed quickly, 8 pupating 37 days post egg-hatch, with 1 adult eclosing 11 days later. Remaining pupae were stored at 4–5 °C for overwintering. Eggs are laid haphazardly, glued lightly on inert surfaces or at the bases of grasses; oviposition is enhanced

L2 tied grass nest

by sunshine. Early instars eat grass blades partially through from the side. Later instars eat through the tender grass tips, causing much of the blade to fall to the ground. L1 construct simple leaf shelters by silking over the edge of a grass blade. L2–L5 construct untidy vertical shelters among host grasses; at night they are often found outside their nests feeding. Frass is stored within the shelters. Throughout development, larvae rest with their heads turned back parallel to their bodies in a tight U shape. Later instars make fewer nests and feed almost entirely on tender leaf tips, resting at the cut-off ends. Prepupal larvae construct a stronger final shelter in which pupation occurs, typically horizontal and near the ground. There are 5 instars; protection is largely based on concealment, particularly in later instars. The pupa overwinters (Scott, 1986a; Pyle, 2002); however, reared larvae usually produce second-brood adults; this does not appear to occur in nature but could affect breeding and reintroduction programs.

Description of Immature Stages

The smooth, greenish white egg turns light orange. L1 is greenish white sparsely covered with short pale setae, becoming greenish tan with small brown spots. The head is dark brown, and there is a dark brown dorsal collar on segment 1. L2 is greenish brown with dark spots, each with a tiny black seta. The head is dark brown-black with 2 parallel lighter stripes. The dark brown collar is edged anteriorly in white. L3 is gray-brown speckled with white and black and has an indistinct middorsal black stripe. L4 is darker with more profuse white-and-black speckling and more contrasting head markings. L5 is dark brown to gray-black with reduced white speckling and a middorsal black stripe. Each segment has 3 transverse folds posteriorly. The head is black with prominent white-tan parallel stripes and smaller stripes laterally. The collar is black, edged anteriorly in white. The posterior segment is white and black dorsally. Secretions from ventral glands between segments 7–8 and 8–9 are minor. The pupa is dark green or blue-black on the head, thorax, and wing cases with a waxy pruinose bloom. The abdomen is yellowish tan with a middorsal dark stripe. Immature stages could be confused with *P. sonora*, which may occur in the same habitats, but

Adult male

Egg @ 1 mm

L5 @ 21 mm (head)

L1 @ 3 mm

Pupa @ 15 mm

L2 @ 5 mm

L3 @ 9 mm

L4 @ 13 mm

L5 @ 21 mm

its larvae are browner (*P. mardon* grayer) and have no white markings on the head. The larvae of *P. sabuleti* most closely resemble those of *P. mardon*, but the pupa is not as dark. Allen et al. (2005) illustrate a late-instar larva of *P. mardon* from CA similar to ours.

Discussion

The biology and ecology of *P. mardon* are receiving attention in Cascadia in light of its endangered listing (e.g., Beyer and Schultz, 2010), although most studies are directed toward the adults. More research is needed on the biology and ecology of immature stages. Our rearing demonstrated 2 broods in captivity, and the possibility of this occurring in populations in some locations should be investigated. Confirmation of the pupa as the overwintering stage is required (Beyer and Schultz, 2010, suggest that larvae overwinter) as well as details of the phenology of all immature stages under natural conditions. Beyer and Schultz (2010) showed that most ovipositions occurred on *F. idahoensis* and the introduced *Poa pratensis* (Kentucky Bluegrass); however, they also showed that *P. mardon* is a generalist in terms of host grass selection, able to use many grass species. Many *P. mardon* sites have some grazing by cattle, and studies are needed to clarify the impact, positive or negative, of grazing on survival of larval populations. The identity and impact of natural enemies on *P. mardon* populations also needs research. The recent discovery of multiple new colonies of *P. mardon* in the Rimrock Lk area of Yakima Co. (James, unpubl; Jepsen et al., 2007) suggests that habitat requirements are not as narrow as previously thought, and populations in central WA appear to be more widespread than previously believed.

Mardon Skipper

Polites mardon (W. H. Edwards)

Tawny-edged Skipper   *Polites themistocles* (Latreille)

Adult Biology

Widely distributed and common in much of E and central North America, *Polites themistocles* is restricted in Cascadia to the S interior of BC and NE WA. This is a lawn skipper and inhabitant of a variety of grassy areas. The habitat of *P. themistocles* in Cascadia is limited to moister areas in higher elevations such as pond and marsh margins, moist meadows, and stream margins. Males perch on low vegetation or short grass, and both sexes are avid flower visitors (e.g., dandelion, vetch, clover, asters). Double brooded or multivoltine in E and S US, *P. themistocles* has a single generation in Cascadia, flying late May–early Aug, peaking in late June–early July. Males eclose ~1 week earlier than females, which dominate toward the end of the 3–4-week flight period. Host plants are a variety of grasses and sedges, including *Panicum* spp. (panic grasses), *Digitaria filiformis* (Slender Crabgrass), *Poa pratensis* (Kentucky Bluegrass), *Glyceria* (mannagrass), *Dichanthelium aciculare* (Needleleaf Witchgrass), *Koeleria macrantha* (Prairie Junegrass), and *Carex* spp. (sedges) (Scott, 1986a; Allen, 1997; Warren, 2005). Eggs are laid singly on or near host grasses (Scott, 1992). Oviposition was observed at Mt Hull, Okanogan Co., WA, on short turf.

Immature Stage Biology

We reared or partially reared this species several times from gravid females. On July 10, a female from Tiger Meadows, Pend Oreille Co., WA, produced a single egg which hatched after 9 days. Reared on *P. pratensis* under natural daylengths (20–22 °C), the larva developed to L3 in 5 weeks, then it constructed a tightly woven multiple-blade vertical grass nest and diapaused head-down for ~1 month. When the grass senesced, the larva moved to an inert surface, where it wove a silken cocoon, but it failed to survive winter. Ten females from Mt Hull on June 23 were confined with *Cynodon dactylon* (Bermuda Grass) in a plastic container (30 × 20 × 10 cm) with a muslin lid. Exposed to partial sunshine at 21–30 °C, they laid ~250 eggs over 9 days. Most were on lower blades and stems, with some (~20) on cage surfaces. Eggs hatched after 9 days (22–27 °C), and L1 fed and built loose silk nests on single grass blades. Larvae were reared on *C. dactylon* under continuous lighting, pupating in 40 days. Second-brood adults eclosed after 10 days. Substantial mortality occurred during L1–L3, mostly due to excessive humidity. Survival during L4–L5 was good, with >50 adults produced. Nest building occurred throughout development, with multiple blades used as larvae aged. Late instars incorporated dead as well as fresh grass into nests. Some larvae rested openly in a J shape. Larvae possess a ventral gland on segment 1 from L3 onward. Pupation occurred in stronger nests formed on the ground or at the base of the grass. Contrary to Minno et al. (2005), no obvious wax glands were present in prepupal larvae. Scott (1986a) and Pyle (2002) reported overwintering as pupae, but Scott (1992) suggested that "half-grown larvae hibernate," as in our findings. There are 5 instars, and protection is based on concealment and perhaps chemicals from the ventral gland and L1 setal droplets.

Description of Immature Stages

The egg is pale green, white with reddish brown dots and dashes as it matures. L1 is pale green with a middorsal dark line and a shiny black head and collar on segment 1. The head and body are sparsely clothed (~20/segment) with short black setae, each bearing a small droplet. The setae on the posterior segment are long and swept backward. L2 is pale olive green with a shiny black head, black collar (bordered anteriorly in white), and indistinct middorsal dark line. L3 is mottled brown-gray with a prominent middorsal dark stripe and many brown-based short black setae. Above the prolegs there is a distinct fold and the posterior segment has a dorsal "flap." The black spiracle on segment 10 is larger and positioned higher than the rest. L4 is similar, but in some individuals, pale orange bars and an inverted V are present frontally on the black head. There is also an orange spot on each side of the head near the mouthparts. L4 has an indistinct dark stripe midlaterally. These markings are prominent in L5, which is mottled gray-brown. The dorsal "flap" is white, sometimes with black stripes. The pupa is greenish white with

Adult male

Egg @ 1.0 mm (fresh, *left*, maturing, *right*)

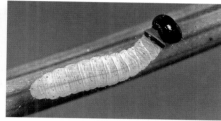

L1 @ 2 mm

Pupa @ 18 mm

L2 @ 4 mm (late)

L3 @ 9 mm

a dark middorsal stripe. The wing cases are dark olive green, and there is a dark orange spot behind each eye. The proboscis case is extended and tipped in orange. The immature stages of *P. themistocles* may be confused with those of other *Polites* spp., although only *P. peckius* and *P. mystic* occur in the same areas. Only *P. themistocles* has an orange-marked head. Images of late-instar *P. themistocles* were published in Allen (1997), Wagner (2005), Allen et al. (2005), and Minno et al. (2005); all are similar to ours, except that none appear to have orange-marked heads. However, Allen (1997) noted that "pale vertical stripes" may occur in mature larvae, and Dornfield (1980), Scott (1986a), and Guppy and Shepard (2001) all reported white-marked heads.

L4 @ 16 mm

Discussion

The oviposition period appears to be short, and young mated females must be obtained for egg laying. Larvae accept a range of grasses and sedges as hosts. Complete development was obtained under continuous lighting; rearing under declining daylengths caused L3 to overwinter. The chemistry and function of the ventral gland in L3–L5 and droplet-bearing setae in L1 should be investigated. The incidence of orange-striped heads in mature *P. themistocles* larvae deserves study to see whether this is a characteristic of Cascadia populations. The apparent restricted habitat requirements of *P. themistocles* in Cascadia compared with those of E US populations is intriguing and should be investigated. The decline of this species in some parts of the US has been blamed on human activities like increased use of urban pesticides (Pyle, 2002; Cech and Tudor, 2005).

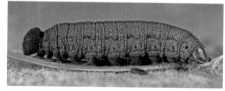

L5 @ 20 mm

L5 @ 20 mm (head)

Tawny-edged Skipper *Polites themistocles* (Latreille)

Long Dash   *Polites mystic* (W. H. Edwards)

Adult Biology

Polites mystic is widely distributed across the N US and Canada but in Cascadia is restricted to SE BC, NE WA, and N ID. It frequents wet meadows, bogs, streamsides, and marshy habitats in Cascadia but occurs in drier areas elsewhere. A single brood flies from early June to late Aug, peaking in June–July. Flight is rapid, and males perch on low vegetation in open areas. Courtship and mating occur in the afternoon, with oviposition observed from late morning to late afternoon (Allen, 1997). Both sexes nectar on a variety of flowers, including Self-heal, dogbane, milkweed, thistles, and clovers. Larval hosts are grasses, with *Poa* (bluegrass) most frequently cited (Shapiro, 1966; Scott, 1986a). Other grasses are likely used as well (e.g., *Agrostis*, *Agropyron*), but accurate records are lacking. Eggs are dropped to the ground. Scott (1992) noted that eggs are poorly glued to substrates and invariably drop off.

Immature Stage Biology

We reared or partially reared this species 6 times (different years) from gravid females obtained from Tiger Meadows and Big Meadow Lk, Pend Oreille Co., WA. One female obtained July 15 and confined with *Setaria glauca* (Yellow Foxtail) in a small plastic cylinder (13 cm dia × 8 cm tall) laid 10 eggs during 3 days. Two hatched after 8 days at 21–27 °C; both died in L1. Eggs from females obtained July 1 hatched and L1 were reared on *Festuca* sp., producing 35 diapausing L3 and L4 on Sep 4, but none survived the winter. On July 16, 2 females produced ~80 eggs; L1 were reared on *Festuca* sp., but all died prior to overwintering. Five females obtained July 10 produced 65 eggs; L1 were reared on *Poa pratensis* (Kentucky Bluegrass), 8 diapaused, but none survived the winter. Females obtained June 26 and July 22, confined with *Bromus carinatus* (California Brome) laid ~300 eggs during 4 days. Most oviposition occurred late afternoon,

eggs scattered loose. Eggs hatched after 8 days; L1 consumed eggshells and began feeding on and rolling edges of grass blades for shelter (see L1 image). Grasses consumed included *S. glauca*, *B. carinatus*, *Poa annua* (Annual Bluegrass), and *Lolium multiflorum* (Italian Ryegrass). Development was rapid in L1 and L2, occupying 8 and 9 days, respectively, at 21–27 °C under natural daylengths. Development slowed in L3, with some molting to L4 after 28–35 days (Aug 20–30), but most diapaused and remained in L3 for overwintering. Exposure of larvae to continuous lighting prevented diapause, with 13, 13, and 10 days spent in L3–L5, respectively, pupation occurring Sep 20 onward. Overwintering larvae constructed silk-lined loosely tied nests and were placed at 5 °C on Sep 10. Forty-four days later, 3 L3 and 1 L4 were removed and placed at 27 °C under continuous lighting. One L4 pupated 27 days later. Remaining overwintering larvae (6 L3) were placed at 20–25 °C under 15 hr light on Jan 15. All fed and molted to L4 and L5 after 16 and 27 days, respectively. L5 ($n = 3$) pupated 34–40 days post overwintering, and adults eclosed 11–14 days later. All instars silked grass blades together to form vertical, aerial tubular nests, with construction more extensive and untidy in successive instars. Frass scattered at night indicated nocturnal activity and frass shooting. Pupation occurred in a silken cocoon within the final nest or on nearby inert surfaces. The cocoon was liberally lined with flocculent material produced by ventral abdominal glands. Larval defense is based on concealment, and there are 4 or 5 instars. Overwintering occurs as L3 (~90%) or L4.

Description of Immature Stages

The greenish white egg is shiny and smooth. L1 is white becoming yellowish white, sparsely covered with short brown setae, and has a shiny black head. A black collar with a tan-brown anterior edge is present dorsally on segment 1. L2 is yellowish green peppered with tiny black setae and has a middorsal dark stripe. Each segment has 5 transverse folds posteriorly. The head is black, and the black collar is edged in white. L3 is yellowish green peppered with black setae, becoming tan-brown with the onset of winter dormancy. L4 is dark brown speckled with irregular small white patches and tiny black setae. The head is black with short pale setae, and the middorsal dark stripe is prominent. L5 is purplish brown with reddish intersegmental areas, and the middorsal stripe is indistinct. Each segment has 4 transverse folds posteriorly. Ventrally there are 2 pairs of patches of a white flocculent secretion between abdominal

Adult

Eggs @ 0.8 mm

L5 @ 25 mm (head)

Pupa @ 18 mm

L1 @ 2.5 mm (*below*) and shelter (*above*)

L2 @ 6 mm

L3 @ 13 mm (*above*) and L3 @ 15 mm (diapause, *below*)

L4 @ 22 mm

L5 @ 25 mm

Long Dash *Polites mystic* (W. H. Edwards)

segments 7–8 and 8–9. The pupa is black with brown intersegmental bands on the abdomen and brown spiracles. The white flocculent secretion adheres strongly to the abdomen. Most of the pupa except the wing covers is covered with pale medium-length setae. The immature stages of *P. mystic* may be confused with those of other *Polites* spp. *Polites themistocles* and *P. peckius* frequently coexist in the same areas and their larvae are similar, although *P. peckius* has pale side stripes and *P. themistocles* has a more distinct dorsal stripe. Mature larvae of *P. sabuleti* and *P. mardon* have white-marked heads.. Allen et al. (2005) illustrated a late-instar larva from WV similar to ours, although a little lighter.

Discussion

Polites mystic females readily produce eggs in captivity and larvae are fairly easily reared to diapause, but overwintering can be difficult. Our only successful overwintering occurred when we artificially shortened "winter" to 6 weeks. Overwintering may be avoided by rearing larvae under continuous lighting. We used 3 unrecorded grass species as hosts in our rearings, and the natural host range is likely to be more extensive than currently known. Soggy rearing conditions or open water were detrimental to larval survival in captivity. Larvae tolerate each other well but should be reared in uncrowded conditions with ample live grass, as they respond poorly to nest disruption when host plants are changed. Variation in instar number has not previously been reported for *Polites* spp. and deserves further study. This species has a fairly restricted moist habitat range in Cascadia, yet is less restricted farther E. The environmental factors and natural enemies limiting *P. mystic* populations in Cascadia need study.

Sonora Skipper *Polites sonora* (Scudder)

Adult Biology

Polites sonora occurs in disjunct populations throughout the montane W US from S BC to N Baja and E to CO. In Cascadia it occurs on the slopes of the Cascade, Olympic, Blue, and Wallowa Mts, also at lower elevations in coastal and SW WA and in the Willamette Valley of OR. A variety of habitats are used, including flowery or wet meadows, prairies, roadsides, stream banks, forest clearings, and occasionally even disturbed habitats like lawns (Warren, 2005). *Polites sonora* is univoltine, flying from early June to early Aug in WA, longer in S OR. Males are territorial and conspicuously guard low perches while females are more reclusive, often remaining concealed in low vegetation. Both sexes visit flowers, including thistles, clovers, and asters. Larval hosts are grasses, including *Festuca idahoensis* (Idaho Fescue) and *Poa pratensis* (Kentucky Bluegrass) (Newcomer, 1966; Scott, 1986a). Eggs are laid singly and, according to Scott (1992), lack "glue" and fall to the ground. A female near Rimrock Lk in Yakima Co., WA, on July 15 was observed crawling and ovipositing at the bases of grass clumps.

Immature Stage Biology

We reared or partially reared this species 5 times from gravid females, in 4 different years. Seven females obtained from Rimrock Lk July 15–Aug 1 and confined with *Setaria glauca* (Yellow Foxtail) in a small plastic cylinder (13 cm dia × 8 cm tall) with muslin ends laid 30 eggs, loose at the bottom of the cage, except for 1 "glued" to a grass blade. All larvae died in L1. On July 30, 7 females collected at American R, Yakima Co., produced many eggs; L1 were reared on *Elytrigia repens* (Quackgrass), switched to a fescue mix they rejected, then to a lawn mixture of bluegrass and fescue, which they accepted. Eleven larvae developed to L3 and diapaused, but none survived to spring. On July 31, 4 females collected at American R produced 35 eggs. L1–L2 were reared on *P. pratensis*, but none survived to L3. On July 30, 1 female from Rimrock Lk confined with *S. glauca* laid 8 eggs on the cage bottom. On *S. glauca*, development of L1, L2, and L3 took 10, 11, and 14 days, respectively, at 18–25 °C (natural daylengths), reaching L4 by mid-Sep. Most L4 fed sparingly and within a few days constructed strongly silked shelters in which they became dormant. On Aug 5, 6 females from Washington Pass, Chelan Co., WA laid 25 eggs. Larvae were reared on Timothy Grass then on a bluegrass-fescue mixture, developing to L4 by early–mid-Sep. At temperatures of 18–25 °C and natural daylengths, L4 wandered for a few days, then constructed strongly silked shelters (cocoons) and entered winter dormancy. One larva continued feeding and molted to L5 on Sep 22, pupating on Oct 18 under paper toweling, fastened to the cage bottom with silk. Egg-hatch took 7–8 days at 18–27 °C, and L1 consumed shells. Larvae fed by eating grass blades from the edges and near the tips. L1 built light silken shelters. L2–L5 constructed vertical, aerial tubular nests in the upper half of the grass, more extensive and untidier in each instar. Scott (1992) did not find aerial nests associated with *P. sonora* in CO and thought larvae tunneled into the soil. The pupa is liberally decorated with flocculent material from ventral abdominal glands. Protection is based on concealment and a ventral gland on segment 1 that likely emits defensive chemicals. There are 5 instars, and L4 appears to be the overwintering stage.

Description of Immature Stages

The greenish white egg is shiny and smooth with an indistinct micropyle. L1 is white maturing to yellowish green with many tiny black setae. The head is shiny black, and there is a white or orange-edged black dorsal collar on segment 1. The posterior segment is slightly sclerotized dorsally and bears a pair of long setae. L2 is yellowish green becoming greener with maturity, particularly dorsally. Ventrally, L2 is yellowish, and there is an indistinct middorsal green stripe. The body is peppered with tiny black setae, and each segment has 5 transverse folds posteriorly. The head is black clothed with tiny pale setae. L3 is greenish tan peppered with tiny black setae and has a middorsal dark stripe. The head and collar are black, the latter edged in white, and this coloration persists to L5. The spiracles are black and largest on segment 11. L4 is dark brown-black speckled with irregular

Adult

Egg @ 1.0 mm

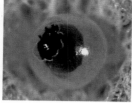

Egg @ 1.0 mm (hatching)

L1 @ 2.5 mm

L2 @ 4 mm (late)

Pupa @ 17 mm

L3 @ 8 mm

L4 @ 14 mm

olive patches on a gray ground color, and bears tiny black setae and a prominent middorsal dark stripe. The spiracles are black circled in orange. L5 is dark purplish brown with irregular orange tints and a prominent middorsal dark stripe. The posterior segment is speckled with black dots and dashes. Each segment has 4–5 transverse folds posteriorly, and the spiracles are black. Ventrally there are 2 pairs of patches of a white flocculent secretion between abdominal segments 7–8 and 8–9. The pupa is black with brown intersegmental bands on the abdomen and has brown spiracles. Most of the pupa except the wing case is covered with pale, short–medium-length setae. The immature stages of *P. sonora* may be confused with those of other *Polites* spp., particularly *P. mystic*, which is very similar, but the 2 species do not occur together. Mature larvae of *P. sabuleti* and *P. mardon* have white-marked heads. Our images appear to be the first published of immature stages of *P. sonora*.

Discussion

Polites sonora is not an easily reared species; obtaining fertile eggs can be difficult, and mortality of early instars is frequently high. Under natural daylengths, most larvae overwinter in L4, but in our rearings 1 individual avoided dormancy and developed to adulthood in late Oct. Rearing larvae under continuous lighting might prevent diapause and overwintering, as it does with *P. mystic*. *Setaria glauca* was used successfully as a host in our rearing, but information on grasses used by this species in Cascadia is needed. The habitat requirements of *P. sonora* are unclear and need research. Natural mortality factors are also unknown and deserve study.

L5 @ 20 mm (head)

L5 @ 20 mm

Woodland Skipper   *Ochlodes sylvanoides* (Boisduval)

Adult Biology

Ochlodes sylvanoides is abundant in Cascadia, flying late June–mid-Oct, and can sometimes be seen by the hundreds visiting late-season nectar sources such as rabbitbrush and knapweed. It occurs in nearly every county in WA and OR, also throughout S BC and ID. Its range extends from BC to Baja and from the Pacific Coast to E CO. Most open grassy habitats are used, including forest roadsides, meadows, yards, and shrub-steppe habitats from sea level to 7,000 ft (Warren, 2005). Larval hosts are grasses, especially large coarse species. Pyle (2002) listed *Cynodon dactylon* (Bermuda Grass), *Agropyron spicatum* (Bluebunch Wheatgrass), *A. caninum* (Bearded Wheatgrass), *Agrostis tenuis* (Colonial Bentgrass), *Avena fatua* (Common Wild Oat), *Leymus giganteus* (Siberian Wild Rye), and *L. cinereus* (Giant Wildrye). Warren (2005) added *Leymus* spp., *Pseudoroegneria spicta* (Bluebunch Wheatgrass), *Agrostis capillaris* (Colonial Bentgrass), and *Danthonia spicta* (Poverty Oatgrass). We found that *Phalaris arundinacea* (Reed Canarygrass) is also widely used in WA. Adults nectar readily, often on flowers not used by other butterflies (Pyle, 2002; Shapiro and Manolis, 2007), and males mud-puddle or perch while awaiting females. Eggs are laid singly in partial shade on the undersides of dead leaves, 20–40 cm above ground on wide-leafed, usually tall "hay" grasses (Scott, 1986a). *Ochlodes sylvanoides* is single brooded in Cascadia, but there are 2 or more broods in parts of CA (Scott, 1986a; Shapiro and Manolis, 2007).

Immature Stage Biology

We reared this species once from gravid females, and on several occasions from postdiapause larvae. Females collected Aug 14 from Yakima Co., WA and confined with cut *C. dactylon* produced ~50 eggs during 3 days. L1 hatched after 7–18 days, according to temperature (21–30 °C), consumed eggshells, and constructed overwintering shelters without feeding.

Overwintering shelters are constructed by tying the edges of a grass blade together with 7 silk cords, according to Scott (1992), but we found this number to vary from 4 to 10. The cohort was overwintered at 5 °C on Sep 15. After 4 months, larvae were transferred to 26 °C and continuous lighting and provided cut *C. dactylon*. Larvae fed immediately, developing to pupation in 44 days, with 9, 8, 7, 5, and 15 days spent in L1–L5, respectively. Adults eclosed 6–10 days later. Overwintered larvae exposed to 15–21 °C and 12 hr light fed briefly, then reentered dormancy. Overwintered L2–L4 were collected May 11 and June 5 at Bellevue, King Co., WA, and May 30 in Whatcom Co., WA, and pupae were found in Chelan Co., WA, on June 9 and July 29. Nests are constructed by postdiapause L1–L5 pulling together the edges of a grass blade and tying it into a tube with several cross-stitches of silk; the larva leaves the nest to feed nocturnally, eating the tip and sides. When the blade has been consumed, the larva moves to another site and builds a new nest. As larvae mature, more time is spent outside nests, and some feeding occurs by day. Frass is forcibly expelled from the nest, presumably to avoid attracting predators. Larvae collected in May–June developed rapidly from L2 to L4 at ~21–22 °C and natural daylengths but spent ~30 days in L5, supporting the claim by Scott (1986a) that mature larvae aestivate for about a month before pupation. Pupation occurs in a new tied-leaf nest with much flocculent material; the pupa hangs with its ventral side up in the nest, attached to the top by a girdle thread and cremaster (Scott, 1992). We found 1 pupa in a tied-leaf nest on *Eriogonum elatum* (Tall Buckwheat). There are 5 instars, and protection is based on concealment, nocturnal feeding, camouflage, and frass expulsion.

Description of Immature Stages

The white egg is faintly textured, sometimes with 1–3 short spikes. L1 is whitish (during diapause) turning dull orange with green dorsal infusion postdiapause. The head and collar are shiny black, the latter edged anteriorly in white, persisting until pupation. Each segment has 4 small green spots dorsally, and there is a narrow dorsal green stripe. L2 is greenish yellow, blue-green dorsally, bearing numerous small dark speckles (~100/segment). There is a dull green dorsal stripe and a whitish dorsolateral stripe, and numerous tiny white (head) or black (body) setae. L3 is similar but darker dorsally, paler laterally and posteriorly. L4 is similar but more olive green with a whitish ventral periphery. The dorsal stripe is distinct, and there are brown or white patches on the head.

Adult female

Egg @ 1.0 mm   **L5 @ 30 mm (head)**

Pupa @ 16 mm

L1 @ 2.8 mm

L2 @ 8.5 mm

L3 @ 12 mm

L4 @ 20 mm

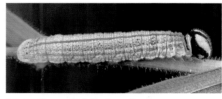

L5 @ 20 mm (white head)

L5 is olive green to cinnamon brown with a dark dorsal stripe. Laterally there are 1 or 2 olive brown stripes bordered with white. The head is bifurcated dorsally, initially white with a black line from the bifurcation halfway to the mouth, where it divides into an inverted Y; there is also a dark cheek spot. The head in mature L5 is orange-brown with black markings. The pupa is slightly to moderately waxy, whitish to cinnamon brown with dark brown spots dorsally on the abdomen. The head and thorax are darker, with 2 pairs of dark spots and a V shape dorsally. There is a distinct red "false eye" behind each true eye. A number of grass-feeding skippers are similar, but none, except the closely related *Ochlodes yuma*, have similar head capsule markings. The rare *O. yuma* is found only on *Phragmites australis* (Common Reed); *O. sylvanoides* has not been recorded on this plant. Allen et al. (2005) published an image of a larva in UT, and Layberry et al. (1998) showed a pupa; both of these are similar to ours.

Discussion
This species can be difficult to rear, as its preferred host grasses are large and hard to keep alive. Overwintering L1 is also difficult. Exposing eggs and L1 to increasing daylengths might prevent diapause and allow a second generation to develop, as occurs in CA. All instars can be found in spring by searching for tied-leaf nests, or grasses can be beaten onto a sheet at dusk, when the larvae become active. Scott (1992) found that many larvae and pupae were victims of unidentified parasitoids. The incidence and function of the spikes found on some eggs deserves study.

Woodland Skipper   *Ochlodes sylvanoides* (Boisduval)

Yuma Skipper   *Ochlodes yuma* (W. H. Edwards)

Adult Biology

Ochlodes yuma has a restricted and disjunct range in Cascadia, occurring in isolated populations E of the Cascades. In WA, there are discrete populations in Grant and Klickitat Cos., while in OR it is found in Hood River, Wasco, and Sherman Cos. along the Columbia R, and in Wallowa and Lake Cos. in NE and S OR (Warren, 2005). Beyond Cascadia the range includes NV, UT, W CO, and the central valley of CA, with smaller colonies in WY, AZ, NM (Allen et al., 2005), and ID (Warren, 2005). *Phragmites australis* (Common Reed) appears to be the sole natural host, although Pyle (2002) reported finding males associated with the introduced ornamental *Miscanthus* (eulalia grass) in Klickitat Co. *Phragmites australis* is a plant of marshy areas, alkaline seeps, and lake edges in hot, dry desert areas, and this defines the habitat of *O. yuma*. The flight period is late June–late July in WA and mid-July–mid-Aug in OR. Males eclose ~7–10 days before females; they perch at eye level at marsh edges or fly slowly through dense vegetation at the bases of reeds in search of mates. In Grant Co., females are found low in the dense vegetation under *Phragmites*; when disturbed, they move deeper into tangles where Poison Oak abounds. Later, when males are scarce, females nectar openly on flowers near reeds. Eggs are laid on or near host leaves (Scott, 1986a). *Ochlodes yuma* is single brooded in Cascadia, but 2 broods occur in NV, W UT, and CA (Scott, 1986a; Allen et al., 2005).

Immature Stage Biology

We partially reared this species 4 times. On July 21, females from Grant Co. oviposited on *P. australis*, and larvae were reared to postdiapause L3. On July 22 (different year), females from the same location oviposited on the introduced subspecies of *P. australis*, producing ~70 eggs during 5 days. Eggs were attached to reed stems and leaves. Overwintered larvae (L1–L5) were collected in Grant Co. 3 times (mid-May–early June). One L5 pupated and produced an adult. Eggs hatched after 9–10 days at 25–30 °C, and L1 immediately formed a shelter by bending over the

edge of a blade and tying it with silk cross-strands. L1 did not feed and entered diapause, remaining in the nest. The cohort was overwintered at 5 °C from Aug 17. After 4.5 months, larvae were exposed to 15–21 °C and 15 hr light but did not leave their shelters and feed for 21–28 days. Larvae removed from overwintering after 7 months began feeding after 9 days. Postdiapause L1 fed on *Miscanthus sinensis* (Chinese Silvergrass), rolling and silking tight tubes from blades. L1 and L2 left their shelters at night to feed and wander. Postdiapause development from first feeding to pupation occupied 45–55 days. Adults eclose after ~14 days. L3–L5 live in rolled blade tubes with silk ties, hidden or partly exposed, venturing out nocturnally to feed. The side or tip of the shelter blade is eaten from the edge, leaving scalloped patterns. When the blade end has been consumed, the larva moves to another blade and creates a new shelter. A high percentage of empty shelters on consumed leaves are normally found, suggesting that larvae move repeatedly. Pupation occurred within blade nests in captivity, but many empty nests were examined in the wild prior to the flight period and no pupae were found, suggesting that larvae may wander before pupation. Larvae are solitary throughout development; smaller larvae often wandered off host reeds, whereas larger larvae occupied prime sites, suggesting competition and intolerance between individuals. Protection is based on concealment (in nests) and nocturnal activity to avoid predators. There are 5 instars.

Description of Immature Stages

The white hemispherical egg has a faintly dimpled surface and obscure micropyle. L1 is creamy white, becoming brownish orange after a few days with a shiny dimpled black head and collar. The body and head are sparsely clothed with short pale setae, except for 1 long rear-projecting pair on the posterior segment. Each segment has 5 lateral folds on the posterior half, a character that persists to pupation. Postdiapause L1 become green with feeding. L2 is mostly greenish and has numerous tiny indistinct brown spots bearing tiny pale setae. The head is dark chestnut brown tending to black around the perimeter. L3 is green anteriorly, pale whitish posteriorly, and there is a distinct dark green dorsal stripe and numerous small green spots. There are 2 indistinct dark green stripes laterally that persist to pupation. The head is brown, and the collar is white with a black edge posteriorly. The head has a black perimeter and a bold vertical black line from the top to the middle-front, where it splits into an inverted Y. L4 is similar; large rounded prolegs are prominent below the posterior tip in this and all instars. L5 is light brown, and the dark body spots

Adult male

Egg @ 1.1 mm

L5 @ 24 mm (head)

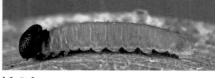

L1 @ 4 mm

Pupa @ 25 mm

L2 @ 7 mm

L3 @ 15 mm (late, *above*) and L3 @ 9 mm (in nest)

L4 @ 19 mm

L5 @ 35 mm

are irregular and splotchy. The black inverted-Y head capsule marking has a contrasting narrow pale line down the middle. The pupa is encased in a silk sleeve with white waxy secretions. The body is light brown with random dark brown spots on the abdomen, and the wing cases are dark brown. The head and anterior-ventral parts are blackish, the proboscis extending beyond the wing cases. *Ochlodes sylvanoides* is similar but differs by being browner with different head capsule markings in L3–L5. Allen et al. (2005) published an image of a mature larva from UT similar to ours, and Emmel and Emmel (1973) provided black-and-white images of an egg and pupae.

Discussion

Larvae of *O. yuma* are readily found in occupied habitats by searching for rolled and tied nests high on *Phragmites* reeds. Eggs can be obtained from females, but timing is critical, with oviposition occurring during a fairly brief period. Cut *Phragmites* is difficult to keep fresh for larvae, which grow very slowly; however, *Miscanthus* spp. provide more readily available and convenient alternative hosts that appear acceptable to *O. yuma* larvae. The exotic subspecies of *P. australis*, invasive and widely abundant in Cascadia, was accepted by ovipositing females in captivity, but there are no records of this plant being utilized by *O. yuma* under natural conditions. It is not well understood why this species is found in isolated remote colonies, nor do we understand how it spreads to remote areas. These issues, as well as the natural factors controlling abundance and distribution, deserve study.

Yuma Skipper   *Ochlodes yuma* (W. H. Edwards)

Dun Skipper   *Euphyes vestris* (Boisduval)

Adult Biology

The Dun Skipper ranges throughout most of E US and S Canada. In W US it occurs in coastal areas from S BC to CA. In Cascadia it is found in and W of the Cascades, with a few outlier populations in NE and central WA. Favored habitats are moist to wet areas where sedges grow near a nectar source. It flies in a single brood from early June to Sep but has 2 or more flights in S US. Favored habitats in OR include riparian corridors, weedy roadsides, and fields from sea level to 5,000 ft (Warren, 2005). The flight is rapid, darting, and inconspicuous. Males perch on grasses or other vegetation 2–3 ft above the ground and visit mud, scat, and carrion. Both sexes visit a number of flowers, including Self-heal, mints, dogbane, and thistle. Mating generally occurs in the afternoon (Allen, 1997). Reported larval hosts include *Carex filifolia* (Thread-leaf Sedge), *Carex heliophila* (Sun Sedge), and *Cyperus esculentus* (Yellow Nutsedge) (Scott, 1986a). *Carex stricta* is used in WV (Allen, 1997), and Cech and Tudor (2005) reported 5 additional *Carex* spp. used as hosts in E US. Oviposition occurs during late morning into midafternoon (Allen, 1997). Eggs are typically placed in narrow spaces between leaves, and laid singly or in loose groups of 2 or 3.

Immature Stage Biology

We reared this species 3 times from field-collected larvae and gravid females. A female obtained from Kittitas Co., WA on June 27 laid ~15 eggs at the base of cut blades of a sedge (*Carex* sp.). Two hatched after 10 days, and L1 tied sedge blades together with silk stitching, secreting a silk sleeve lining inside the nest. The nest was silked closed at the lower end but left open at the head end. One larva wandered and

Adult

ceased feeding in L3 and was lost. The other larva reached L5 in early Aug. By Aug 14 it ceased feeding and constructed a tubular nest sealed at both ends. This nest was heavily lined inside with a waxy white, powdery material produced by ventral abdominal glands, likely a waterproofing adaptation (Dethier, 1942). On Aug 17 the larva pupated in this nest, 42 days post egg-hatch, and an adult eclosed 16 days later. Others report pupation at the bases of host plants (Allen, 1997; Cech and Tudor, 2005). An L2 collected Aug 12 from the Kittitas Co. site developed to L3 before wandering and disappearing in mid-Aug. Four females from Klickitat Co., WA, on July 11 and confined with *Stipa* sp. laid ~30 eggs. Eggs hatched in 9 days, and L1 were fed on *Carex macloviana* (Thickhead Sedge) and an unidentified coarse grass species, developing to L2 in 12 days at 22–26 °C. *Scirpus americanus* (American Bulrush) was also supplied as a potential host but rejected by the larvae. Twenty days were spent in L2, with L3 entering dormancy at the end of Aug. Overwintering occurred in L3 (75%) or L2, but all died. Cech and Tudor (2005) reported that overwintering occurs in L3 in E US. Tubular nests are used throughout development, with larvae resting in them during the day and feeding at night on edges and tips of sedge blades. Protection is based on concealment.

Description of Immature Stages

The egg is conical, initially white but becoming green. When partially mature, the egg develops a distinct bright red ring around the middle and a red "bulls-eye" spot on the top. L1 is smooth, beige, and sparsely covered with short brown setae, more obvious posteriorly. The head is shiny black and bifurcated dorsally, with a black collar. L2 is bluish green with ~8 transverse folds on each segment. The black head capsule is brownish dorsally with a distinct black-and-cream eye-shaped pattern. L3 is similar but greener, with the head more strongly marked with black and white bands. Setae are numerous, tiny, and indistinct. L4 and L5 are similar, green with fine black speckling and more distinct transverse folds. There is an indistinct middorsal dark stripe. The head capsule has a distinctive complex pattern of black, white, and brown lines and streaks, and the black collar is anteriorly edged in white. Pulsating paired organs are visible dorsally. The prepupal larva has large white ventral spots, 1 on each side of the posterior part of the abdomen. The pupa is slender, black, and cylindrical, gently tapering to a larger head and heavily coated with a waxy white secretion. Larvae

Egg @ 1.4 mm (midstage, *left*, mature, *right*)

L1 @ 4 mm

Pupa @ 20 mm

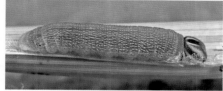

L2 @ 9 mm

L3 @ 15 mm

L4 @ 26 mm

could be confused with other green skipper larvae, but none have the distinctive head capsule pattern of *E. vestris*. The "bull's-eye" egg is unique. Larvae shown in Allen (1997), Wagner (2005), Allen et al. (2005), and Tveten and Tveten (1996) agree closely with ours. The FL larva in Minno et al. (2005) has a black crescent dorsally on the posterior segment. Heitzman (1964) described the immature stages of E US *E. vestris*.

Discussion

Euphyes vestris eggs are readily obtained from gravid females by caging them with clippings of their sedge host plants or coarse grasses. Gravid females generally contain relatively limited numbers of the large eggs. Dissection of an unmated female revealed 18 eggs. Females collected early in the flight period may have some infertile eggs but appear to be more reliable sources than older, worn adults. Larvae may be found by diligently searching for silked blade nests, especially during Aug, when larvae are larger. Eaten leaf tips and edges and sedge leaves stitched into tubular nests provide good search patterns. It is advisable to separate larvae during rearing as they do not tolerate each other well. Overwintering of larvae is difficult, excessive humidity causing mold problems. Provision of long daylengths or continuous lighting for larvae may allow complete development, avoiding overwintering. The full range of hosts used in Cascadia needs study and may include coarse grasses (e.g., *Stipa* spp.) as well as sedges (Layberry et al., 1998).

L5 @ 34 mm

L5 @ 34 mm (head)

Dun Skipper *Euphyes vestris* (Boisduval)

Common Roadside Skipper   *Amblyscirtes vialis* (W. H. Edwards)

Adult Biology

Amblyscirtes vialis occurs in every US state except AK, NV, AZ, and HI but is rarely abundant anywhere. Its small size, rapid darting flight, and general inconspicuousness make it a much overlooked species, and it is probably more common than generally realized. In Cascadia it occurs in S BC, also in WA in a narrow strip E of the Cascades, and in the NE. In OR it occurs closer to the coast and in the NE. It is found in all the major mountain ranges in Cascadia, mostly in wooded areas, along trails, roads, and watercourses, as well as in meadows and gullies and on hillsides. Males perch all day on or very near the ground, scouting for females and taking moisture. They fly short distances when disturbed. Many low-growing flowers are visited, including phlox, clover, wild strawberry, ground ivy, and violets. There is a single annual brood from mid-April to early July, depending on elevation and seasonal conditions, although 2 broods may occur elsewhere (Allen, 1997; Minno et al., 2005). Larval hosts are various grasses; Scott (1992) reported *Elymus canadiensis* (Canada Wild Rye), *E. trachycaulis* (Slender Wheatgrass), *Bromus lanatipes* (Woolly Brome), *B. inermis* (Smooth Brome), and *Phleum pratense* (Timothy) as hosts. Pyle (1981) recorded *Poa pratensis* (Kentucky Bluegrass), *Avena striata* (Striped Oats), *Agrostis* spp. (bent grass), and *Cynodon dactylon* (Bermuda Grass). Eggs are strongly cemented singly on host grasses, usually on the undersides of blades (Scott, 1992).

Immature Stage Biology

We reared this species once and partially on another occasion. Three females collected July 2, Tiger Meadows, Pend Oreille Co., WA, produced 25 eggs that were reared to L2 on *Festuca* sp., dying after ~2 weeks. One female obtained from Cummings Creek, Columbia Co., WA on June 19 and confined with *Setaria viridis* (Green Foxtail) in a plastic box with a muslin lid (28 × 16 × 7 cm) laid ~20 eggs during 4 days. Eggs hatched after 6 days at 22–26 °C. Larvae fed on *S. viridis* and *Panicum capillare* (Witch Grass) but rejected *Digitaria sanguinalis* (Large Crabgrass). Development was rapid, with L5 reached after 26 days (July 19). Thereafter, development slowed, with 22 days in L5 before pupation (Aug 10). Adults eclosed 8–9 days later. Overwintering is reported to occur as a larva (Scott, 1986a; Pyle, 2002; Minno et al., 2005) or as a pupa (Pyle, 1981; Allen, 1997; Schlicht et al., 2007). Larvae build silk-tied shelters during L1–L4, with L5 mostly resting exposed on blades. L1 attach silk strands across blades and rest underneath; sometimes blade edges are silked over to form shelters. L2–L4 form tubular shelters from grass blades tied in place by equally spaced silk strands. When L5 use shelters, they usually consist only of a partially rolled blade tied with a single silk strand (see image). Feeding occurs along the edges of blades and beyond tubular shelters, mostly at night. Mature L5 silk together dead grass blades or other material (e.g., paper toweling), usually on the ground, for pupation. Survival is based on concealment, although young larvae are vulnerable to small predators like Minute Pirate Bugs that can enter nests.

Description of Immature Stages

The white egg is smooth with a faintly textured surface comprising tiny polygons. L1 is whitish green dorsally, yellowish laterally. The head is slightly triangular, shiny black with a black collar anteriorly edged in white. Short, dark setae sparsely cover the body (~12/segment); 6 long pale setae project backward from the posterior segment. L2 is similar, green dorsally, whitish yellow laterally, with a shiny black head and white-and-black collar. Setae are more profuse, short, pale, and inconspicuous. L3 is whitish green, with >250 tiny white dots/segment, each carrying a tiny white seta. There is a middorsal dark stripe and a less distinct midlateral stripe. L4 is whitish gray, resulting from increased density of white, flakelike setae, giving the larva a flocculent, powdery bloom appearance. The brown-marked black head is also covered with white flakelike setae, and there is a prominent middorsal dark stripe. L5 is white with a middorsal dark stripe and increased density of white flakelike setae on the body and head. The head has fewer flakes centrally, producing dark brown markings. The pupa is pale greenish white with a light brown head, thorax, and posterior segment. The proboscis casing extends to the posterior segment and is light brown beyond

Adult

Egg @ 0.9 mm (fresh)   Egg @ 1.1 mm (mature)

L1 @ 2.0 mm

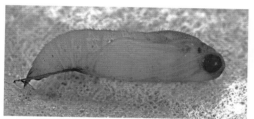

Pupa @ 14 mm

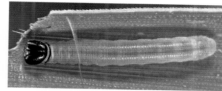

L2 @ 8 mm

L3 @ 10 mm

L4 @ 16 mm

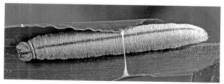

L5 @ 24 mm

L5 @ 24 mm
(head)

Common Roadside Skipper *Amblyscirtes vialis* (W. H. Edwards)

the wing cases. Short dark brown setae cover the pupa dorsally, and there is a red spot behind each eye. A few days prior to eclosion, the eyes become red, with the pupa turning black before adult emergence. The white "powdery bloom" appearance of mature larvae prevents confusion with other grass-feeding species, although early instars could be confused with other skippers. An image of a mature larva from MI appears in Allen et al. (2005) and is very similar to ours. An image of a mature larva from FL appears in Minno et al. (2005) but lacks the white "powdery bloom" of our larvae.

Discussion

This widespread but inconspicuous species is easily reared on a number of grass species and in captivity produces a second generation. Research is needed on basic aspects of biology, including development and overwintering. The overwintering stage requires determination, but is likely to be a fully fed L5. Minno et al. (2005) reported that mature larvae clip the final larval shelter from the host and overwinter in it on the ground. A similar behavior was observed in some of our prepupal L5, which formed tight aerial nests but then left them for pupation in shelters formed on the ground. The delay seen in development of L5 in our rearing may have been due to some "indecision" whether to enter dormancy or continue development. Much of the L5 period was spent inactive. The influence of temperature and photoperiod on development and induction of dormancy needs study.

Glossary

abdomen (entomology)—The hindmost of the three main body divisions of an insect.

aberrant—Deviating in important characters from the typical condition.

abiotic—Nonbiological factors, e.g., climate, daylength.

adult (entomology)—The mature breeding stage of an insect; in Lepidoptera, the winged, flying stage.

aestivation—A period of metabolic and behavioral inactivity during the summer months.

allochronic—Two species not occurring at the same time.

allopatric—Two species whose ranges do not overlap.

anal—Near or toward the anus.

antennae, s **antenna**—Paired sensory appendages on the head of an adult insect.

anterior—Toward the front.

apex, pl **apices**—The highest point; the tip.

aposematic—Brightly colored, usually as a warning defense by toxic organisms.

basal—Toward the base.

biotic—Biological factors, e.g., predators, host plants.

bivoltine—Having two generations per year.

brood—A group of young (larvae) hatched at one time from the same mother.

bulla, pl **bullae**—A blister or blisterlike protuberance on early-instar larvae, usually bearing a seta.

cancellate—Cross-barred, latticed; a pattern with longitudinal lines or grooves crossed by transverse ones.

catkin (botany)—A cylindrical, often drooping cluster of unisexual flowers found in willows, birches, and oaks.

chitin—A tough, protective substance composed of nitrogen compounds, forming arthropod exoskeletons.

chorion (entomology)—The outer shell or membrane of the insect egg.

chrysalis, pl **chrysalids** or **chrysalides** (butterflies)—The pupal (resting) stage prior to adulthood.

class (biology)—A broad taxonomic category ranking below phylum and above order, generally consisting of a group of species exhibiting similar characteristics.

cocoon—The silken sheath or envelope surrounding a pupa or chrysalis.

cohort—A group of individuals with something in common, such as larvae from the same batch of eggs.

congener—An organism that is a member of the same genus as another organism.

conspecific—Belonging to the same species.

cremaster—A support hook or a cluster of small hooks at the terminal end of a pupa.

crepuscular—Active at dusk or dawn.

cryptic—Concealed, camouflaged, or not easily recognized.

diapause—A physiological condition in which development or activity is suspended; nonresponsive to normally favorable stimuli (e.g., as in hibernation, aestivation).

dimorphic—Occurring in two different forms; used especially for sexual differences within a species.

distal—Away from the body.

diurnal—Active during daylight hours.

dormant—Temporary and readily reversible suspension of development/activity.

dorsal—On or near the back.

dorsolateral—Located between the side and the back.

dorsum—The midline of the back.

double brooded—Having two generations per year. See **bivoltine**.

ecdysis (entomology)—The process of molting, in which the old growth-limiting insect exoskeleton splits, allowing a larger stage to emerge and grow.

eclose, eclosure (entomology)—The emergence of an adult insect from a pupa, or a larva from an egg.

endemic—Restricted to a particular locality.

eversible (entomology)—Able to be turned inside out, in regard to a larval **osmeterium**.

exoskeleton—The hard outer structure or shell of an insect.

exotic—Nonnative; an introduced species.

family (biology)—A fairly narrow taxonomic category ranking below order and above genus, consisting of a group of species exhibiting similar characteristics.

frass—Caterpillar fecal material.

frass pier—A narrow extension of a leaf vein or small stick, composed of frass and silk produced by a larva, especially in admiral butterflies, upon which the larva may rest.

genotype—The entire genetic identity of an individual.

genus, pl **genera** (biology) A narrow taxonomic category ranking below family and above species, consisting of a group of species exhibiting similar characteristics.

girdle thread—A silk thread a mature larva places around its middle before pupating, to help secure the pupa in place.

glabrous—Having a waxy or powdery coating.

glaucous—Not hairy.

gravid—Pertaining to a female bearing eggs. This term usually implies the eggs are fertilized.

head (entomology)—The frontmost of three insect body segments, bearing the eyes, mouth, and antennae.

hibernaculum, pl **hibernaculae**—A rolled leaf tied with silk, used by some larvae as an overwintering shelter.

hibernation—A torpid or resting state for passing the winter.

host plant—A plant eaten by larvae of Lepidoptera.

hydric—Tolerant of or adapted to wet conditions.

immaculate—Not spotted or otherwise marked.

instar (entomology)—In insect larvae, a growth stage between **molts**. The second instar is the stage following the first larval molt but before the second larval molt. See also **ecdysis**.

larva, pl **larvae**—The immature stage between the egg and pupa of an insect; in Lepidoptera, called a caterpillar.

lateral—On the side.

lithosol—Thin, rocky soil.

maculation—Spotting, usually contrasting.

median, mesial—Toward the middle.

mesic—Pertaining to sites with intermediate amounts of moisture; not decidedly wet or dry.

metamorphosis (entomology)—Morphological and physiological changes that occur from egg to adult in a butterfly.

micropyle—A tiny opening or pore in an insect's egg, through which sperm enters during fertilization.

mimicry—Imitating behavior or appearance.

molt—The shedding of an exoskeleton or skin to allow room for growth. See **ecdysis**.

monophagous (entomology)—Among Lepidoptera, a species that uses only one kind of plant as a larval host plant.

morph—A local population of a species distinguishable from other populations but capable of interbreeding with them.

multibrooded—Having more than two generations per year. See **multivoltine**.

multivoltine—Having more than two generations per year.

myrmecophilous—Living symbiotically with ants.

nocturnal—Active during the night.

obligate—Having no choice or options.

ocellus, pl **ocelli**—A simple eye; sometimes used to describe a superficial pattern resembling an eye, as on the wings of some butterflies.

oligophagous—A species that uses several closely related host plants.

omnivorous—An organism that eats both plant and animal matter.

order (biology)—A broad taxonomic category ranking below class and above family and generally consisting of a group of organisms exhibiting similar characteristics.

osmeterium—A Y-shaped gland located behind the head of caterpillars in the family Papilionidae that can be everted to emit a chemical disagreeable to potential predators.

oviposit—To lay eggs.

ovoid—Egg shaped.

ovum, pl **ova**—A mature egg.

palpate—To examine by touch or feel.

parapatric—Two species living in adjacent but nonoverlapping areas.

parasite—An animal that lives in or on another animal, obtaining nourishment from the host without benefiting or killing the host.

parasitoid—An animal which develops within a host and ultimately kills it.

penultimate—Next to last.

pharate—Describes a pupa near to adult eclosion showing wing markings of the adult.

phenology—The study of the seasonal timing of life cycle events.

phenotype—The observable physical characteristics of an organism, determined by genetics and environmental influences.

pheromone—A chemical produced by an insect or animal that mediates the behavior of other individuals of the same species.

photoperiod—The relationship between the length of light and dark in a 24-hour period.

phototactic—The movement of an organism toward or away from a source of light.

pier—See **frass pier**.

polymorphic—The occurrence of different forms or types in organisms of the same species, independent of sexual variations.

polyphagous—Feeding on a wide variety of host plants, including different plant families.

posterior—Toward the back.

predator—An animal that lives by killing and eating other animals.

proboscis (entomology)—The slender tubular feeding organ of adult butterflies.

proleg—The fleshy unjointed structures found on the ventral surface of the abdomen of lepidopteran larvae; used for larval movement but generally lacking musculature.

protandry (Lepidoptera)—Males eclosing or arriving before females. This term does not refer to sex change during the life cycle in the context of Lepidoptera.

proximal—Toward the body.

pruinose—Having a bluish powdery coating, as on the skin of a prune.

pubescent—Hairy.

punctate—Having small holes, spots, points, or depressions (on the surface).

pupa, pl **pupae**—A generally inactive developmental stage of butterflies, intermediate between the larva and the adult.

pygidium, pl **pygidia**—The terminal posterior segment(s) of an insect.

salique—A seedpod of a mustard plant.

sclerite—One of the individual pieces (plates) of an insect's exoskeleton.

segment (entomology)—A visible division of the thorax or abdomen, often marked by a constriction.

segregate—A group of organisms isolated from others.

senesce (botany)—To decline with maturity, often hastened by environmental stress; to dry out seasonally.

setae, s **seta**—Bristles or hairs.

single brooded—Having only one generation per year. See also **univoltine**.

skeletonize (entomology)—To eat the tissue of a leaf, leaving only the stem and veins.

species—A narrow taxonomic category below genus, whose members can interbreed. (Note: A universally accepted definition of "species" has eluded scientists for decades.)

spermatophore—Small packet of sperm.

sphragis—A waxy secretion containing sperm, produced during mating by a male parnassian butterfly and placed at the tip of his mate's abdomen, preventing her from mating again.

spinneret—A tubelike structure on a larva's head that contains the silk spinning apparatus.

spiracle—An external opening of the tracheal respiratory system; a breathing pore in the side of the body.

symbiotic, **symbiosis**—An association between two organisms of different kinds, resulting in benefits to both.

sympatric—Different species living in the same geographical area.

synchronic—Occurring at the same time.

temporal—Related to or limited by time.

thorax (entomology)—The portion of the body between the head and the abdomen and to which the legs and wings are attached.

true leg—Jointed walking structure on the thorax of insect larvae and adults, having musculature and five segments.

tubercle—A small knoblike or rounded protuberance on larvae that sometimes bears spines.

univoltine—Having a single generation per year.

vagility—The ability of a species to spread within its environment.

venter—The midline of the belly.

ventral—The underside; the belly.

ventrolateral—Located between the side and the belly.

voltinism (biology)—The number of broods or generations of an organism in a year.

xeric—Characterized by or tolerant of dry conditions.

Bibliography

Acorn, J., and I. Sheldon. 2006. *Butterflies of British Columbia*. Lone Pine Publishing, Alberta, Canada. 360 pp.

Aiello, A. 1984. *Adelpha* (Nymphalidae): Deception on the wing. *Psyche* 91(1–2): 1–45.

Albanese, G., M. Nelson, P. Vickery, and P. Sievert. 2007. Larval feeding behavior and ant association in frosted elfin, *Callophrys irus* (Lycaenidae). *J. Lepid. Soc.* 61(2): 61–66.

Alcock, J., and K. M. O'Neill. 1986. Density-dependent mating tactics in the gray hairstreak, *Strymon melinus* (Lepidoptera: Lycaenidae*). J. Zool. Lond.* 209: 105–113.

Alcock, J., and K. M. O'Neill. 1987. Territory preferences and intensity of competition in the gray hairstreak, *Strymon melinus* (Lepidoptera: Lycaenidae) and the tarantula hawk, *Hemipepsis ustulata* (Hymenoptera: Pompilidae). *Am. Midl. Nat.* 118(1): 128–138.

Allen, T. J. 1997. *The Butterflies of West Virginia and Their Caterpillars*. University of Pittsburgh Press. 388 pp.

Allen, T. J., J. P. Brock, and J. Glassberg. 2005. *Caterpillars in the Field and Garden: A Field Guide to the Butterfly Caterpillars of North America*. Oxford University Press. 232 pp.

Anweiler, G. G., and B. C. Schmidt. 2003. Edith's copper, *Lycaena editha* (Lycaenidae), confirmed for Canada. *J. Lepid. Soc.* 57(3): 249–250.

Auckland, J. N., D. M. Debinski, and W. R. Clark. 2004. Survival, movement and resource use of the butterfly, *Parnassius clodius. Ecol. Entomol.* 29: 139–149.

Austin, G. T., and P. J. Leary. 2008. Larval hostplants of butterflies in Nevada. *Holarct. Lepid.* 12(1–2): 1–134.

Ballmer, G. R., and G. F. Pratt. 1988. A survey of the last instar larvae of the Lycaenidae of California. *J. Res. Lepid.* 27: 1–80.

Ballmer, G. R., and G. F. Pratt. 1989. Instar number and larval development in *Lycaena phlaes hypophlaeas* (Boisduval) (Lepidoptera: Lycaenidae). *J. Lepid. Soc.* 43: 59–65.

Ballmer, G. R., and G. F. Pratt. 1991. Quantification of ant attendance (Myrmecophily) of lycaenid larvae. *J. Res. Lepid.* 30(1–2): 95–112.

Beckage, N. E., and L. M. Riddiford. 1978. Developmental interactions between the tobacco hornworm, *Manduca sexta* and its braconid parasite, *Apanteles congregatus. Entomol. Exp. Appl.* 23: 139–151.

Berkhousen, A. E., and A. M. Shapiro. 1994. Persistent pollen as a tracer for hibernating butterflies: the case of *Hesperia juba* (Lepidoptera: Hesperiidae). *Great Basin Nat.* 54(1): 71–78.

Betros, B. 2008. *A Photographic Field Guide to the Butterflies in the Kansas City Region*. Kansas City Star Books, Kansas City, MO. 407 pp.

Beyer, L. J., and S. H. Black. 2007. Site utilization by adults and larvae of Mardon skipper butterfly (*Polites mardon*) at four sites in Washington and Oregon. *Final Report to the Forest Service and BLM from the Xerces Society.*

Beyer, L. J., and C. B. Schultz. 2010. Oviposition selection by a rare montane skipper *Polites mardon* in montane habitats: advancing ecological understanding to develop conservation strategies. *Biol. Conserv.* 143: 862–872.

Boggs, C. L. 1987. Demography of the unsilvered morph of *Speyeria mormonia* in Colorado. *J. Lepid. Soc.* 41(2): 94–97.

Boggs, C. L., and C. L. Ross. 1993. The effect of adult food limitation on life history traits in *Speyeria mormonia* (Lepidoptera: Nymphalidae). *Ecology* 74(2): 433–441.

Bower, H. M. 1911. Early stages of *Lycaena lygdamus* Doubleday (Lepid.). *Entomol. News* 22(8): 359–362.

Bowers, M. D. 1984. Iridoid glycosides and host-plant specificity in larvae of the buckeye butterfly, *Junonia coenia* (Nymphalidae). *J. Chem. Ecol.* 10(11): 1567–1577.

Bowers, M. D. 1988. Chemistry and co-evolution: iridoid glycosides, plants and herbivorous insects. In K. C. Spencer, ed., *Chemical Mediation of Co-evolution*, pp. 133–165. Academic Press, San Diego, CA.

Brower, A.V. Z., and K. R. Sime. 1998. A reconsideration of mimicry and aposematism in caterpillars of the *Papilio machaon* group. *J. Lepid. Soc.* 52(2): 206–212.

Brower, L. P., P. B. McEvoy, K. L. Williamson, and M. A. Flannery. 1972. Variation in cardiac glycoside content of monarch butterflies from natural populations in eastern North America. *Science* 177: 426–429.

Brussard, P. F., and P. R. Erhlich. 1970a. The population structure of *Erebia epipsodea* (Lepidoptera: Satyrinae). *Ecology* 51: 119–129.

Brussard. P. F., and P. R. Erhlich. 1970b. Adult behavior and population structure in *Erebia epipsodea* (Lepidoptera: Satyridae). *Ecology* 51: 880–885.

Carey, D. B. 2004. Patch dynamics of *Glaucopsyche lygdamus* (Lycaenidae): Correlations between butterfly density and host species diversity. *Oecologia* 99(3–4): 337–342.

Cech, R., and G. Tudor. 2005. *Butterflies of the East Coast: An Observer's Guide.* Princeton University Press. 345 pp.

Clark, S. H., and A. P. Platt. 1969. Influence of photoperiod on development and larval diapause in the viceroy butterfly, *Limenitis archippus*. *J. Insect Physiol.* 15(10): 1951–1957.

Comstock, J. A., and C. Henne. 1965. Notes on the life history of *Philotes enoptes dammersi* Hew. (Lepidoptera, Theclinae). *Bull. South. Calif. Acad. Sci.* 64(3): 153–156.

Crozier, L. 2003. Winter warming facilitates range expansion: Cold tolerance of the butterfly, *Atalopedes campestris*. *Oecologia* 135: 648–656.

Crozier, L. 2004. Warmer winters drive butterfly range expansion by increasing survivorship. *Ecology* 85: 231–241.

Davies, Z. G., R. J. Wilson, T. M. Brereton, and C. D. Thomas. 2005. The re-expansion and improving status of the silver-spotted skipper butterfly (*Hesperia comma*) in Britain: A metapopulation success story. *Biol. Conserv.* 124: 189–198.

Dethier, V. G. 1942. Abdominal glands of Hesperiinae. *J. N. Y. Entomol. Soc.* 50(2): 203–206.

Dimock, T. E. 1978. Notes on the life cycle and natural history of *Vanessa annabella* (Nymphalidae). *J. Lepid. Soc.* 32(2): 88–96.

Dornfield, E. J. 1980. *The Butterflies of Oregon.* Timber Press, Forest Grove, OR. 276 pp.

Douglas, M. M., and J. M. Douglas. 2005. *Butterflies of the Great Lakes Region.* University of Michigan Press, Ann Arbor. 345 pp.

Ehrlich, P. R. 1961. Intrinsic barriers to dispersal in a checkerspot butterfly. *Science* 134: 108–109.

Ehrlich, P. R., and I. Hanski. 2004. *On the Wings of Checkerspots: A Model System for Population Biology.* Oxford University Press. 371 pp.

Emmel, J. F., and B. M. Griffin. 1998. A new *Papilio indra* subspecies (Lepidoptera: Papilionidae) from the Muddy Mountains of southern Nevada. In T. C. Emmel, ed., *Systematics of Western North American Butterflies*, pp. 707–710. Mariposa Press, Gainesville, FL. 878 pp.

Emmel, J. F., and T. C. Emmel. 1998. A new subspecies of *Papilio indra* (Lepidoptera: Papilionidae) from northwestern California. In T. C. Emmel, ed., *Systematics of Western North American Butterflies*, pp. 701–706. Mariposa Press, Gainesville, FL. 878 pp.

Emmel, J. F., T. C. Emmel, and B. M. Griffin. 1998. A new dwarfed subspecies of *Papilio indra* (Lepidoptera: Nymphalidae) from the Dead Mountains of southeastern California. In T. C. Emmel, ed., *Systematics of Western North American Butterflies*, pp. 711–716. Mariposa Press, Gainesville, FL. 878 pp.

Emmel, T. C. 1969. Taxonomy, distribution and biology of the genus *Cercyonis* (Satyridae). 1. Characteristics of the genus. *J. Lepid. Soc.* 23(3): 165–175.

Emmel, T. C., and J. F. Emmel. 1973. *The Butterflies of Southern California.* Natural History Museum of Los Angeles County, Science Series 26. 148 pp.

Emmel, T. C., and S. O. Mattoon. 1972. *Cercyonis pegala blanca*, a "missing type" in the evolution of the genus *Cercyonis* (Satyridae). *J. Lepid. Soc.* 20(3): 140–149.

Emmel, T. C., M. Minno, and B. Drummond. 1992. *Florissant Butterflies: A Guide to the Fossil and Present-day Species of Central Colorado.* Stanford University Press. 118 pp.

Ferris, C. D., and P. M. Brown. 1981. *Butterflies of the Rocky Mountain States.* University of Oklahoma Press, Norman. 442 pp.

Fiedler, K., P. Seufert, N. E. Pierce, J. G. Pearson, and H. T. Baumgarten. 1992. Exploitation of lycaenid-ant mutualisms by braconid parasitoids. *J. Res. Lepid.* 31(3–4): 153–168.

Forister, M. L., A. Fordyce, and A. M. Shapiro. 2004. Geological barriers and restricted gene flow in the Holarctic skipper *Hesperia comma* (Hesperiidae). *Mol. Ecol.* 13: 3489–3499.

Fox, B. W. 2005. The larva of the white admiral butterfly, *Limenitis camilla* (Linnaeus, 1764) – a master builder. *Entomol. Gaz.* 56: 225–236.

Frankfater, C., M. R. Tellez, and M. Slattery. 2009. The scent of alarm: Ontogenetic and genetic variation in the osmeterial gland chemistry of *Papilio glaucus* (Papilionidae) caterpillars. *Chemoecology* 19: 81–96.

Fraser, A. M., A. H. Axén, and N. E. Pierce. 2001. Assessing the quality of different ant species as partners of a myrmecophilous butterfly. *Oecologia* 129: 452–460.

Freitas, A. V. L. 2006. Immature stages of *Adelpha malea goyama* Schaus (Lepidoptera: Nymphalidae, Limenitidinae). *Neotrop. Entomol.* 35(5): 625–628.

Frohawk, F. W. 1892. Life history of *Carterocephalus palaemon. The Entomologist* 25: 225–256.

Funk, R. S. 1975. Association of ants with ovipositing *Lycaena rubidus* (Lycaenidae). *J. Lepid. Soc.* 29(4): 261–262.

Fuxa, J. R., J. S. Sun, E. H. Weidner, and L. R. LaMotte. 1999. Stressors and rearing diseases of *Trichoplusia ni*: Evidence of vertical transmission of NPV and CPV. *J. Invertebr. Pathol.* 74: 149–155.

Garcia-Barros, E. 2006. Number of larval instars and sex-specific plasticity in the development of the small heath butterfly, *Coenonympha pamphilus* (Lepidoptera: Nymphalidae). *Eur. J. Entomol.* 103(1): 47–53.

Glassberg, J. 2002. *Butterflies of North America.* Michael Friedman Publishing. 201 pp.

Guppy, C. S., and J. H. Shepard. 2001. *Butterflies of British Columbia.* UBC Press and Royal British Columbia Museum. 414 pp.

Hardy, G. A. 1954. Notes on the life history of *Hesperia comma* L. *manitoba* Scud. (Lepidoptera, Rhopalocera) on Vancouver Island. *Proc. Entomol. Soc. B. C.* 51: 40–43.

Harry, J. 2009. Natural life histories of Alaska *Colias* (Lepidoptera: Pieridae). *Taxon. Rep. Int. Lepid. Surv.* 7(2): 1–20.

Hayes, J. L. 1981. Some aspects of the biology of the developmental stages of *Colias alexandra* (Pieridae). *J. Lepid. Soc.* 34(4): 345–352.

Heitzman, J. R. 1964. The early stages of *Euphyes vestris. J. Res. Lepid.* 3(3): 151–153.

Hellmann, J. J., S. L. Pelini, K. Prior, and J. D. K. Dzurisin. 2008. The response of two butterfly species to climatic variation at the edge of their range and the implications for poleward range shifts. *Oecologia* 157: 583–592.

Herman, W. S., and S. H. Dallman. 1981. Endocrine biology of the painted lady butterfly, *Vanessa cardui. J. Insect Physiol.* 27(3): 163–168.

Hill, J. K., C. D. Thomas, and O. T. Lewis. 1996. Effects of habitat patch size and isolation on dispersal by *Hesperia comma* butterflies: Implications for metapopulation structure. *J. Appl. Ecol.* 65: 725–735.

Hinchliff, J. 1994. *An Atlas of Oregon Butterflies.* The Evergreen Aurelians and Oregon State University. 176 pp.

Hinchliff, J. 1996. *An Atlas of Washington Butterflies.* The Evergreen Aurelians and Oregon State University. 162 pp.

Hiruma, K., J. Pelham, and H. Bouhin. 1997. Termination of pupal diapause in *Callophrys sheridanii* (Lycaenidae). *J. Lepid. Soc.* 51(1): 75–82.

Hitchcock, C. L., and A. Cronquist. 1973. *Flora of the Pacific Northwest – An Illustrated Manual.* University of Washington Press. 730 pp.

Hoffman, R. J. 1978. Environmental uncertainty and evolution of physiological adaptation in *Colias* butterflies. *Am. Midl. Nat.* 112: 999–1015.

Honda, K. 1981. Larval osmeterial secretions of the swallowtails (*Papilio*). *J. Chem. Ecol.* 7: 1089–1113.

Honda, K., and N. Hayashi. 1995. Chemical nature of larval osmeterial secretions of papilionid butterflies in the genera, *Parnassius, Sericinus* and *Pachliopta. J. Chem. Ecol.* 21: 859–867.

James, D. G. 1986. Thermoregulation in *Danaus plexippus* (L.) (Lepidoptera: Nymphalidae): Two cool climate adaptations. *Gen. Appl. Entomol.* 18: 43–47.

James, D. G. 1987. Effects of temperature and photoperiod on the development of *Vanessa kershawi* McCoy and *Junonia villida* Godart (Lepidoptera: Nymphalidae). *J. Aust. Entomol. Soc.* 26: 289–292.

James, D. G. 1988. Notes on the adult summer diapause of *Heteronympha merope merope* Fabricius (Lepidoptera: Nymphalidae). *Aust. Entomol. Mag.* 15: 67–72.

James, D. G. 1993. Migration biology of the monarch butterfly in Australia. In S. Malcolm and M. Zalucki, eds., *Biology and Conservation of the Monarch Butterfly*, pp. 189–200. Natural History Museum of Los Angeles County.

James, D. G. 1999. Larval development in *Heteronympha merope merope* (Fabricius) (Lepidoptera: Nymphalidae). *Aust. Entomol.* 26: 97–102.

James, D. G. 2000. Feeding on larvae of *Danaus plexippus* (L.) (Lepidoptera: Nymphalidae) causes mortality in the assassin bug, *Pristhesancus plagipennis* Walker (Hemiptera: Reduviidae). *Aust. Entomol.* 27: 5–8.

James, D. G. 2008a. Comparative studies on the immature stages and developmental biology of five *Argynnis* spp. (subgenus *Speyeria*) (Nymphalidae) from Washington. *J. Lepid. Soc.* 62(2): 61–66.

James, D. G. 2008b. Two heads better than one? Sclerotization of the posterior abdominal segments in *Nymphalis californica* larvae creates the illusion of a second head. *News Bull. Lepid. Soc.* 50(2): 35–36.

James, D. G. 2009a. A lesson in self defense from your California sister. *News Bull. Lepid. Soc.* 51(2): 47–52.

James, D. G. 2009b. Comparative studies on the immature stages and developmental biology of *Hesperia colorado idaho* and *Hesperia juba* (Hesperiidae). *J. Lepid. Soc.* 63: 129–136.

Janz, N. 2005. The relationship between habitat selection and preference for adult and larval food resources in the polyphagous butterfly *Vanessa cardui* (Lepidoptera: Nymphalidae). *J. Insect Behav.* 18(6): 767–780.

Janz, N., S. Nylin, and N. Wedell. 2001. Evolutionary dynamics of host-plant specialization: A case study of the tribe Nymphalini. *Evolution* 55: 783–796.

Jepsen, S., S. H. Black, and L. Lauvray. 2007. *Xerces Society Surveys for* Polites mardon mardon *in the Naches Ranger District of Washington (Summer 2007)*. Report to the US Forest Service. 56 pp.

Jones, R. E., V. G. Nealis, P. M. Ives, and E. Scheermeyer. 1987. Seasonal and spatial variation in juvenile survival of the cabbage butterfly, *Pieris rapae*: Evidence for patch density-dependence. *J. Anim. Ecol.* 56: 723–737.

Klots, A. B. 1951. *A Field Guide to the Butterflies of North America, East of the Great Plains*. Houghton Mifflin. 349 pp + 40 pls.

Kuussaari, M., S. V. Nouhuys, J. J. Hellman, and M. C. Singer. 2004. Larval biology of checkerspots. In P. R. Ehrlich and I. Hanski, eds., *On the Wings of Checkerspots*. Oxford University Press. 371 pp.

Lawrence, D. A., and J. C. Downey. 1966. Morphology of the immature stages of *Everes comyntas* Godart (Lycaenidae). *J. Res. Lepid.* 5: 61–96.

Layberry, R., P. Hall, and D. LaFontaine. 1998. *The Butterflies of Canada*. University of Toronto Press. 280 pp.

Lazri, B., and E. M. Barrows. 1984. Flower visiting and pollen transport by the imported cabbage butterfly (Lepidoptera: Pieridae) in a highly disturbed urban environment. *Environ. Entomol.* 12: 166–170.

Lokkers, C., and R. E. Jones. 1999. The cabbage white, *Pieris rapae* (Pieridae). In R. L. Kitching, E. Scheermeyer, R. E. Jones, and N. E. Pierce, eds., *Biology of Australian Butterflies*, pp. 153–172. CSIRO Publishing, Canberra. 395 pp.

Lyman, H. H. 1896. Notes on the preparatory stages of *Erebia epipsodea* (Butler). *Can. Entomol.* 28(11): 274–278.

Lyman, H. H. 1897. Notes on the life history of *Colias interior*, Scud. *Can. Entomol.* 29(11): 249–258.

MacNeill, C. D. 1964. The skippers of the genus *Hesperia* in western North America with special reference to California (Lepidoptera: Hesperiidae). *Univ. Calif. Publ. Entomol.* 35: 1–230.

Matter, S. F., and J. Roland. 2002. An experimental examination of the effects of habitat quality on the dispersal and local abundance of the butterfly *Parnassius smintheus*. *Ecol. Entomol.* 27(3): 308–313.

McCorkle, D. V., and P. C. Hammond. 1988. Biology of *Speyeria zerene hippolyta* (Nymphalidae) in a marine-modified environment. *J. Lepid. Soc.* 42(3): 184–192.

McDonald, A. K., and H. F. Nijhout. 1996. The effect of environmental conditions on mating activity of the buckeye butterfly, *Precis coenia*. *J. Res. Lepid.* 35: 22–28.

McNeil, J. N., and P. G. Fields. 1985. Seasonal diapause development and diapause termination in the European skipper, *Thymelicus lineola* (Ochs). *J. Insect Physiol.* 31: 467–470.

Miller, J. C. 1995. *Caterpillars of Pacific Northwest Forests and Woodlands*. USDA, Tech Transfer FHM-NC-06-05. 80 pp.

Miller, J. C., and P. C. Hammond. 2003. *Lepidoptera of the Pacific Northwest: Caterpillars and Adults*. USDA, Forest Health Technology Enterprise Team, FHTET-2003-03. 324 pp.

Miller, J. C., and P. C. Hammond. 2007. *Butterflies and Moths of Pacific Northwest Forests and Woodlands: Rare, Endangered and Management-sensitive Species*. USDA Forest Health Technology Enterprise Team, FHTET-2006-07. 234 pp.

Minno, M. C., J. F. Butler, and D. W. Hall. 2005. *Florida Butterfly Caterpillars and their Host Plants*. University Press of Florida, Gainesville. 342 pp.

Murphy, D.D., and P. Feeny. 2006. Chemical facilitation of a naturally occurring host shift by *Papilio machaon* butterflies (Papilionidae). *Ecol. Monogr.* 76(3): 399–414.

Muyshondt, A., and A. Muyshondt. 1976. Notes on the life cycle and natural history of butterflies of El Salvador. 1C. – *Colobura diice* L. (Lepidoptera: Coloburinae). *J. N. Y. Entomol. Soc.* 84: 23–33.

Neill, W. 2001. *The Guide to Butterflies of Oregon and Washington*. Westcliffe, Englewood, CO. 160 pp.

Neill, W. 2007. *Butterflies of the Pacific Northwest*. Mountain Press, Missoula, MT. 192 pp.

Nelson, S. M. 2003. The western viceroy butterfly (Nymphalidae: *Limenitis archippus obsoleta*): An indicator for riparian restoration in the arid southwestern United States. *Ecol. Indicators* 3: 203–211.

Newcomer, E. J. 1963. The synonymy, variability and biology of *Lycaena nivalis*. *J. Res. Lepid.* 2(4): 271–280.

Newcomer, E. J. 1964. Life histories of *Papilio indra* and *P. oregonius*. *J. Res. Lepid.* 3(1): 49–62.

Newcomer, E. J. 1966. Life histories of three western species of *Polites*. *J. Res. Lepid.* 5(4): 243–247.

Newcomer, E. J. 1967. Early stages of *Chlosyne hoffmanni manchada* (Nymphalidae). *J. Lepid. Soc.* 21(1): 71–73.

Newcomer, E. J. 1973. Notes on life histories and habits of some western Theclinae. *J. Lepid. Soc.* 27: 13–15.

Nielsen, M. C. 1999. *Michigan Butterflies and Skippers*. Michigan State University Extension. 248 pp.

Nylin, S., K. Nyblom, F. Ronquist, N. Janz, J. Belicek, and M. Kallersjo. 2001. Phylogeny of *Polygonia, Nymphalis* and related butterflies (Lepidoptera: Nymphalidae): A total evidence analysis. *Zool. J. Linn. Soc.* 137(4): 441–468.

Opler, P. A. 1974. Studies on Nearctic *Euchloe*. Part 7. Comparative life histories, hosts and the morphology of immature stages. *J. Res. Lepid.* 13(1): 1–20.

Otero, L. D., and A. Aiello. 1996. Description of the immature stages of *Adelpha alala* (Nymphalidae). *J. Lepid. Soc.* 50(4): 329–336.

Pelham, J. P. 2008. A Catalogue of the Butterflies of the United States and Canada. *J. Res. Lepid.* 40. 658 pp.

Pengelly, D. H. 1960. *Thymelicus lineola* (Ochs) (Lepidoptera: Hesperiidae) a pest of hay and pasture grasses in southern Ontario. *Proc. Entomol. Soc. Ont.* 91: 189–197.

Peterson, M. A. 1993. The nature of ant attendance and the survival of larval *Icaricia acmon* (Lycaenidae). *J. Lepid. Soc.* 47(1): 8–16.

Peterson, M. A. 1997. Host plant phenology and butterfly dispersal: Causes and consequences of uphill movement. *Ecology* 78(1): 167–180.

Pratt, G. F., and G. R. Ballmer. 1991. Acceptance of *Lotus scoparius* (Fabaceae) by larvae of Lycaenidae. *J. Lepid. Soc.* 45(3): 188–196.

Pratt, G. F., and J. F. Emmel. 1998. Revision of the *Euphilotes enoptes* and *E. battoides* complexes (Lepidoptera: Lycaenidae). In T. C. Emmel, ed., *Systematics of Western North American Butterflies*, pp. 207–270. Mariposa Press, Gainesville, FL. 878 pp.

Pratt, G. F., D. M. Wright, and G. R. Ballmer 1993. Multivariate and phylogenetic analyses of larval and adult characters of the *editha* complex of the genus *Lycaena* (Lepidoptera: Lycaenidae). *J. Res. Lepid.* 30(3–4): 175–195.

Prudic, K. L., S. Keara, A. Solyom, and B. N. Timmermann. 2007. Isolation, identification, and quantification of potential defensive compounds in the viceroy butterfly and its larval hostplant, Carolina willow. *J. Chem. Ecol.* 33: 1149–1159.

Prudic, K. L., A. M. Shapiro, and N. S. Clayton. 2002. Evaluating a putative mimetic relationship between two butterflies, *Adelpha bredowii* and *Limenitis lorquini*. *Ecol. Entomol.* 27: 68–75.

Prudic, K. L., A. D. Warren, and J. Llorente-Bousquets. 2008. Molecular and morphological evidence reveals three species within the California sister butterfly, *Adelpha bredowii* (Lepidoptera: Nymphalidae: Limenitidinae). *Zootaxa* 1819: 1–24.

Pyle, R. M. 1981. *National Audubon Society Field Guide to North American Butterflies.* Chanticleer Press, New York. 924 pp.

Pyle, R. M. 1999. *Chasing Monarchs: Migrating with Butterflies of Passage.* Houghton Mifflin, New York. 307 pp.

Pyle, R. M. 2002. *The Butterflies of Cascadia: A Field Guide to All the Species of Washington, Oregon, and Surrounding Territories.* Seattle Audubon Society. 420 pp.

Ravenscroft, N. O. M. 1992. "The ecology and conservation of the chequered skipper butterfly, *Carterocephalus palaemon* (Pallas)." Unpublished PhD thesis, University of Aberdeen, Scotland.

Rawlins, J. E., and R. C. Lederhouse. 1981. Developmental influences of thermal behavior on monarch caterpillars (*Danaus plexippus*): An adaptation for migration. *J. Kans. Entomol. Soc.* 54: 387–407.

Reinhard, H. V. 1981. Notes on early stages of *Nymphalis californica* (Nymphalidae). *J. Lepid. Soc.* 35(3): 243–244.

Remington, C. L. 1952. The biology of Nearctic Lepidoptera I. Foodplants and life histories of Colorado papilionids. *Psyche* 59(2): 61–70.

Roland, J. 2006. Effect of melanism of alpine *Colias nastes* butterflies (Lepidoptera: Pieridae) on activity and predation. *Can. Entomol.* 138: 52–58.

Schlicht, D. W., J. C. Downey, and J. C. Nekola. 2007. *The Butterflies of Iowa.* University of Iowa Press, Iowa City. 233 pp.

Schultz, C. B., P. C. Hammond, and M. V. Wilson. 2003. Biology of the Fender's blue butterfly (*Icaricia icarioides fenderi* Macy), an endangered species of western Oregon native prairies. *Nat. Areas J.* 23(1): 61–71.

Scott, J. A. 1975. Movements of *Precis coenia*, a "pseudoterritorial" submigrant (Lepidoptera: Nymphalidae). *J. Anim. Ecol.* 44(3): 843–850.

Scott, J. A. 1986a. *The Butterflies of North America: A Natural History and Field Guide.* Stanford University Press. 583 pp.

Scott, J. A. 1986b. Larval hostplant records for butterflies and skippers (mainly from western U.S.), with notes on their natural history. *Papilio* (New Series) 4. 37 pp.

Scott, J. A. 1992. Hostplant records for butterflies and skippers (mostly from Colorado) 1959–1992, with new life histories and notes on oviposition, immatures and ecology. *Papilio* (New Series) 6. 185 pp.

Scott, J. A. 2006. Butterfly hostplant records, 1992–2005, with a treatise on the evolution of *Erynnis*, and a note on new terminology for mate-locating behavior. *Papilio* (New Series) 14. 74 pp.

Shapiro, A. M. 1966. *Butterflies of the Delaware Valley.* Entomological Society of America, Ann Arbor, MI. 79 pp.

Shapiro, A. M. 1975. Notes on the biology of a "weedy" butterfly *Pieris occidentalis* (Lepidoptera: Pieridae) at Fairbanks, Alaska. *Arct. Alp. Res.* 7(3): 273–278.

Shapiro, A. M. 1976. The biological status of Nearctic taxa in the *Pieris protodice-occidentalis* group (Pieridae). *J. Lepid. Soc.* 30(4): 289–300.

Shapiro, A. M. 1981. Egg-mimics of *Streptanthus* (Cruciferae) deter oviposition by *Pieris sisymbrii* (Lepidoptera: Pieridae). *Oecologia* 48(1): 142–143.

Shapiro, A. M. 1986. Seasonal phenology and possible migration of the mourning cloak butterfly, *Nymphalis antiopa* (Lepidoptera: Nymphalidae) in California. *Great Basin Nat.* 46(1): 112–116.

Shapiro, A. M., and T. D. Manolis. 2007. *Field Guide to Butterflies of the San Francisco Bay and Sacramento Valley Regions.* University of California Press, Berkeley. 346 pp.

Shigeru, A. A. 1957. Effects of photoperiod on *Colias eurytheme. Lepid. News* 11(6): 207–214.

Shigeru, A. A. 1958. Comparative studies of developmental rates, hibernation and food plants in North American *Colias* (Lepidoptera: Pieridae). *Am. Midl. Nat.* 60(1): 84–96.

Shiojiri, K., and J. Takabayashi. 2005. Effects of oil droplets by *Pieris* caterpillars against generalist and specialist carnivores. *Ecol. Res.* 20: 695–700.

Silberglied, R. E., and O. R. Taylor, Jr. 1978. Ultraviolet reflection and its behavioral role in the courtship of the sulphur butterfies *Colias eurytheme* and *C. philodice* (Lepidoptera: Pieridae). *Behav. Ecol. Sociobiol.* 3(3): 203–243.

Sims, S. R. 1983. Prolonged diapause and pupal survival of *Papilio zelicaon* Lucas (Lepidoptera: Papilionidae). *J. Lepid. Soc.* 37(1): 29–37.

Sims, S. R. 1984. Reproductive diapause in *Speyeria* (Lepidoptera: Nymphalidae). *J. Res. Lepid.* 23: 211–216.

Singer, M. C., and P. R. Erhlich. 1979. Population dynamics of the checkerspot butterfly, *Euphydryas editha. Fortschr. Zool.* 25(2): 53–60.

Singer, M. C., C. D. Thomas, H. L. Billington, and C. Parmesan. 1994. Correlates of speed of evolution of host preference in a set of twelve populations of the butterfly *Euphydryas editha.* Ecoscience 1(2): 107–114.

Slansky, F. 1973. "Energetic and nutritional interactions between larvae of the imported cabbage butterfly, *Pieris rapae* L. and cruciferous food plants." PhD thesis, Cornell University.

Smedley, S. R., F. C. Schroeder, D. B. Weibel, J. Meinwald, K. A. Lafleur, J. A. Renwick, R. Rutowski, and T. Eisner. 2002. Mayolenes: labile defensive lipids from the glandular hairs of a caterpillar (*Pieris rapae*). *Proc. Natl. Acad. Sci. USA* 99: 6822–6827.

Smith, K. C. 1991. The effects of temperature and daylength on the *Rosa* polyphenism in the buckeye butterfly, *Precis coenia* (Lepidoptera: Nymphalidae). *J. Res. Lepid.* 30(3–4): 225–236.

Sourakov, A. 1995. Systematics, evolutionary biology and population genetics of the *Cercyonis pegala* group (Lepidoptera: Nymphalidae: Satyrinae). *Holarct. Lepid.* 2(1): 1–20.

Spencer, L. 2006. *Arkansas Butterflies and Moths.* Ozark Society Foundation, Little Rock, AR. 314 pp.

Spomer, S. M., and W. W. Hoback. 1998. New ant associations for *Glaucopsyche lygdamus* Doubleday (Lycaenidae). *J. Lepid. Soc.* 52(2): 216–217.

Stamp, N. E. 1984. Interactions of parasitoids and checkerspot caterpillars *Euphydryas* spp. (Nymphalidae). *J. Res. Lepid.* 23: 2–18.

Tabashnik, B. E. 1983. Host range evolution: The shift from native legume hosts to alfalfa by the butterfly, *Colias philodice eriphyle. Evolution* 37(1): 150–162.

Takabayashi, J., Y. Sato, S. Yano, and N. Ohsaki. 2000. Presence of oily droplets from the dorsal setae of *Pieris rapae* larvae (Lepidoptera: Pieridae). *Appl. Entomol. Zool.* 35: 115–118.

Taylor, O. R., J. W. Grula, and J. L. Hayes. 1981. Artificial diets and continuous rearing methods for the sulphur butterflies *Colias eurytheme* and *C. philodice* (Pieridae). *J. Lepid. Soc.* 35(3): 281–289.

Thomas, J. A., C. D. Thomas, D. J. Simcox, and R. T. Clarke. 1986. Ecology and declining status of the silver-spotted skipper butterfly (*Hesperia comma*) in Britain. *J. Appl. Ecol.* 23: 365–380.

Tveten, J., and G. Tveten. 1996. *Butterflies of Houston and Southeast Texas.* University of Texas Press, Austin. 292 pp.

Wagner, D. L. 2005. *Caterpillars of Eastern North America.* Princeton University Press. 512 pp.

Wagner, D. L., V. Giles, R. Reardon, and M. McManus. 1997. *Caterpillars of Eastern Forests.* US Dept. of Agriculture, Tech Transfer FHTET-96-34. 113 pp.

Warren, A. D. 1998. Greater fritillaries—lesser frustration: A guide to the species at Rabbit Ears Pass. *Am. Butterflies* 6(1): 16–27.

Warren, A. D. 2005. *Butterflies of Oregon: Their Taxonomy, Distribution, and Biology.* Lepidoptera of North America 6. C. P. Gillette Museum of Arthropod Diversity, Dept. of Bioagricultural Sciences and Pest Mgmt, Colorado State University, Fort Collins. 408 pp.

Warren, L. O., and J. E. Roberts. 1956. A hesperiid, *Atalopedes campestris* (Bdv.) as a pest of Bermuda grass pastures. *J. Kans. Entomol. Soc.* 29(4): 139–141.

Warren, M. S. 1987. The ecology and conservation of the heath fritillary, *Melitaea athalia*: Host selection and phenology. *J. Appl. Ecol.* 24: 467–482.

Webster, R. P. 1998. The life history of the maritime ringlet, *Coenonympha tullia nipisiquit* (Satyridae). *J. Lepid. Soc.* 52(4): 345–355.

Wehling, W. F., and J. N. Thompson. 1997. Evolutionary conservatism of oviposition preference in a widespread polyphagous insect herbivore, *Papilio zelicaon*. *Oecologia* 111(2): 209–215.

Weiss, M. R. 2003. Good housekeeping: Why do shelter-dwelling caterpillars fling their frass? *Ecol. Lett.* 361–370.

Weiss, S. B., R. R. White, D. D. Murphy, and P. R. Ehrlich. 1987. Sun, slope and butterflies: Topographic determinants of habitat quality for *Euphydryas editha*. *Ecology* 69: 1486–1496.

Williams, E. H., C. E. Holdren, and P. R. Ehrlich. 1984. The life history and ecology of *Euphydryas gillettii* Barnes (Nymphalidae). *J. Lepid. Soc.* 38: 1–12.

Woodward, M. K. 2005. *Butterflies and Butterfly Gardening in the Pacific Northwest.* Whitecap Books, Vancouver, B.C. 104 pp.

Zalucki, M. P., and R. L. Kitching. 1982. Temporal and spatial variation of mortality in field populations of *Danaus plexippus* L. and *D. chrysippus* L. larvae (Lepidoptera: Nymphalidae). *Oecologia* 53: 201–207.

Photo Credits and Data

Key to Photographers

DJ	David James	KW	Keith Wolfe
DN	David Nunnallee	MP	Merrill Peterson
IU	Idie Ulsh	ND	Nicky Davis

Each photograph is identified below by its page number and a descriptor (the growth stage and/or page position). Locality information is given for the original source of the specimen or its wild progenitor. Abbreviations used: E = Egg(s); L1–L6 = first to sixth instar larvae; P = Pupa; Ad = Adult; All = all stages on the page. Bot = bottom, C = center, CL = center left, CR = center right, LL = lower left, LR = lower right, UC = upper center, UL = upper left, UR = upper right.

Pg. 12: Arctic Fritillary *Boloria chariclea* + Rhiannon James, DJ, Apex Mt, BC.

Pg. 13: *Euphyes vestris* eggs, DJ, Satus Pass, Klickitat Co., WA.

Pg. 14: *Euphilotes* sp. ovipositing, DN, Elk Heights, Kittitas Co., WA.; *Adelpha* + *Brephidium* egg comparison, DJ, California

Pg. 15: *Pyrgus centaureae* hatching egg, DJ, Slate Peak, Okanogan Co., WA.; *Atalopedes campestris* premolt L4, DJ, Benton City, Benton Co., WA.

Pg. 16: *Argynnis egleis* in J shape, DJ, Diamond Peak, Garfield Co., WA.; *Celastrina echo* larval molt series, DJ, Satus Pass, Klickitat Co., WA.

Pg. 17: *Papilio multicaudata* pupation series, DJ, Naches Hts, Yakima Co., WA.; *Polygonia satyrus* pupa, DJ, Tucannon R, Columbia Co., WA.

Pg. 18: *Aglais milberti* eggs, DJ, Waterworks Canyon, Yakima Co., WA.; *Vanessa cardui* larva on thistle, DN, Bellevue, King Co., WA.

Pg. 19: *Euphilotes* on *E. thymoides*, DN, Schnebley Coulee, Kittitas Co., WA.

Pg. 20: *Hesperia juba* nest, DJ; Shrub-steppe habitat, DJ, Benton Co., WA.

Pg. 21: *Vanessa virginiensis* eggs, DJ, Feather R, CA.

Pg. 22: *Lotus crassifolius*, DN, Tahuya, Mason Co., WA.

Pg. 23: *Celastrina echo* on *Cornus*, DJ, Satus Pass, Klickitat Co., WA.

Pg. 24: *Chlosyne hoffmanni* nest, DN, Quartz Mt, Kittitas Co., WA.

Pg. 25: *Erynnis propertius* L5, DJ; *Cupido amyntula* larva in vetch pea, DN, Swauk Pass, Kittitas Co., WA.

Pg. 26: *Euphydryas colon* L2 in nest, DN, Reecer Canyon, Kittitas Co., WA.; *Argynnis zerene* L1 in violet seed, DJ, Mt Hull, Okanogan Co., WA.

Pg. 27: *Euphydryas colon* nest on penstemon, DN, Reecer Canyon, Kittitas Co., WA.

Pg. 28: Dormant L1 *Hesperia colorado*, DJ, Bear Canyon, Yakima Co., WA.; *Plebejus icarioides* L2, DJ, Waterworks Canyon, Yakima Co., WA.

Pg. 29: *Euchloe ausonides* eggs and thrips, DJ, Waterworks Canyon, Yakima Co., WA.; *Nymphalis californica* eggs, DJ, Naches Hts, Yakima Co., WA.

Pg. 30: Predatory ambush bug, DJ, Naches Hts, Yakima Co., WA.; *Orius tristicolor*, DJ, Naches Hts, Yakima Co., WA.

Pg. 31: *Hesperia juba* being stalked, DJ, Cummings Creek, Columbia Co., WA.

Pg. 32:   *Euphilotes columbiae* and parasitoid, DN, Chumstick Mt, Chelan Co, WA.; *Papilio rutulus* larva and parasitoid, DJ, Cummings Creek, Columbia Co., WA.

Pg. 33:   *Glaucopsyche lygdamus* and tachinid fly, IU, Haney Mdws, Kittitas Co., WA; *Nymphalis l-album* larva killed by virus, DJ, Sinlahekin R, Okanogan Co., WA.

Pg. 34:   *Erynnis propertius* L1 shelters on oak, DJ, Bear Canyon, Yakima Co., WA.

Pg. 35:   *Adelpha californica* L1, DJ, Feather R, CA; *Euphilotes columbiae* on buckwheat, DN, Reecer Canyon, Kittitas Co., WA.

Pg. 36:   *Callophrys gryneus* L on cedar, DN, Nooksack R, Whatcom Co., WA.

Pg. 37:   *Adelpha californica* larval head, DJ, California; *Nymphalis antiopa* L4, DJ, Schnebley Coulee, Kittitas Co., WA;.

Pg. 38:   *Nymphalis californica* larva, DJ, Tucannon R, Columbia Co., WA; *Argynnis coronis* larval head, DJ, Umtanum Canyon, Yakima Co., WA.

Pg. 39:   *Argynnis zerene* larval head, DJ, Mt Misery, Garfield Co., WA.; *Neophasia menapia* larval head, DJ, Little Naches R, Yakima Co., WA.

Pg. 40:   *Pontia occidentalis* larva, DJ, Steens Mt, OR; Ant on *Glaucopsyche lygdamus*, DN, Haney Mdws, Kittitas Co., WA.

Pg. 44:   Scrub-heath habitat, DN, Mason Co., WA.

Pg. 46:   Mountain habitat, DJ, Bear Creek Mt, Yakima Co., WA.

Pg. 47:   Shrub-steppe habitat, DJ, Waterworks Canyon, Yakima Co., WA.

Pages   Quick Photo Guides. All photos shown in the guides are also found in their
62–67:   respective species accounts, for which photo credits are listed separately.

Pg. 69:   Swallowtails and Parnassians Montage: UR, *Parnassius smintheus* P, DN, Reecer Canyon, Kittitas Co., WA; UL, *Papilio multicaudata* E, DJ, Naches Hts, Yakima Co, WA.; C, *Papilio zelicaon* prepupa, DJ, Red Mt, Benton Co., WA; CR, *Papilio rutulus* P, DJ, Bethel Ridge, Yakima Co., WA; CL + Bot, *Papilio machaon* L5+Ad, DJ, Snake R Jct, Franklin Co., WA.

Pg. 71:   (*Parnassius clodius*) E, DN, Carnation, WA; L1–L4+P, DJ, Bear Creek Mt, Yakima Co., WA; L5, DN, Snoqualmie Pass, Kittitas Co., WA.; Ad, IU, Reecer Canyon, Kittitas Co., WA.

Pg. 73:   (*Parnassius smintheus*) E, DN, Quartz Mt, Kittitas Co., WA; L1–L2, DJ, Diamond Peak, Garfield Co., WA; L3, DJ, Apex Mt, BC; L4+P, DN, Reecer Canyon, Kittitas Co., WA; L5, DJ, Umatilla NF, WA; Ad, DN, Chumstick Mt, Chelan Co., WA.

Pg. 75:   (*Papilio machaon*) All, DJ, Snake R Jct, Franklin Co., WA.

Pg. 77:   (*Papilio zelicaon*) E(banded), DN, Umtanum Canyon, Kittitas Co., WA; E(plain)+ L1–L3+L5(white)+P, DJ, Bear Canyon, Yakima Co., WA; L4, DN, Schnebley Coulee, Kittitas Co., WA; L5(green), DN, Chumstick Mt, Chelan Co., WA; Ad, DJ, Red Mt, Benton Co., WA.

Pg. 79:   (*Papilio indra*) L5(solid black), DJ, Columbia R Gorge, Klickitat Co., WA; All others, DJ, Klickitat, WA.

Pg. 81:   (*Papilio rutulus*) E, DJ, Tucannon R, Columbia Co., WA; L1, DJ, Cummings Creek, Columbia Co., WA; L2–L3+L4(35 mm), DJ, Bethel Ridge, Yakima

Co., WA; L4(22 mm), DN, Colockum, Kittitas Co., WA; L5(both), DN, Umtanum Canyon, Kittitas Co., WA; P, DJ, Cowiche, Yakima Co., WA; Ad, DN, Swakane Canyon, Chelan Co., WA.

Pg. 83: (*Papilio multicaudata*) E–L3+P, DJ, Naches, Yakima Co., WA; L4, DJ, Waterworks Canyon, Yakima Co, WA; L5(both), DN, Spokane, Spokane Co., WA; Ad, DN, Sinlahekin, Okanogan Co., WA.

Pg. 85: (*Papilio eurymedon*) E(green), DN, Swakane Canyon, Chelan Co., WA; E(brown)+L1–L2, DJ, Cummings Creek, Columbia Co., WA; L3–P, DN, Chumstick Mt, Chelan Co., WA; Ad, DJ, Tucannon, Columbia Co., WA.

Pg. 87: Whites and Sulphurs Montage: UL, *Euchloe ausonides* prepupa, DJ, Red Mt, Benton Co., WA; UC, *Euchloe lotta* P, DN, Schnebley Coulee, Kittitas Co., WA; UR, *Colias interior*, DJ, Tiger Mdws, Pend Oreille Co., WA; C, *Pontia beckerii* L5, DJ, Grand Ronde R, Garfield Co., WA; Inset R, *Colias interior* E, DJ, Tiger Mdws, Pend Oreille Co., WA; Bot, *Pieris marginalis*, DJ, Cummings Creek, Columbia Co., WA.

Pg. 89: (*Neophasia menapia*) Ad, DN, Mineral Sprgs, Kittitas Co., WA; All others, DJ, Little Naches R, Yakima Co., WA.

Pg. 91: (*Pontia beckerii*) E+L2–P, DJ, Grand Ronde R, Asotin Co., WA; L1, DN, Colockum, Kittitas Co., WA; Ad, DJ, Waterworks Canyon, Yakima Co., WA.

Pg. 93: (*Pontia sisymbrii*) E(yellow)+L4–L5, DN, Durr Rd, Kittitas Co., WA; E(red)+L1–L3, DJ, Umtanum Canyon, Kittitas Co., WA; P, DN, Whiskey Dick, Kittitas Co., WA; Ad, DN, Swakane Canyon, Chelan Co., WA.

Pg. 95: (*Pontia protodice*) E–L2, DJ, LA Co., CA; all others, DJ, Plumas Co., CA.

Pg. 97: (*Pontia occidentalis*) E–P, DJ, Steens Mt, Harney Co., OR; Ad, DJ, Bear Creek Mt, Yakima Co., WA.

Pg. 99: (*Pieris marginalis*) E+L5+Ad, DJ, Cummings Creek, Columbia Co., WA; L1–L4, DJ, Sawtooth Ridge, Columbia Co., WA; P, DN, Fall City, King Co., WA.

Pg. 101: (*Pieris rapae*) E+L4(side)+P(brown)+Ad, DN, Stossel Creek, King Co., WA; L1+L3, DN, Dry Falls, Grant Co., WA; L2+L4(dorsal)+P(green), DJ, Horn Rapids, Benton Co., WA; L5, DN, Sammamish, King Co., WA.

Pg. 103: (*Euchloe ausonides*) E–L4, DJ, Umtanum Canyon, Kittitas Co., WA; L5–P, Red Mt, Benton Co., WA; A, DN, Reecer Canyon, Kittitas Co., WA.

Pg. 105: (*Euchloe lotta*) E(white)+L3+P(dark), DN, Schnebley Coulee, Kittitas Co., WA; E(red)+L1–L2+L4–L5+P(light), DJ, Red Mt, Benton Co., WA; Ad, DN, Priest Rapids, Yakima Co., WA.

Pg. 107: (*Anthocharis sara*) E+L2, DN, Durr Rd, Kittitas Co, WA; L1+L3+L5+P, DN, Corral Pass, Pierce Co., WA; L4, DJ, Waterworks Canyon, Yakima Co.; Ad, DJ, Umatilla NF, WA.

Pg. 109: (*Colias philodice*) Ad, DN, Wanapum Dam, Kittitas Co., WA; All others, DJ, Apex Mt, BC.

Pg. 111: (*Colias eurytheme*) E–L1, DN, Winchester Wasteway, Grant Co., WA; L2–P, DJ, Umatilla NF, WA; Ad, DN, Colorado Sprgs, CO.

Pg. 113: (*Colias occidentalis*) E–L1, DJ, Bear Creek, Yakima Co., WA; L2+L4–P, DJ, Satus Pass, Klickitat Co., WA; L3, DN, Taneum Creek, Kittitas Co., WA; Ad, DN, Haney Mdws, Kittitas Co., WA.

Pg. 115: (*Colias alexandra*) E+L3–P+Ad DN, Moses Mdws, Okanogan Co., WA; L1–L2, DJ, Bunchgrass Mdws, Pend Oreille Co., WA.

Pg. 117: (*Colias nastes*) E+Ad+scene, DJ, Apex Mt BC; All others, ND, Baffin Is, Nunavut Terr., Can.

Pg. 119: (*Colias pelidne*) E–L3, DJ, Steens Mt, OR; L5–P, ND, North Slope, AK.

Pg. 121: (*Colias interior*) All, DJ, Tiger Mdws, Pend Oreille Co., WA.

Pg. 123: (*Nathalis iole*) Ad, DJ, San Diego Co., CA; All others, DJ, Pima Co., AZ.

Pg. 125: Coppers Montage: UL, *Lycaena rubidus* L3, DJ, Juniper Dunes, Franklin Co., WA; UR, *Lycaena editha* E, DJ, Tucannon R Columbia Co., WA; CL, *Lycaena mariposa* P, DJ, Tiger Mdws, Pend Oreille Co., WA; CR, *Lycaena helloides* Ad, DN, Haney Mdws, Kittitas Co., WA; Bot, *Lycaena mariposa* L4, DN, Moses Mdws,Okanogan Co., WA.

Pg. 127: (*Lycaena cupreus*) L4, MP, Slate Peak, Okanogan Co., WA; all others, DN, Slate Peak, Okanogan Co., WA.

Pg. 129: (*Lycaena editha*) E, DN, Blue Mts, Garfield Co., WA; All others, DJ, Tucannon R, Columbia Co., WA.

Pg. 131: (*Lycaena rubidus*) L1+L4(pair), DN, Juniper Dunes, Franklin Co., WA; All others, DJ, same loc.

Pg. 133: (*Lycaena heteronea*) E(white)+L4(13 mm)+P(green)+Ad, DN, Reecer Canyon, Kittitas Co., WA; All others, DJ, Tucannon R, Columbia Co., WA.

Pg. 135: (*Lycaena helloides*) E+L2, DJ, Satus Pass, Klickitat Co, WA; L1+L4(side)+P+Ad, DN, Wenas Well, Kittitas Co., WA; L3+L4(dorsal), DJ, Conrad Mdws, Yakima Co., WA.

Pg. 137: (*Lycaena nivalis*) L1+Ad, DJ, Blue Mts, Garfield Co.,WA; All others, DN, Reecer Canyon., Kittitas Co., WA.

Pg. 139: (*Lycaena mariposa*) E–L3, DJ, Tiger Mdws, Pend Oreille Co., WA; L4–P, DN, Moses Mdws, Okanogan Co., WA; Ad, DJ, Bear Cr Mt, Yakima Co., WA.

Pg. 141: Hairstreaks Montage: UL, *Callophrys polios* L4, DN, Moses Mdws, Okanogan Co., WA; UR, *Callophrys affinis* pupae, DJ, Umtanum Canyon, Kittitas Co., WA; CR, *Callophrys gryneus* nr *chalcosiva* E, DJ, Juniper Dunes, Franklin Co., WA; LC, *Callophrys augustinus* L4, DN, Reecer Canyon, Kittitas Co., WA; Bot, *Satyrium californica* adults, DN, Reecer Canyon, Kittitas Co., WA.

Pg. 143: (*Habrodais grunus*) All, DN, Willard, Skamania Co., WA.

Pg. 145: (*Satyrium titus*) All, DN, Colockum Rd, Kittitas Co., WA.

Pg. 147: (*Satyrium behrii*) E, DN, Colockum Rd, Kittitas Co., WA; L1–L4, DN, Cooke Canyon, Kittitas Co., WA; P, DN, Umtanum Ridge, Kittitas Co., WA; Ad, DN, Similkameen, Okanogan Co., WA.

Pg. 149: (*Satyrium semiluna*) E+Ad, DN, Wenas Lk, Yakima Co., WA; L1–L2+P, DJ, Conrad Mdws, Yakima Co., WA; L3, DJ, Bear Cr Mt, Yakima Co., WA; L4, DN, Whiskey Dick, Kittitas Co., WA.

Pg. 151: (*Satyrium californica*) L1–L2, DN, Chumstick, Chelan Co.,WA; All others, DN, Reecer Canyon, Kittitas Co., WA.

Pg. 153: (*Satyrium sylvinus sylvinus*) All, DN, Hwy 97 nr Reecer Canyon, Kittitas Co., WA.

Pg. 155: (*Satyrium sylvinus nootka*) All, DN, Big Meadow Lk, Pend Oreille Co., WA.

Pg. 157: (*Satyrium saepium*) E–L2, DJ, Satus Pass, Klickitat Co., WA; L3–P, DN, Satus Pass, Klickitat Co., WA; Ad, DN, Chumstick, Chelan Co., WA.

Pg. 159: (*Callophrys perplexa*) E+L2(2.5 mm)+L3(8 mm)+Ad, DN, Tahuya Penin, Mason Co., WA; All others, DJ, Satus Pass, Klickitat Co., WA.

Pg. 161: (*Callophrys affinis*) E–L2+L4(pair)+Ad, DJ, Umtanum, Kittitas Co., WA; L3+L4(single)+P, DN, Umtanum Ridge, Kittitas Co., WA.

Pg. 163: (*Callophrys sheridanii*) E+L3(7.5 mm), DN, Schnebley Coulee, Kittitas Co., WA; L1–L2+L4(16 mm)+P, DN, Cowiche Canyon, Yakima Co., WA; L3(early)+Ad, DJ, Waterworks Canyon, Yakima Co., WA; other L4s, DN, Chumstick Mt, Chelan Co., WA.

Pg. 165: (*Callophrys johnsoni*) All, DN, Lk Cushman, Mason Co., WA.

Pg. 167: (*Callophrys spinetorum*) E–L1+L4–P+Ad, DJ, Ashnola R, BC; L2–L3+ plant, DN, Mt Hull, Okanogan Co., WA.

Pg. 169: (*Callophrys gryneus plicataria*) Ad, DN, Stossel Creek, King Co., WA; All others, DN, Nooksack R, Whatcom Co., WA.

Pg. 171: (*Callophrys gryneus* nr *chalcosiva*) All, DJ, Juniper Dunes, Franklin Co., WA.

Pg. 173: (*Callophrys augustinus*) E–L3+L4(pink), DN, Reecer Canyon, Kittitas Co., WA; L4(green)+Ad, DJ, Satus Pass, Klickitat Co., WA; P, DN, Moses Mdws, Okanogan Co., WA.

Pg. 175: (*Callophrys mossii*) All, DN, Reecer Canyon, Kittitas Co., WA.

Pg. 177: (*Callophrys polios*) All, DN, Tahuya Peninsula, Mason Co., WA.

Pg. 179: (*Callophrys eryphon*) E–L2+L4(side), DJ, Tucannon R, Columbia Co., WA; L3+L4(dorsal)+P, DN, Moses Mdws, Okanogan Co., WA; Ad, DN, Tahuya Penin, Mason Co., WA.

Pg. 181: (*Strymon melinus*) E–L1+L3(10 mm)+L4(17 mm), DN, L Granite Dam, Whitman Co, WA; L2+L3(6 mm)+L4, DJ, Juniper Dunes, Franklin Co., WA; P, DN, Schnebley Coulee, Kittitas Co., WA; Ad, IU, Tahuya, Kitsap Co., WA.

Pg. 183: Blues Montage: Top, *Celastrina echo nigrescens* L4, DN, Black Canyon, Okanogan Co., WA; CL, *Euphilotes enoptes* L4, DN, Rimrock Lk, Yakima Co., WA; UCR, *Plebejus lupini* E, DJ, Bear Canyon, Yakima Co., WA; LCR, *Brephidium exilis* P, DJ, Fairfield, CA; Bot, *Plebejus icarioides* adults, DJ, Umtanum Canyon, Kittitas Co., WA.

Pg. 185: (*Brephidium exilis*) E(top)+L3(side)+L4(dorsal), DN, Indio, Riverside Co., CA; All others, DJ, Fairfield, Solano Co., CA.

Pg. 187: (*Cupido amyntula*) E+L2+L4(side), DJ, Mt Hull, Okanogan Co., WA; L1+P(pale), DN, Liberty, Kittitas Co., WA; L3, DJ, Umtanum Canyon, Kittitas Co., WA; L4(in pod)+P(dark), DN, Hurley Creek, Kittitas Co., WA; Ad, IU, central WA.

Pg. 189: (*Cupido comyntas*) All, DN, Adair Village, Benton Co., OR.

Pg. 191: (*Celastrina echo echo*) E, DJ, Bear Canyon, Yakima Co., WA; L4(dorsal), DN, Mercer Island, King Co., WA; Ad, DN, Reecer Canyon, Kittitas Co., WA; All others, DJ, Satus Pass, Klickitat Co., WA.

Pg. 193: (*Celastrina echo nigrescens*) All, DN, Black Canyon, Okanogan Co., WA.

Pg. 195: (*Celastrina lucia*) All, DJ, Cowiche Canyon, Yakima Co., WA.

Pg. 197: (*Euphilotes enoptes*) All, DN, Rimrock Lk, Yakima Co., WA.

Pg. 199: (*Euphilotes columbiae*) E–L1, DN, Elk Heights, Kittitas Co., WA; L2–L3+L4(11 mm)+P+Ad, DN, Dallesport, Klickitat Co., WA; L4(6 mm), DN, Chumstick Mt, Chelan Co., WA.

Pg. 201: (*Euphilotes* on *E. heracleoides*) E–L1, DN, Swakane Canyon, Chelan Co., WA; All others, DN, Chumstick Mt, Chelan Co., WA.

Pg. 203: (*Euphilotes glaucon*) All, DN, Chumstick Mt, Chelan Co., WA.

Pg. 205: (*Euphilotes* on *E. sphaerocephalum*) E, DN, Wild Horses Mon., Grant Co., WA; All others, DN, Ellensburg viewpoint, Kittitas Co., WA.

Pg. 207: (*Glaucopsyche lygdamus*) E+L3+L4(green), DN, Coffin Butte, Benton Co., OR; L1–L2+L4(brown)+P, DJ, Tucannon R, Columbia Co., WA; Ad, DN, Schnebley Coulee, Kittitas Co., WA.

Pg. 209: (*Glaucopsyche piasus*) L1, DJ, Snake R, Walla Walla Co., WA; All others, DN, Number 2 Canyon, Wenatchee, Chelan Co., WA.

Pg. 211: (*Plebejus idas*) E–L3, DJ, Apex Mt, BC; All others, DN, Moses Meadows, Okanogan Co., WA.

Pg. 213: (*Plebejus anna*) E, DN, Mt St Helens, Skamania Co., WA; L3, DJ, Apex Mt, BC; All others, DJ, Chinook Pass, Yakima Co., WA.

Pg. 215: (*Plebejus melissa*) L4(12 mm)+Ad, DN, Wenas Lk, Yakima Co., WA; All others, DJ, Waterworks Canyon, Yakima Co., WA.

Pg. 217: (*Plebejus saepiolus*) E(side)–L2, DJ, Tucannon R, Columbia Co., WA; All others, DN, Taneum Creek, Kittitas Co., WA.

Pg. 219: (*Plebejus icarioides*) E–L2+L4(dorsal), DJ, Waterworks Canyon, Yakima Co., WA; Ad, DJ, Umtanum Canyon, Yakima Co., WA; All others, DN, Whiskey Dick Ridge, Kittitas Co., WA.

Pg. 221: (*Plebejus acmon*) E, DN, Dallesport, Klickitat Co., WA; All others, DN, Rimrock Lk, Yakima Co., WA.

Pg. 223: (*Plebejus lupini*) E, DJ, Bear Canyon, Yakima Co., WA; L1+L3, DJ, Snake R Jct, Franklin Co., WA; L2+P, DN, Ellensburg Viewpoint, Kittitas Co., WA; L4+Ad, DN, Yakima R nr Thorp, Kittitas Co., WA.

Pg. 225: (*Plebejus glandon*) E+L2–L3+L4(9 mm)+Ad, DJ, Slate Peak, Okanogan Co., WA; All others, DN, same loc.

Pg. 227: (*Apodemia mormo*) E+L2, DJ, Umtanum Canyon, Kittitas Co., WA; L1, DJ, Waterworks Canyon, Yakima Co, WA; L3–L4+L5(dorsal), DN, Spokane, WA; L5(side)+P, DN, Swakane Canyon, Chelan Co., WA; Ad, DN, Ellensburg, Kittitas Co., WA.

Pg. 229: Fritillaries Montage: Top + CL, *Argynnis zerene* L6+P, DJ, Diamond Peak, Garfield Co., WA; UC, *Argynnis atlantis* E, DJ, Bunchgrass Mdws, Pend Oreille Co., WA; C Bot, *Argynnis mormonia erinna* L1, DN, Tiger Mdws, Pend Oreille Co., WA; CR, *Argynnis cybele leto* L6, DN, Hurley Creek, Kittitas Co., WA; Bot, *Argynnis* mixed species, DN, Reecer Canyon, Kittitas Co., WA.

Pg. 231: (*Euptoieta claudia*) All, DN, nr Golden, Jefferson Co., CO.

Pg. 233: (*Argynnis cybele*) E+L2–L5+L6(side)+P, DN, Hurley Creek, Kittitas Co., WA; L1+L6(dorsal), DJ, Grande Ronde R, Asotin Co., WA; Ad, DN, Mineral Sprgs, Kittitas Co., WA.

Pg. 235: (*Argynnis coronis*) L2, DJ, Naches Hts, Yakima Co., WA; L5+P, DJ, Bird Creek Mdws, Klickitat Co., WA; Ad, DN, Ellensburg, Kittitas Co., WA; All others, DJ, Bear Creek, Mt, Yakima Co., WA.

Pg. 237: (*Argynnis zerene*) Ad, DN, Mineral Sprgs, Kittitas Co., WA; All others, DJ, Diamond Peak, Garfield Co., WA.

Pg. 239: (*Argynnis callippe*) E, DN, Reecer Canyon., Kittitas Co., WA; Ad, DN, Mineral Sprgs, Kittitas Co., WA; All others, DJ, Satus Pass, Klickitat Co., WA.

Pg. 241: (*Argynnis egleis*) All, DJ, Diamond Peak, Garfield Co., WA.

Pg. 243: (*Argynnis hesperis dodgei*) L5+P, DJ, Wallowa Mts, OR; All others, DJ, Blue Mts, Garfield Co., WA.

Pg. 245: (*Argynnis hesperis brico*) All, DN, Tiger Mdws., Pend Oreille Co., WA.

Pg. 247: (*Argynnis atlantis*) E(mature)+L1, DN, Bunchgrass Mdws, Pend Oreille Co., WA; E(fresh)+L3+P, DJ, Tiger Mdws, Pend Oreille Co., WA; All others, DJ, Bunchgrass Mdws, Pend Oreille Co., WA.

Pg. 249: (*Argynnis hydaspe*) L3+L6(side), DN, Quartz Mt, Kittitas Co., WA; All others, DJ, Bear Cr Mt, Yakima Co., WA.

Pg. 251: (*Argynnis mormonia erinna*) E+L1+Ad, DN, Tiger Mdws, Pend Oreille Co., WA; All others, DJ, Tiger Mdws, Pend Oreille Co., WA.

Pg. 253: (*Argynnis mormonia washingtonia*) L6(light), DN, Lion Rock, Kittitas Co., WA; All others, DJ, Bear Creek Mt, Yakima Co., WA.

Pg. 255: (*Boloria selene*) L4, DN, Mary Ann Creek, Okanogan Co., WA; Ad, DN, Stewart Mdws, Pend Oreille Co., WA; All others, DJ, Tiger Mdws, Pend Oreille Co., WA.

Pg. 257: (*Boloria bellona*) P+Ad, DJ, Moses Mdws, Okanogan Co., WA; All others, DN, Moses Mdws, Okanogan Co., WA.

Pg. 259: (*Boloria epithore*) E+L5+P, DN, Haney Mdws, Kittitas Co., WA; L1, DJ, Swauk Pass, Kittitas Co., WA; L2–L4, DJ, Bear Creek Mt, Yakima Co., WA; Ad, DN, Liberty, Kittitas Co., WA.

Pg. 261: (*Boloria freija*) L5(dorsal), DN, Ashnola R, BC; All others, DJ, Ashnola R, BC.

Pg. 263: (*Boloria astarte*) E–L3+Ad, DN, Slate Peak, Okanogan Co., WA; L4–P, DJ, Slate Peak, Okanogan Co., WA.

Pg. 265: (*Boloria chariclea*) E+L3+L4(13 mm), DJ, Apex Mt, BC; L1+L2+L4(12 mm), Chinook Pass, Yakima Co, WA; Ad, DN, Harts Pass, Okanogan Co., WA.

Pg. 267: Checkerspots and Crescents Montage: UL, *Phyciodes pallidus* L5, DN, Similkameen R, Okanogan Co., WA; UR, *Euphydryas colon* P, DJ, Cummings Creek, Columbia Co., WA; UCL, *Euphydryas colon* E, DN, Reecer Canyon, Kittias Co., WA; C, *Euphydryas editha* L4, DN, Quartz Mt, Kittitas Co., WA; Bot, *Euphydryas anicia* Ad, DN, Moses Mdws, Okanogan Co., WA.

Pg. 269: (*Chlosyne hoffmanni*) E–L2+L4+nest, DN, Reecer Canyon, Kittitas Co., WA; L3, DN, Quartz Mt, Kittitas Co., WA; L5+P, DN, Liberty, Kittitas Co., WA; Ad, DN, Derby Canyon, Chelan Co., WA.

Pg. 271: (*Chlosyne acastus*) E–L4+Ad, DN, Schnebley Coulee, Kittitas Co., WA; L5+P, DJ, Red Mt, Benton Co., WA.

Pg. 273: (*Chlosyne palla*) All, DN, Moses Mdws, Okanogan Co., WA.

Pg. 275: (*Phyciodes cocyta*) All, DJ, Cummings Creek, Columbia Co., WA.

Pg. 277: (*Phyciodes pulchella*) E–L1+L5, DJ, Satus Pass, Klickitat Co., WA; L2+L4+P, DN, Haney Mdws, Kittitas Co., WA; L3, DJ, Rimrock Lk, Yakima Co., WA; Ad, DN, Whistler Mdws, Chelan Co., WA.

Pg. 279: (*Phyciodes pallida*) E–L1+L5+Ad, DJ, Similkameen R, Okanogan Co., WA; L2–L4+P, DN, Similkameen R, Okanogan Co., WA.

Pg. 281: (*Phyciodes mylitta*) E+L4, DJ, Umatilla NF, Garfield Co., WA; L1–L3+L5+P, DJ, Naches Hts, Yakima Co., WA; Ad, DN, Cowiche Canyon, Yakima Co., WA.

Pg. 283: (*Euphydryas anicia*) E–L5(pale)+P, DN, Moses Mdws, Okanogan Co., WA; L5(dark), DN, Colockum Pass, Kittitas Co., WA; Ad, DN, Chumstick Mt, Chelan Co., WA.

Pg. 285: (*Euphydryas colon*) E–L1+L3–L5, DN, Reecer Canyon, Kittitas Co., WA; L2+P, DJ, Cummings Creek, Columbia Co., WA; L6+Ad, DJ, Tucannon R, Columbia Co., WA.

Pg. 287: (*Euphydryas editha*) L1, DJ, Diamond Peak, Garfield Co., WA; L2, DN, Colockum Pass, Kittitas Co., WA; All others, DN, Quartz Mt, Kittitas Co., WA.

Pg. 289: Brushfoots Montage: UL, *Nymphalis l-album* E, DN, Sinlahekin R, Okanogan Co., WA; UR, *Polygonia satyrus* E, DN, Reecer Canyon, Kittitas Co., WA; C, *Limenitis archippus* L5, DJ, Horn Rapids, Benton Co., WA; LL, *Polygonia satyrus* P, DJ, Tucannon R, Columbia Co., WA; LR, Jasmine James and *Limenitis lorquini*, DN, Columbia Co., WA.

Pg. 291: (*Polygonia satyrus*) E+L2+L3+P(21 mm), DN, Reecer Canyon, Kittitas Co., WA; L1+L4–P(20 mm)+Ad, DJ, Tucannon R, Columbia Co., WA.

Pg. 293: (*Polygonia faunus*) E(single)+L1, DJ, Tucannon R, Columbia Co., WA; Ad, DN, Snoqualmie Pass, Kittitas Co., WA; All others, DN, Reecer Canyon, Kittitas Co., WA.

Pg. 295: (*Polygonia gracilis*) E, DN, Ashnola R, BC; L1–L2+Ad, DJ, Umatilla NF, WA; L3–L5(23 mm), DJ, Ashnola R, BC; L5(33 mm)+P, DN, Mt Hull, Okanogan Co.,WA.

Pg. 297: (*Polygonia oreas*) All, DN, Reecer Canyon, Kittitas Co., WA.

Pg. 299: (*Nymphalis l-album*) E–L5, DN, Sinlahekin R, Okanogan Co., WA; P, DN, Mt Hull, Okanogan Co, WA; Ad, DN, Ione, Pend Oreille Co., WA.

Pg. 301: (*Nymphalis californica*) E+L2–P, DJ, Tucannon R, Columbia Co., WA; L1, DJ, Sawtooth Ridge, Columbia Co., WA; Ad, DN, Tiger Mdws, Pend Oreille Co., WA.

Pg. 303: (*Nymphalis antiopa*) E–L3+Ad, DN, Schnebley Coulee, Kittitas Co., WA; L4–P, DJ, Schnebley Coulee, Kittitas Co., WA.

Pg. 305: (*Aglais milberti*) E+L4, DJ, Waterworks Canyon, Yakima Co., WA; L1–L2, DN, Reecer Canyon, Kittitas Co., WA; L3+L5(26 mm)+P, DN, Stossel Creek, King Co., WA; L5(25 mm), DJ, Diamond Pk, Garfield Co., WA; Ad, DJ, Mt Misery, OR.

Pg. 307: (*Vanessa virginiensis*) All, DJ, Feather R, CA.

Pg. 309: (*Vanessa annabella*) L2, DN, West Seattle, King Co., WA; Ad+L5+P, DJ, Mt Misery, Garfield Co., WA; All others, DJ, Benton City, Benton Co., WA.

Pg. 311: (*Vanessa cardui*) E+L1+L3+L4, DJ, Commercial stock; L2, DN, Bellevue, King Co, WA; L5+P+Ad, DN, Cle Elum, Kittitas Co, WA.

Pg. 313: (*Vanessa atalanta*) E(fresh), DN, Soos Creek, King Co., WA; E(mature)+L1–L4+P, DJ, Soos Creek, King Co., WA; L5(both)+Ad, DN, Stossel Creek, King Co., WA

Pg. 315: (*Junonia coenia*) All, DJ, Feather R, CA.

Pg. 317: (*Limenitis lorquini*) P, DN, Winthrop, Okanogan Co., WA; All others, DJ, Tucannon R, Columbia Co., WA.

Pg. 319: (*Limenitis archippus*) All, DJ, Horn Rapids, Yakima R, Benton Co., WA.

Pg. 321: (*Adelpha californica*) All, DJ, Feather R, CA.

Pg. 323: (*Danaus plexippus*) E(fresh)+L1(2.5 mm)+L2+L4, DJ, Waterworks Canyon, Yakima Co., WA; E(mature)+L1(3.4 mm)+Ad, DN, Chicago, IL; L3+L5+P, DJ, Trinity Co., CA.

Pg. 325: Satyrs Montage: Top, *Cercyonis sthenele* L3, DJ, Grant Co., WA; UL, *Coenonympha tullia* P, DJ, Napa, CA.; LL, *Oeneis chryxus* E, DJ, Mt Hull, Okanogan Co., WA; CR, *Cercyonis pegala* prepupa, DJ, Tucannon R, Columbia Co., WA; Bot, *Erebia vidleri* Ad, DN, Slate Peak, Okanogan Co., WA.

Pg. 327: (*Coenonympha tullia*) E–L2, DJ, Klickitat Co., WA; L3, DJ, Waterworks Canyon, Yakima Co., WA; L4(23 mm), DN, Durr Rd, Kittitas Co., WA; L4(22 mm)+P, DJ, Napa, CA; Ad, DN, Elk Hts, Kittitas Co., WA.

Pg. 329: (*Cercyonis pegala*) E, DJ, Leavenworth, Chelan Co., WA; All others, DJ, Tucannon R, Columbia Co., WA.

Pg. 331: (*Cercyonis sthenele*) E(mature)+L1(pink)+Ad, DN, Dry Falls, Grant Co., WA; All others, DJ, Dry Falls, Grant Co., WA.

Pg. 333: (*Cercyonis oetus*) E(mature), DN, Reecer Canyon, Kittitas Co., WA; E(fresh)–P, DJ, Tiger Mdws, Pend Oreille Co., WA; Ad, DN, Quartz Mt, Kittitas Co., WA.

Pg. 335: (*Erebia epipsodea*) E(both)+L1(pink)+L2–L3, DN, Sinlahekin R, Okanogan Co., WA.; L1(brown), DJ, Cummings Creek, Columbia Co., WA; L4, DJ, Mt Hull, Okanogan Co., WA; L5, DJ, Apex Mt, BC; Ad, DN, Fields Spring, Asotin Co., WA.

Pg. 337: (*Erebia vidleri*) E(mature), DJ, Apex Mt, B.C.; All others, DN, Slate Peak, Okanogan Co., WA.

Pg. 339: (*Oeneis nevadensis*) E(pink), DN, Rimrock Lk, Yakima Co., WA; E(gray)+L2+L3, DJ, Rimrock Lk, Yakima Co., WA; L1(both), DN, Liberty, Kittitas Co., WA; L5(both), KW, Branscomb Rd, Mendocino Co., CA; Ad, DN, Lk Chelan, Chelan Co., WA.

Pg. 341: (*Oeneis chryxus*) E(both)+L1(3.2 mm), DN, Mt Hull, Okanogan Co., WA; L1(6 mm), DJ, Moses Mdws, Okanogan Co., WA; L5+P, KW, Saddlebag Lk, Mono Co., CA; Ad, DJ, Ashnola R, BC.

Pg. 343: Spreadwing Skippers Montage: Top, *Pholisora catullus* L5, DJ, Benton City, Benton Co., WA; center UL, *Pyrgus communis* P, DN, Swakane Canyon, Chelan Co., WA;

Pg. 343 (cont.): center LL + CR, *Erynnis propertius* L5+E, DJ, Bear Canyon, Yakima Co., WA; LL, *Erynnis persius* Ad, DN, Winthrop, Okanogan Co., WA; LR, *Epargyreus clarus* L5, DN, Tahuya, Kitsap Co., WA.

Pg. 345: (*Epargyreus clarus*) L5(37 mm)+Ad, DJ, Tahuya, Kitsap Co., WA; All others, DN, Tahuya, Kitsap Co., WA.

Pg. 347: (*Thorybes pylades*) E–L3+L5, DN, Satus Pass, Klickitat Co., WA; L4+P+Ad, DJ, Satus Pass, Klickitat Co., WA.

Pg. 349: (*Erynnis icelus*) E–L2, DJ, Satus Pass, Klickitat Co., WA; All others DN, Oak Creek Canyon, Yakima Co., WA.

Pg. 351: (*Erynnis propertius*) L1, DN, Albany, OR; All others, DJ, Bear Canyon, Yakima Co., WA.

Pg. 353: (*Erynnis pacuvius*) L2, DJ, Swakane Canyon, Chelan Co., WA; All others, DN, Swakane Canyon, Chelan Co., WA.

Pg. 355: (*Erynnis persius*) E(top), DN, Reecer Canyon, Kittitas Co., WA; E(side), DN, Sinlahekin R, Okanogan Co., WA; L1, DN, Haney Mdws, Kittitas Co., WA; L3–P, DN, Tahuya, Kitsap Co., WA; Ad, DN, Winthrop, Okanogan Co., WA.

Pg. 357: (*Pyrgus centaureae*) All, DJ, Slate Peak, Okanogan Co., WA.

Pg. 359: (*Pyrgus ruralis*) E(fresh)+L1-L3, DJ, Bear Creek Mt, Yakima Co., WA; E(mature)+L4-P, DN, Nooksack Falls, Whatcom Co., WA; Ad, DN, Tahuya, Kitsap Co., WA.

Pg. 361: (*Pyrgus communis*) E(0.5 mm)+L1–L3, DJ, Naches Hts, Yakima Co., WA; All others, DN, Swakane Canyon, Chelan Co., WA.

Pg. 363: (*Heliopetes ericetorum*) L1+L4+P(19 mm)+Ad, DN, Swakane Canyon, Chelan Co., WA; All others, DJ, Benton City, Benton Co., WA.

Pg. 365: (*Pholisora catullus*) L3(10 mm)+Ad, DJ, Benton City, Benton Co., WA; All others, DN, Swakane Canyon, Chelan Co., WA.

Pg. 367: Monocot (Grass) Skippers Montage: Top, *Hesperia colorado* L5, DJ, Bear Canyon, Yakima Co., WA; CL, *Polites mystic* Ad, DJ, Tiger Mdws, Pend Oreille Co., WA; center UR, DJ, *Euphyes vestris* E, Satus Pass, Klickitat Co., WA; center LR, *Ochlodes yuma* L5, DN, Grant Co., WA; LL, *Polites sonora* P, DJ, Rimrock Lk, Yakima Co., WA; LR, *Hesperia colorado* E, DN, Reecer Canyon, Kittitas Co., WA.

Pg. 369: (*Carterocephalus palaemon*) E–L4, DN, Liberty, Kittitas Co., WA; L5–P, DN, Tumwater Canyon, Chelan Co., WA; Ad, DJ, Slate Peak, Okanogan Co., WA.

Pg. 371: (*Oarisma garita*) L2, DN, Tiger Mdws, Pend Oreille Co., WA; All others, DJ, Tiger Mdws, Pend Oreille Co., WA.

Pg. 373: (*Thymelicus lineola*) E+Ad, DN, Blaine, Whatcom Co., WA; All others, DN, Lk Terrell, Whatcom Co., WA.

Pg. 375: (*Hesperia juba*) E–L1, DN, Lower Granite Dam, Whitman Co., WA; All others, DJ, Cummings Creek, Columbia Co., WA.

Pg. 377: (*Hesperia comma*) Ad, DN, Parachute Mdws, Okanogan Co., WA; All others, DJ, Apex Mt, BC.

Pg. 379: (*Hesperia colorado*) E–L1, DN, Reecer Canyon, Kittitas Co., WA; L2–P, DJ, Bear Canyon, Yakima Co., WA; Ad, DJ, Wallowa Mts, OR.

Pg. 381: (*Hesperia nevada*) All, DJ, Umtanum Ridge, Kittitas Co., WA.

Pg. 383: (*Atalopedes campestris*) L5, DN, Richland, Benton Co., WA; All others, DJ, Benton City, Benton Co., WA.

Pg. 385: (*Polites peckius*) E, DJ, Tiger Mdws, Pend Oreille Co., WA; Ad(ventral), DN, Mt Hull, Okanogan Co., WA; All others, DN, Tiger Mdws, Pend Oreille Co., WA.

Pg. 387: (*Polites sabuleti*) All, DJ, Hat Rock State Park, Umatilla Co., OR.

Pg. 389: (*Polites mardon*) All, DN, Scatter Creek, Thurston Co., WA.

Pg. 391: (*Polites themistocles*) E(mature), DN, Mt Hull, Okanogan Co., WA; L3, DN, Tiger Mdws, Pend Oreille Co., WA; All others, DJ, Mt Hull, Okanogan Co., WA.

Pg. 393: (*Polites mystic*) E+L1, DN, Tiger Mdws, Pend Oreille Co., WA; All others, DJ, Tiger Mdws, Pend Oreille Co., WA.

Pg. 395: (*Polites sonora*) E+L1, DN, American R, Yakima Co., WA; All others, DJ, Rimrock Lk, Yakima Co., WA.

Pg. 397: (*Ochlodes sylvanoides*) E, DJ, Cummings Creek, Columbia Co., WA; L1, DN, Umtanum Canyon, Kittitas Co., WA; L2-L4, DN, Lk Terrell, Whatcom Co., WA; L5+Ad, DJ, Indian Flat, Yakima Co., WA; P, DN, Sedro Woolley, Skagit Co., WA.

Pg. 399: (*Ochlodes yuma*) E+L1+L3(15 mm), DJ, Dry Falls, Grant Co., WA; All others, DN, Grant Co., WA.

Pg. 401: (*Euphyes vestris*) E(hatching)+L2, DJ, Satus Pass, Klickitat Co., WA; All others, DN, nr Ellensburg, Kittitas Co., WA.

Pg. 403: (*Amblyscirtes vialis*) E(fresh)+L1, DN, Tiger Mdws, Pend Oreille Co., WA; All others, DJ, Cummings Creek, Columbia Co., WA.

Acknowledgments

A book of this scope necessarily draws on a host of people for technical assistance, advice, and information as well as for the less tangible but equally important inspiration, energy, and encouragement. First and foremost, this book would not have happened had it not been for the considerable assistance of Jonathan Pelham. Throughout the years of this work we drew heavily and frequently on Jon's amazing knowledge of Pacific Northwest Lepidoptera; it was Jon who inspired one of us (DN) to begin this work. We are likewise indebted to Robert Michael Pyle for his untiring dedication to Cascadia butterflies and for his strong support and encouragement of our work. His monumental *Butterflies of Cascadia* (2002) has been a major inspiration for the format and content of this work. Jon and Bob both had a hand in introducing David to David, and both gentlemen inspired us to work on this book.

Andrew Warren is another inspiration and technical giant on whose talents we have drawn, and his *Butterflies of Oregon: Their Taxonomy, Distribution, and Biology* (2005) provided a wealth of information we used frequently. Idie Ulsh, founding president of the Washington Butterfly Association, was tireless in support of finding a publisher, assisting in the field, and constantly encouraging our efforts.

The members of the Washington Butterfly Association have been patiently and strongly supportive for more than a decade, and a number of members, too many to name here, with their keen field-trip eyes, provided important observations and specimens on a number of occasions.

Paul Opler, John Acorn, and Kiyoshi Hiruma provided important letters of support that are greatly appreciated. Livestock and rearing assistance were graciously provided for several species by Martha Robinson and her INSTARS (children's butterfly group) as well as by Jim Brock, Gary Pearson, Bob Hardwick, Dennis St. John, David Droppers, and Chris Hanslik. Tora Brooks, Victor Reyna, and Larry Wright (Washington State University) helped DJ rear a number of species during his frequent absences, as did Tanya James. The WSU Department of Entomology and Irrigated Agriculture Research and Extension Center at Prosser, WA, are thanked for their support of this project. Photos and rearing notes for several species difficult to find in our area were generously provided by Keith Wolfe and Nicky Davis, and additional photos were provided by Idie Ulsh and Merrill Peterson.

Technical review, advice, field observations, and field techniques were graciously provided by Gordon Pratt, Dave McCorkle, Kiyoshi Hiruma, Todd Stout, Harold Greeney, Keith Wolfe, and Caitlin LaBar. Field suggestions by Cris Guppy, Norbert Kondla, Dennis St. John, Art Shapiro, Erik Runquist, Don Rolfs, Mark Walker, and Bob Hardwick were very helpful. Dale Swedberg of the Washington Department of Fish and Wildlife provided much appreciated hospitality in the Sinlahekin area of Washington, and Ann Potter of the same agency assisted in obtaining permits and specimens for the endangered Mardon Skipper.

We are of course deeply indebted to the fine staff of the Oregon State University Press, particularly Mary Braun, for having faith in our project and accepting this book for publication. Other OSU Press staff who have made important contributions include Tom Booth (associate director), Jo Alexander (managing editor), Micki Reaman (editorial and marketing), and Judy Radovsky (editorial and production). We also want to acknowledge the meticulous and painstaking editing of freelance copy editor Lucinda Treadwell, proofing by Susan Campbell, and index preparation by Mary Harper.

And to our long-suffering wives, Jo Nunnallee and Tanya James, and to our families, we are indebted more than we can express. Their patience with years of postponed summer vacations when livestock required constant attention, for tolerating rearing cages and equipment in our homes, for providing companionship and assistance on hundreds of sometimes frustrating field sojourns, for enduring the many expenses associated with our travels, rearing equipment, photographic and computer equipment, and host plants, as well as the thousands of hours during which we communicated so poorly while engrossed in the book, has been remarkable and is deeply appreciated. DJ particularly thanks Tanya for her entomological insights and field work enthusiasm, also Dick and Susan Price for their generous enthusiasm and support of this project. The James "vanessids," Jasmine Vanessa and Rhiannon Vanessa, spent their first 4–6 years of life involved in the field side of this project, and we are indebted to Jasmine in particular, who at 5 years of age caught the American Lady female that enabled our coverage of *Vanessa virginiensis*!

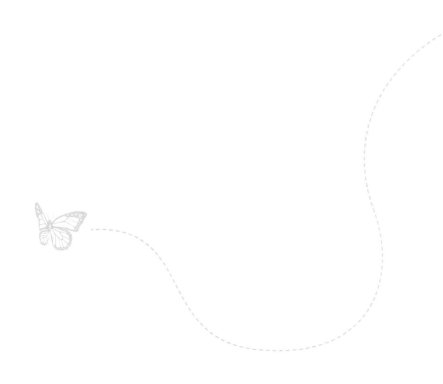

Index

Page numbers in **bold** mark the main entry for featured species.

David Nunnallee has studied butterflies in the Pacific Northwest for 15 years. Two hundred of his photographs are published in field guides, online, and in permanent public displays. He is cofounder of the Washington Butterfly Association.

David G. James is Associate Professor of Entomology at Washington State University, specializing in integrated pest management, biological control, and chemical ecology. He has more than 650 publications, including 166 scientific papers (30 on Lepidoptera).

Field Notes: